MW00843514

GEOLOGIC FUNDAMENTALS OF GEOTHERMAL ENERGY

ENERGY AND THE ENVIRONMENT

SERIES EDITOR

Abbas Ghassemi

New Mexico State University

PUBLISHED TITLES

Geologic Fundamentals of Geothermal Energy
David R. Boden

Introduction to Bioenergy
Vaughn Nelson and Kenneth Starcher

Introduction to Renewable Energy, Second Edition
Vaughn Nelson and Kenneth Starcher

Environmental Impacts of Renewable Energy
Frank R. Spellman

Geothermal Energy: Renewable Energy and the Environment, Second Edition
William E. Glassley

Energy Resources: Availability, Management, and Environmental Impacts
Kenneth J. Skipka and Louis Theodore

Finance Policy for Renewable Energy and a Sustainable Environment
Michael Curley

Wind Energy: Renewable Energy and the Environment, Second Edition
Vaughn Nelson

Solar Radiation: Practical Modeling for Renewable Energy Applications
Daryl R. Myers

Solar and Infrared Radiation Measurements
Frank Vignola, Joseph Michalsky, and Thomas Stoffel

Forest-Based Biomass Energy: Concepts and Applications
Frank Spellman

Solar Energy: Renewable Energy and the Environment
Robert Foster, Majid Ghassemi, Alma Cota,
Jeanette Moore, and Vaughn Nelson

GEOLOGIC FUNDAMENTALS OF GEOTHERMAL ENERGY

David R. Boden

CRC Press is an imprint of the
Taylor & Francis Group, an **informa** business

CRC Press
Taylor & Francis Group
6000 Broken Sound Parkway NW, Suite 300
Boca Raton, FL 33487-2742

Printed and bound in India by Replika Press Pvt. Ltd.

Printed on acid-free paper
Version Date: 20160411

International Standard Book Number-13: 978-1-4987-0877-7 (Hardback)

Library of Congress Cataloging-in-Publication Data

Names: Boden, David R., author.
Title: Geologic fundamentals of geothermal energy / author, David R. Boden.
Description: Boca Raton : Taylor & Francis, CRC Press, 2016. | Series: Energy and the environment | Includes bibliographical references and index.
Identifiers: LCCN 2016016324
Subjects: LCSH: Geothermal resources.
Classification: LCC GB1199.5 .B63 2016 | DDC 551.2/3--dc23
LC record available at https://lccn.loc.gov/2016016324

Visit the Taylor & Francis Web site at
http://www.taylorandfrancis.com

and the CRC Press Web site at
http://www.crcpress.com

Dedication

The author dedicates this book to his wife, Mimi Grunder,
for her encouragement, patience, and love throughout.

Contents

Series Preface

By 2050 the demand for energy could double or even triple as the global population rises and developing countries expand their economies. According to data from the United Nations, it is projected that the world population will increase from 7.2 billion to more than 9 billion by 2050. This increase, coupled with continued demand for the same limited natural resources, will cause a significant increase in the consumption of energy. All life on Earth depends on energy and the cycling of carbon. Affordable energy resources are essential for economic and social development as well as food production, water supply availability, and sustainable, healthy living. To avoid the long-term adverse and potentially irreversible impact of harvesting energy resources, we must explore all aspects of energy production and consumption, including energy efficiency, clean energy, the global carbon cycle, carbon sources and sinks, and biomass, as well as their relationships to climate and natural resource issues. Knowledge of how to utilize energy has allowed humans to flourish in numbers unimaginable to our ancestors. The world's dependence on fossil fuels began approximately 200 years ago. Are we running out of oil? No, but we are certainly running out of the affordable oil that has powered the world economy since the 1950s. We know how to recover fossil fuels and harvest their energy to operate power plants, planes, trains, and automobiles, but doing so has modified the carbon cycle and amplified greenhouse gas emissions. This has resulted in debates on the availability of fossil energy resources, the concept of peak oil, when the era of fossil fuel might end, energy pricing, and environmental impacts vs. what the various renewable resources offer with regard to reduced carbon footprints and emissions, in addition to necessary controls (i.e., cap and trade) and the emergence of "green power."

Our current consumption has largely relied on oil for mobile applications and on coal, natural gas, nuclear, and water power for stationary applications. To address the energy issues in a comprehensive manner, it is vital to consider the complexity of energy. Any energy resource—oil, gas, coal, wind, biomass, etc.—is an element of a complex supply chain and must be considered in the entirety as a system from production through consumption. All of the elements of the system are interrelated and interdependent. The use of oil, for example requires the interlinking of many elements, including exploration, drilling, production, transportation, water usage and production, refining, refinery products and byproducts, waste, environmental impacts, distribution, consumption/application, and finally emissions. Inefficiency in any part of the system has an impact on the overall system, and disruption of any one of these elements can have a significant cost impact. As we have experienced in the past, interrupted exploration results in disruptions in production, restricted refining and distribution, and consumption shortages; therefore, any proposed energy solution requires careful, extensive evaluation, which can prove to be an important barrier to implementing alternative resources, such as hydrogen as a mobile fuel.

Even though an admirable level of effort has gone into improving the efficiency of fuel sources for the delivery and use of energy, we are faced with severe challenges on many fronts, including population growth, emerging economies, new and expanded usage, and limited natural resources. All energy solutions include some level of risk, such as technology snafus and changes in market demand and economic drivers, among others. This is particularly true for energy solutions involving the implementation of untested alternative energy technologies.

There are concerns that emissions from fossil fuels are resulting in climate changes with possibly disastrous consequences. Over the past five decades, the world's collective greenhouse gas emissions have increased significantly even as efficiency has increased and extended energy benefits to more of the population. Many propose that we improve the efficiency of energy use and conserve resources to lessen greenhouse gas emissions and avoid a climate catastrophe; however, using fossil fuels more efficiently has not reduced overall greenhouse gas emissions for various reasons, and it is unlikely that such initiatives will have a perceptible effect on atmospheric greenhouse gases. Despite the debatable correlation between energy use and greenhouse gas emissions, there are effective means to produce energy, even from fossil fuels, while controlling emissions. There are also emerging technologies and engineered alternatives that can actually control the composition of the atmosphere but will require significant understanding and careful use of energy.

We need to step back and reconsider our knowledge of energy use. The traditional approach of micromanaging greenhouse gas emissions is not feasible or functional over a long period of time. More assertive methods to influence the carbon cycle are needed and will be emerging in the coming years. To modify the carbon cycle requires looking at all options for managing atmospheric greenhouse gases, including considering various ways to produce and consume energy in a more environmentally friendly way. We need to be willing to face reality and search in earnest for alternative energy solutions. Technologies that could assist may not all be viable. The proposed solutions must not be quick fixes but must be more of a comprehensive, long-term (10, 25, 50 years) approach that is science based and utilizes aggressive research and development. The proposed solutions must be capable of being retrofitted into our existing energy chain. In the meantime, we must continually seek to increase the efficiency of converting energy into heat and power.

The concept of sustainable development addresses the long-term, affordable availability of limited resources, including energy. Foremost among the many potential constraints to sustainable development is the competition for water use among energy production, manufacturing, farming, and others in light of a limited supply of fresh water for consumption and development. Sustainable development is also dependent on the Earth's limited amount of productive soil. In the not too distant future, it is anticipated that we will have to restore and build soil as a part of sustainable development. We need to focus our discussions on the motives, economics, and benefits of natural resource conservation, as well as on the limited ability of improvements in technology to impact sustainability; that is, how many fish we can catch from the ocean is limited by the number of fish available, not by the size of our boat or design of our net. Hence, possible sustainable solutions must not be based solely on enhancing and improving the technology used to obtain fossil fuel resources, but instead must be comprehensive and based on integrating our energy use with nature's management of carbon, water, and life on Earth as represented by the carbon and hydrogeological cycles. The challenges presented by the need to control atmospheric greenhouse gases are enormous, and to achieve sustainable development requires "out of the box" thinking, innovative approaches, imagination, and bold engineering initiatives. We must ingeniously exploit even more sources of energy and integrate their use with effective control of atmospheric greenhouse gases.

The continued development and application of energy are essential to the sustainable advancement of society. We must consider all aspects of our energy options, including performance against known criteria, basic economics and benefits, efficiency, processing and utilization requirements, infrastructure requirements, subsidies and credits, waste disposal, and effects on ecosystems, as well as unintended consequences such as impacts on natural resources and the environment. Additionally, we must view the emerging energy picture in light of current and future efforts in developing renewable alternatives and modifying and enhancing the use of fossil fuels. We must also evaluate the energy return for the investment of funds and the use of other natural resources such as water. Water is a precious commodity that has a significant impact on energy production, including alternative sources, due to the nexus between energy and water and issues related to the environment and sustainability.

A significant driver for creating a book series focused on alternative energy and the environment was my lecturing around the country and in the classroom on the subject of energy, the environment, and natural resources such as water. Although the correlations among the many relevant elements, how they relate to each other, and the impact of one on the other are fairly well understood, they are not always considered when it comes to integrating alternative energy resources into the energy matrix. Additionally, as renewable technology implementation continues to grow, the need for informed and trained human resources has resulted in universities, community colleges, and trade schools offering minors, certificate programs, and even in some cases majors in renewable energy and sustainability. As the field grows, so too is the demand for trained operators, engineers, designers, and architects. A deluge of flyers, e-mails, and texts promotes various short courses in solar, wind, geothermal, biomass, etc. under the umbrella of retooling an individual's career and providing the trained resources necessary to interact with financial, governmental, and industrial organizations.

Throughout all my years in this field, I have devoted significant effort to locating integrated textbooks that explain alternative energy resources in a suitable manner, that would complement a syllabus for potential courses to be taught at the university and that would provide good reference material for interested parties getting involved in this field. I have been able to locate a number of books related to energy, energy systems, energy conversion, and energy sources such as fossil, nuclear, and renewable, as well as specific books on the subjects of natural resource availability, use, and impacts as related to energy and environment. However, specific books that are correlated and present the various subjects in detail are few and far between, which is why this series of texts addressing specific technology fields in the renewable energy arena has been created. This series so far includes texts on wind, solar, geothermal, biomass, and hydro power, with others yet to be developed. These texts are intended for upper-level undergraduate and graduate students and for informed readers who have a solid fundamental understanding of science and mathematics, as well as individuals and organizations involved in design development of the renewable energy field, entities that are interested in having reference material available to their scientists and engineers, consulting organizations, and reference libraries. Each book presents fundamentals as well as numerical and conceptual problems designed to stimulate creative thinking and problem solving.

The series author wishes to express his deep gratitude to his wife Maryam, who has served as a motivator and intellectual companion and too often was victim of this effort. Her support, encouragement, patience, and involvement have been essential to completion of this series.

Abbas Ghassemi, PhD
Las Cruces, New Mexico

Preface

The heat or thermal energy of the Earth (geothermal) is enormous. It exceeds (by many orders of magnitude) the energy of all known resources of coal, oil, and natural gas (the fossil fuels). With such a tremendous energy resource, why are we still using fossil fuels? The answer is complex but basically boils down to this: Although the flow of heat from the Earth to the surface is continuous and steadfast, it is not uniformly distributed. Some regions are characterized by elevated heat flow, whereas in other areas the flow of energy is average or below average. To use Earth's heat directly, such as for space heating, or indirectly in the generation of electrical power requires finding regions that have above-average heat flow. This unequal distribution of heat flow at the Earth's surface is controlled by variable geologic environments and forces. In most places, the necessary geologic conditions are not adequate to concentrate the heat sufficiently for us to harness it for the benefit of society. In this way, geothermal resources are not unlike oil and gas plays or mineral deposits, as they are not found everywhere and require a special orchestration or convergence of geologic processes. The principal intent of this book is to explore those processes that help form geothermal resources. Studying geologic conditions and forces is central to finding potentially developable geothermal resources. These conditions and forces also strongly influence the type of geothermal facility that might be constructed to utilize a discovered resource.

Although much has been researched and written on the geology of geothermal systems, this information is mainly scattered among professional journals, government publications, and conference proceedings. Most of the existing reference books covering geothermal energy focus on engineering and/or thermodynamic elements (energy analysis), power plant designs, and applications of geothermal energy; geology, in such cases, is condensed into a chapter or two. Glassley's book on geothermal energy (*Geothermal Energy: Renewable Energy and the Environment*, 2nd ed., CRC Press, 2015) discusses more geology than other reference books on the topic, and as such this book complements and draws upon Glassley's book. Additional material presented here includes central information on Earth materials, rock structures, and plate tectonics. Other topics include hydrothermal alteration processes, the role between active geothermal systems and the possible formation of mineral deposits, and pioneering efforts to explore and develop deep supercritical water systems and regions of hot dry rock (also known as petra-heat). Successful realization of the latter exploratory efforts may someday greatly expand development of geothermal energy.

For college geology students and instructors, the book can serve as a foundation for learning how geologic fundamentals can be applied toward developing geothermal energy and as a springboard for more detailed treatments. To assist in pursuing more in-depth reviews on select topics, references on many seminal articles are provided at the end of each chapter. Furthermore, because background information on Earth materials, rock structures, and plate tectonics is provided, the contents of this book should also be accessible to energy engineers, environmental scientists, and energy policymakers who would like to learn more about geologic controls of geothermal systems and in doing so better develop and manage geothermal resources. For similar reasons, the book should be of interest to the educated layperson who is broadly interested in energy issues, in how geothermal relates to other sources of renewable energy, and in the role geology plays in defining the resource. It is hoped that the reader will find that information in the book complements and supplements extant information on geothermal energy in a useful way and thereby will gain an improved understanding of the geologic foundations and societal benefits of geothermal energy.

Acknowledgments

This book would not have been possible without the help of many people. I thank Bill Glassley for his suggestions about the book and for facilitating finding a publisher. The following people reviewed select chapters of the book: Peter Schiffman, David John, Patrick Dobson, Mark Walters, Pete Stelling, Don Hudson, Mark Coolbaugh, Dick Benoit, Stefan Arnorsson, Jim Stimac, and Lisa Shevenell. I am grateful for their efforts, which certainly improved scientific accuracy and clarity and helped curtail my tendency to make overextended statements or dubious interpretations. Nonetheless, any errors and/or omissions that may remain are solely my own.

I also thank students in Professor Rob Zierenberg's class on geothermal energy at the University of California–Davis for reviewing select chapters of the book and offering meaningful comments on content and clarity. In particular, these include W. Rodrigues, V. Manthos, C. McHugh, and C. Rousset.

Jon Price and Chris Henry of the Nevada Bureau of Mines and Geology kindly reviewed and supported my book proposal for potential publishers. Their encouragement helped push me over the edge to proceed.

I thank Stuart Simmons for his suggestions and his quick replies to my e-mails seeking yet another of his many publications on geothermal energy and epithermal mineral deposits. I am grateful to Mariana Eneva for enlightening me on using InSAR for monitoring developed geothermal systems and as a possible exploration tool.

Special thanks to Joe Clements, CRC Press senior editor, for his help in soliciting potential reviewers and his patience for a project that took much longer than expected. Furthermore, I am grateful to the many publishers of books and journals who kindly gave permission to reuse published illustrations, variably modified herein, that certainly help clarify and reinforce concepts discussed in the text. For any of those I may have missed, I regret the oversight, but credit was assigned to all previously published figures, modified or not, used in this book.

My wife, Mimi Grunder, and our daughter, Kate Boden, came to the rescue in the final preparation of the manuscript and helped edit and improve writing lucidity. They discovered and corrected more grammatical errors and obtuse sentence structure than I would like to admit. Kate used her experience from earning her physics degree to make equations in the text easier to read. I also thank Kate's friend and fellow UC Berkeley graduate Anna B. Dimitruk for diving in and helping edit a few chapters of the manuscript when she thought she came to visit to relax.

Many times while putting this book together I felt overwhelmed and thought it somewhat presumptuous that one author can write a book on such a multifaceted topic as geothermal energy. Focusing on just the geoscience aspects of geothermal energy includes not only geology (involving the roles mainly of Earth materials and rock structures) but also geochemistry (involving hydrothermal fluid chemistry, stable isotope behaviors, equilibrium and disequilibrium chemical reactions, etc.) and geophysics (involving studies on resistivity, aeromagnetism, gravity, and seismicity). All of these factors, and of course drilling, are used to help identify and characterize geothermal reservoirs. Indeed, many people having considerable direct geothermal experience (either researching active systems or working in the operations of finding and developing geothermal systems) might be better at writing a book such as this one. For better or worse, however, I took the plunge with the intent not only to emphasize the geologic underpinnings of geothermal energy but also to consider the environmental implications and societal benefits of using Earth's internal heat. If parts of the book may seem uneven, I ask for your forbearance and welcome your comments.

Series Editor

Abbas Ghassemi, PhD, is the director of the Institute for Energy and Environment (IEE) and professor of chemical engineering at New Mexico State University. In his role as IEE director, he is the chief operating officer for programs in education and research, as well as outreach in energy resources including renewable energy, water quality and quantity, and environmental issues. He is responsible for the budget and operation of the program. Dr. Ghassemi has authored and edited several textbooks, and his published works include the areas of energy, water, carbon cycle (including carbon generation and management), process control, thermodynamics, transport phenomena, education management, and innovative teaching methods. His research areas of interest include risk-based decision making, renewable energy and water, carbon management and sequestration, energy efficiency and pollution prevention, multiphase flow, and process control. Dr. Ghassemi serves on a number of public and private boards, editorial boards, and peer review panels. He earned his master's degree and doctorate in chemical engineering, with minors in statistics and mathematics, from New Mexico State University, and his bachelor of science degree in chemical engineering, with a minor in mathematics, from the University of Oklahoma.

Author

Dave Boden is currently professor of geoscience at Truckee Meadows Community College (TMCC) in Reno, Nevada. He teaches courses on physical geology, natural hazards, geological field methods, and, of course, the geology of geothermal energy. Dr. Boden served as chair of the physical sciences department at TMCC from 2009 to 2012. He is also an adjunct instructor at the University of Nevada, Reno, where he teaches a course on geothermal energy as part of the Graduate Renewable Energy Certificate program. As a result of a grant for developing a renewable energy technology program at TMCC, Dr. Boden developed the course on the geology of geothermal energy in 2007. This book is an outgrowth of teaching that class for the last 9 years and was further inspired from visits to select geothermal systems in New Zealand, Germany, and Iceland and discussions there with Mr. Ted Montegue of Contact Energy, New Zealand; Drs. Greg Bignall and Andrew Rae of GNS Science, New Zealand; and Dr. Omar Fridleifsson of HS Orka, Iceland. These visits were made possible by the author's participation on an National Science Foundation-sponsored CREATE project, which focuses on the training and education of renewable energy technologies at technical colleges and is headed by Dr. Kathy Alfano.

Prior to accepting a teaching position at Truckee Meadows Community College in 2004, the author worked as a minerals exploration geologist looking for base and precious metal deposits (the fossil analogs of modern geothermal systems) in the western United States, Alaska, and South America for almost 20 years. In that role, he worked for many companies (most of which no longer exist), including Anaconda, Phelps Dodge, Echo Bay Exploration, and Corona Gold. He also served as a consultant for Homestake, Lac Minerals, Kennecott, and Andean Silver Corporation. He has worked with the Nevada Bureau of Mines and Geology, helping map the geology, with Dr. Chris Henry, of the Tuscarora mining district and vicinity in northeast Nevada and the Talapoosa mining district and vicinity in west-central Nevada. As a result of his work in minerals exploration and geologic mapping, Dr. Boden has authored or is a coauthor on several scientific articles and geologic maps, including the volcanic geology of the giant Round Mountain gold deposit in central Nevada, the Bullfrog gold deposit in southern Nevada, and mineralization and volcanic geology of the Tuscarora area in northeastern Nevada.

Dr. Boden has degrees in geology and geological engineering consisting of a bachelor of science degree earned from the University of California–Davis, a master's degree from the Colorado School of Mines, and a doctorate from Stanford University. He was recently elected to the board of directors of the Geothermal Resources Council, where he would like to further promote education about geothermal energy in schools and with the general public. In his spare time, the author enjoys hiking and skiing to backcountry hot springs with family and friends.

1 An Overview of Energy

KEY CHAPTER OBJECTIVES

- Describe and contrast nonrenewable and renewable sources of energy.
- Identify characteristics that make geothermal energy distinctive from other forms of renewable energy and describe how temperature affects how geothermal energy is utilized.
- Recognize the difference between energy and power and apply the terms in the correct context.
- Discuss the attributes of geothermal energy in terms of fuel source, emissions, and baseload.

Succinctly, geothermal energy is heat from the Earth that can be harnessed and used for the benefit of society. Geothermal energy is below us everywhere and is available all the time, unlike other forms of renewable or alternative energy, such as solar and wind. And, yet, in many ways geothermal is overlooked because people are not able to see it like sunshine or feel it like wind. Geothermal, unlike solar and wind energy, is a baseload energy resource capable of providing power 24 hours a day all year long, similar to traditional fossil-fuel-fired power plants. This chapter provides a cursory overview of all forms of energy to provide a perspective of how geothermal energy fits into the energy milieu. Also, key concepts on energy and power are reviewed so the reader understands how energy and power are related and measured.

BASIC TERMINOLOGY OF ENERGY AND POWER

Energy comes in many forms, including kinetic (energy of motion), potential (the ability to deliver energy), chemical (energy in fossil fuels, such as gasoline and natural gas), and of course thermal or heat energy. The heat energy of the Earth is enormous and so is its ability to do work. Examples of Earth's work include moving huge pieces of the Earth's crust and uppermost mantle a few centimeters every year, the eruption of volcanoes, and the episodic lurching and shaking during an earthquake. Tapping just 1% of the thermal energy contained in the Earth's uppermost 10 km would produce 1000 times the annual energy used in the United States (Moore and Simmons, 2013). In other words, tapping just 1/1000 of that 1% would equal all the energy used in the United States annually. The work of geothermal energy ranges from electrical power production for high-temperature resources (generally >100°C) to space heating and cooling for lower temperature resources. The basic unit for measuring energy is the Joule (or newton-meter), which is also the SI unit for work, which is defined as the product of the force required (newton) to move a mass a specified distance (meter).

Power, on the other hand, is the rate at which the energy is delivered. The SI unit for power is joule per second (J/s). One joule per second is equal to 1 watt (W), and 1 kilowatt (kW) equals 1000 J/s. Power and energy are related simply as power (P) = energy (E)/time (t), or $E = P \times t$. Energy, then, is a quantity (how much) and power is a rate (how fast).

The U.S. power industry uses the hour instead of a second as the basic time measurement. So, 1 kilowatt-hour (kWh) equals 1000 J/s × 3600 s/hour, or 3.6 million J/hour; that is, 1 kWh equals 3.6 million joules of energy. To make use of these relationships, let's determine the energy used for a 100-W light bulb over a given amount of time. For example, if a 100-W light bulb is left on for 1 hour per day, how much energy is used in a 30-day month? That would be 100 W × 1 hour/day × 30 days/month, which equals 3000 Wh or 3 kWh. At $0.15/kWh, the cost would be $0.45 to have the light bulb turned on for 1 hour/day for 30 days. Most homes use about 500 to 1000 kWh of energy per month, depending on the size of the home and time of the year, which translates to energy bills of $75 to $150 per month.

Power plant size is typically rated using megawatts electric (MWe).* Large fossil fuel and nuclear power plants are on the order of 500 to 2000 MWe or 0.5 to 2 gigawatts electric (GWe). Individual geothermal power plants typically range between 10 and 100 MWe. Some geothermal power plans, such as Hellisheidi in Iceland, are combined heat and power facilities and provide both electrical and thermal power, measured as megawatts electric (MWe) and megawatts thermal (MWt), respectively. To provide some perspective, 1 megawatt is enough power to serve the needs of about 1000 homes in the United States. Thus, a 100-MWe power plant would serve the residential power needs of about 100,000 homes, or about 300,000 to 400,000 people.

Some government documents and articles report energy produced or consumed in the form of British thermal units (Btu) (such as for many home appliances). A Btu is a measure of energy, not power, and is equal to 1055 joules or 1.055 kJ. Due to the unfortunate mixing of units from one publication to another, the ability to convert between units so comparisons can be made is necessary. For example, how are Btu and kWh related? Using the above equality of 1 Btu equals 1.055 kJ and the already noted equality that 1 kWh is 3.6 million joules or 3600 kJ, a Btu then equals 1.055 kJ × 1 kWh/3600 kJ, or 2.9×10^{-4} kWh; the inverse of 1 kWh is roughly 3413 Btu. A final energy unit used is the *quad*, which is shorthand for 1 quadrillion Btu (a quadrillion is 1×10^{15}). One quad is about the energy consumed by 5.5 million U.S. households in a given year, and the annual energy consumption of the entire United States is about 98 quads (EIA, 2015a).

As a practical example, let's determine the power plant rating (in MWe) for a plant that produces 6.5×10^{9} kWh of energy in a year. We need to be able to convert from energy to power or *vice versa* so equivalent comparisons can be made, a practice that is necessary because data are commonly reported differently depending on the source. To return to our example above, we need to convert kWh to MWe. To do so, we realize that power is a rate and energy a quantity so we need to cancel out the time term as follows:

* The prefix M (mega) represents 1 million, the prefix G (giga) 1 billion, and the prefix T (terra) 1 trillion. The small "e" in MWe stands for electric power, and the small "t" in MWt stands for thermal power.

$$(6.5 \times 10^9 \text{ kWh}) \times (1 \text{ year}/365 \text{ days})$$
$$\times (1 \text{ day}/24 \text{ hours}) \times (1 \text{ MWe}/1000 \text{ kWe}) = 74 \text{ MWe}$$

This calculation assumes that the plant runs 24 hours per day, every day of the year, requiring a capacity factor of 100%.

CURRENT SOURCES OF ENERGY

Civilization has developed three main sources of energy. The most commonly used source includes the fossil fuels of coal, oil, and natural gas, which account for about 82% of the energy used in the United States (Figure 1.1). The other two sources of energy are nuclear energy (~8%) and renewable energy (~10%). On a global basis, the proportion of fossil fuels used is comparable to that in the United States (81.7%), but the use of nuclear energy is less (4.8%) and the use of renewable sources is slightly greater (13.5%), if biofuels are included (IEA, 2015).

Nonrenewable Sources of Energy

Nonrenewable sources of energy include the fossil fuels and nuclear energy. The term *nonrenewable* means the resources are finite and require tens of thousands to millions of years to form and thus cannot be renewed on the time scale of human demands. Although the energy obtained from fossil fuels and nuclear processes is nonrenewable, these sources of energy can be sustainable and made available to future generations if used wisely and efficiently (sustainable and renewable resources are discussed further in Chapter 12).

Fossil Fuels

The fossil fuels of oil, natural gas, and coal make up about 85% of the energy used in the United States (EIA, 2015a). With the current boom in shale oil and natural gas and our current abundance of coal resources, these fuel sources could serve energy

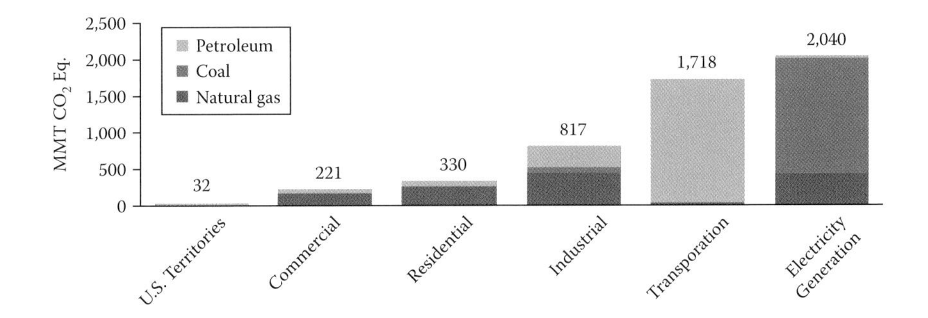

FIGURE 1.1 Amount of carbon dioxide emissions (in million metric tons) according to sector. The proportions have been normalized to include only the fossil fuels and not renewable sources of energy. (From USEPA, *Inventory of U.S. Greenhouse Gas Emissions and Sinks: 1990–2013*, U.S. Environmental Protection Agency, Washington, DC, 2015, Chapter 3.)

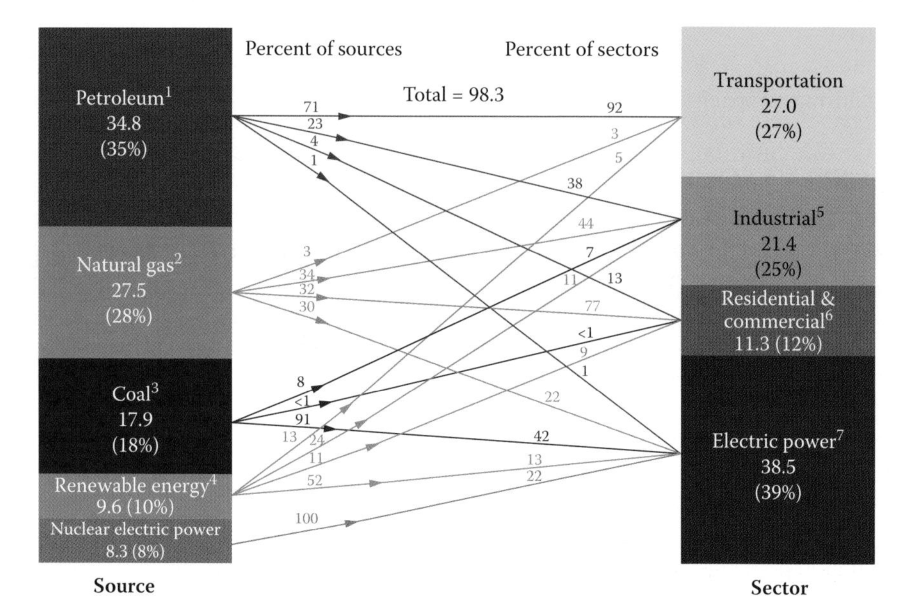

[1]Does not include biofuels that have been blended with petroleum–biofuels are included in "Renewable Energy".

[2]Excludes supplemental gaseous fuels.

[3]Includes less than –0.1 quadrillion Btu of coal coke net imports.

[4]Conventional hydroelectric power, geothermal, solar/photovoltaic, wind, and biomass.

[5]Includes industrial combined-heat-and-power (CHP) and industrial electricity-only plants.

[6]Includes commercial combined-heat-and-power (CHP) and commercial electricity-only plants.

[7]Electricity-only and combined-heat-and-power (CHP) plants whose primary business is to sell electricity, or electricity and heat, to the public. Includes 0.2 quadrillion Btu of electricity net imports not shown under "Source."

Notes: Primary energy in the form that it is first accounted for in a statistical energy balance, before any transformation to secondary or tertiary forms of energy (for example, coal is used to generate electricity). Sum of components may not equal total due to independent rounding.

FIGURE 1.2 Primary energy consumption by source and sector in 2014. Note that electric power is the largest user of energy and burning of coal is the largest source of energy (42%) used in making power. (From EIA, *Primary Energy Consumption by Source and Sector*, U.S. Energy Information Administration, Washington, DC, 2014.)

demands for the United States for decades to come, if not longer. The downside to the continued use or reliance on fossil fuels, however, is their continued contribution of greenhouse gases to our atmosphere which are now considered by an overwhelming majority of scientists to be accelerating climate change (Stocker et al., 2013). In the United States, for example, power generation is the largest source of CO_2 emissions, with coal-fired power plants being the largest contributor (Figure 1.1).

In the United States, coal is the third most common resource for the production of fossil fuel energy, but it is the number one source of energy for electrical power production, which makes up the largest sector of energy used (Figure 1.2). At today's mining rates, the United States has about a 200-year supply of coal; 92% of the coal mined is used for electrical power generation, and about 42% of all electric power came from coal in 2014 (Figure 1.1). Coal-fueled power generation, however, peaked in the mid-2000s at about 2 billion kWh (52%) and has since declined to 1.75 billion kWh. During the same period, the percentage of electricity generated by natural gas has increased from 15% to 22% (EIA, 2014).

Natural gas is the second largest source of energy used in the United States (Figure 1.2). About 31% of natural gas produced is used for electrical power generation, and a comparable proportion is used in the residential/commercial and industrial sectors. Currently, only a small proportion of natural gas is used for transportation, but utilization of natural gas for transportation will likely increase in the future if the supply continues to grow, keeping the price attractive. Shipping and trucking companies will be encouraged to shift from diesel to natural gas as domestic natural gas production increases. The United States is expected to become a net exporter of natural gas (EIA, 2015a).

Most of our currently consumed energy comes from oil, and 71% of that is used for transportation. Currently, we import about 45% of the oil we consume, with Canada and Mexico being our largest foreign suppliers. Our dependence on foreign oil has declined since it peaked in 2005, due in part to increased domestic production and manufacturing and the use of more fuel-efficient vehicles. As a result of increased domestic oil production and deployment of increasingly fuel-efficient vehicles, the U.S. Energy Information Administration projects that the United States will no longer need to import oil by the early 2030s and could become a net exporter of oil (EIA, 2015a).

Nuclear Energy

Nuclear energy represents about 8% of our energy base, and all of it is used for electrical power generation. About 19% of U.S.-generated electricity (101 GWe) came from nuclear power plants in 2011. Currently, there are 104 operating nuclear reactors in the United States. Nuclear power is projected to grow by about 14% over the next 30 years via both upgrades at existing plants and new construction (EIA, 2015a). Nonetheless, nuclear power's overall share is projected to decline from 19% to about 17% during that period due to increases in power generation from natural gas and renewable sources. The production of nuclear energy generates no greenhouse gas emissions but suffers from the challenges of waste disposal, negative public perspectives, high upfront costs, and protracted permitting and construction time frames* that hamper further development of this source of power. As a result, the development of small modular reactor (SMR) technology that can produce somewhere between 45 and 250 MWe is being pursued. SMR units could be fabricated in factories in 2 years or less and then delivered to power sites via barge, train, or truck. Nonetheless, the Nuclear Regulatory Commission (NRC) has not yet licensed SMRs because the regulatory infrastructure to support licensing review of SMR designs has yet to be developed. In 2013, the U.S. Department of Energy announced its intention to provide up to $450M to assist in further development of SMR technology as a joint effort with private industry (USDOE, 2013). A private contractor was selected that same year with which development costs will be shared on a one-to-one basis, with the goals of developing SMR designs for licensing by the NRC and beginning commercial operation by 2025 (USDOE, 2013).

* It takes about $12 billion to build a conventional, twin-unit, 2.2-GWe nuclear power plant and usually more than a decade to permit and construct such a plant.

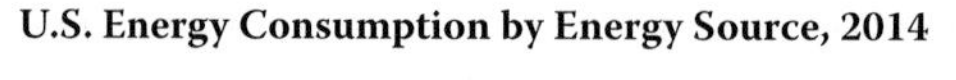

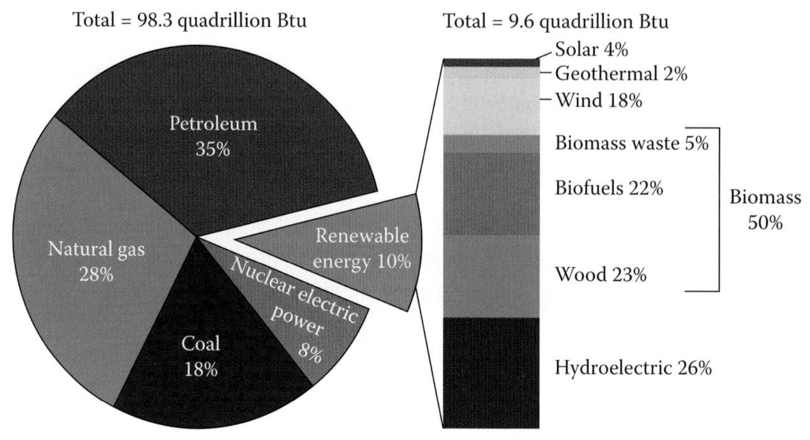

FIGURE 1.3 Energy consumption according to source in 2014. It is unclear whether the 2% geothermal value includes the energy contribution of geothermal heat pumps (see text under geothermal subheading). If not, the 2% value would be low by more than a factor of about 4 using data from Lund and Taylor (2015). (From EIA, *Renewable Energy Explained*, U.S. Energy Information Administration, Washington, DC, 2015.)

Renewable Sources of Energy

Renewable energy has now surpassed nuclear, comprising about 10% of our energy used. About half of that percentage comes from biomass, such as biofuels (Figure 1.3) (EIA, 2014), and over half of the renewable energy produced is used for producing electrical power. The remaining 46% is distributed among the transportation, industrial, and residential/commercial sectors. The principal types of renewable energy are, in order of decreasing contribution, biomass, hydroelectric, wind, solar photovoltaic (PV) and thermal, and geothermal (Figure 1.3). The 2% value for geothermal, however, could be low by a factor of about 4 if the thermal energy provided by geothermal heat pumps (as discussed below under the geothermal subheading) is not included, because about 88% of the energy from geothermal direct use comes from geothermal heat pumps (Lund and Boyd, 2015).

Hydropower

Representing a little over a third of that 10%, hydropower accounts for the largest proportion of renewable energy (but see following discussion on biomass energy), on average about 35 GWe of power or about 3 quadrillion Btu of energy annually. As the graph in Figure 1.4 illustrates, hydropower consumption is spikey, reflecting its sensitivity to variations in amounts of seasonal precipitation. Also evident is that the average amount of hydropower used has remained relatively flat for the last 40 years, because no new dams have been built over that period. Hydropower production will likely decrease in the future as more aging dams become decommissioned without replacement.

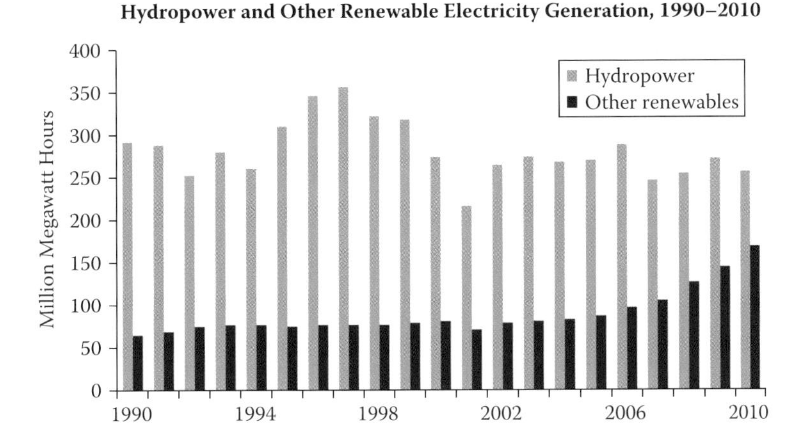

FIGURE 1.4 Variation of hydropower and other renewables from 1990 to 2010. Note that hydropower output varies considerably from year to year. (From EIA, *Today in Energy*, U.S. Energy Information Administration, Washington, DC, 2011.)

Biomass

Wood and biofuels, if considered together under biomass, comprise the largest share of renewable energy, at about 50% (Figure 1.3). Wood is largely used for space heating, such as in pellet stoves and rarely in power generation. Biofuels consist of biodiesel, largely made from recycled cooking oils or animal fat, and ethanol, made from the processing of plant starch, such as corn, or agricultural feedstocks. Both are primarily used for purposes of transportation but can also be used for heating oils. Although both biodiesel and ethanol are renewable because their source material can be grown and replenished, fossil fuels are commonly used in the processing to yield the biofuel. Moreover, some controversy has developed over whether food crops should be used in the production of fuel when hunger is still a serious worldwide problem.

Wind

Wind energy has seen remarkable growth over the last decade. Installed wind power capacity in the United States has increased sixfold in the last 7 years (5-year average annual growth rate of 29%). This represents the greatest capacity addition of any renewable energy technology. As of the end of 2012, installed wind capacity in the United States was slightly more than 60 GWe. The world leader of installed wind power capacity is China, at about 75 GWe. Total world installed wind capacity is about 283 GWe. These numbers are significant, but when load or capacity factor is considered the actual power realized is about 25 to 30% of the total installed capacity, or 70 GWe. The load or capacity factor is a ratio of the actual time of production to the total possible power that can be produced. Thus, power is being generated only 8 to 10 days per month on average. As the map in Figure 1.5 shows, most of the wind power in the United States comes from the Midwest.

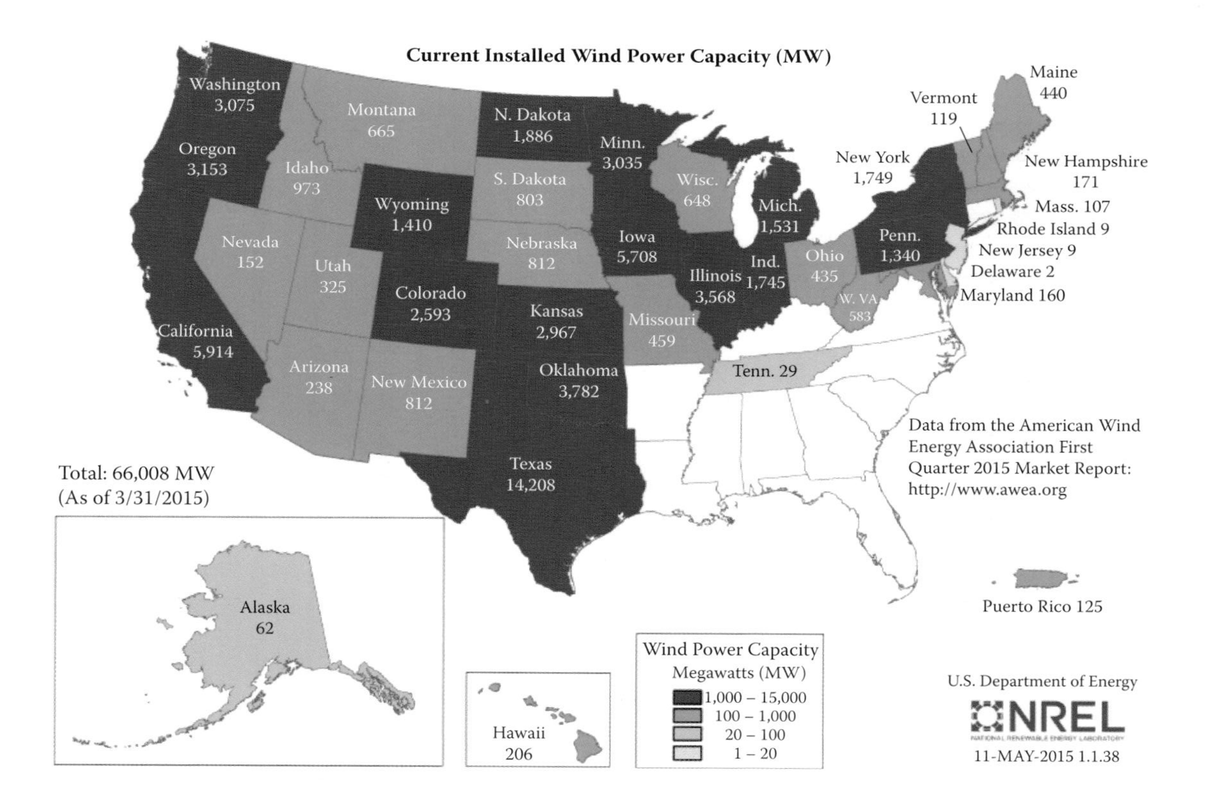

FIGURE 1.5 Map showing wind power capacity per state as of March 2015. Note that most of the wind power comes from the Midwest, reflecting not only the number of wind turbine installations but also a region typified by fairly consistent wind frequency. Although the wind power capacity is high, the actual power realized is about one fourth to one third of the stated power capacity of 66 GWe because the wind does not blow all the time. (From EERE, *Installed Wind Capacity*, Office of Energy Efficiency & Renewable Energy, U.S. Department of Energy, Washington, DC, 2016.)

Solar

Solar energy can be divided among three forms of technology—photovoltaic (PV), thermal, and concentrating solar power (CSP). Solar PV and CSP are used for power production, whereas thermal is used mainly for space heating. Solar CSP typically involves focusing the sun's rays using special mirrors that range from parabolic to Fresnel collectors (curved linearly aligned mirrors) to special relatively flat mirrors called *heliostats* that track the sun's path across the sky. The mirrors focus the sunlight to heat water or some other fluid mixture to produce steam that drives a turbogenerator.

One type of CSP involves the use of a solar power tower (~600 feet high) that is surrounded by heliostats that focus the sun's heat to the top of a tower containing molten salt heated to more than 500°C. The molten salt then passes through a heat exchanger to boil water. The resulting steam drives a turbine connected to an electrical generator. Heat from the molten salt can be stored overnight so power production can continue without sunlight. Utilization of the solar power tower technology, however, is currently the most expensive of all producing renewable sources of energy and requires a higher level of solar insolation, compared to solar PV, to be operational.

Solar PV has sustained the second fastest growth rate of renewable technologies, after wind power, in the United States. Globally, however, solar PV is the fastest growing renewable technology, with operating capacity increasing 58% annually from 2006 to 2011 (REN21, 2012). In the United States, as of the end of the first quarter of 2015, 19.6 GWe of solar PV and 1.7 GWe of solar CSP had been installed, for a total of 21.3 GWe (SEIA, 2015). The capacity of concentrated solar power will further increase due to the recent completion of SolarReserve's Crescent Dunes Solar Energy Project (110 MWe), a solar tower facility, near Tonopah, Nevada (NREL, 2016).

Solar thermal is used for residential and commercial space heating (and cooling) and water heating. It usually consists of flat panels filled with potable water or a water/glycol mixture contained in tubes that heat potable water via a heat exchanger. About 66% of the energy used in the average home is for space and water heating (Figure 1.6).

Geothermal

Unlike solar- and wind-produced energy, geothermal is primarily a *baseload* energy resource that produces electrical or thermal energy throughout the day, all year long. Hydropower is also a baseload resource, but its energy output can vary considerably depending on levels of precipitation in the drainage basin. Geothermal energy can be used in three principal ways—power generation, direct use, and geothermal heat pumps (also known as geoexchange). The pressure and temperature ranges of these three applications are shown in Figure 1.7. Electrical power generation requires the highest temperature resources (generally >100°C) and is the least widespread, being restricted to geologically favorable regions, such as along or near the boundaries of the Earth's tectonic plates (as discussed in Chapter 4). As of 2015, total worldwide installed geothermal power capacity was 12.6 GWe (Figure 1.8) (Bertani, 2015).

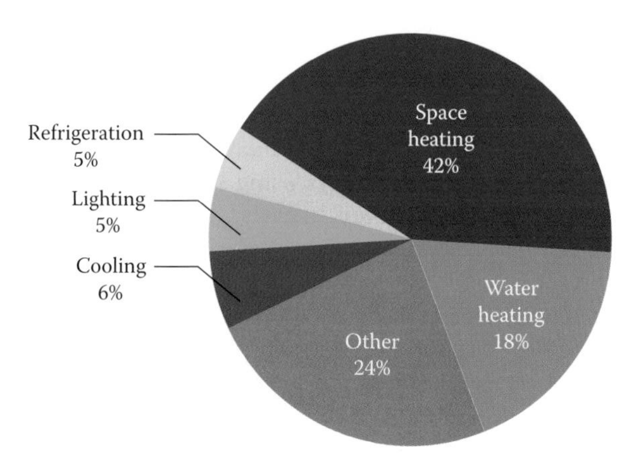

FIGURE 1.6 Pie chart showing how energy is used in the average U.S. household. (From USDOE, *Tips: Your Home's Energy Use*, Office of Energy Efficiency & Renewable Energy, U.S. Department of Energy, Washington, DC, 2016.)

The second form is direct use, which requires elevated but lower temperature fluids (generally ~50° to >100°C), used mainly in the heating (and sometimes cooling) of buildings, but also for a variety of other needs, including fish farming (aquaculture), fruit and vegetable drying, lumber processing, and, of course, bathing and

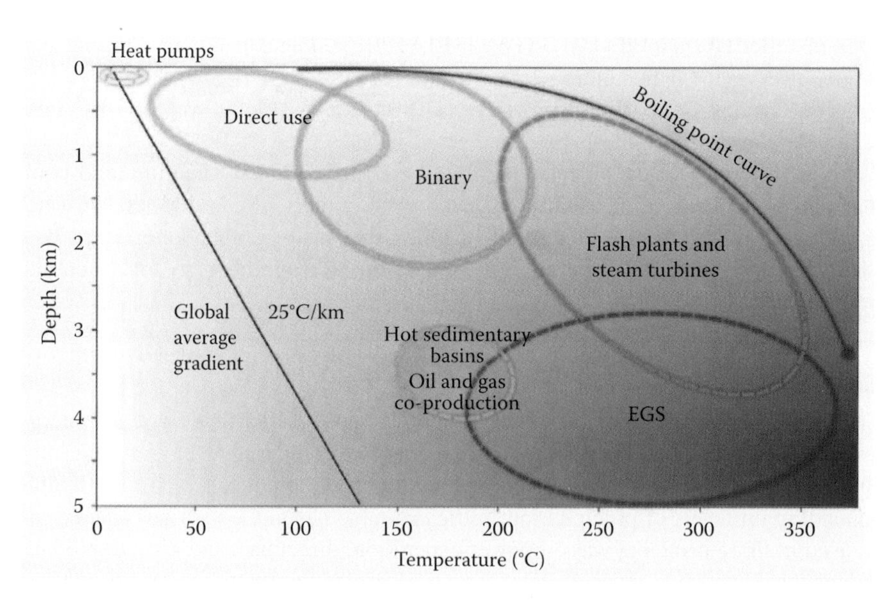

FIGURE 1.7 Temperature with depth graph showing the temperature–depth regions of different types of geothermal energy from near-surface geothermal heat pumps to the yet largely undeveloped enhanced geothermal systems (EGSs) occurring at depths mainly below currently developed power-generating reservoirs. (Adapted from Moore, J.N. and Simmons, S.F., *Science*, 340(6135), 933–934, 2013.)

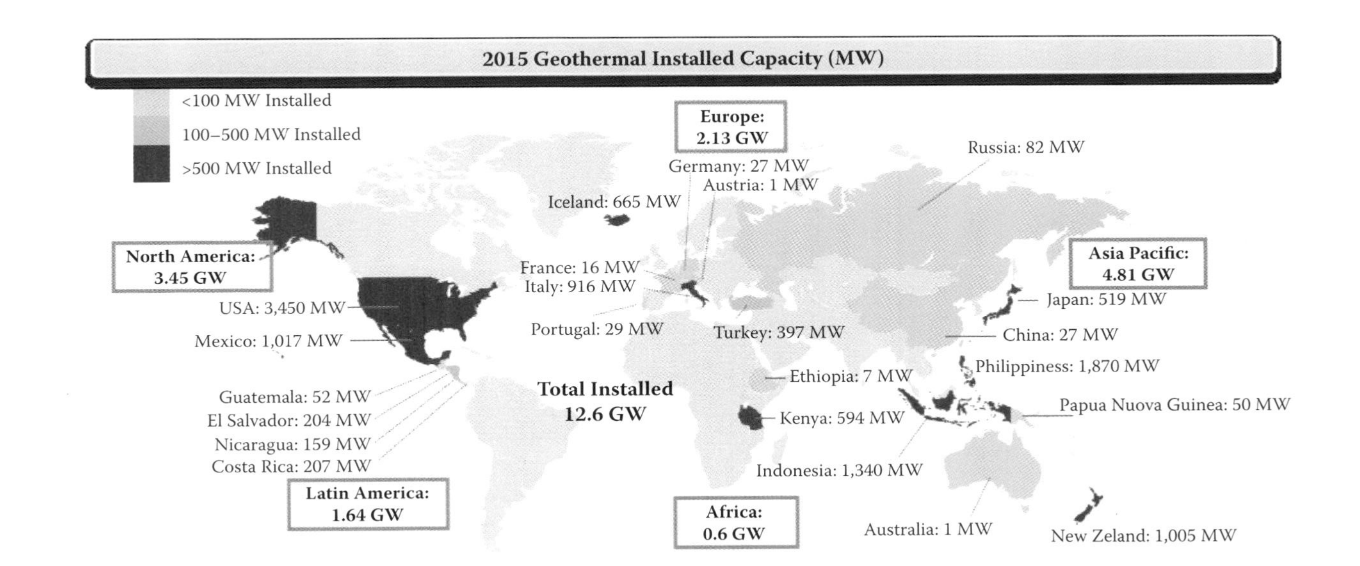

FIGURE 1.8 Total planet-wide installed geothermal electrical capacity as of beginning of 2015. The United States continues to have the largest installed capacity for geothermal power. (Adapted from Bertani, R., in *Proceedings of World Geothermal Congress 2015*, Melbourne, Australia, April 19–24, 2015.)

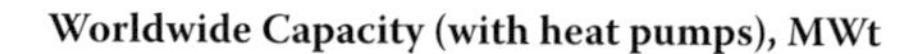

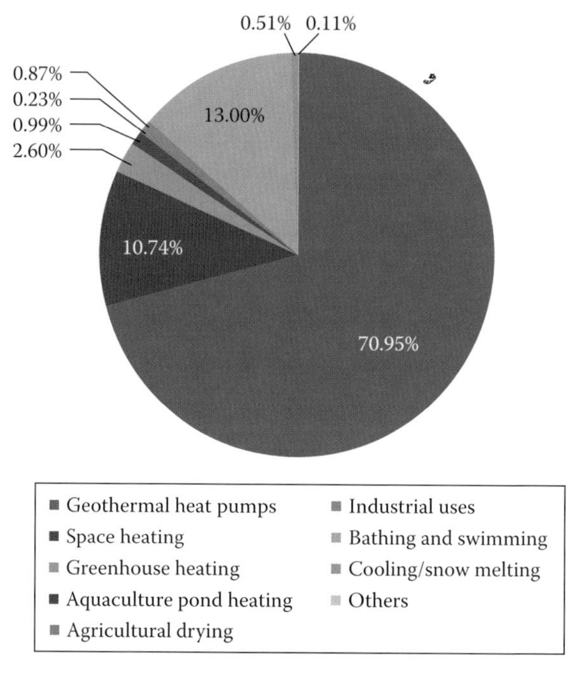

FIGURE 1.9 Pie chart showing the distribution of direct-use applications of geothermal energy. Geothermal heat pumps comprise the largest proportion by far, reflecting their wide geographic applicability. (Adapted from Lund, J. and Boyd, T., in *Proceedings of World Geothermal Congress 2015*, Melbourne, Australia, April 19–24, 2015.)

soaking. Direct-use geothermal fluids also are restricted to geologically favorable areas but are more widespread because modest temperature fluids are more common than high-temperature fluids required for electrical power generation.

The third form is geoexchange or ground-source (geothermal) heat pumps, which can be utilized virtually anywhere on the planet. Geoexchange is also a form of geothermal direct use, but at the low-temperature end. Rather than utilize geothermally heated fluids, geoexchange uses the Earth as a thermal bank in which heat is deposited in summer and withdrawn in winter. This is because the average temperature of the Earth a few meters below its surface is within the limited range of 10 to 15°C year-round over much of the planet. If geothermally heated fluids are unavailable, as is the case in most places, geoexchange systems are the most efficient way to heat and cool buildings for two main reasons—it is easier to move heat than create it and only a modicum of electricity is used to power circulating pumps and a compressor. Indeed, geothermal heat pumps comprise about 71% of worldwide installed direct-use capacity, followed by swimming and bathing and space heating from warm geothermal fluids (Figure 1.9). When all three forms of geothermal energy are considered, they currently offset about 300 million barrels of oil per year globally (Moore and Simmons, 2013). To put that in perspective, world daily oil consumption is estimated at 93 million barrels of oil, and the United States uses about 19.05 million barrels of oil per day (EIA, 2016; IEA, 2016). The environmental aspects of geothermal energy including both advantages and challenges are explored in Chapter 9.

The annual growth rate of geothermal energy development of a few percent has paled in comparison to that of wind and solar, which have grown at tens of percent per year over the last half dozen years or so. There are several reasons for this, including the high upfront capital costs and associated high risk of developing geothermal resources. Moreover, high-temperature geothermal systems require good permeability (the ability of water to flow through rock, as discussed in Chapter 5), hot rock, and an ample supply of circulating fluid at accessible depths (generally ≤3 km), but such areas are geologically rare and challenging to find. Another resource that may come into play at some point in the future is hot but relatively dry or impermeable rock. This currently undeveloped resource falls within the realm of enhanced or engineered geothermal systems (EGSs) and is discussed in more detail in Chapters 11 and 12. According to a seminal report on the future of geothermal energy (Tester et al., 2006), EGSs could provide 100 MWe or more of baseload, cost-competitive generating capacity in the United States by the middle of this century. For perspective, the United States has a current geothermal power-producing capacity of about 3500 MWe, which is less than 0.5% of the U.S. power capacity. Although EGSs offer the possibility of significantly expanding the development of geothermal energy, especially in the western United States, they will be expensive and will require additional water—water that is in short supply in the semi-arid but geothermally promising western United States. Nonetheless, developing EGS resources could move geothermal energy from a marginal to significant resource for baseload power and thermal energy and is being promoted by the Geothermal Technology Office of the U.S. Department of Energy as part of their new Frontier Observatory for Research in Geothermal Energy (FORGE) initiative.

With more renewable, yet intermittent, energy sources of solar and wind power coming onto the power grid, flexible power delivery may become the new "baseload." This is evident in the so-called *duck curve* (Figure 1.10), which shows a possible overgeneration of power during the middle of the day. This happens because solar generation is greatest in the middle of the day, which can force the curtailment of precontracted renewable power in order to maintain inflexible baseload sources of power, such as nuclear and natural gas power plants. The latter facilities are designed to operate most efficiently in a stable mode. Geothermal power plants, however, can have the ability to run as either a firm supplier of power (baseload) or a flexible supplier of power (such as load following—the ability to respond to changes in power demand during the course of the day), depending on how they are engineered and managed. For example, and as discussed further in Chapter 12, the Puna geothermal power plant on the Big Island of Hawaii is managed as an auxiliary power source and is able to increase or decrease power (Nordquist et al., 2013). In light of the growing use of intermittent power sources to satisfy environmental requirements for reduction in carbon emissions, geothermal has the potential to provide both baseload and flexible power, but to do so comes at a higher price and it must compete with power from ramping natural gas-fueled power plants.

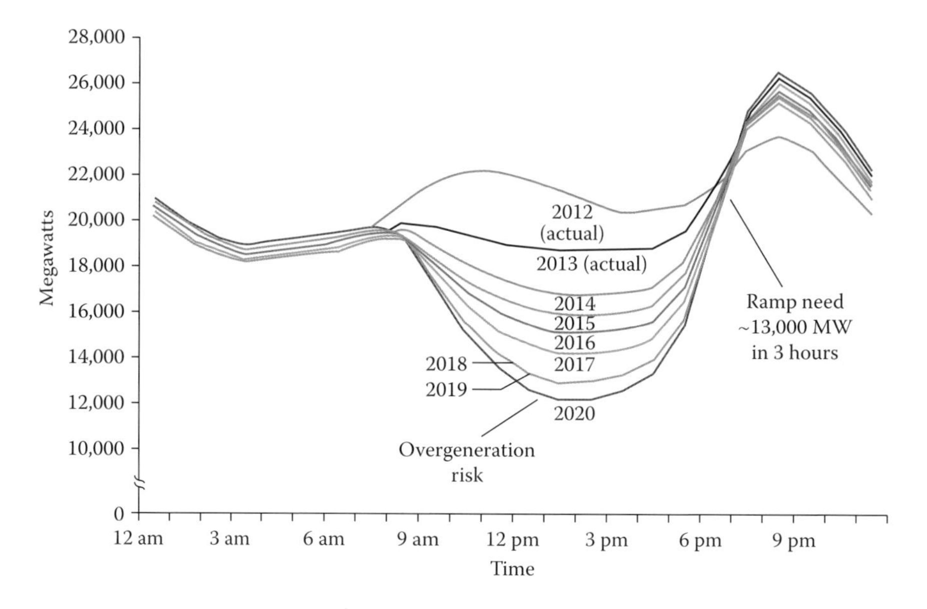

FIGURE 1.10 The so-called "duck curve" illustrating the reduction in power needs during the middle of the day (when solar power generation is greatest) and the ramp of power needed (about 13 GW by 2020) from about 5 p.m. to 9 p.m., when solar power generation drops off quickly. (From California ISO, *Fast Facts: What the Duck Curve Tells Us about Managing a Green Grid*, California Independent System Operator, Folsom, 2013.)

ORGANIZATION OF BOOK

This book focuses on how geology impacts finding, developing, and operating geothermal resources from high to low temperatures. The book is intended as an introduction to the subject, suitable for an undergraduate course in most geoscience departments. Some background in geologic principles is helpful but not required, as key geologic elements are discussed herein (see Chapter 4) to ensure understanding of the geology and geothermal phenomena. Accordingly, the book should be of interest and accessible to a wide audience, including engineers, environmental scientists, or the educated layperson interested in renewable/geothermal energy. Nonetheless, although information is presented at a fundamental level, the book can serve as a springboard to more advanced topics or in-depth treatment of the presented material. References are provided to help serve the needs of specialists.

Beginning in the next chapter, the various ways in which geothermal systems are classified are discussed. Classification not only helps in understanding the geologic processes behind the different types of geothermal systems but also aids in determining how geothermal resources might be used, whether for power generation (using either binary or flash technology, as discussed in Chapter 2) or for direct use. Chapter 3 explores the geology and heat architecture of Earth's interior which in turn affect how heat is moved, whether conductively or convectively. The roles of plate tectonics, earth materials (rocks and minerals), and geologic forces in the production of structures that serve as pathways and traps for geothermal fluid flow are discussed in Chapter 4. Chapter 5 explains how whether or not a geothermal system can be developed easily is strongly dependent on the ability of rocks to store and move heated geothermal fluids, reflecting the roles of porosity and permeability. How geothermal systems may be eventually developed depends in large measure on the physical and chemical characteristics of geothermal systems, as discussed in Chapter 6. In this light, the thermodynamic considerations of water are reviewed in preparation for looking at the two main physicochemical types of geothermal systems—the more common liquid-dominated systems and the rare, but more energy efficient, vapor-dominated systems.

The earlier chapters provide the foundation for analyzing the geologic and tectonic settings of select geothermal systems in Chapter 7. One important criterion is to distinguish geothermal systems that are powered by more deeply underlying magma from those that derive their heat energy from deeply circulating groundwater in thinned, hot crust (amagmatic). Furthermore, the type of geothermal resource is also impacted by the tectonic setting, such as divergent, convergent, transform, hot spot, continental rifting, or stable craton, and select examples of geothermal systems in different tectonic settings are discussed. Chapter 8 examines the discovery of geothermal systems, a process that involves geologic, geochemical, geophysical, remote sensing, and drilling of bore holes. One of the greatest benefits of developing and using geothermal energy is the environmental impact. Emissions are low or nil depending on the type of geothermal facility, and, compared to solar and wind power facilities, geothermal provides much more power per area of land used. This and other environmental attributes, along with potential problems, such as subsidence, induced seismicity, and potential degradation of geothermal surface manifestations, are presented in Chapter 9.

As described in Chapter 10, active geothermal systems are the modern analogs of those that formed mineral deposits, such as gold and silver, in the geologic past, with some rich enough to be mined. The potential deposition of minerals due to changes in temperature and pressure as geothermal fluids rise from the reservoir to the power plant, however, is to be avoided as it retards fluid flow and power production. An examination of how to avoid the deposition of minerals affords insights into the mechanisms involved in the formation of mineral deposits, helping in their exploration and potential development. Also, studying mineral deposits (fossil geothermal systems) aids in developing a better understanding of the thermal and pressure conditions of geothermal systems (by studying the mineralogy of the rocks hosting mineralization) without the complications of dealing with hot water. Indeed, some developed geothermal systems began as mineral exploration plays, and some mined mineral deposits still contain hot water that can help power some of the mining operation.

The final two chapters explore the next generation of geothermal systems (Chapter 11) and the future utilization of geothermal systems (Chapter 12). The next generation of geothermal systems includes engineered and enhanced geothermal systems (rock is hot but fluid is absent or is unable to flow well naturally), the potential use of supercritical fluids (water and CO_2) as a motive fluid, and deep, hot sedimentary aquifers that can provide a much larger target and volume of fluid than currently developed geothermal systems. The future development of geothermal energy depends as much on improved technical expertise as on political direction and economic factors, such as the country's commitment to reduce greenhouse gas emissions and the market price of fossil fuels. Natural gas, which is the cleanest of the fossil fuels, is currently inexpensive and abundant, and electricity from natural-gas-fueled power plants is possibly the greatest challenge facing continued development of geothermal power production. To help development of geothermal energy prosper in the future and compete economically, a leveling of the playing field of energy prices will be necessary. Competitive pricing for geothermal energy can be achieved as a result of either market forces causing the price of natural gas to increase or governmental sponsored programs, such as tax incentives or production tax credits for non- or low-carbon producing forms of energy. The future of geothermal energy is ripe with promise; time will tell if promise becomes reality.

SUMMARY

Fossil fuels, nuclear energy, and renewable energy are the main sources of energy, although fossil fuels provide more than 80% of the energy in the United States. Nuclear energy represents about 8%, and renewable sources have climbed to about 10%. The renewable energy sector is dominated by biomass and hydroelectric. Although solar, wind, and geothermal comprise only about 25% of the renewable energy sector, they are the fastest growing components, particularly wind and solar. The growth of geothermal energy, mainly for electrical power and direct use, is challenged by its restricted geographic availability (as dictated by geologic controls), high upfront costs of drilling, and risk of not finding a viable resource after incurring possible significant expenditures of time and money.

Energy and power are related but reflect different characteristics. Energy is a fundamental property of all objects and comes in many forms, such as kinetic, potential, nuclear, and thermal. Energy is measured in joules in the SI system and reflects the potential of doing work through applying a force over a given distance. A related term is *enthalpy*, which is discussed in greater detail in Chapters 2 and 6. Enthalpy reflects the internal energy of a system and its potential to do work. Power, on the other hand, is the rate at which energy is delivered or the rate at which work gets done and is measured in joules per second. One joule per second equals 1 watt. Power is related to energy by the following equation: power (P) = energy (E)/time (t); that is, $P = E/t$, or $E = P \times t$. Energy in the power industry is typically measured by the kilowatt-hour (kWh) or megawatt-hour (MWh). Units of power, then, are kW or MW, as the time term cancels.

Although development of geothermal energy has challenges, it also has distinctive attributes compared to other forms of renewable energy. For one, geothermal energy is a baseload resource and is not dependent on the sun shining and wind blowing, because the flow of heat (thermal energy) from the Earth is continuous. Hydroelectric power is also a baseload resource of energy but is sensitive to climatic conditions, such as drought, and energy and power output can vary from one season to the next or from one year to another. Other benefits of using geothermal energy include a small land footprint for the energy produced, low to nil gaseous and particulate emissions, and immunity to availability and price volatility of fuel sources.

Also, geothermal energy can be used over a range of temperatures. Electrical power generation can occur at high temperatures; direct use, such as for space and water heating, can take advantage of moderate temperatures; and geothermal heat pumps can operate at ambient Earth temperatures.

SUGGESTED PROBLEMS

1. How much energy can a 50-MWe geothermal power plant produce in a year (assuming the plant is online for 90% of the time)? Please show your work and report your results in units of kWh, Btu, and quad.
2. Compare the benefits and disadvantages, from an environmental view, of geothermal power plants and fossil-fuel power plants. Also, how are the development and use of geothermal energy different compared to energy produced by solar and wind power?

REFERENCES AND RECOMMENDED READING

Bertani, R. (2015). Geothermal power generation in the world: 2010–2015 update report. In: *Proceedings of World Geothermal Congress 2015*, Melbourne, Australia, April 19–24 (http://www.geothermal-energy.org/pdf/IGAstandard/WGC/2015/01001.pdf).

California ISO. (2013). *Fast Facts: What the Duck Curve Tells Us about Managing a Green Grid*. Folsom: California Independent System Operator (http://www.caiso.com/documents/flexibleresourceshelprenewables_fastfacts.pdf).

Denholm, P. and Margolis, R.M. (2008). Land-use requirements and the per-capita solar footprint for photovoltaic generation in the United States. *Energy Policy*, 36(9): 3531–3543.

Denholm, P., Hand, M., Jackson, M., and Ong, S. (2009). *Land Use Requirements of Modern Wind Power Plants in the United States*, Technical Report NREL/TP-6A2-45834. Golden, CO: National Renewable Energy Laboratory (http://www.nrel.gov/docs/fy09osti/45834.pdf).

EERE. (2016). *Installed Wind Capacity*. Washington, DC: Office of Energy Efficiency & Renewable Energy, U.S. Department of Energy (http://apps2.eere.energy.gov/wind/windexchange/wind_installed_capacity.asp).

EIA. (2011). *Today in Energy*. Washington, DC: U.S. Energy Information Administration (http://www.eia.gov/todayinenergy/detail.cfm?id=2650).

EIA. (2014). *Primary Energy Consumption by Source and Sector*. Washington, DC: U.S. Energy Information Administration (http://www.eia.gov/totalenergy/data/monthly/pdf/flow/css_2014_energy.pdf).

EIA. (2015a). *Annual Energy Outlook 2015*. Washington, DC: U.S. Energy Information Administration (http://www.eia.gov/forecasts/aeo/).

EIA. (2015b). *Renewable Energy Explained*. Washington, DC: U.S. Energy Information Administration (http://www.eia.gov/energyexplained/index.cfm?page=renewable_home).

EIA. (2016). *Frequently Asked Questions*. Washington, DC: U.S. Energy Information Administration (http://www.eia.gov/tools/faqs/faq.cfm?id=33&t=6).

Holm, A., Jennejohn, D., and Blodgett, L. (2012). *Geothermal Energy and Greenhouse Gas Emissions*. Washington, DC: Geothermal Energy Association.

IEA. (2015). *Key World Energy Statistics*. Washington, DC: International Energy Agency (https://www.iea.org/publications/freepublications/publication/KeyWorld_Statistics_2015.pdf).

IEA. (2016). *Oil*. Washington, DC: International Energy Agency (http://www.iea.org/aboutus/faqs/oil/).

Lund, J. and Boyd, T. (2015). Direct utilization of geothermal energy: 2015 worldwide review. In: *Proceedings of World Geothermal Congress 2015*, Melbourne, Australia, April 19–24 (http://www.geothermal-energy.org/pdf/IGAstandard/WGC/2015/01000.pdf).

Moore, J.N. and Simmons, S.F. (2013). More power from below. *Science*, 340(6135): 933–934.

Nordquist, J., Buchanan, T., and Kaleikini, M. (2013). Automatic generation control and ancillary services. *Geothermal Resources Council Transactions*, 37: 761–766.

NREL. (2016). *Concentrating Solar Power Projects*. Washington, DC: National Renewable Energy Laboratory (http://www.nrel.gov/csp/solarpaces/project_detail.cfm/projectID=60).

REN21. (2012). *Renewables 2012: Global Status Report*. Paris: Renewable Energy Policy Network for the 21st Century (http://www.ren21.net/resources/publications/).

SEIA. (2015). *Solar Market Insight Report 2015 Q1*. Washington, DC: Solar Energy Industries Association (http://www.seia.org/research-resources/solar-market-insight-report-2015-q1).

Stocker, T.F., Dahe, Q., Plattner, G.-K. et al. (2013). Technical summary. In: *Climate Change 2013: The Physical Science Basis. Contribution of Working Group I to the Fifth Assessment Report of the Intergovernmental Panel on Climate Change* (Stocker, T.F. et al., Eds.), pp. 33–115. New York: Cambridge University Press (http://www.climatechange2013.org/images/report/WG1AR5_TS_FINAL.pdf).

Tester, J.W., Anderson, B., Batchelor, A. et al. (2006). *The Future of Geothermal Energy: Impact of Enhanced Geothermal Systems (EGS) on the United States in the 21st Century*. Cambridge, MA: Massachusetts Institute of Technology (https://mitei.mit.edu/system/files/geothermal-energy-full.pdf).

USDOE. (2013). *Energy Department Announces New Funding Opportunity for Innovative Small Modular Reactors*. Washington, DC: U.S. Department of Energy (http://energy.gov/articles/energy-department-announces-new-funding-opportunity-innovative-small-modular-reactors).

USDOE. (2016). *Tips: Your Home's Energy Use*. Washington, DC: U.S. Department of Energy (http://energy.gov/energysaver/articles/tips-your-homes-energy-use).

USEPA. (2015). *Inventory of U.S. Greenhouse Gas Emissions and Sinks: 1990–2013*. Washington, DC: U.S. Environmental Protection Agency (https://www.iea.org/publications/freepublications/publication/KeyWorld_Statistics_2015.pdf).

2 Classification and Uses of Geothermal Systems

KEY CHAPTER OBJECTIVES

- Characterize the different types of geothermal systems in terms of physical characteristics such as temperature, porosity and permeability, and phase state of fluids.
- Identify what variables are used to determine how a geothermal system is used, such as for power generation, direct use, or ground-source heat pumps.
- Describe the three main types of geothermal power plants and identify the physical conditions that determine the type of power plant built.
- Identify the main components of the different types of power plants.
- Relate how the energy produced from power plants is a function of the difference in temperature of the input and output fluids and describe what is done to help maximize the difference in temperature.

CLASSIFICATION SCHEMES

A variety of methods or criteria have been applied to analyze geothermal systems, reflecting their complex and multidisciplinary nature. Some of the main schemes employed and discussed below include the following:

- How heat is transferred (conductive systems vs. convective systems)
- The types of heat sources (presence or absence of underlying molten rock or magma)
- The geologic or tectonic settings (location along or near plate boundaries or within the interior portions of continents)
- Environments of low, moderate, and high enthalpy or heat content
- The type of fluid medium present in the geothermal reservoir (liquid- or vapor-dominated)
- The uses of geothermal systems (e.g., power production, direct use of geothermal fluids, or geoexchange, also known as ground-source heat pumps or geothermal heat pumps)

Conductive vs. Convective Systems

In the Earth, heat is transferred mainly by either conduction or convection. *Conduction* is the transfer of heat by contact, and *convection* involves the transfer of heat via motion, mainly of a fluid, such as a liquid or gas. Convection is initiated from a thermal gradient that causes changes in the density of a fluid. In the presence of a gravitational field, the changes in density cause hot material to rise (become more buoyant) and cold material to sink (become less buoyant). Convection, therefore, requires good permeability and porosity* in order for fluid to flow and transfer heat. Where porosity and permeability are low, on the other hand, heat is more slowly transferred by conduction. Conductive and convective heat flow processes are discussed in further detail in Chapter 3. The difference in rates of heat transport between conduction and convection has implications for the sustainability of developed geothermal systems, as explored in Chapter 12.

Conductive Systems

Most of the heat transferred from Earth's interior to its surface occurs by conduction. The gradient at which this heat is transferred averages about 25 to 30°C per kilometer in the upper crust, but it can be higher or lower depending on the geologic setting and thermal conductivity of the rocks. In most geothermally active areas, the geothermal gradient is above 50°C/km. The geothermal gradient is commonly determined by measuring the temperature variations in bore holes, such as water wells or deeper wells of oil and gas that sometimes encounter temperatures >100°C at depths of 3 to 5 km. Indeed, bottom-hole temperatures from wellbores can be used to identify areas of elevated heat flow when discharge of geothermal fluids at the surface is lacking, as is the case for "blind" geothermal systems.

One type of conductive geothermal system consists of deep (3 to >6 km) *sedimentary aquifers*. These systems are heated by conduction from below. They consist of a sandstone or carbonate aquifer with modest permeability and are blanketed by rocks, such as a shale, that have low permeability and low thermal conductivity (attributes of a good insulator). A good example is the Paris Basin, in which a simple production and injection well couplet is used for heating. Deep sedimentary basins in the Great Basin of Nevada and Utah may represent an attractive target for possible power generation due to the region's anomalously high heat flow which allows temperatures of 175° to 200°C at depths of 4 to 5 km (Allis et al., 2012). Those depths are well within the limit of established drilling technology. Deep, hot sedimentary aquifers are discussed in more detail in Chapter 11.

Another type of conductive system is the *geopressured reservoir*, which forms where fluids are trapped in permeable horizons and are rapidly buried and isolated by impermeable rock layers. With time and with further burial, the pore fluid transitions from hydrostatic to lithostatic head (weight of the overlying rock rather than weight of an overlying column of water) (Figure 2.1). Geopressured reservoirs thus differ from the above-described deep sedimentary aquifers, whose pressure conditions are largely hydrostatic head as indicated from deep exploratory oil and gas wells (Allis, 2014). Geopressured reservoirs also commonly contain dissolved methane and sometimes oil. Because of deep burial (typically >3 km) the fluid is heated

---

* *Porosity* is the ratio of open space to solid material and is reported as a percentage. *Permeability* is a measure of the ability of a fluid to flow through a material and is commonly measured in darcy or millidarcy, after Henry Darcy, who studied the flow of fluids through porous media. Porosity and permeability are discussed in greater detail in Chapter 5.

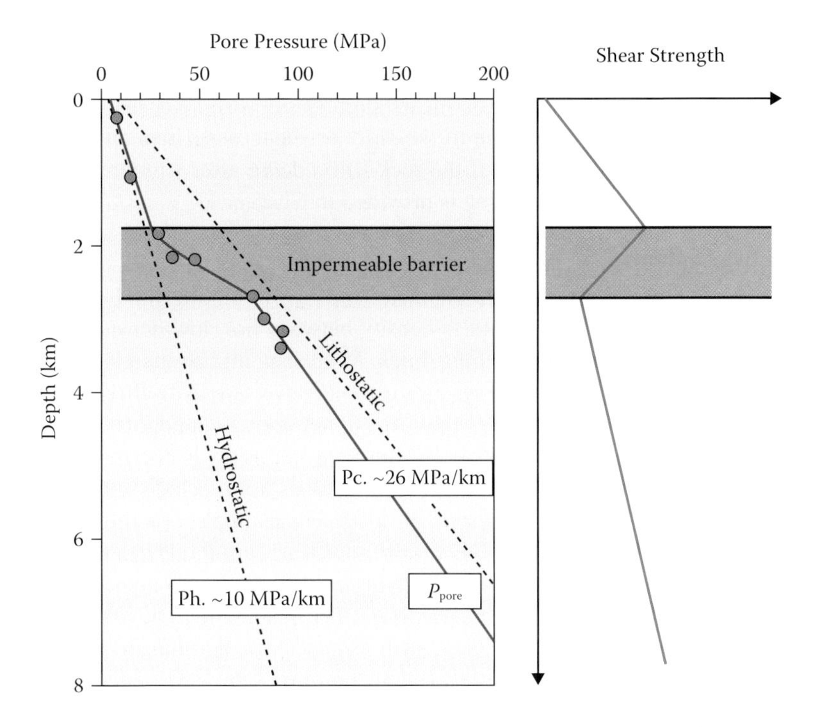

FIGURE 2.1 Graph showing the change in pressure with depth for hydrostatic and overburden (lithostatic) conditions. Geopressured reservoir conditions begin when pore fluid pressure (P_{pore}) exceeds hydrostatic pressure, as shown within and below the impermeable barrier by a red line. In a fully geopressurized reservoir, pore fluid pressure essentially equals lithostatic head, as noted below the impermeable barrier, and because pore pressure is high rock shear strength is reduced (green line). (From Geosciences, *Hydraulic Fracturing*, University of Sydney, School of Geosciences, New South Wales, 2016.)

to temperatures between 100° and 150°C. Thus, the fluid contains thermal energy, but it also contains chemical energy (from dissolved methane or oil) and mechanical energy resulting from the high pressure. Most known geopressured reservoirs are located along the Gulf Coast of the United States, where an experimental pilot geothermal plant operated in the 1980s. Results, however, were disappointing, as the plant was plagued by operational problems due to the high salinity of the fluids and carbon dioxide (CO_2); no further work has been done (Griggs, 2004).

A final type of conductive system is *hot dry rock*, in which high temperatures are encountered (typically >200°C), but there is little water or permeability. As such, the flow of heat results from conduction, not convection. The goal is to produce an artificial convecting hydrothermal system, or engineered geothermal system (EGS), by injecting cold fluids under pressure to fracture the rock. This creates pathways for fluids to flow and pick up heat that can then be channeled to the surface via a production well. This procedure is currently underway at Newberry Volcano in central Oregon and has met with modest success through a process called *hydroshearing*. Hydroshearing is different than *hydrofracturing*, or "fracking," in that hydroshearing uses no chemicals or proppants (solid particles to keep fractures open) and is conducted under relatively low fluid pressures. As cold fluids encounter hot rock, the rock contracts, breaks, and shifts or shears slightly sideways. Because sides of the fracture are no longer aligned, due to the shift or shear, asperities on the fracture walls prop the fracture open even if the rock shifts down after stimulation. A more thorough treatment of hydroshearing is provided in Chapter 11.

Convective Systems

Convective geothermal systems are characterized by circulating fluid due to buoyancy forces produced by differences in density between hot (low-density) upwelling fluids and cool (more dense) descending fluids. At present, all commercial geothermal power stations and most direct-use systems exploit convective hydrothermal systems. Surface manifestations of fumaroles (gas vents), hot springs, mudpots, geysers, and warm springs are characteristic of underlying convecting hydrothermal reservoirs. Nonetheless, the absence of such surface features does not preclude the existence of a blind convecting hydrothermal system at depth. For example, if a thick blanket of impermeable rock should cap the reservoir or upwelling, hot fluids can be channeled laterally away in the subsurface by an overlying cold groundwater aquifer. For fluids to circulate, high rock permeability is necessary. Permeability can be either an intrinsic property of the reservoir rock itself, such as sandstone that has primary permeability, or induced secondarily through fracturing of rock by Earth (tectonic) forces. As discussed in Chapter 5, fractures formed in otherwise impermeable rock, such as granite, can greatly enhance fluid movement and abstraction of heat from hot rocks.

From his years of research of the Steamboat Springs geothermal system, near Reno, Nevada, White (1973) proposed a model of how a convecting geothermal system operates (Figure 2.2). Surface and near-surface cool groundwater percolates downward along fractures in otherwise impermeable rock to a considerable depth (2 to 6 km), where it is heated from below by magma or anomalously hot rock. The heated water may flow laterally along a permeable rock horizon capped by relatively impermeable rock, or the heated water may rise to the surface along possible fractures in the cap rock due to buoyancy forces between cool descending fluid and hot rising fluid. Otherwise, the fluid will circulate in the permeable rock layer (reservoir), rising where hotter in the middle and sinking where cooler along the sides of the reservoir. The graph in Figure 2.2 illustrates changes in the temperature of fluid at different depths that correspond to the cross-section in the figure. A characteristic feature of a convecting system is the isothermal profile of temperature with depth (see points C and D on the graph and the corresponding cross-section illustration in Figure 2.2). By contrast, conductive heat flow is indicated for points C, F, and G (Figure 2.2), where the temperature steadily increases with depth because conductive heat flow is much slower than convective heat flow.

The other point to note in the graph in Figure 2.2 is the curve from E to D and its projection (dashed line of curve 1). This is the boiling point-to-depth curve, which shows that the boiling point increases with depth due to increased pressure. The curve shown here is for pure water, but it would shift to higher temperatures for a given depth with increased dissolved solids or if a component of lithostatic pressure is present in addition to the hydrostatic head.

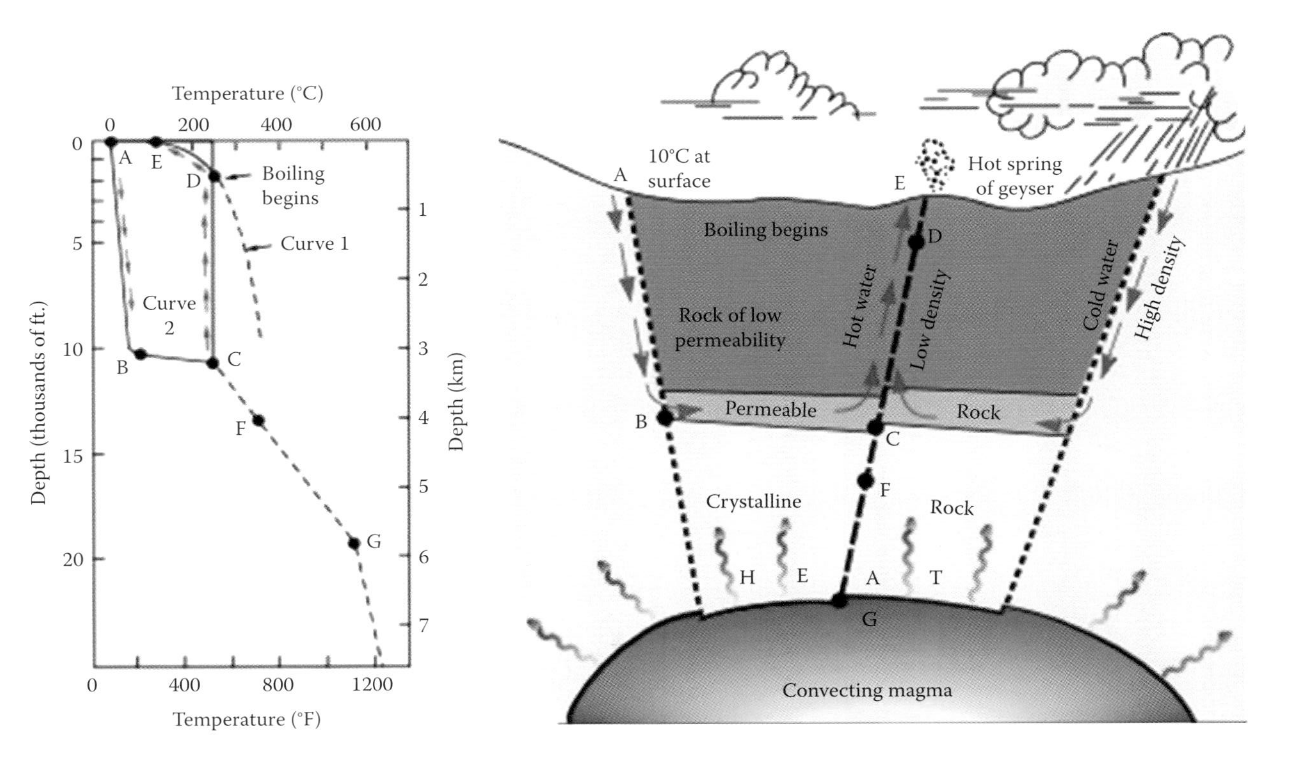

FIGURE 2.2 Graph and corresponding schematic sketch showing change in temperature of geothermal fluids above a heat source (convecting magma). Blue arrows show the path of recharging, cool, dense groundwater that is heated conductively from below. Red arrows show the path of buoyant hot water that rises convectively. Letters A to G in the graph denote positions shown in the cross-sectional sketch. See text for further details. (Adapted from White, D.E., in *Geothermal Energy, Resources, Production, and Stimulation*, Kruger, P. and Otte, C., Eds., Stanford University Press, Stanford, CA, 1973, pp. 69–94.)

Liquid- and Vapor-Dominated Systems

All convective systems can be divided into two types—liquid-dominated or vapor-dominated—depending on whether the mobile phase is liquid or vapor (steam). Most reservoirs are of the liquid type and have a vertical pressure distribution that is close to hydrostatic. In vapor-dominated systems, the pressure distribution is steam static, meaning that these systems are actually underpressured relative to the hydrostatically to lithostatically pressured rocks surrounding the steam reservoir. This, in part, explains the rarity of vapor-dominated systems, because if the vapor barrier is breached, the reservoir would be flooded and revert to a liquid-dominated system. White (1973) estimated that only about 5% of all geothermal systems are vapor dominated.

Vapor-Dominated Systems

The two largest vapor-dominated systems are The Geysers in northern California and in Larderello, Italy, where geothermal power was first demonstrated in 1904. These two systems have a combined capacity of about 2200 MWe, which represents about 17% of the total worldwide installed geothermal capacity (Bertani, 2015; DiPippo, 2012). Vapor-dominated systems are found in regions underlain by long-lived (>100,000 years), potent sources of heat (typically heat from underlying magma) and generally require temperatures of at least 240°C to form. At The Geysers, for example, the geothermal system was generated by the intrusion of magma into the upper crust about a million years ago. This led to the formation of an overlying liquid-dominated geothermal reservoir that persisted to about 0.25 million years ago (Hulen et al., 1997; Shook, 1995). The current vapor-dominated reservoir then developed upon abrupt depressurization, perhaps due to earthquake rupture, and consequent boiling and lowering of the liquid-dominated reservoir. In order for the steam cap to grow with time, some leakage must occur above the reservoir; otherwise, the system would pressurize, slowing or stopping evaporation from the underlying boiling brine. Therefore, unless the escaping steam is intercepted by a shallow aquifer and absorbed or carried away, blind vapor-dominated systems (i.e., having no visible surface expression) are unlikely (White et al., 1971). Also, the natural recharge must be slow enough to boil; otherwise, the vapor zone would be flooded by liquid. Despite wells being located as deep as nearly 4 km at The Geysers, a deep, boiling liquid brine has yet to be intercepted (Calpine, 2016).

Liquid-Dominated Systems

Liquid-dominated systems, also known as *conventional hydrothermal systems*, are the most common type of geothermal reservoir. Except for The Geysers and Larderello systems, and a handful of other much smaller examples of vapor-dominated systems, the rest of the world's geothermal power production comes from liquid-dominated reservoirs. The fluid remains a liquid due to the increase of the boiling temperature with depth, known as the boiling point-to-depth curve (BPD), reflecting the increased hydrostatic head (Figure 2.3). The BPD is also sensitive to dissolved solutes, further increasing the temperature of boiling. On the other hand, the presence of dissolved gases, such as CO_2, can lower the temperature of boiling. The first liquid-dominated system developed for geothermal power production was Wairakei in New Zealand

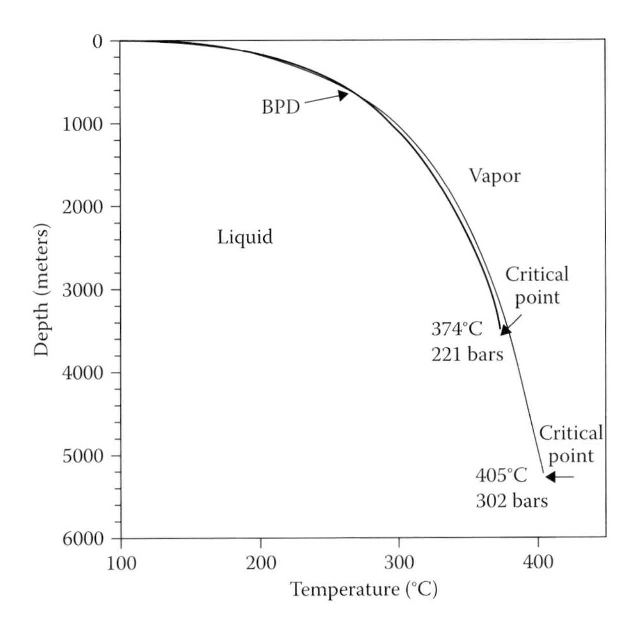

FIGURE 2.3 Graph illustrating the increase in boiling point temperature of water with depth or pressure. The bold line is for pure water and the thin line is for seawater (3.3% NaCl). The critical points indicate the temperature and pressure above which the fluid exhibits both liquid-like and vapor-like properties. Supercritical fluids are discussed further in Chapters 6 and 11. (From Arnorsson, S. et al., *Reviews in Mineralogy and Geochemistry*, 65(1), 259–312, 2007.)

in 1958. It recently celebrated 50 years of continuous production. In a typical liquid-dominated system developed for power production, the fluid will begin to flash, or boil, as it rises up a well due to the decrease in pressure. Only the steam fraction is sent to the power plant, which represents typically about 25 to 30% of the total fluid mass. The residual brine (fluid left over after flashing) is commonly reinjected back into the reservoir. Sometimes it is sent to an adjoining power plant (binary plant, as discussed later in this chapter) designed to handle lower temperature liquid before it is reinjected. On occasion, continued production from a liquid-dominated reservoir over time can lead to the formation of an artificial steam or vapor cap as the liquid level is drawn down from producing wells. This happened at Matsukawa in Japan and in parts of the Wairakei field in New Zealand (Poihipi). The steam forms due to persistent high temperatures and a reduction in hydrostatic head as fluid is withdrawn at a rate faster than recharge. The steam can become trapped and cannot escape, mainly because of an impermeable clay cap; otherwise, any formed steam escapes and fluid pressure is reduced, typically resulting diminished power production.

Temperature and Uses

Another straightforward means of classifying geothermal systems is based on temperature, which certainly impacts potential uses of the systems. This form of classification is most commonly used by engineers who are keen to see how a given geothermal system might best be utilized, because the greater the heat content (or *enthalpy*[*]), the greater the ability to do work. Using this scheme, systems are ranked as low, moderate, or high enthalpy. Temperature divisions separating the different levels of enthalpy are somewhat arbitrary but are mainly based on use. For purposes here, the following approximate divisional boundaries will be employed: low enthalpy (<~100°C), moderate enthalpy (~100° to ~175°C), and high enthalpy (>~175°C).

Low-Enthalpy Systems

Low-enthalpy systems would involve using geothermal fluids directly (typically referred to as *direct use of direct geothermal energy*) and geoexchange systems (also known as *ground-source heat pumps* or *geothermal heat pumps*), which utilize the Earth's ambient temperature at depths of a few meters to a hundred meters or so. Direct-use fluids are still the type most commonly used (since antiquity, in fact) for swimming and bathing, followed by space and water heating (Lund and Boyd, 2015; Lund et al., 2004). Other important direct uses of geothermal fluids include fish or alligator farming (aquaculture), vegetable and fruit drying, commercial greenhouse production, and processing of ore in certain mining operations. If direct-use fluids are hot enough, about 90°C or higher, they can also be used for cooling, as in absorption chillers, which use thermal rather than electrical energy. Cooling in this case works on the same principle as a gas refrigerator that uses heat from burning natural gas to evaporate a refrigerant (commonly a mixture of water and ammonia or lithium bromide). Rather than using a gas flame as the heat source, an absorption chiller uses the hot geothermal fluid to promote evaporation of the refrigerant. The process of evaporation cools the residual liquid and surrounding air.

Lowest enthalpy systems consist of geoexchange systems, also known as geothermal or ground-source heat pumps. Rather than being strictly a form of heat mining, geoexchange uses the Earth as a thermal bank to both withdraw (heat source) and deposit (heat sink) energy to minimize heating and cooling requirements achieved by other means, such as burning fossil fuels. Geothermal heat pumps provide the highest efficiency (and hence lowest costs) in heating and cooling compared to other technologies (USDOE, 2016). Because of their widespread applicability, geoexchange systems currently offer the greatest potential for growth (>10% annually), compared to other enthalpy levels of geothermal energy (Lund and Taylor, 2015; Lund et al., 2004).

Moderate- and High-Enthalpy Systems

Moderate- and high-enthalpy systems can be used for both electrical power generation and direct use, depending on temperature and need. Moderate-enthalpy systems (<~175°C) are generally too cool to power a steam turbine directly from the flashing of water because the steam fraction is less than about 20% of the total fluid mass (Glassley, 2015). A way to address this problem is to use a secondary or binary fluid (typically a hydrocarbon) that has a lower boiling point than that of water. A larger

[*] Enthalpy is a thermodynamic property that reflects the internal energy, pressure, and volume of a system. In general, as temperature increases so does the enthalpy and the potential to do work. The SI unit for enthalpy is the joule, but other units include the British thermal unit (Btu) and calorie.

mass of steam (and hence higher pressure) is produced, compared to using water alone, to drive a steam turbine generator. Thus, two fluids are used, as a primary geothermal fluid heats and vaporizes a secondary or working (motive) fluid via a heat exchanger. Appropriately, geothermal systems of moderate enthalpy are exploited by such binary geothermal power facilities.

In high-enthalpy systems the produced geothermal fluid is used directly in power generation, removing the need for a heat exchanger as used in a binary geothermal power plant. For a liquid-dominated reservoir, enough fluid flashes to steam to drive a steam turbine in what are referred to as *flash power plants*. The higher the enthalpy (temperature) of the fluid, the greater the proportion of steam produced and the greater amount of power generated (see Chapter 6 for details on this process). In the rare vapor-dominated systems, all the fluid goes to the power plant, resulting in higher power per fluid mass produced as there is no partitioning between liquid and vapor phase as in liquid-dominated systems.

Coproduction geothermal facilities, also called *combined heat and power* (CHP) plants, produce both electrical power and thermal energy. Thermal energy comes from fluid left over after producing electrical power. Such plants require geothermal systems of high enthalpy (generally >200°C) so fluids contain enough heat to make a direct-use application possible. A good example of such a facility is the Hellisheidi geothermal plant near Reykjavik, Iceland, that provides approximately 300 MWe of electrical power and 133 MWt of thermal energy for heating homes, businesses, and swimming pools. Interestingly, Iceland has the highest number of heated swimming pools per capita of any country on the planet, thanks to its natural endowment of geothermal resources (Arnórsson et al., 2008).

Geologic and Tectonic Setting

As will be discussed in Chapter 4, most geothermal resources, except for geoexchange systems, are restricted to select geologic or tectonic regions. Most of these geothermally producing or prospective regions are located along or near the boundaries of the planet's dozen or so tectonic plates or widely scattered geologic hot spots that can occur within tectonic plates, such as the Hawaiian Islands (Figure 2.4). In these areas, the flow of heat to the surface is typically much higher than average. Average continental crustal heat flow is about 65 milliwatts per square meter (mW/m^2). Along or near tectonic plate boundaries or geologic hot spots, heat flow can be 100 W/m^2 or greater.

The boundaries of tectonic plates consist of three main types: convergent, divergent, and transform. Convergent boundaries occur where plates collide, divergent where plates separate, and transform where plates slide past each other. Other important geologic settings for moderate- and high-enthalpy geothermal systems are hot spots and deeply buried sedimentary basins in the interiors of continents. Hot spots are the surface manifestations (volcanoes) of localized upwelling plumes of hot material within the interior of a tectonic plate, such as Yellowstone National Park and the Hawaiian Islands. Convergent and divergent boundaries are typically characterized by active volcanoes that develop above reservoirs of magma residing at depths of about 5 to 10 km. Examples of convergent and divergent boundaries are

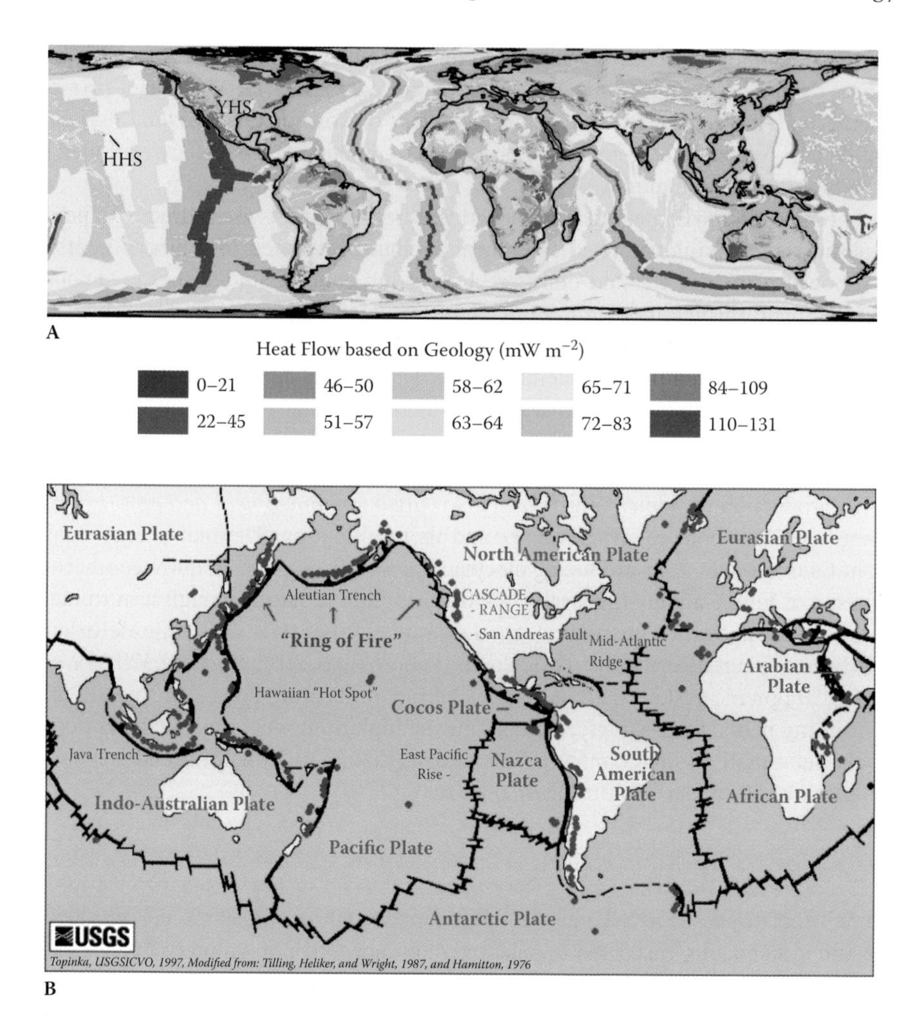

FIGURE 2.4 (A) Heat flow map of the world. HHS and YHS denote Hawaiian hot spot and Yellowstone hot spot, respectively. (Adapted from Davies, J.H., *Geochemistry, Geophysics, Geosystems*, 14, 4608–4622, 2013.) (B) Map of tectonic plates. Plate boundaries are indicated by heavy black lines and active volcanoes by red circles. Note how boundaries of plates coincide with areas having high heat flow and therefore good geothermal energy potential. (Illustration adapted from https://commons.wikimedia.org/wiki/File:Map_plate_tectonics_world.gif.)

the Cascade volcanoes in the Pacific Northwest of the United States and Iceland in the north Atlantic, respectively. Because of its lower density relative to surrounding rock, magma rises upward into the upper crust. Above the pools of magma, heat is conducted upward where it can engender, if fluids and permeability are present, convective geothermal systems within the top few kilometers of the crust. Transform boundaries are more complex, as localized volcanism and crustal extension can contribute to anomalously high heat flow. When the crust is extended or stretched, it is thinned and hot rocks at depth are then closer to the surface; this increases the heat flow and geothermal gradient (change in temperature with depth), which can foster the formation of geothermal systems at accessible depths for potential development. Further details on plate tectonic boundaries, intraplate geologic hot spots, and relationships to geothermal systems are explored in Chapter 4.

The geologic/tectonic setting of a geothermal system impacts the temperature and physicochemical characteristics of geothermal systems, which is important for exploration and development. For example, geothermal systems located on the flanks of volcanoes associated with convergent boundaries typically have high temperatures (>200°C). This is good for power production, but they can contain acidic gases, which can potentially damage equipment, or high dissolved chemical components, which can lead to scaling in piping and reduced flow rates and power production.

Magmatic vs. Amagmatic Systems

Worldwide most commercially producing geothermal regions derive their heat energy from more deeply buried regions of molten rock (magma) and are aptly designated as magmatic. These systems are typically of high enthalpy and consist of both vapor- and liquid-dominated systems, with the latter supporting flash power plants. Due to the presence of underlying magma, heat flow is quite high and drill depths for tapping the geothermal reservoirs are modest (typically between depths of 1 and 2 km). Magmatic geothermal systems are characteristic of active volcanoes related to convergent and divergent boundaries and geologic hot spots. Magmatically heated geothermal systems can also form locally along transform boundaries, such as The Geysers, under proper geological conditions (see Chapters 4 and 7).

In the last 20 years or so, a new class of geothermal system has been exploited for producing geothermal energy. These geothermal systems are heated as a result of crustal extension, which thins the crust, effectively positioning hot rocks of the Earth's mantle closer to the surface. As the crust is stretched, it breaks, producing fractures and faults that provide pathways for near-surface groundwater to move downward. Groundwater becomes heated as it moves through hot rocks at depth and then rises buoyantly upward along other faults and fractures. Molten rock is lacking, but heat flow and the geothermal gradient are elevated. Systems formed under these conditions are termed *amagmatic* or *extension-related*. Many of the systems in the Great Basin of the western United States are amagmatic and generally have temperatures < 200°C at depths of 1 to 2 km. They are most commonly, but not always, developed using binary geothermal plants. Because of the lower geothermal gradient and heat flow compared to typical magmatic systems, wells in many cases must be drilled deeper to access temperatures high enough to support a geothermal power plant.

TYPES OF GEOTHERMAL ENERGY PLANTS

Electricity-producing geothermal power plants consist of three main types: dry steam, flash, and binary. These plant types include variations such as hybrid versions, which typically involve combined or integrated flash and binary systems. Also, some power plants use mixed technologies, such as solar photovoltaic (PV), solar thermal, and geothermal binary. For example, the hybrid Stillwater geothermal facility in western Nevada is the first of its kind to combine solar PV and solar thermal to augment geothermal electrical production. As previously noted, another type of geothermal energy facility is the combined heat and power (CHP) plant, which produces both electrical power and thermal energy for space heating.

Dry Steam Power Plants

For about 45 years, prior to 1958, dry steam power plants were the only form of geothermal electrical power generation. The first commercial geothermal power plant consisted of dry steam and was commissioned in 1913 in Larderello, Italy. Dry steam plants are the simplest and most energy efficient of all types of geothermal plants, as almost all of the fluid mass produced by the wells goes to the power plant to drive

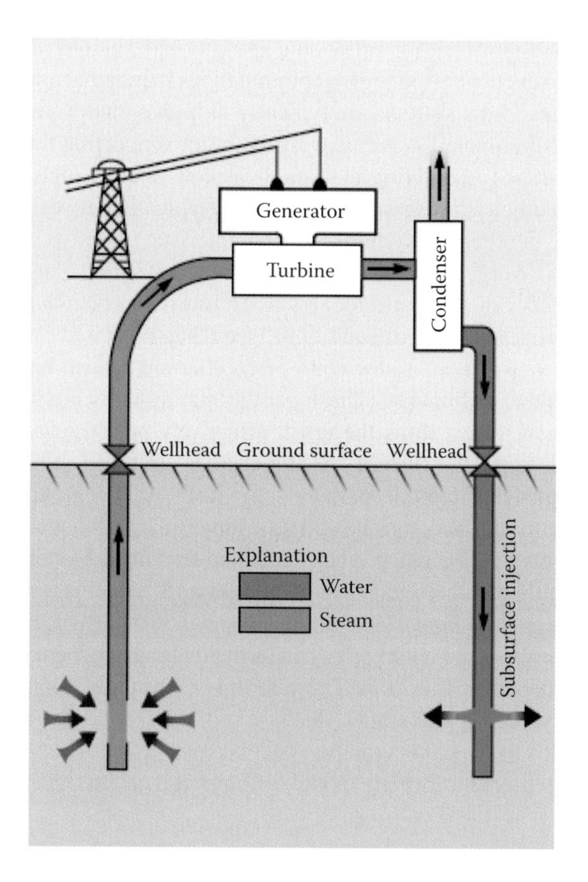

FIGURE 2.5 Simplified diagram showing basic components of a dry-steam geothermal power plant developing a vapor-dominated geothermal system. Steam is drawn directly from the wells and piped to the power plant to drive a turbine generator. After the steam is condensed, the condensate is reinjected to help prolong the productive life of the geothermal system. (Adapted from Duffield, W.A. and Sass, J.H., *Geothermal Energy—Clean Power from the Earth's Heat*, Circular No. 1249, U.S. Geological Survey, Reston, VA, 2003.)

a turbo generator (Figure 2.5). No partitioning of energy between liquid and steam occurs, unlike in flash power plants tapping liquid-dominated systems. This is because this partitioning between vapor and liquid was already achieved in nature via development of a steam cap above a deeply boiling brine (White et al., 1971). Therefore, most of the enthalpy of the fluid is fully available for conversion of heat energy to mechanical energy to ultimately electrical energy. The steam enters the turbine at pressures of 40 to 100 pounds per square inch (3 to 7 bars) and at velocities of several tens of meters per second. Furthermore, because separation of brine and steam is unnecessary, equipment and piping expense is minimized. Rather than a separator as used in flash power plants, only a filter is necessary to remove any solid or liquid particulates that might be present with the steam to prevent damage to the turbine blades.

Individual plant sizes range from about 50 MWe to 100 MWe. This high conversion of geothermal fluid energy to mechanical and electrical energy and the simplicity of power plant construction and operation make dry steam plants the gem of the geothermal power industry. As of mid-2011, about 12% of all geothermal power plants on the planet were dry steam (DiPippo, 2012). By 2015, the percentage of dry steam geothermal power plants had decreased slightly to about 10% due to the construction of other types of power plants (mainly binary). Yet, because of their high enthalpy and large fluid flux rate to the turbine, dry steam plants account for 22% (2863 MWe) of total worldwide geothermal power production (Bertani, 2015).

Flash Power Plants

This type of energy conversion technology is the current mainstay of the geothermal power industry worldwide. As of mid-2011, 228 units of this type were in operation in 17 countries, accounting for almost 40% of all geothermal power plants (DiPippo, 2012). By 2015, the number of flash power plants (including single, double, and triple flash) increased to 237 units, providing 63% of geothermally produced power on the planet (Bertani, 2015). These power plants develop high-enthalpy (typically >175°C), liquid-dominated geothermal systems. A simplified schematic of a flash plant is shown in Figure 2.6. Production wells feeding these power plants deliver two-phase flow as water begins to boil or flash due to lower pressure in the upper parts of the wells. This two-phase fluid goes to a separator in which the steam is separated from the liquid or brine (Figure 2.7). Only the steam is delivered to the powerhouse. The residual brine usually is either reinjected directly or sent to a bottoming binary plant prior to reinjection, or it can also be used to heat domestic water for direct-use applications. Fluid is reinjected ultimately in most cases to maintain reservoir pressure and to minimize potential environmental contamination of shallow groundwater used by municipalities for irrigation and drinking water.

Based on the pressure–enthalpy relations of water, generally only about a third of the total fluid mass delivered in the production wells is converted to steam (Glassley, 2015). The actual proportion of steam available is a function of the difference in temperature of the steam entering and leaving the turbine—the greater the difference in temperature of ingoing and outgoing steam, the larger the proportion of steam separated and the more power that can be generated. The importance of exit temperature, pressure, and produced power is discussed further in the section on condensers.

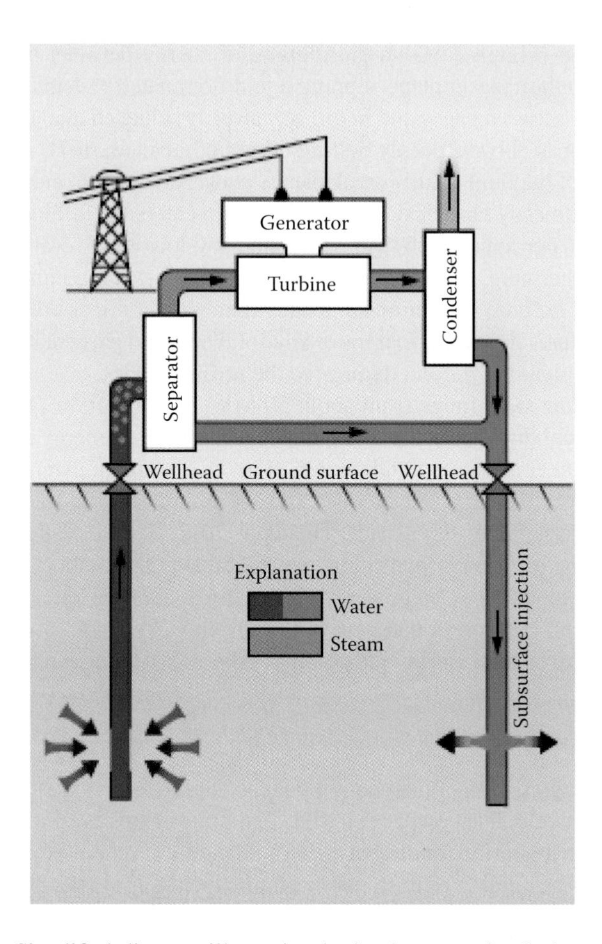

FIGURE 2.6 Simplified diagram illustrating basic elements of a flash geothermal plant developing an underlying liquid-dominated geothermal reservoir. Due to decreasing pressure as the fluid rises in the well, it begins to flash or boil, yielding a two-phase mixture of steam and brine. The steam is separated and piped to the turbine generator. The residual brine is reinjected to help prolong the productive life of the geothermal system. (Adapted from Duffield, W.A. and Sass, J.H., *Geothermal Energy—Clean Power from the Earth's Heat*, Circular No. 1249, U.S. Geological Survey, Reston, VA, 2003.)

Depending on the temperature of the incoming geothermal fluid, the fluid may undergo more than one flash (Figure 2.8). Under such circumstances, flash plants may involve double or even triple flashes. For multiple flashing to be energetically worthwhile, temperature of the fluid must be generally >200°C. Multiple flashing, consisting of a high-pressure flash at about 6 to 7 bar and another at a low pressure of 2 to 3 bar, yields about 15 to 25% more power than a single flash for the same geothermal fluid conditions (DiPippo, 2012). Clearly, such a power plant will be more complex requiring more upfront equipment costs and maintenance, but the added power produced in most cases justifies the added expense. Some of the extra maintenance for double- or triple-flash plants results from possible additional coating (scaling) of

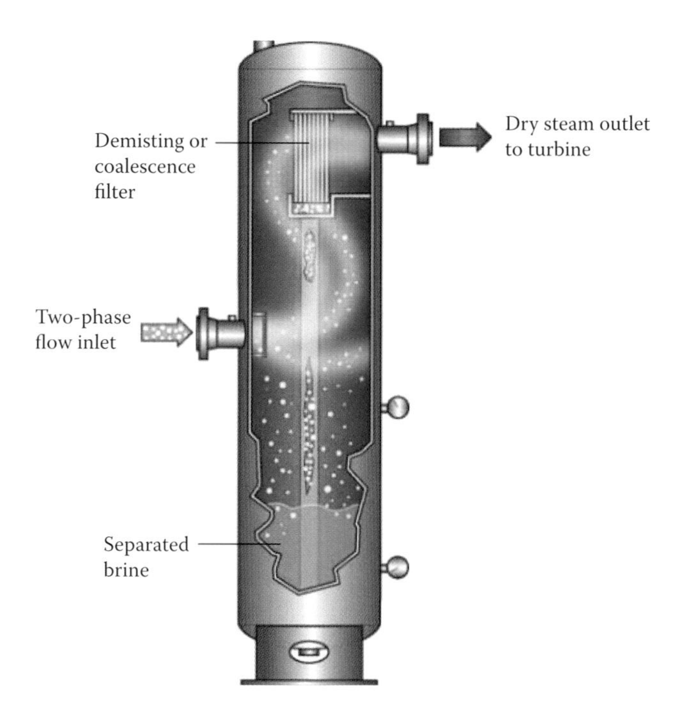

FIGURE 2.7 Schematic view of a cyclone separator. Heavier water particles are thrown to side of vessel and flow down to accumulate as brine at the bottom, while low density steam rises toward the top. (Illustration adapted from www.Peerless-Canada.com.)

pipes from mineral deposition, such as silica, due to the additional cooling and pressure drop of fluids brought about by multiple flashing prior to reinjection (changes in solubilities of dissolved minerals in geothermal fluids are discussed in Chapter 6).

Worldwide, about a quarter of all geothermal flash power plants involve multiple flash cycles, whereas in the United States the majority of flash plants (about 88%) utilize double flash. This may reflect the greater availability of capital in the United States to build and maintain more expensive (but more efficient) multiple-flash units. On a worldwide basis, 63% of geothermal power is produced from flash plants (single, double, and triple), reflecting the importance of this technology (Bertani, 2015). In terms of actual flash plants, 167 are single flash, 68 are double flash, and only 2 are triple flash. Although double-flash plants comprise 29% of total flash plants, they produce 34% of the power (Bertani, 2015), reflecting their more efficient extraction of energy per given mass of fluid produced.

Binary Geothermal Power Plants

Binary geothermal power plants have been the fastest growing type as they develop moderate-enthalpy (generally 120° to 175°C), liquid-dominated resources. These are more common and widespread than high-enthalpy geothermal systems developed using flash plants. As such, they are the most common type of geothermal

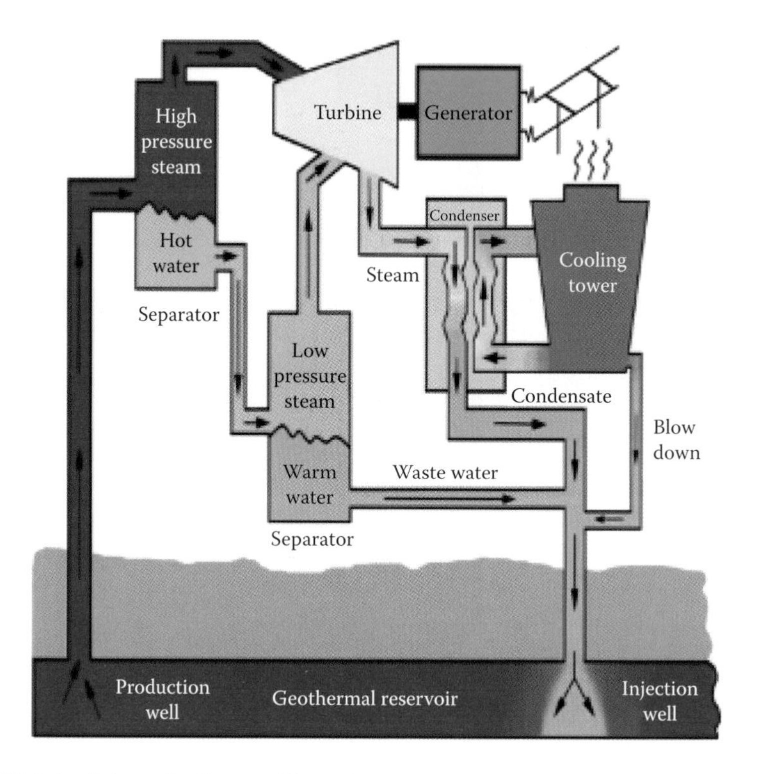

FIGURE 2.8 Schematic diagram illustration a double-flash geothermal power plant. The geothermal fluid contains enough enthalpy to support a high- and low-pressure flash to boost power production. See text for details. (From Gradient Resources, *Geothermal Technology*, Gradient Resources, Reno, NV, 2016.)

power plant. From 2011 to 2015, the percentage of binary geothermal power plants increased from 40% to 47% worldwide (Bertani, 2015; DiPippo, 2012). Although binary geothermal plants are the most common type, they generate only 12% of the total geothermal power (Bertani, 2015), which reflects the smaller energy potential from lower temperature (enthalpy) resources. Accordingly, the average power produced from binary plants (excluding hybrid plants) is only about 6.3 MWe on a planet-wide basis (Bertani, 2015). Nonetheless, the percentage of power produced from binary plants increased from 7% to 12% since 2011 (DiPippo, 2012).

Most binary geothermal power plants built in the United States have an average installed capacity of about 20 MWe (Bertani, 2015; DiPippo, 2012). The largest is the newly operational McGinness Hills facility in central Nevada with 72 MWe installed capacity (two plants at 36 MWe each; B. Delwiche, pers. comm., 2015). The single largest binary geothermal power facility in the world is the recently commissioned Ngatamariki plant in New Zealand with an installed capacity of 100 MWe (Legmann, 2015). Part of the reason for the high power output of the Ngatamariki plant is that the wells produce geothermal fluids of 193°C, which is about 30° to 40°C hotter than typical fluids utilized by binary geothermal power plants. The decision to use binary rather than flash technology at Ngatamariki involved the following:

- One hundred percent of the geothermal fluid is reinjected, thus promoting reservoir stability.
- Zero water consumption is the result of using air-cooled condensers, eliminating the need for make-up water to replace geothermal fluid loss from evaporative cooling.
- There are no air emissions, as geothermal and working fluids are in closed loops and not exposed to the atmosphere.
- Total costs over the life span of the plant are lower compared to those of a comparable flash plant.

As noted, binary geothermal power plants use two fluids; the geothermal fluid heats a secondary or working fluid (commonly a hydrocarbon such as isopentane). The working fluid has a lower boiling point than water so a higher amount of steam is produced than if the geothermal fluid were flashed solely. The two fluids circulate through a heat exchanger (or *vaporizer*), where the heat of the geothermal fluid boils the working fluid, whose steam goes to drive a turbine in the power house. Both the geothermal fluid and working fluid form closed loops, so there are no emissions, making binary geothermal power plants the most environmentally clean of the three types of geothermal power plants. A generalized schematic diagram of a binary geothermal power plant is shown in Figure 2.9.

Binary geothermal plants are the mainstay for moderate-enthalpy geothermal resources typically developed in amagmatic regions of actively extending crust. Examples are the Great Basin of the western United States and basins in western Turkey. They can also occur, in addition to flash plants, in other geologic settings, not only in New Zealand but also in Iceland, Costa Rica, and Kenya, depending on the temperature of the utilized geothermal system at a given location. In many of the above localities, flash plants are paired with binary plants (*bottoming plants*) in which the separated brine, prior to reinjection, is used to heat a secondary working fluid to drive a separate turbine to gain additional power output. Such a facility is a type of hybrid power plant, as discussed below.

At Steamboat Springs, Nevada, the operator, Ormat, uses a dual-fluid binary arrangement in its newer power plants. The incoming geothermal fluid contains sufficient heat to boil the secondary working fluid in two loops—a high-pressure and a low-pressure loop. The steam from the high- and low-pressure loops goes to different turbines connected on each side of a single generator (Figure 2.10).

Hybrid Power Plants

As the name implies, hybrid geothermal power plants combine multiple technologies at a given facility. Some important hybrid plants consist of

- Integrated flash–binary plants
- Combined cycle flash–binary plants
- Integrated combined cycle flash–binary plants
- Combined heat and power (CHP) geothermal plants
- Combined solar and binary geothermal plants

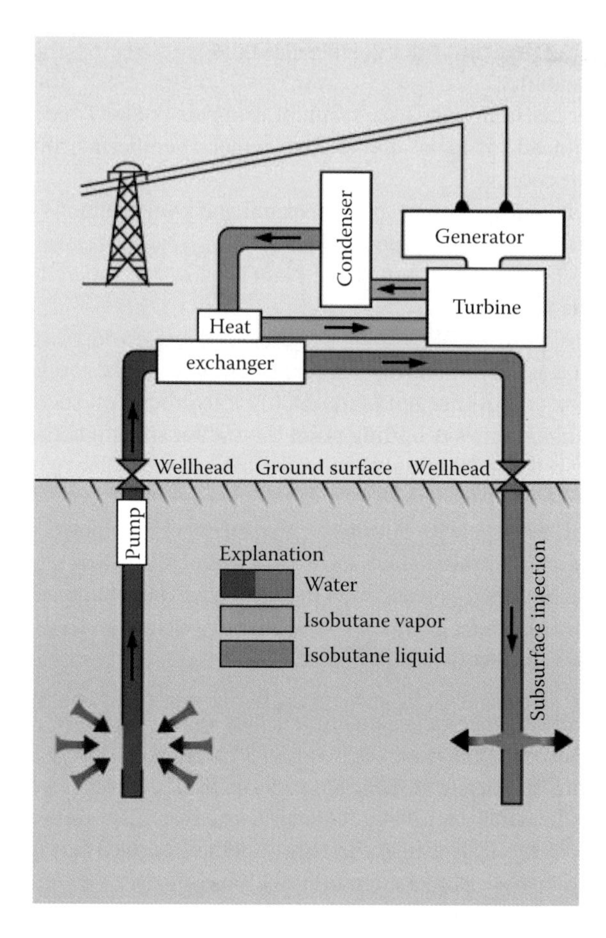

FIGURE 2.9 A diagram showing how electricity is generated from a binary, moderate-temperature hydrothermal system. The geothermal fluid is used to boil a second or working fluid (isobutene, in this example) whose vapor then drives a coupled turbine generator. After heating the secondary working fluid, the geothermal fluid is reinjected to help prolong the life of the hydrothermal system, or it can be used for direct-use heating prior to reinjection. (From Duffield, W.A. and Sass, J.H., *Geothermal Energy—Clean Power from the Earth's Heat*, Circular No. 1249, U.S. Geological Survey, Reston, VA, 2003.)

The integrated flash–binary plant is a common configuration in which the brine from the separator of a flash plant is sent to power a binary plant (bottoming plant) that adjoins the flash plant. Adding a bottoming binary plant to the Steamboat Hills flash plant boosted power output by about 10 to 15% for the combined operation (J. Nordquist, pers. comm., 2015). A further advantage of a binary bottoming plant, rather than having the fluid undergo a second low-pressure flash, is that the possibility of mineral deposition (scaling) in piping is reduced. This is because the silica concentration, in this case, is not increased as there is no secondary flashing in the binary plant (DiPippo, 2012). A potential downside is that the injected brine is now cooler, by approximately 20°C, than if reinjected directly after steam separation. This process

FIGURE 2.10 View of a portion of the Galena III power plant at Steamboat Springs, Nevada. The cylinders on the left contain high-pressure steam that goes to a high-pressure turbine (A), whereas the cylinders on the right contain low-pressure steam that goes to a low-pressure turbine (B). Both turbines are connected to the generator (C). (Adapted from https://en.wikipedia.org/wiki/Ormat_Industries.)

may cool the geothermal reservoir more quickly, resulting in reduced power output over time. To counteract this potential cooling of the reservoir, new wells may need to be drilled in different parts of the reservoir unaffected or less affected by injected fluids. Sustainable development of geothermal resources is explored in Chapter 12.

Another type of hybrid plant is the combined flash–binary plant (Figure 2.11). Rather than having two independent cycles as in the integrated flash–binary plant, the steam, after leaving the steam turbine, is used to evaporate a secondary working fluid to drive a binary turbine. The initial steam turbine in this configuration is also called a *back-pressure turbine* because the steam is not directly condensed upon leaving the turbine as in a typical steam-powered flash plant. Condensation of steam occurs in the process of vaporizing the secondary fluid, thus it occurs at higher pressure than in a typical condenser, hence back pressure. The originally separated brine goes to preheat the secondary working fluid prior to its vaporization.

A variation of the combined cycle power plant is the integrated combined cycle plant, in which the separated brine is not used to preheat the secondary or working fluid as above, but goes to a separate binary bottoming unit to flash a separate working fluid cycle that drives a separate turbine generator. This configuration is used for high-enthalpy resources characterized by high steam pressure, high gas content (noncondensable gases such as CO_2 and H_2S), and a high water fraction. A good example of the integrated combined cycle power plant configuration is Ormat's 38-MWe Puna geothermal facility on the Big Island of Hawaii (Figure 2.12).

As touched on above, in several cases both power generation and direct heat usage are achieved in a single power plant known as a combined heat and power (CHP) plant. By capturing the unused heat in the residual brine prior to reinjection the overall utilization of the resource is optimized. Studies have shown that utilization

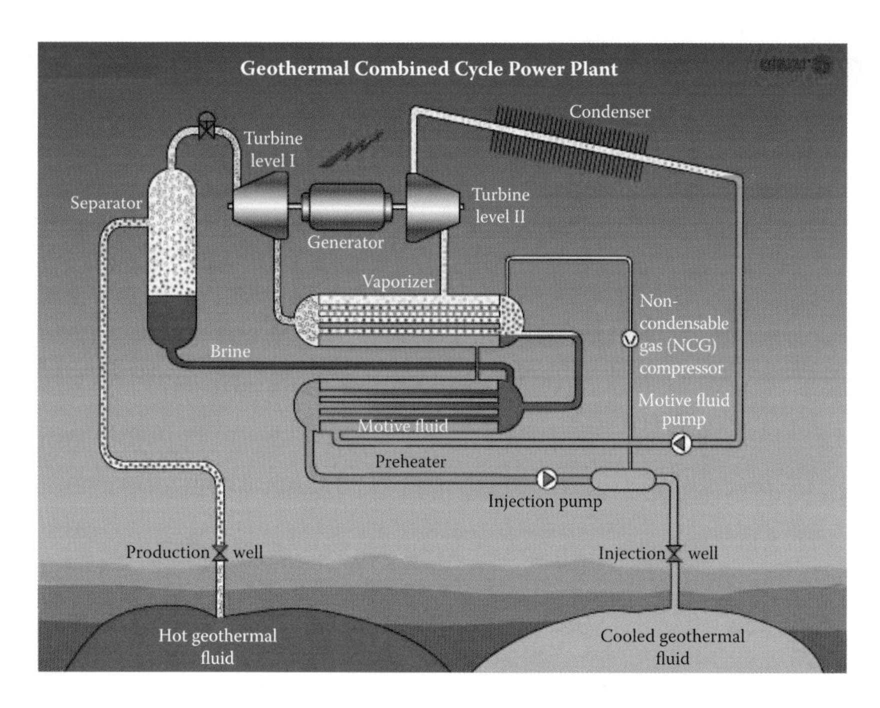

FIGURE 2.11 Schematic illustration of a combined or integrated flash-binary geothermal plant. See text for details. (From Ormat, *Geothermal Power*, Ormat Technologies, Inc., Reno, NV, 2016.)

efficiency of the geothermal resource increases by about 7%, from about 31% for single-flash alone to 38% for CHP, considering reasonable assumptions on reservoir temperature, separator temperature, and condensing temperature (DiPippo, 2012). These plants are used where there is an ample supply of fresh water that can be heated from the residual brine after flashing via a bank of heat exchangers. The heated fresh water is then piped to nearby communities for household uses, space heating, and swimming pools. Good examples of CHP plants include Hellisheidi and Nesjavellir, located about 30 km east of Reykjavik, Iceland. Indeed, the Hellisheidi geothermal plant is one of the largest geothermal energy facilities on the planet, based on total installed capacity of electric (303 MWe) and thermal (133 MWt) energy. It could become the largest when planned thermal additions of 133 MWt for 2020 and another 133 MWt for 2030 are completed (Hallgrimsdottir et al., 2012). Nesjavellir, located about 11 km northeast of Hellisheidi, has an installed capacity of 120 MWe and 200 MWt and delivers about 300 gallons per second of 82° to 85°C water to the greater Reykjavik region for space heating. Due to the high flow rate and pipe insulation, the fluid cools only about 1.5°C upon reaching Reykjavik. Although capturing as much heat as possible prior to reinjection improves efficiency, it may, as noted above, lead to an accelerated decline in reservoir temperature if the heat flux from the earth is inadequate to replace the heat mined, which has implications for sustainability discussed in Chapter 12.

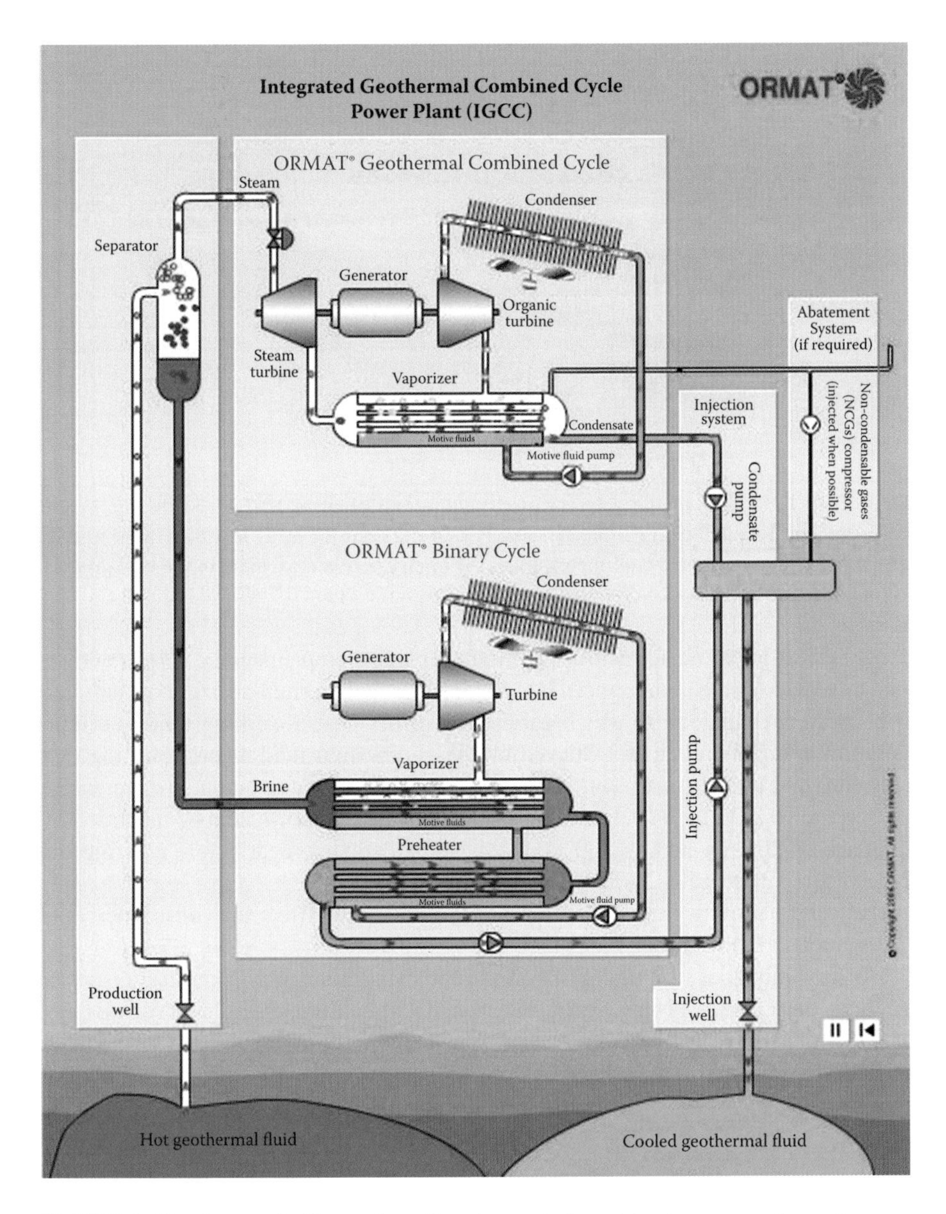

FIGURE 2.12 Schematic illustration of an integrated combined cycle geothermal power plant. Rather than preheating the working fluid, the separated brine directly vaporizes the working fluid in an adjacent binary facility. (From Ormat, *Integrated Combined Cycle Units: Geothermal Power Plants*, Ormat Technologies, Inc., Reno, NV, 2016.)

A final hybrid configuration, discussed here, is the combined use of solar and geothermal energy in the same power plant to achieve a synergistic output (e.g., Allis and Larsen, 2013). There are two main versions: (1) the solar and geothermal energy enhance each other (integrated solar–geothermal), and (2) a solar photovoltaic (PV)

FIGURE 2.13 View of the solar-assisted geothermal plant at ENEL's Stillwater hybrid solar-geothermal plant near Fallon, Nevada. Part of the solar PV array is shown in the middle ground with the air-cooled cooling towers of the binary geothermal plant in the background. (From https://www.flickr.com/photos/geothermalresourcescouncil/7940198440.)

array is positioned adjacent to a geothermal plant (complementary solar–geothermal or solar-assisted geothermal). Integrated solar–geothermal utilizes concentrated solar power technology in which parabolic mirrors heat a working fluid to a temperature hotter than can be achieved by the geothermal fluid alone, resulting in a solar thermal assist. As a result, turbine efficiency and power output are increased. As reported by DiPippo (2012), calculations indicate that power can be increased by as much as 20% by adding 100°C of solar-provided superheat before entering the turbine for a base case 20-MWe geothermal flash plant. Of course, the downside is that solar energy is intermittent, which can adversely affect the consistency (and economics) of power production.

As an example of combining solar and geothermal energy to help offset reservoir decline, Enel Green Power North America (EGP-NA) added a 26-MW solar PV array to its 33-MW Stillwater geothermal power plant (built in 2009 and located east of Fallon, Nevada). In 2012, EGP-NA created the first solar PV-assisted geothermal power facility in the world (Figure 2.13). The solar PV array covers 240 acres (98 ha) and consists of 89,000 solar panels (DiMarzio et al., 2015). The PV array complements the baseload nature of the geothermal resource by providing peak electrical power needs during the afternoon (when electricity can be sold at a higher rate) and during the sunshine-rich summer months, when efficiency and power output of the binary plant are lowest (see discussion on condensers below) (Figure 2.14). Then, in 2014, EGP-NA developed a fully integrated solar–geothermal plant by adding concentrated solar thermal technology and heat storage to augment the temperature of produced geothermal fluids. The solar thermal energy is estimated to add a 2-MWe boost to power production by adding extra heat to the incoming geothermal fluid (DiMarzio et al., 2015). Adding the solar thermal facility cost about $15M, yielding a cost of about $7.5M/MW. Although this amount is about three times the cost of drilling a $5M well that might yield 2 MWe, drilling a new well to make up power loss

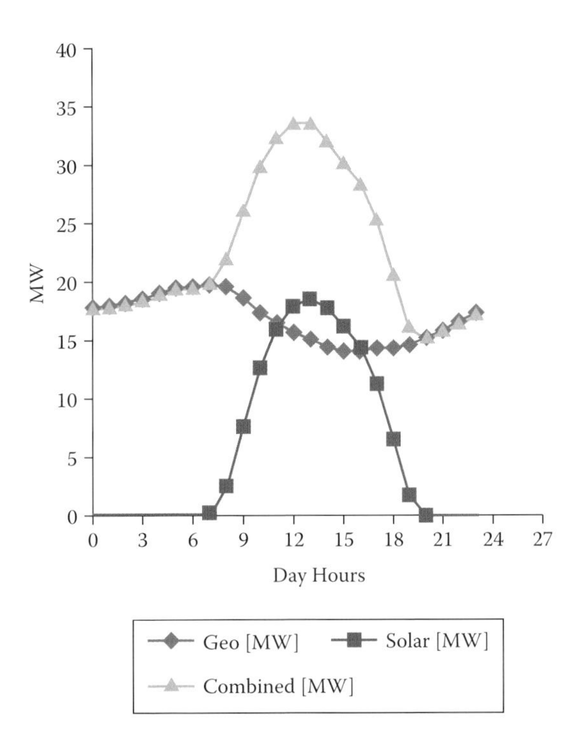

FIGURE 2.14 Distribution of solar and geothermal power for a typical spring day at the Stillwater hybrid geothermal solar power facility. Notice how the solar PV power offsets the dip in geothermal power in the middle of the day due to less efficient cooling of the air condensers. (Adapted from DiMarzio, G. et al., The Stillwater Triple Hybrid Power Plant Integrating Geothermal, Solar Photovoltaic, and Solar Thermal Power Generation. paper presented at World Geothermal Conference 2015, Melbourne, Australia, April 19–24, 2015.)

does not always lead to success. Perhaps two or three wells must be drilled to generate 2 MWe. If so, the economics become comparable and reduce risk. The other key ingredient is that Stillwater is located in a very sunny location, enhancing the capture of solar energy to complement and enhance the existing geothermal energy.

As a result, Stillwater is now the world's first triple-hybrid power plant, integrating geothermal, solar PV, and concentrated solar thermal energy. The storage of solar heated fluids allows continued elevated heating of geothermal fluids during the night, assuming of course that the preceding day was sunny. For solar PV-assisted geothermal or solar thermal-assisted geothermal or their combination to be viable, overlap of an *in situ* geothermal resource and an arid, sunny climate, is necessary.

Importance of Condensers and Power Output

An easily overlooked but critical process of any geothermal plant involves the condensation of the steam back to a liquid. Because steam occupies about 1600 times the space as an equivalent mass of liquid water, the condenser is a region of low pressure. This contrast between high pressure at the inlet of the turbine (~6 to 8 bars) and low pressure of the condenser (~80 millibars or 0.008 bar) promotes the expansion of the steam through the turbine—the greater the pressure difference, the greater the turbine efficiency and the greater the power output. For example, at the Wairakei geothermal power facility in New Zealand, for every 10-millibar increase in condenser pressure, the power output goes down by 1 MWe (T. Montague, pers. comm., 2013). The amount of pressure drop depends on how efficiently the steam is cooled. There are two main ways to cool the steam. The most traditional method uses evaporative cooling towers in which the hot steam condensate undergoes evaporative cooling as it cascades down a series of baffles that line the side of a cooling tower. This is a very efficient cooling process, but on average about 60 to 70% of the water is lost to the atmosphere from evaporation. At The Geysers in northern California in the summer, as much as 90% to 100% of the condensed steam is lost from the evaporative cooling towers (M. Walters, pers. comm., 2015). Such loss underscores the importance of make-up water, and the problem of where it comes from, to maintain fluid pressure in the geothermal reservoir via reinjection.

Another means to cool and condense steam is using air. This process is typically used in arid climates where surface or ground water is in short supply and in many binary geothermal plants. Unlike evaporative cooling, which is most efficient during dry summer months, air cooling is least efficient at that time. This explains why the power output from binary geothermal facilities can fluctuate significantly both for the time of day and time of the year—being highest at night and during the winter months when air cooling is most efficient. There are times, however, when electrical demand can be lower, resulting in correspondingly lower rates paid by utilities. An example of variable power production is the Steamboat geothermal complex near Reno, Nevada. The power output varies from about 75 to 80 MWe during the summer months but can increase to about 125 to 130 MWe during the coldest winter months, reflecting the more efficient cooling and condensing of steam by cold air. To help even out the power production, Ormat is experimenting with using misters associated with the air cooling fans at their Steamboat facility (Galena III plant). Ormat has seen a 5 to 10% increase of power output when misters are engaged at air temperatures exceeding about 85°F (J. Nordquist, pers. comm., 2015).

Condensation of the working fluid steam is a measure of the thermodynamic efficiency of the system. Thermodynamic efficiency (e) is simply the difference between the initial temperature (T_1) and the final temperature (T_2) divided by the initial temperature: $e = (T_1 - T_2)/T_1$ (°K). For example, the thermodynamic efficiency of cooling steam at 200°C to 65°C is about 0.29, but cooling the same fluid to 20°C is 0.39, or about a 34% increase in efficiency (Figure 2.15). Although this increase does not translate into an equal increase in power output, due to losses from energy conversion and friction, power output will be increased with increased cooling at the end-state condensers.

DIRECT USE OF GEOTHERMAL ENERGY

Direct-use geothermal systems use the heat directly from fluids in the Earth, or its near-surface thermal mass, for geothermal heat pumps for heating or cooling. Direct use is also the most efficient way to use geothermal energy as losses from energy conversion do not apply, as happens in converting heat energy into mechanical energy (driving a

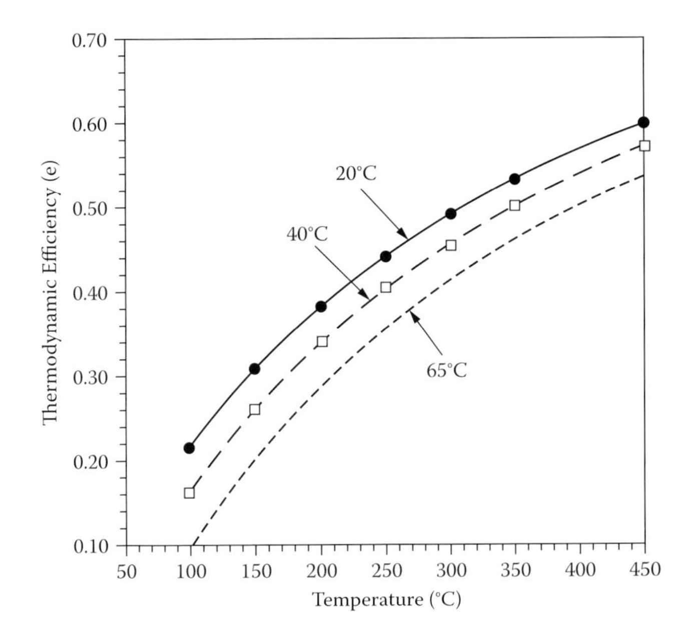

FIGURE 2.15 Graph illustrating changes in thermodynamic efficiency as a function of different exit temperatures as noted by the arrows. See text for discussion. (Adapted from Glassley, W.E., *Geothermal Energy: Renewable Energy and the Environment*, 2nd ed., CRC Press, Boca Raton, FL, 2015.)

turbine) and finally to electrical energy via a generator. As such, producing electricity is an indirect use of geothermal energy. Although not widely recognized, about half of the planet's utilized renewable energy, on a baseload basis, comes from direct-use geothermal (Lund, 2005; Lund and Boyd, 2015). As noted previously, solar and wind energy have captured most of the public's attention with regard to renewable energy. Geothermal energy, on the other hand, either direct or indirect use, is largely overlooked or underappreciated because it is basically invisible (excluding occasional visible plumes of condensed water vapor above a geothermal power plant's cooling towers on a cool day). Surficial geothermal features, such as hot springs, mudpots, fumaroles, and rare geysers, are uncommon and considered mainly as geologic wonders by the public (which they are) and not as expressions of a potential source of energy.

Direct geothermal use in 2015 was being utilized in 82 countries (Lund and Boyd, 2015), and growth of the direct use of geothermal energy grew from about 15,000 MWt in 2000 to about 50,000 MWt in 2010 (IGA, 2015a)—an increase of 230% for the 10-year period and an average annual growth rate of about 14%. By comparison, installed worldwide geothermal electric power capacity over the same period increased from about 8000 MWe to just over 10,700 MWe—an increase of 34% and an average annual growth rate of 3.3% (IGA, 2015b). From 2010 to 2015, direct-use geothermal increased to slightly more than 70,000 MWt for an increase of about 45% (Lund and Boyd, 2015), or an annual growth rate of 8.8%. Geothermal electric power output from 2010 to 2015 increased to 12,600 MWe, an approximately 18% increase over 5 years, yielding an annual growth rate of about 4.2%. For the 15-year period from 2000 to 2015, utilization of direct-use geothermal energy grew about three times faster than geothermal electrical power development. This difference is due mainly to the restricted geographic availability of the key geologic conditions required to produce geothermal fluids of high enough temperatures to run a geothermal power plant. Direct use of geothermal fluids, on the other hand, is found over a larger area because geothermal warm water is more common than geothermal hot water. And, geothermal heat pumps, the lowest thermal level of direct use, can be applied almost anywhere, especially in regions with strong seasonal contrasts in temperature.

Direct Use of Geothermal Fluids

Direct use typically involves geothermal fluids at temperatures between 35°C and 95°C but can also at times access higher temperature fluids (>100°C). Bathing, space heating, and greenhouse use are the three main direct-use applications of geothermal fluids. Aquaculture and industrial uses (such as lumber drying or in the manufacturing of paper and textiles) can also benefit from direct use (Lund and Boyd, 2015). In Klammath Falls, Oregon, geothermal fluids are used to de-ice streets, bridges, and sidewalks. The Klamath Basin Brewing Company uses geothermal fluids to heat water to brew beer. According to the GeoHeat Center at the Oregon Institute of Technology in Klamath Falls, Oregon, 271 locations in the western United States have geothermal fluids of at least 50°C within 8 km of a town (Oregon Tech, 2016).

Lund et al. (2010) reported that 20 facilities in the United States provide space heating as part of privately or government-operated geothermal districts or utilities, including about 2000 buildings in 17 states that use geothermal fluids from individual shallow wells for space heating.. Such systems distribute geothermal energy from one or more production wells to homes and businesses within the district. In addition to district-operated geothermal districts, individual geothermal systems also exist, sometimes hundreds at some locations, in which each structure has its own geothermal well. Geothermal district heating systems can save consumers 30 to 50% of the cost of natural gas heating (EERE, 2016).

Worldwide, direct utilization of geothermal energy has displaced about 350 million barrels of equivalent oil per year and prevented the emission of 148 million tonnes of CO_2 into the atmosphere (Lund and Boyd, 2015). For comparison, total carbon dioxide emissions from the consumption of energy for the planet amounted to 32,310 million metric tons in 2012 (EIA, 2016).

Case Study: Moana Geothermal Field in Reno, Nevada

In Reno, Nevada, Nevada Geothermal Utility Company serves about 110 homes located in the Moana geothermal resource area (Trexler, 2008). The Nevada Geothermal Utility Company produces about 95°C fluid from two production wells and has one injection well for spent fluid. Flow rates range from about 200 gallons per minute (gpm) in summer (mainly to heat swimming pools) to about 400 gpm in winter. In addition, over 250 geothermal wells have been drilled in the Moana geothermal zone that support, at least in part, the heating needs of an estimated 200 additional homes not part of the district served by the Nevada Geothermal Utility Company (Flynn, 2001). A single very large user of direct-use geothermal fluids from the Moana geothermal field is the Peppermill Resort Hotel and Casino. The Peppermill Resort Hotel constructed a 4400-foot-deep production well that delivers 1200 gpm (75 kg/s) of 79°C fluid (Spampanato et al., 2010). Production from this well serves the hot water and space heating needs of the entire hotel campus (a 2.2-million-square-foot complex). This installation has displaced the use of four 25,000,000-Btu natural-gas fired boilers that consumed about $2.2M of natural gas per year. Drilling of the new production and injection well and laying of new piping cost about $9.7M, yielding a payback period of less than 5 years (D. Parker, pers. comm., 2015).

Ground-Source Heat Pumps (Geoexchange or Geothermal Heat Pumps)

A heat pump is an amazingly simple but efficient device that moves heat that already exists. For ground-source heat pumps, that heat is provided by the immense thermal mass of the Earth. The Earth acts like a thermal bank in which heat is deposited during the summer and withdrawn in the winter. This is because at depths of only 3 to 5 m the temperature of the Earth is a consistent 10 to 13°C year-round over much of the planet where people live. The basic configuration consists of either a closed or open loop in which fluid is circulated between the ground and the building (Figure 2.16). In a closed loop, the heat-exchanging fluid, such as antifreeze, is fully contained in the piping and heat is exchanged via conduction. In an open loop, the exchange fluid (in this case, water) is transferred between a surface body of water (such as the pond shown in Figure 2.16) or with groundwater. In closed loop systems, the geometry of the loop can be vertical, horizontal, or a combination of both in a slinky style. Which geometric configuration is best depends on the soil or rock conditions and space limitations. For instance, if the bedrock is relatively shallow and hard, then a horizontal configuration is probably best due to the higher costs of drilling rock. On the other hand, if space is a factor and the bedrock is relatively soft or deep, then a vertical configuration generally makes more efficient use of the Earth's thermal mass. In a closed loop, the circulating fluid is confined to the piping and isolated from the environment physically but not thermally. In an open loop, however, new fluid is continually introduced from, say, a shallow aquifer or pond when operating and then reinjected back into the reservoir to thermally and physically mix and re-equilibrate. Heat is transferred from the ground to the building in the winter and transferred from the building to the ground in the summer.

To provide an idea of the effectiveness of geothermal heat pumps or the heat transfer process, indices called the *coefficient of performance* (COP) and *energy efficiency ratio* (EER) are used. The COP is a ratio of the net heat (difference between output heat and input heat) to the energy used by the heat pump (a small electric motor in the compressor) to move the heat. For example, net heat input values are typically on the order of 5000 to 7000 Wt; dividing these values by the 1500 We necessary to drive or move the heat yields COP values of between 3 and 5. This means that the energy delivered is 300 to 500% of the energy required to move the energy. For comparison, the most energy efficient gas-fired furnaces convert about 90 to 95% of the energy available to usable heat for heating, resulting in a

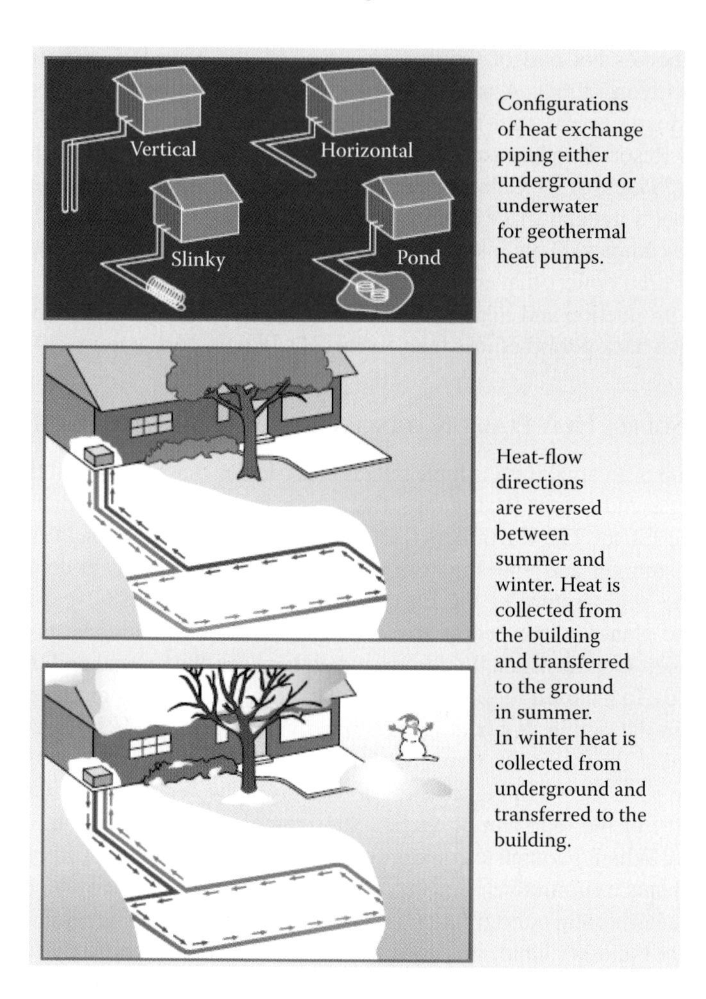

FIGURE 2.16 Diagram showing the different piping configurations and the change in the flow of fluids from winter to summer for geothermal heat pumps. (Adapted from Duffield, W.A. and Sass, J.H., *Geothermal Energy—Clean Power from the Earth's Heat*, Circular No. 1249, U.S. Geological Survey, Reston, VA, 2003.)

COP of 0.9 to 0.95. The greater the contrast between the input and output temperature, the greater the COP; therefore, highest COP values are achieved in areas that have hot summers and cold winters, such as the upper Midwest of the United States. In areas such as San Diego, on the other hand, the COP for geothermal heat pumps would be notably less, reflecting that region's mild climate and lower difference between input and output heat.

Lund and Boyd (2015) reported that there were more than 4 million geothermal heat pump systems worldwide, yielding almost 50 GWt of thermal capacity, which is about a 50% increase from 2010. Geothermal heat pumps comprise almost 71% of the total installed capacity of direct-use applications of geothermal energy (Lund and Boyd, 2015) (Figure 2.17).

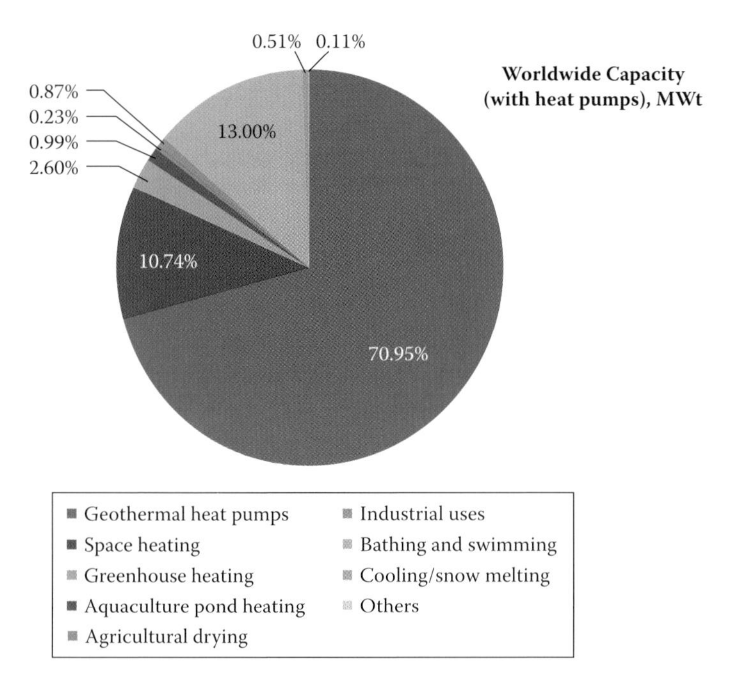

FIGURE 2.17 Pie chart showing distribution of applications of direct uses of geothermal energy. Notice that geothermal heat pumps comprise by far the largest proportion of installed capacity. (Adapted from Lund, J. and Boyd, T., in *Proceedings of World Geothermal Congress 2015*, Melbourne, Australia, April 19–24, 2015.)

Cooling efficiency is measured using the energy efficiency ratio (EER), which is the cooling capacity (in Btu/hour) divided by the electrical input in watts. EER values for geothermal heat pumps are commonly in the range of 15 to 25 Btu/Wh. Again, the wide range reflects the difference in input and output temperatures. Where the temperature difference is high, so is the EER. For comparison, standard room air conditioners have EER values typically around 10 to 12.

Case Study: Kendyl DePoali Middle School in Reno, Nevada

Kendyl DePoali Middle School is a state-of-the-art, energy-efficient building of nearly 200,000 square feet. The school opened in 2009 and cost about $40M to construct (just over $200 per square foot). The backbone of the school's energy efficiency is a closed-loop geoexchange system consisting of 373 300-foot deep wells that underlie the school's athletic field. The wells circulate water at a constant 64°F that is used to heat and cool air. In association with other temperature-efficient monitoring controls, such as variable air volume control valves on the heat pumps, the school saves about 60% on utilities compared to comparable sized schools built in the 1990s. Savings are due in part to the geothermal heat pump moving four units of heat (Btu) for every one unit of electricity (kWh), resulting in greater than 400% efficiency (or a coefficient of performance of about 4). The extra cost for installing the geoexchange system is expected to pay for itself in about 5 years (Alerton, 2010).

SUMMARY

Geothermal systems can be classified by a variety of criteria, such as the nature of heat transfer (conductive vs. convective), the presence or absence of recent magmatism or volcanic activity, the particular geologic setting (e.g., type of volcanic environment) or tectonic setting (e.g., type of plate boundary or intraplate geologic hot spot), and, of course, temperature (low-, moderate-, and high-enthalpy systems). Other criteria include fluid chemistry, such as acidic or near pH-neutral systems; vapor- vs. liquid-dominated systems; and how the system is used (power, direct use, or geoexchange). It is not uncommon for the different classifications to overlap. For example, hot-spot tectonic settings, such as Yellowstone or Hawaii, are typically magmatic systems, whereas those in extended crust, such as in much of the Basin and Range Province of Nevada, are mainly amagmatic. Ultimately, the classification of any geothermal system is based on study and plays a strong role in how that system might be developed. For example, a liquid-dominated system, based on temperature and mass flow rate confirmed by drilling, can be used for flash or binary power generation, combined power and heat, or direct use; whereas, the geologically rare vapor-dominated systems are used for power generation, as that is the most economic and efficient use of such a resource.

Although power generation captures much of the attention of the geothermal energy field, direct use is much more widely applied as sub-power generating temperatures of fluids are more widespread and can be developed with much less expense than building a power plant. Indeed, one of the important attributes of geothermal systems is their wide range of use over a cascading range of temperatures. Even where no hot fluids or rocks are present, the Earth acts like a thermal bank, where heat can be deposited during the summer and withdrawn during the winter. In regions characterized by hot summers and cool to cold winters, geoexchange systems can significantly reduce energy consumption using traditional fossil fuel sources.

SUGGESTED PROBLEMS

1. The graph below shows the boiling point of water (fresh and with 3.2% NaCl) with depth. If fluid in a geothermal reservoir is 250°C at a depth of 1 km, will it be boiling? If not, at what depth will boiling occur? Explain your reasoning.

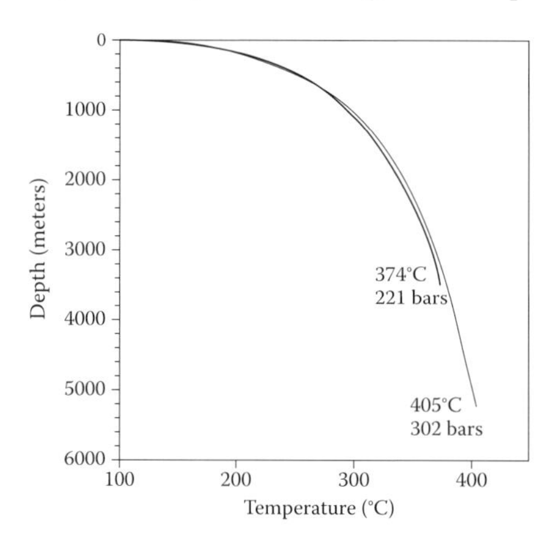

2. Consider a drill that intercepts fluid at 240°C at a depth of 1 km with good flow rates. What kind(s) of geothermal power plant would best make use of these reservoir conditions? Please be as specific as possible in justifying your decision.
3. Will heat movement in a vapor-dominated geothermal reservoir be dominantly convective or conductive? Why?
4. Will a vapor-dominated system also be geopressured? Why or why not?

REFERENCES AND RECOMMENDED READING

Alerton. (2010). *Kendyl DePoali Middle School: Case Study*. Redmond, WA: Alerton (http://alerton.com/en-US/solutions/k-12schools/Case%20Studies/MK-CS-KENDYL DEPOALI.pdf).

Allis, R.G. (2014). Formation pressure as a potential indicator of high stratigraphic permeability. In: *Proceedings of the 39th Workshop on Geothermal Reservoir Engineering*, Stanford, CA, February 24–26 (http://www.geothermal-energy.org/pdf/IGAstandard/SGW/2014/Allis.pdf).

Allis, R., Blackett, B., Gwynn, M. et al. (2012). Stratigraphic reservoirs in the Great Basin—the bridge to development of enhanced geothermal systems in the U.S. *Geothermal Resources Council Transactions*, 36: 351–357.

Arnórsson, S., Axelsson, G., and Samundsson, K. (2008). Geothermal systems in Iceland. *Jokull*, 58: 269–302.

Bertani, R. (2015). Geothermal power generation in the world: 2010–2015 update report. In: *Proceedings of World Geothermal Congress 2015*, Melbourne, Australia, April 19–24 (http://www.geothermal-energy.org/pdf/IGAstandard/WGC/2015/01001.pdf).

Calpine. (2016). *The Geysers*. Middletown, CA: Calpine Corporation (http://www.geysers.com/numbers.aspx).

Davies, J.H. (2013). Global map of solid Earth surface heat flow. *Geochemistry, Geophysics, Geosystems*, 14: 4608–4622 (http://www.mantleplumes.org/WebDocuments/Davies2013.pdf).

DiMarzio, G., Angelini, L., Price, W., Chin, C., and Harris, S. (2015). The Stillwater Triple Hybrid Power Plant Integrating Geothermal, Solar Photovoltaic, and Solar Thermal Power Generation, paper presented at World Geothermal Conference 2015, Melbourne, Australia, April 19–24 (https://pangea.stanford.edu/ERE/db/WGC/papers/WGC/2015/38001.pdf).

DiPippo, R. (2012). *Geothermal Power Plants: Principles, Applications, Case Studies, and Environmental Impacts*, 3rd ed. Waltham, MA: Butterworth-Heinemann.

Duffield, W.A. and Sass, J.H. (2003). *Geothermal Energy—Clean Power from the Earth's Heat*, Circular No. 1249. Reston, VA: U.S. Geological Survey (http://pubs.usgs.gov/circ/2004/c1249/c1249.pdf).

EERE. (2016). *Geothermal Energy at the U.S. Department of Energy*. Washington, DC: Office of Energy Efficiency & Renewable Energy, U.S. Department of Energy (http://www1.eere.energy.gov/geothermal/directuse.html).

EIA. (2016). *International Energy Statistics*. Washington, DC: U.S. Energy Information Administration (http://www.eia.gov/cfapps/ipdbproject/iedindex3.cfm?tid=90&pid=44&aid=8).

Flynn, T. (2001). Moana geothermal area Reno, Nevada: 2001 update. *GeoHeat Center Bulletin*, 22(3): 1–7.

GEA. (2012). *Geothermal Basics: Q&A*. Washington, DC: Geothermal Energy Association, (http://geo-energy.org/reports/Gea-GeothermalBasicsQandA-Sept2012_final.pdf).

Geosciences. (2016). *Hydraulic Fracturing*. New South Wales: University of Sydney, School of Geosciences (http://www.geosci.usyd.edu.au/users/prey/Teaching/Geos-2111GIS/Faults/Sld004c.html).

Glassley, W.E. (2015). *Geothermal Energy: Renewable Energy and the Environment*, 2nd ed. Boca Raton, FL: CRC Press.

Gradient Resources. (2016). *Geothermal Technology*. Reno, NV: Gradient Resources (http://www.gradient.com/geothermal-power/geothermal-technology/).

Griggs, J. (2004). A Re-Evaluation of Geopressured-Geothermal Aquifers as an Energy Source, master's thesis, Louisiana State University, Baton Rouge.

Haas, Jr., J.L. (1971). The effect of salinity on the maximum thermal gradient of a hydrothermal system at hydrostatic pressure. *Economic Geology*, 66(6): 940–946.

Hallgrimsdottir, E., Ballzus, C., and Hrolfsson, I. (2012). The geothermal power plant at Hellisheioi, Iceland. *Geothermal Resources Council Transactions*, 36: 1067–1072.

Hulen, J.B., Quick, J.C., and Moore, J.N. (1997). Converging evidence for fluid overpressures at peak temperature in the pre-vapor-dominated Geysers hydrothermal system. *Geothermal Resources Council Transactions*, 21: 623–628.

IGA. (2015a). *Geothermal Energy: Direct Uses*. Bochum, Germany: International Geothermal Association (http://www.geothermal-energy.org/geothermal_energy/direct_uses.html).

IGA. (2015b). *Geothermal Energy: Electricity Generation*. Bochum, Germany: International Geothermal Association (http://www.geothermal-energy.org/geothermal_energy/electricity_generation.html).

Legmann, H. (2015). The 100-MW Ngatamariki Geothermal Power Station: A Purpose-Built Plant for High Temperature, High Enthalpy Resource, paper presented at World Geothermal Conference 2015, Melbourne, Australia, April 19–24 (http://www.geothermal-energy.org/pdf/IGAstandard/WGC/2015/06023.pdf).

Lund, J. and Boyd, T. (2015). Direct utilization of geothermal energy: 2015 worldwide review. In: *Proceedings of World Geothermal Congress 2015*, Melbourne, Australia, April 19–24 (http://www.geothermal-energy.org/pdf/IGAstandard/WGC/2015/01000.pdf).

Lund, J., Sanner, B., Ryback, L., Curtis, G., and Hallstrom, G. (2004). Geothermal (ground-source) heat pumps: a world overview. *Geo Heat Center Quarterly Bulletin*, 25(3): 1–10.

Lund, J.W., Gawell, K., Boyd, T.L., and Dennajohn, D. (2010). The United States of America update 2010. *Geo-Heat Center Quarterly Bulletin*, 29(1): 2–11.

Oregon Tech. (2016). *Geo-Heat Center*. Klamath Falls: Oregon Institute of Technology (http://geoheat.oit.edu/colres.htm).

Ormat. (2016a). *Geothermal Power*. Reno, NV: Ormat Technologies, Inc. (http://www.ormat.com/geothermal-power).

Ormat. (2016b). *Integrated Combined Cycle Units: Geothermal Power Plants*. Reno, NV: Ormat Technologies, Inc. (http://www.ormat.com/solutions/Geothermal_Integrated_Combined_ Cycle).

Shook, G.M. (1995). Development of a vapor-dominated reservoir with a "high-temperature" component. *Geothermics*, 24(4): 489–505.

Spampanato, T., Parker, D., Bailey, A., Ehni, W., and Walker, J. (2010). *Overview of the Deep Geothermal Production at the Peppermill Resort*. Palm Desert, CA: Geothermal Resource Group (http://geothermalresourcegroup.com/wp-content/uploads/2011/03/Deep-Geothermal-Production-at-the-Peppermill-Resort.pdf).

Trexler, D.T. (2008). Nevada Geothermal Utility Company: Nevada's largest privately owned geothermal space heating district. *GeoHeat Center Bulletin*, 28(4): 13–18.

USDOE. (2016). *Geothermal Heat Pumps*. Washington, DC: U.S. Department of Energy (http://energy.gov/energysaver/articles/geothermal-heat-pumps).

White, D.E. (1973). Characteristics of geothermal resources. In: *Geothermal Energy, Resources, Production, and Stimulation* (Kruger, P. and Otte, C., Eds.), pp. 69–94. Stanford, CA: Stanford University Press.

White, D.E., Muffler, L.J.P., and Truesdell, A.H. (1971). Vapor-dominated hydrothermal systems compared with hot-water systems. *Economic Geology*, 66(1): 75–97.

3 Geology and Heat Architecture of the Earth's Interior

KEY CHAPTER OBJECTIVES

- Distinguish between the Earth's compositional and physical mechanical layers.
- Identify the sources of Earth's internal heat.
- Compare and contrast conductive and convective heat flow.
- Recognize conductive and convective zones of heat transfer from drill-hole temperature profiles.
- Explain the significance of heat flow maps and temperature-at-depth maps.

To more completely understand geothermal resources and their distribution on the planet, a review of the Earth's compositional and physical make-up is necessary. The Earth is compositionally inhomogeneous, consisting of an iron–nickel core, a dense rocky mantle, and a thin, comparatively low-density rocky crust. This compositional diversification developed shortly after our planet formed when more dense material sank to the center and low-density material rose toward the surface. Furthermore, because of this compositional diversity, differences in physical or mechanical properties exist (liquid or molten vs. solid; brittle vs. ductile deformation). Brittle behavior means breaking or fracturing after a threshold level of stress is applied, such as what happens when a glass vase is dropped on a hard surface. Ductile deformation, on the other hand, reflects bending without breaking after a material's yield strength is exceeded, such as bending a metal wire or molding clay. Understanding both the compositional and physical characteristics of the Earth's interior lays the groundwork for the discussion about plate tectonics in Chapter 4; plate tectonics exerts a fundamental control on the distribution of Earth's mineral, fossil fuel, and geothermal resources.

EARTH'S COMPOSITIONAL AND RHEOLOGICAL LAYERS

The Earth's radius is just under 6400 km. Extending outward from Earth's center, systematic changes occur in both composition and rheological behavior (physical or mechanical properties of a material, such as changes from solid to liquid or brittle to ductile). We will begin with compositional changes.

Earth's Compositional Layers

The area extending from the center of the Earth to a depth of about 2900 km is known as the Earth's *core*. The core consists of both solid and molten iron and nickel, and its temperature is comparable to the surface of the sun, or about 6000°C. Overlying the core is the *mantle*, which extends from a depth of about 2900 km to less than 100 km. Volumetrically, the mantle makes up the largest part of Earth's interior. The mantle consists of dense iron- and magnesium-rich rock, whose temperature decreases progressively upward from about 5000°C to less than 1500°C. The third and last layer is the Earth's *crust*, which consists of a thin shell, varying from 70 to 80 km thick under parts of continents to less than a few kilometers thick under parts of the ocean floor. A useful analogy of the compositional layers of the Earth is a peach. The size of the pit would be proportional to the Earth's core, the pulp (the edible part) proportionally represents the mantle, and the fuzzy skin would have the proportional thickness of the crust. Earth's compositional layers are illustrated in Figure 3.1.

Unlike the more compositionally homogeneous core and mantle, the crust consists of two types: oceanic and continental (Figure 3.1). Oceanic crust underlies the ocean basins and consists of a dark-colored, moderately dense rock called *basalt*. It is relatively thin, reaching a maximum of 7 km and a minimum of less than a

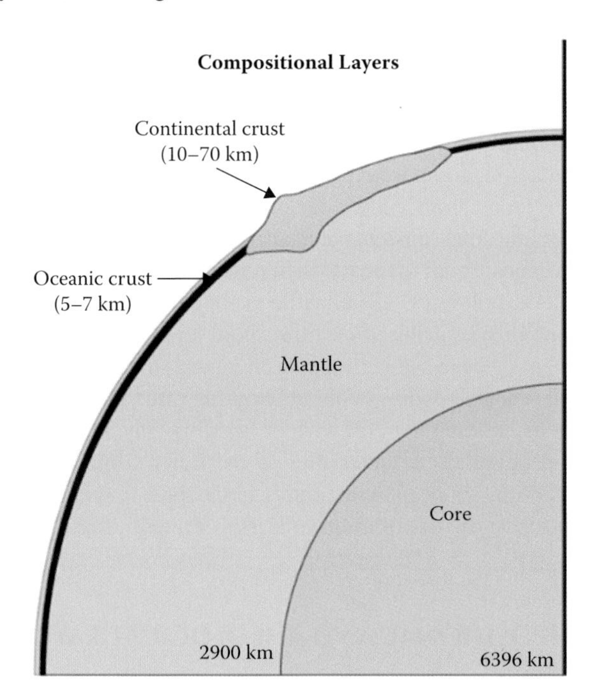

FIGURE 3.1 Cross-sectional view of Earth's compositional layers. Note that the crust consists of thin oceanic crust and thick continental crust. (Adapted from Visionlearning®, http://www.visionlearning.com/img/library/large_images/image_4859.gif.)

kilometer below mid-ocean ridges. Continental crust is comprised mainly of lower density, lighter colored igneous and metamorphic rocks, such as granite and gneiss (discussed in more detail in Chapter 4). These igneous and metamorphic rocks of continental crusts are capped in places by a veneer of sedimentary rocks, including sandstone and limestone. Because continental crust is less dense than oceanic crust, it sits higher compared to oceanic crust, explaining why continents for the most part lie above sea level.

Earth's Rheological (Physical) Layers

In response to changes in pressure and temperature, a material's physical nature (known as rheology) can change, such as from solid to liquid with rising temperature or the reverse with falling temperature or rising pressure. The composition of the material, however, remains essentially unchanged despite changes in the physical state. Another change in rheology would be the change from brittle breaking, forming fractures under low temperature and pressure, to ductile bending under high temperature and pressure prior to actual melting. In other words, a ductile substance is a solid that has the ability to flow, and within the Earth ductile materials flow at rates of a few centimeters per year in response to pressure differences and convection. The Earth is comprised of five main rheological layers, moving from the surface downward: the lithosphere, asthenosphere, mesosphere, outer core, and inner core. The relationship between Earth's compositional and physical or mechanical layers is illustrated in Figure 3.2.

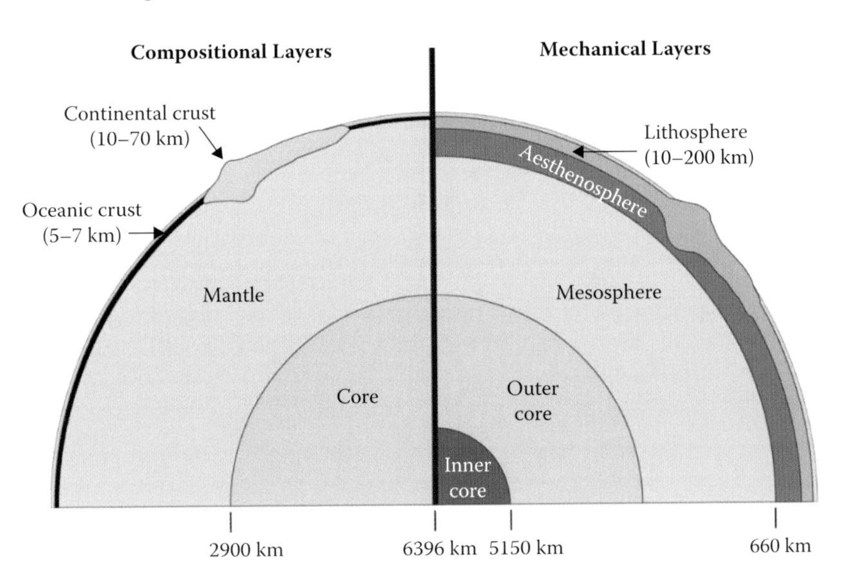

FIGURE 3.2 Cross-sectional view of Earth's compositional and mechanical layers. For details, see text. (Adapted from Visionlearning®, http://www.visionlearning.com/img/library/large_images/image_4859.gif.)

Lithosphere

The lithosphere represents the strong, relatively brittle outermost layer and averages about 100 km thick. It is compositionally diverse as it embraces both the crust and uppermost mantle because both compositional layers behave similarly from a rheological standpoint—relatively strong and brittle. The lithosphere will be discussed more in Chapter 4 because it makes up the Earth's tectonic plates, great chunks of rock that are continually moving with respect to each other.

Asthenosphere

Underlying the lithosphere, between 100 km and about 300 km, is a weak zone of rock called the asthenosphere, which is part of the upper mantle. The rock in the asthenosphere is weak because it is close to its melting point but still mainly a solid (Figure 3.3). However, because of the high heat, the rock is mechanically weakened and has the ability to flow (ductile behavior) in response to thermal and pressure gradients. Motion in the asthenosphere contributes to movement of the overlying lithosphere or tectonic plates.

Mesosphere

Below the asthenosphere, the behavior of the rest of the mantle, referred to as the mesosphere, is mechanically similar. The mesosphere consists of the lower and middle parts of the mantle. Because of the added pressure with depth, the rocks

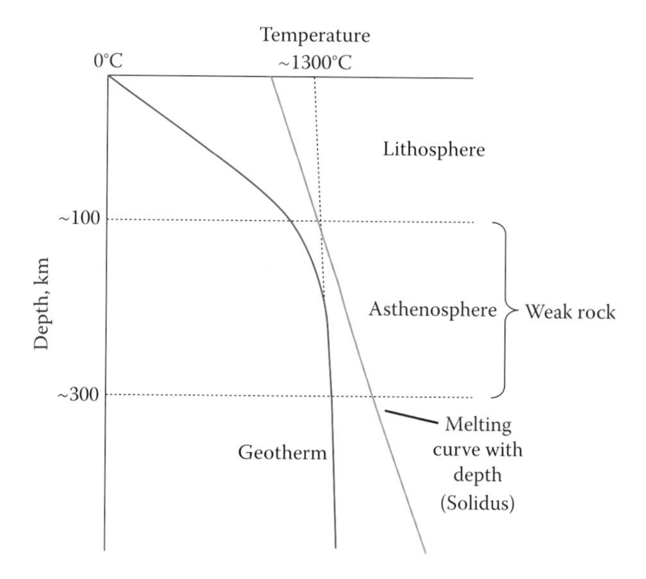

FIGURE 3.3 Depth and temperature plot showing the geothermal gradient (red line) and melting curve with depth of rock (blue line). The melting point of rocks increases with depth because increasing pressure favors the denser, solid phase. Note that the rocks are close to their melting point in the asthenosphere and therefore mechanically weak. As the geotherm and melting point curves diverge below the asthenosphere, the rocks become less weak. The 1300°C marks the approximate temperature at which basaltic rocks begin to melt near the Earth's surface.

are not as close to where they would begin melting as in the asthenosphere and are therefore stronger and less ductile (Figure 3.3). Nonetheless, because of the increasing temperature with depth, rocks of the mesosphere are not as strong or brittle as in the lithosphere and still have the ability to flow but at a slower rate than in the asthenosphere.

Outer Core

At the base of the mantle or mesosphere, temperature increases abruptly across the mesosphere–outer core boundary, reflecting the presence of molten iron and nickel. In response to gravitational and thermal gradients, the molten iron and nickel are convecting or circulating, promoting heat flow into the overlying mantle (resulting in abrupt temperature increases across the boundary). This circulation of molten iron in conjunction with Earth's rotation produces a geodynamo that gives rise to the planet's magnetic field. The liquid nature of the outer core is deduced from seismic wave data (see discussion below). Receiving stations on the opposite side of the planet from which an earthquake occurs will not receive any S-waves (also known as shear or secondary waves), which are attenuated when they encounter liquid material.

Inner Core

The inner core is compositionally the same as the outer core but is a solid rather than a liquid even though the temperature has risen to about 6000°C (depending on the model used). The transition from liquid in the outer core to solid in the inner core results from the extreme pressure at these depths. The radius of the inner core is about 1300 km.

Evidence of Earth's Compositional and Rheological Layers

Our understanding of Earth's compositional and rheological layers is not known from drilling, as the deepest drill hole is about 12 km deep, which is a mere pinprick into the Earth's interior. Rather, our understanding comes from several sources, including meteorites, material erupted from volcanoes, and the nature of Earth's rotation and precession (or wobble) of Earth's axis. Primarily, though, most of what we know of the Earth's internal compositional make-up stems from the study of seismic waves. These waves image the interior of the Earth, much like a computerized axial tomography (CAT) scan discloses internal components of the human body. The speed, direction, and propagation of these waves change in response to the density and composition of the material traversed. By collecting seismic wave data from receiving stations across the planet, the Earth's internal compositional layers can be successfully modeled and imaged (Figure 3.4). Earthquakes generate two types of waves that travel through the interior of the Earth: P-waves, or primary (compressional) waves, and S-waves, or secondary (shear) waves. P-waves travel through solids, liquids, and gases, but S-waves travel only through solids, because liquids and gases have no elasticity to support shear stresses. Therefore, the liquid nature of the outer core is indicated because seismic receiving stations on the opposite side of the Earth from which an earthquake occurs receives no S-wave signal, only a P-wave response. The size of the resulting S-wave shadow zone is a direct indication of the diameter of the core (Figure 3.4).

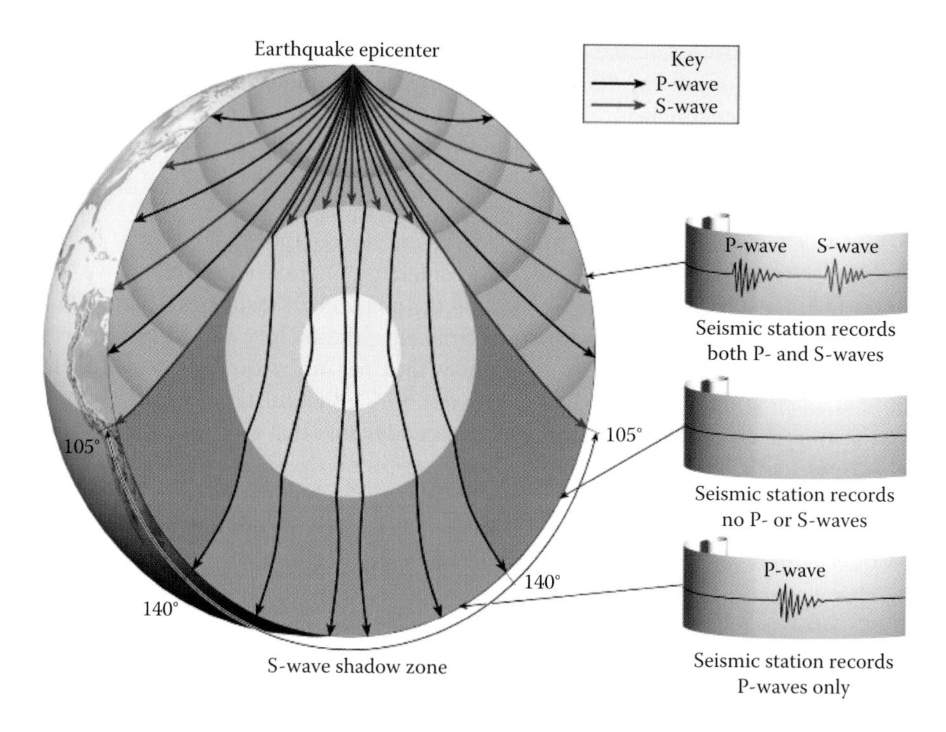

FIGURE 3.4 Cross-sectional view of seismic waves as they traverse Earth's interior. The size of the compositional layers can be determined by the refraction or attenuation of select seismic waves. For example, the liquid outer core is detected by the attenuation of seismic S- or shear waves that cannot travel through liquids, resulting in a shadow zone whose size reflects the diameter of the outer core. The size of the solid, inner core is determined by noting the location of received P- or compressional waves reflected off the sides of the inner core. (Adapted from Tarbuck, E.J. et al., *Earth: An Introduction to Physical Geology*, Prentice Hall, Upper Saddle River, NJ, 2005.)

SOURCES OF EARTH'S HEAT

There are three main sources of Earth's internal heat. First is residual heat left over from the formation of the planet (primordial heat) about 4.6 billion years ago. This heat is a product of the first law of thermodynamics, which states that energy is conserved. Our planet formed by accretion of colliding meteorites or larger chunks of space debris called *planetisimals.* of movement was converted to heat energy after collision, resulting in a largely molten proto-Earth, leading to the eventual gravitational separation of heavy and light elements to form the core, mantle, and crust as described above. Because rock is a good insulator, the deep interior of our planet has stayed hot, with heat flowing outward toward the surface. This outward flow of heat, while fairly uniform at depth from the core through the mantle, becomes irregularly distributed as it flows through the crust, being concentrated in select zones due to plate tectonics (discussed in Chapter 4) and influencing the distribution of areas having high and low geothermal heat flow at the Earth's surface.

A second source of heat comes from the radioactive decay of select elements, principally from uranium, thorium, rubidium, and potassium. These elements are largely concentrated in the crust because their large atomic radii are less compatible in mineral structures in the mantle due to the high pressures there, favoring dense mineral species. As a result, about 60% of the heat in continental crust is due to radioactive decay of these four elements (Glassley, 2015). Nonetheless, these radioactive elements are present in the mantle, and even though their concentration is low there, the large volume of the mantle makes up for the low concentration, indicating that a significant amount of heat coming from the mantle is due to radioactive decay. Recent studies of Earth's internal heat flow budget indicate that the proportion of primordial heat and radiogenic heat to total heat flow is about equal and in total amounts to about 47 terrawatts (TW) (Davies and Davies, 2010; Gando et al., 2011; Korenga, 2011). For comparison, the total installed world power capacity in 2012 was 5.55 TWe (EIA, 2016). The takeaway, clearly, is that Earth's internal heat energy can provide a significant contribution toward supplying the energy needs of civilization. Over 50% of the total heat flow is contributed by convection in Earth's mantle, with about 24% coming from the crust and supplied by a mixture of conduction, hydrothermal convection, and vertical and horizontal movement (advection) of localized zones of magma (Figure 3.5).

A third, albeit minor, source of heat is from gravitational pressure. When something is squeezed it heats up, and when expanded it cools. For gases, this behavior is described by Charles' law; a similar process happens with solids, except that the changes in volume are much smaller for a given increase in pressure. Again, because rocks are good insulators, the escape of heat from Earth's surface is less than the heat generated from internal gravitational attraction or squeezing of rock, so heat builds up with depth.

Other local sources of heat include frictional heating along earthquake faults. This frictional heating can be sufficiently intense to actually partially melt the rock, producing what is called *pseudotachylite*. Indeed, a small amount of heat tapped by geothermal power plants located along major active faults, such as the San Andreas fault in California or active faults in Nevada, probably comes from frictional heating as rocks grind past on either side of the fault.

HEAT TRANSFER MECHANISMS IN THE EARTH

A flux of heat is emitted from every square meter on Earth's surface; in some places it is notably higher, particularly near the boundaries of the tectonic plates, than in other places. Overall, however, the average heat flux or flow for the Earth is about 87 milliwatts per square meter (mW/m^2). Multiplying this value by the total global surface area of 5.2×10^{11} m^2 yields a total heat or power output of about 4.7×10^{13} W or 47 TW (thermal) as noted above. The heat flow for continents averages 65 mW/m^2, and the average heat flux for oceanic crust is 101 mW/m^2. The difference reflects the thinner character of oceanic crust with hot mantle rocks at comparatively shallow depths and the insulating nature of thicker continental

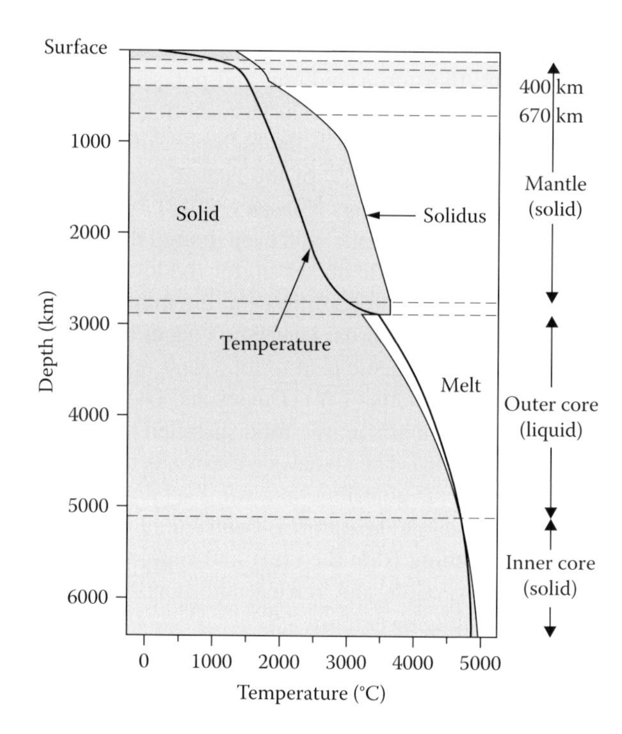

FIGURE 3.5 Graph showing the change in temperature (heavy solid line) of the Earth from its surface to the core (Earth's geothermal gradient). Also shown is the solidus, or the temperature at which rocks begin to melt. Note that the geothermal gradient is highest near the surface, indicative of conductive heat flow, but becomes more gradual with depth, indicating a combination of convective and conductive heat flow. The highlighted yellow layer marks the asthenosphere where the temperature of the solidus and that of the Earth are close, resulting in rheologically weak rock. The layer above the asthenosphere is the lithosphere where the temperatures of the Earth and solidus are further apart, making for rheologically strong rock: (Adapted from Ammon, C.J., *Earth's Origin and Composition*, SLU EAS-A193 Class Notes, Penn State Department of Geosciences, University Park, 2016, http://eqseis.geosc.psu.edu/~cammon/HTML/Classes/IntroQuakes/Notes/earth_origin_lecture.html.)

crust. Indeed, if it were not for the ocean, whose depth averages about 3.7 km, much of the oceanic crust would have the potential for harnessing geothermal energy. But then again, without the oceans there would be a dearth of water, which is the primary vehicle for transferring heat energy from hot rocks at depth to the surface (see later discussion). Heat can be transferred by three main mechanisms: conduction, convection/advection, and radiation. The first two are relevant for the solid Earth, as radiation applies mainly to the transfer of electromagnetic radiation through space, such as sensing heat from a campfire or the transfer of light from the Sun.

Conductive Heat Flow

Conduction is the transfer of heat by contact (transfer of energy from one atom to the next) and is an important means of heat transfer within the Earth. The overall geothermal gradient of Earth—the change in temperature with depth—is largely governed by conductive heat transfer. This gradient is high or changes rapidly near the surface but becomes more gradual at depth (Figure 3.5). This rapid change in temperature with depth is indicative of conductive heat flow, because, in the absence of circulating fluids, rocks are good insulators. The geothermal gradient averages about 25 to 30°C per km for the upper crust (top 10 km or so), whereas in geothermal areas the geothermal gradient is about double to perhaps three times that of non-geothermal regions. In active volcanic regions, the geothermal gradient can be as high as 150°C per km, such as at Yellowstone National Park, and the heat flux can be 500 mW/m^2 or even more.

Heat flux is governed by Fourier's law, which states that the flow of heat (Q) depends directly on the thermal conductivity (k, in units of watts per meter kelvin, or W/m·K) of the material and the geothermal gradient ($\Delta T/\Delta x$ or ∇T). This gives us the equation $Q = k \times \nabla T$. For example, if an exploration well is drilled in granite and encountered a temperature of 200°C at a depth of 1500 m, what is the heat flux at the site?

$$Q = \frac{k_{granite} \times (473\ \text{K} - 298\ \text{K})}{1500\ \text{m}} \tag{3.1}$$

Thermal conductivity itself is modestly sensitive to temperature and generally decreases as temperature increases for Earth materials (Clauser and Huenges, 1995). An average value of granite over this temperature range would be about 2.4 W/m·K (Glassley, 2015). Substituting these values into the equation yields the following:

$$Q = 2.4\ \text{W/m·K} \times 175\ \text{K}/1500\ \text{m} = 0.280\ \text{W/m}^2 \text{ or } 280\ \text{mW/m}^2$$

which would be a very promising heat flow for developing geothermal energy.

For continental crust, the minerals feldspar and quartz are the most common, yet there is a significant difference in the thermal conductivity of quartz and feldspar (Glassley, 2015), such that the thermal conductivity of quartz averages about twice that of alkali feldspar. Thus, the thermal conductivity of a rock will be strongly dependent on the proportion of these two minerals, which in turn will directly influence the heat flow.

Related to thermal conductivity is thermal diffusivity, which measures how quickly an object changes temperature in the presence of a thermal gradient. Thermal diffusivity has the units of square meters per second (m^2/s). Thermal diffusivity is defined by the ratio of thermal conductivity to the heat capacity, by volume, of a material. Heat capacity measures how much heat is required to raise the temperature of a unit volume of material by 1 K. Minerals have thermal diffusivity values of 1×10^{-6} to 10×10^{-6} m^2/s, whereas most metals have diffusivity values in the range of 1×10^{-4} to 5×10^{-4} m^2/s, or about 100 times the diffusivity of minerals. Also affecting conductivity and diffusivity is the porosity or open space in rocks (porosity is discussed in more detail in Chapter 5). Pores in rocks can be filled with water or air or a mixture of the two. Because water conducts heat more readily than air, the thermal conductivity of a water-saturated rock will be 3 to 4 times that of its dry equivalent. Furthermore, conductivity is also dependent on pore size such that larger pores have a lower conductivity for a given water content (Glassley, 2015).

As a result, part of the accurate characterization of the geothermal energy potential of a given region requires measuring and understanding the properties of the geological materials in which the system is developed. How is this important for geothermal power production? Imagine a site having high heat flow but also characterized by quartz-rich rocks, which have relatively high thermal conductivity. Although heat is transferred efficiently to water for production, the cooler injection water could unfavorably cool the reservoir rocks, which would lower the system's enthalpy and power/energy potential. Thus, the rates of production and injection must be such that the system is not adversely perturbed, and determining the production and injection rates requires accurate characterization of the thermal properties of the geological materials.

Examples of conductive geothermal systems include some deep sedimentary basins and geopressured reservoirs, such as those found along the Gulf Coast of the United States. The Paris Basin is an example of a deep sedimentary aquifer whose geothermal fluids are used directly for space heating. The flow of water is slow enough that there is enough time to be heated by the conductive heat flow from the rocks. This happens because there is a general reduction in permeability (flow of water through rock, as discussed in Chapter 5) with depth, which retards the fluid's ability to circulate easily. In geopressured reservoirs, permeable water-bearing horizons are deeply buried (generally >2 km) and are isolated by surrounding impermeable rock. These are self-contained systems in which the pore water was trapped with the sediments at the time of deposition. Because they are isolated from the surface, the pore water is under the weight of the overlying rock (lithostatic) rather than a column of water (hydrostatic). The water is thus pretty much stagnant and is heated conductively in response to the region's geothermal gradient of about 50°C/km.

A final example of conductively heated geothermal systems consists of engineered geothermal systems (EGSs) in which hot rocks exist but permeability or water content is sufficient to produce a circulating hydrothermal system. These conductive systems are being explored in places to artificially produce convective systems (see next section) through controlled fracturing of the rock. An EGS project, at Newberry Volcano in central Oregon, has proved encouraging with regard to developing improved permeability in hot rocks through the injection of cold water. (EGSs, deep sedimentary aquifers, and geopressured reservoirs are discussed in Chapter 11.)

Convective (Advective) Heat Flow

Technically, the movement of heat by bulk fluid flow is advection; however, convection is the more widely used and general term and embraces both advection and conduction, meaning that as heat is transferred by moving material some heat is also transferred by conduction through contact with surrounding material. The slower the movement of the material, the greater the proportion of heat transferred by conduction. For simplicity, we will use the more widely used term convection, understanding that the bulk of heat is transferred by movement of material and a lesser amount by conduction. Because convection involves both movement of material (advection) and thermal diffusion (conduction), it is the most effective means of energy transfer within the Earth.

Convection develops in response to buoyancy forces in the presence of a gravitational field. As material is heated it becomes less dense and will begin to rise. To replace the rising material, cooler (and more dense) material sinks, where it too might be heated and also begin to rise, resulting in a convection cycle. Without convection, a body of water, for example, can become thermally and density stratified, such that warm, less dense water lies near the surface and cooler more dense water at depth. If fluids are convecting, however, they are mixing; thus, temperature changes little with depth over the convecting interval. Recognizing zones of convection from drilling can be an effective exploration tool for identifying prospective geothermal reservoirs (discussed in Chapter 8).

As established, the solid Earth is overall density stratified with a dense iron-rich core and a low-density, outer crust; however, it is not static because the hot, liquid outer core is a potent source of energy. Although the overlying mantle is solid, it has the ability to flow slowly but significantly on the order of the geologic time scale. The rate of flow is controlled in part by the strength of the energy source but also in part on the viscosity of the material. Viscosity is a property that measures the resistance to flow of a material when stressed. For example, molasses is more viscous than water. For most materials, viscosity is inversely related to temperature; as temperature increases, the viscosity decreases, similar to heating honey. Thermally disturbed portions of the lower mantle, perhaps situated above focused zones of upwelling in the underlying and convecting molten outer core, will be gravitationally unstable relative to overlying (and adjacent) cooler and denser mantle and will begin to rise buoyantly upward, producing a system of convection cells (Figure 3.6).

Rayleigh Number

Factors that promote convection are low viscosity, thermally induced expansion, a gravitational field to exert buoyancy forces, and low thermal conductivity to create a strong thermal gradient and drive buoyancy forces. Quantitatively, conditions that promote convection can be represented by the ratio between buoyant and viscous forces, or what is termed the *Rayleigh number* (Ra), which is represented quantitatively below:

$$\mathrm{Ra} = \left(\frac{g \times \alpha \times d^3}{v \times \kappa}\right) \times \Delta T \tag{3.2}$$

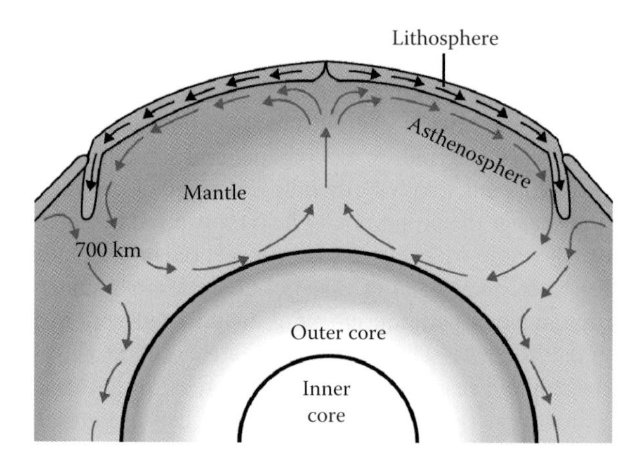

FIGURE 3.6 Cross-sectional view of Earth's interior illustrating convection in the mantle and asthenosphere. Convection is induced by heat transfer and by convection in the liquid outer core. Note that motion of the lithosphere is in part due to convection in the underlying lithosphere. (Adapted from USGS, *Some Unanswered Questions*, U.S. Geological Survey Reston, VA, 1999, http://pubs.usgs.gov/gip/dynamic/unanswered.html.)

where

g = Acceleration of gravity (9.8 m/s^2).
α = Coefficient of thermal expansion (1/K).
d = Depth interval over which the temperature change occurs (m).
v = Kinematic viscosity (m^2/s).
κ = Thermal diffusivity (m^2/s).
ΔT = Vertical temperature change (K).

As a result, Ra is a dimensionless number that provides an indication of whether convection will occur or not and therefore indicates whether the dominant form of heat flow will be by conduction or convection. When Ra is >1000, convection is the dominant heat transfer mechanism; when Ra is <1000, conduction is the dominant form of heat flow.

The Rayleigh number for the mantle ranges between 10^5 and 10^7, indicating that the mantle is mobile and that the main form of heat transfer to the Earth's surface is by convection. The movement of mantle material is a principal driver for plate tectonics, which accounts for much of the distribution and types of geothermal resources across the planet (discussed in Chapter 4).

Convection in the Upper Crust

Currently exploited geothermal systems are typically less than 3 km deep and consist of convective hydrothermal systems, either liquid or vapor dominated. The fluid must reside in rock reservoirs that allow fluids to circulate, which requires connected open space or permeability (discussed in Chapter 5); otherwise, they would be mainly undeveloped conductive reservoirs. When a convective hydrothermal

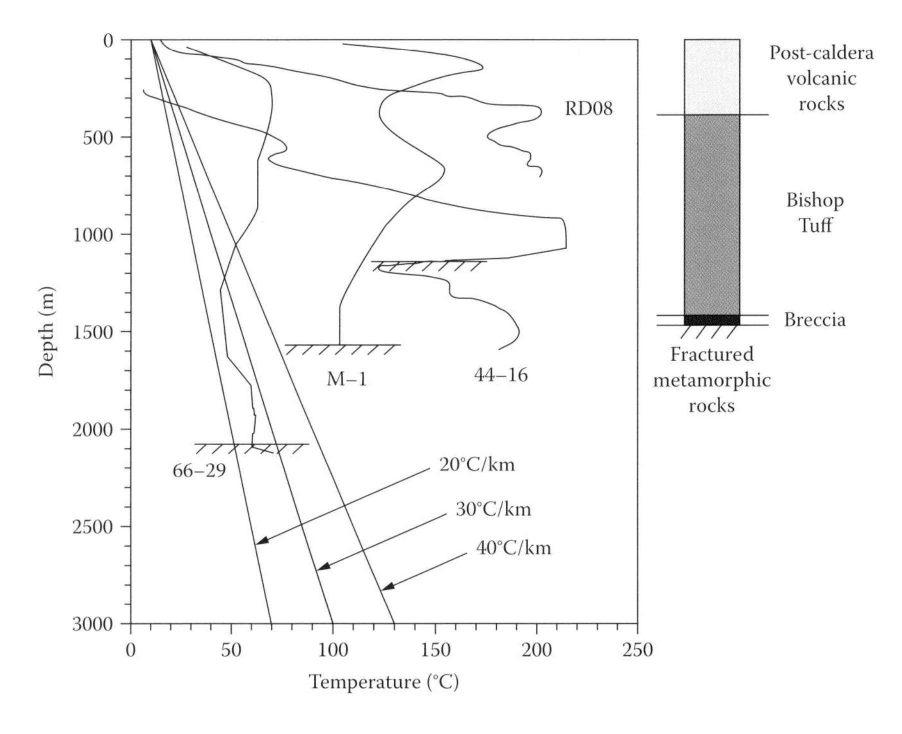

FIGURE 3.7 Temperature profiles of four drill holes into the Casa Diablo geothermal system in the Long Valley caldera, California. Note the areas of conductive heat flow at shallow depths of all drill holes where there is a rapid change of temperature with depth. The convective zones are characterized by little change in temperature with depth, reflecting circulation and thermal mixing. Note that in all holes, one or more temperature reversals occur, reflecting deeper, cooler aquifers. (Adapted from Glassley, W.E., *Geothermal Energy: Renewable Energy and the Environment*, 2nd ed., CRC Press, Boca Raton, FL, 2015.)

reservoir is intercepted by drilling, the geothermal gradient will decrease suddenly, reflecting the thermal mixing of fluid circulation. This is different than thermal stratification, which indicates conductive heat flow zones that commonly bound the tops and bottoms of geothermal reservoirs. In some cases, the geothermal gradient increases again below the convecting hydrothermal reservoir, whereas in other cases the gradient can decrease below the reservoir, reflecting lateral outflow of hydrothermal fluids above cooler (and more dense) groundwater recharge zones. The Casa Diablo geothermal field in the Long Valley caldera in California is an example of the latter (Figure 3.7). Most of the heat flow through the upper crust occurs by conduction (Figure 3.8), and convecting hydrothermal systems require special geologic characteristics. Such characteristics include a source of water, good permeability, properly positioned impermeable cap rocks, and a focused source of heat, such as a body of magma in the upper crust. These conditions are not met everywhere, and just because heat flow may be high in a region does not indicate whether or not potentially developable convecting geothermal reservoirs are present.

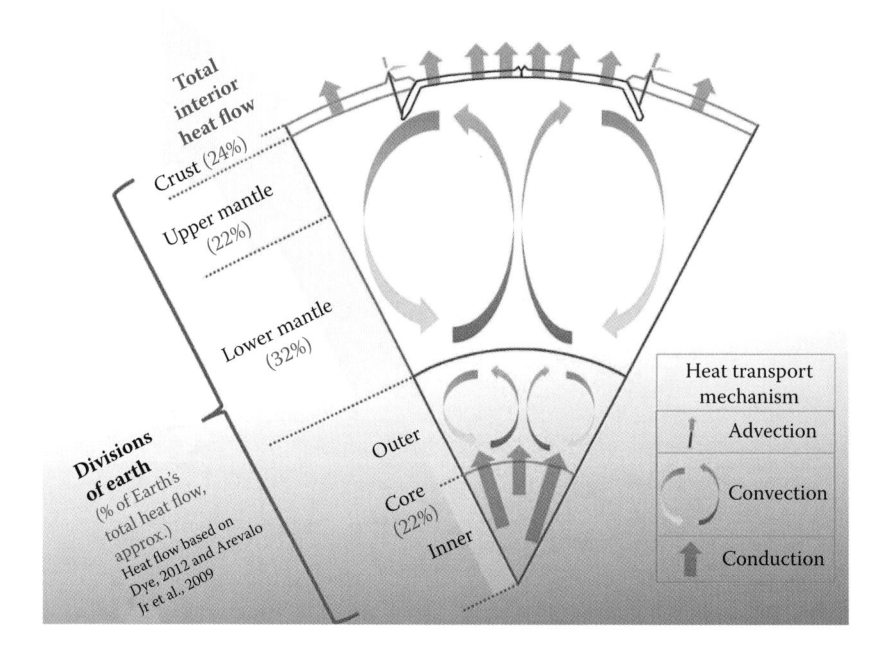

FIGURE 3.8 Pie slice through Earth's interior showing major compositional and rheological divisions and the relative proportion and type of heat flow from each division. Although conduction is the major form of heat flow in the crust, some heat flow occurs by convection in areas of circulating crustal fluids and by advection with the rise of magma below active volcanoes. (From Dye, S.T., *Reviews of Geophysics*, 50(3), RG3007, 2012.)

HEAT FLOW MAPS

The Southern Methodist University Geothermal Laboratory (SMUGL) has been instrumental in compiling drill-hole data and generating and updating a series of maps showing heat flow in the United States. The team of researchers there has also developed, from the compiled heat flow data, a series of temperature-at-depth maps from 3.5 to 9.5 km. These maps help illustrate prospective regions for developing geothermal energy for power and direct use, most of which are located in the western United States (Figures 3.9 and 3.10). Researchers at SMUGL have also developed maps showing the potential for engineered geothermal systems (EGSs) for each of the states based on the heat flow information. For example, Nevada, which has an installed current geothermal power capacity of 580 MWe (as of 2015) from conventional convective geothermal systems, could increase it geothermal power output to 41k MWe if only 2% of its EGS potential is recovered. That number could swell to 288,000 MWe if just 14% of its EGS potential is realized (SMU, 2016). Although such potential is impressive, EGSs are hampered by their still experimental nature and associated high costs which must compete with currently inexpensive natural gas—the fossil fuel of choice for power generation because carbon emissions are about half of those of coal (see Chapter 9). However, as described by Allis et al.

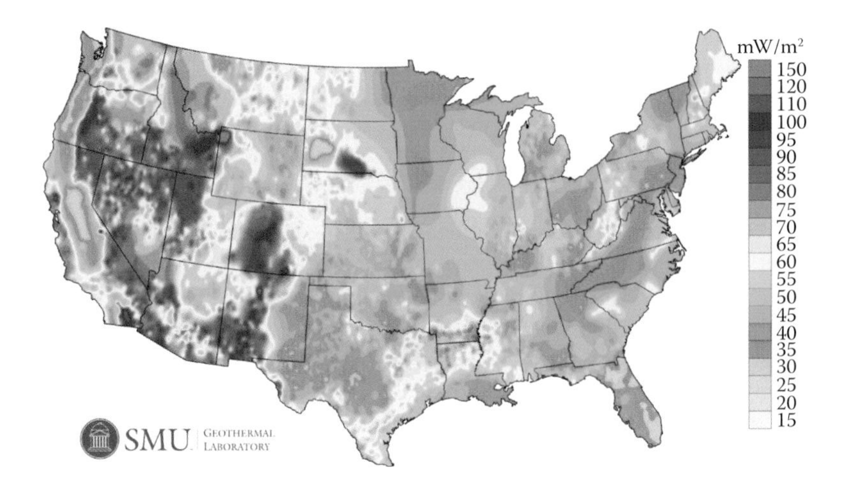

FIGURE 3.9 Heat flow map of the United States for 2011. Ochreous orange to more deeply red indicates heat flow values in excess of 80 mW/m^2. The highest value on the map is the pinkish red, which is >150 mW/m^2, which would be at Yellowstone National Park. This and the following figure illustrate the large area of geothermal energy potential covering much of the western United States. (Adapted from Blackwell, D.M. et al., *Geothermal Resources Council Transactions*, 35, 1545–1550, 2011.)

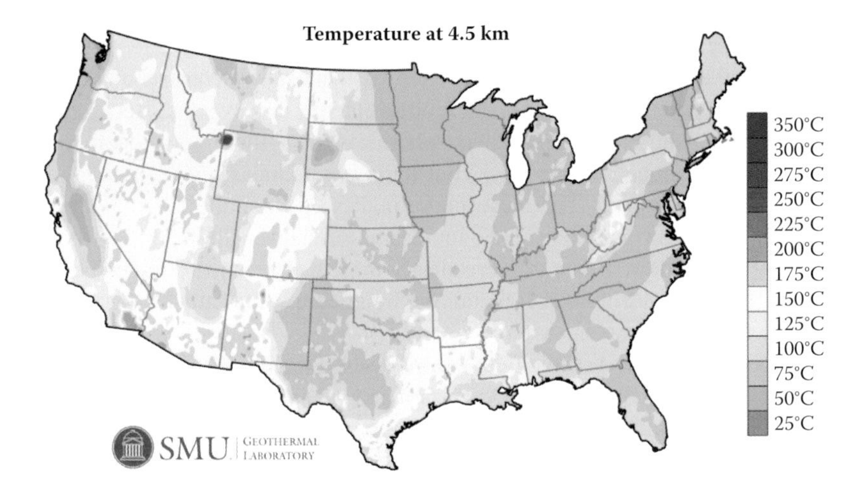

FIGURE 3.10 Map showing temperatures at a depth of 4.5 km. Notice the large area of temperatures of 150°C and higher across much of the western United States. Almost all of northern Nevada has temperatures that are >150°C, including numerous scattered regions with temperatures between 175 and 200°C. This depth level is largely the realm of engineered geothermal systems (EGSs), which if developed (just in small part, ~10%) could greatly expand (by one to two orders of magnitude) geothermal power production. However, accessing this potential energy resource would be expensive, mainly due to the deep levels of drilling required, making it difficult for this technology to compete economically with currently producing, more shallow geothermal reservoirs and natural gas-fired power plants. (Adapted from Blackwell, D.M. et al., *Geothermal Resources Council Transactions*, 35, 1545–1550, 2011.)

(2012), if sedimentary aquifers exist at these depths and are permeable, they might serve as a bridge to actual development of EGSs, offer large potential flow rates, and be cost competitive if current low energy prices were to rise modestly.

SUMMARY

The interior of the Earth is compositionally and rheologically partitioned into distinct layers. Compositionally, the Earth's interior consists of an iron-rich metal core, a mantle, and a thin crust. The mantle makes up the largest volume of the Earth's interior and consists of dense, iron- and magnesium-rich rocks. The crust consists of two types: oceanic and continental. Oceanic crust consists of more dense basalt and is relatively thin (<1 to 7 km thick). Continental crust is made of less dense granitic and metamorphic rocks and can be as much as 70 km thick underneath mountain belts. This compositional division developed very early in the Earth's history, when the planet was still largely molten. Dense constituents settled to the center to form the core, and less dense elements rose toward the surface to form the mantle and crust.

Due to changes in temperature and pressure within the Earth, the compositional layers develop different rheological (mechanical) properties, ranging from solid (brittle and strong), to weak and ductile, to molten. These different mechanical layers are the lithosphere, asthenosphere, mesosphere, outer core, and inner core. The lithosphere consists of the crust and uppermost mantle and is strong rock that when stressed to a certain limit will break (brittle behavior) rather than bend (ductile). On average, the lithosphere is about 100 km thick and comprises the tectonic plates discussed further in Chapter 4. The underlying asthenosphere consists of weak rock near its melting point; however, it is not molten but still largely solid. Because of the hot temperatures, the rock in the asthenosphere has the ability to flow ductilely. The asthenosphere is about 200 km thick, but its lower boundary with the underlying mesosphere is gradational. The mesosphere makes up the bulk of the mantle, and the rocks there are stronger due to the increasing pressure with depth, but they still have the ability to flow, although more slowly than in the asthenosphere. The outer core consists of liquid iron–nickel metal; the inner core is the same composition but a solid due to the extreme pressure.

The compositional and rheological nature of the Earth's interior is largely based on the study of seismic waves whose direction and speed of propagation are based on compositional and physical properties of the material through which they pass. For example, one type of seismic wave does not travel through liquids, and as such seismic receiving stations on the opposite side of the planet from where an earthquake occurs will not detect that wave, creating a shadow zone. The size of the shadow zone is a direct reflection of the size of the liquid, outer core.

Earth's internal heat has two primary sources. The first is heat left over from the tumultuous formation of the Earth, when kinetic energy of celestial collisions was transformed to heat energy (primordial heat). The second major source is radioactive decay of select elements, mainly uranium (U), thorium (Th), rubidium (Rb), and potassium (K). The contribution of each source is about equal. Heat flows from the Earth's interior toward the surface via two main mechanisms: conduction and convection. Conductive heat flow is transfer of energy by contact, also known as thermal diffusion. Conductive heat flow is mainly operative in the Earth's core and crust. Convective heat flow is heat transferred by motion, with subsidiary contribution by conduction. Convective motion is induced by buoyant forces that arise from thermal gradients in a gravitational field. If material becomes hotter than its surroundings, its density is reduced, causing the heated material to rise. Conversely, the surrounding cooler material is more dense and sinks to replace the rising hotter material. Convective heat transfer occurs in the liquid outer core and the rheologically ductile mesosphere and asthenosphere, where buoyant forces exceed viscous forces as measured by the Rayleigh number.

Geothermal heat flow and temperature-at-depth maps illustrate that much of geothermal resources developed and yet to be developed occur in the western United States. For example, most of northern Nevada has a heat flow of >80 mW/m^2, in places >100 mW/m^2. At a depth of 4.5 km, the temperature of crustal rocks in northern Nevada is >150°C and in places as high as 200°C. Although this environment (realm of engineered geothermal systems) represents a vast reservoir of heat and potential source of energy development, it is expensive to access and cannot compete economically with currently developed sources of geothermal energy or natural gas-fired power plants. However, if sedimentary aquifers exist at these depths and are permeable, they could serve as major sources of available geothermal energy if energy prices rise modestly.

SUGGESTED PROBLEMS

1. Explain what factors control whether heat flow will be conductive or convective? What type offers the greatest potential for geothermal energy development and why?
2. Will the Rayleigh number of material affect the heat flow measured at the surface? Why or why not?
3. Assume that a well is drilled in dry sand to a depth of 2500 m and the temperature measured at the bottom is 150°C. For simplicity, assume that the thermal conductivity of dry sand is a constant between 10°C and 200°C. Is there likely to be a geothermal resource? Explain why or why not.
4. Imagine you are a geologist and you have drilled hole RD08 whose temperature profile with depth is shown in Figure 3.7. Using the temperature–depth profiles of the three other wells shown in Figure 3.7, should you continue drilling deeper or stop at the current depth? Justify your position.

REFERENCES AND RECOMMENDED READING

Allis, R., Blackett, B., Gwynn, M. et al. (2012). Stratigraphic reservoirs in the Great Basin—the bridge to development of enhanced geothermal systems in the U.S. *Geothermal Resources Council Transactions*, 36: 351–357.

Ammon, C.J. (2016). *Earth's Origin and Composition*, SLU EAS-A193 Class Notes, University Park: Penn State Department of Geosciences (http://eqseis.geosc.psu.edu/~cammon/HTML/Classes/IntroQuakes/Notes/earth_origin_lecture.html).

Arevalo, Jr., R., McDonough, W.F., and Luong, M. (2009). The K/U ratio of the silicate Earth: insights into mantle composition, structure and thermal evolution. *Earth and Planetary Science Letters*, 278(3–4): 361–369.

Blackwell, D.M., Richards, Z.F., Batir, J., Ruzo, A., Dingwall, R., and Williams, M. (2011). Temperature at depth maps for the conterminous U.S. and geothermal resource estimates. *Geothermal Resources Council Transactions*, 35: 1545–1550.

Clauser, C. and Huenges, E. (1995). Thermal conductivity of rocks and minerals. In: *Rock Physics and Phase Relationships: A Handbook of Physical Constants* (Ahrens, T.J., Ed.), pp. 105–126. Washington, DC: American Geophysical Union.

Davies, J.H. and Davies, D.R. (2010). Earth's surface heat flux. *Solid Earth*, 1(1): 5–24.

Dye, S.T. (2012). Geoneutrinos and the radioactive power of the Earth. *Reviews of Geophysics*, 50(3): RG3007.

EIA. (2016). *International Energy Statistics*. Washington, DC: U.S. Energy Information Administration (http://www.eia.gov/cfapps/ipdbproject/IEDIndex3.cfm?tid=2&pid=2&aid=7).

Gando, A., Gando, Y., Ichimura, K. et al. (2011). Partial radiogenic heat model for Earth revealed by geoneutrino measurements. *Nature Geoscience*, 4(9): 647–651.

Glassley, W.E. (2015). *Geothermal Energy: Renewable Energy and the Environment*, 2nd ed. Boca Raton, FL: CRC Press.

Korenaga, J. (2011). Earth's heat budget: clairvoyant geoneutrinos. *Nature Geoscience*, 4(9): 581–582.

SMU. (2016). Southern Methodist University Geothermal Laboratory website, http://www.smu.edu/dedman/academics/programs/geothermallab.

Tarbuck, E.J., Lutgens, F.K., and Tasa, D. (2005). *Earth: An Introduction to Physical Geology*. Upper Saddle River, NJ: Prentice Hall.

USGS. (1999). *Some Unanswered Questions*. Reston, VA: U.S. Geological Survey (http://pubs.usgs.gov/gip/dynamic/unanswered.html).

4 Fundamental Geologic Elements of Geothermal Systems

KEY CHAPTER OBJECTIVES

- Describe and recognize the three main types of plate tectonic boundaries and their implications for geothermal potential.
- Explain how plate tectonics affects the distribution of prospective geothermal regions on the planet.
- Evaluate intraplate tectonic settings for geothermal potential in light of plate tectonic concepts.
- Describe the three main rock groups and how they impact geothermal resources.
- Relate the different types of geologic structures to the stresses or forces that produced them.
- Evaluate the geothermal potential of a region based on the type of structures exposed.

As discussed in Chapter 3, a tremendous amount of internal heat energy escapes across the surface of the Earth. This escaping heat energy is not uniformly distributed. Just like the distribution of earthquakes and volcanoes, regions of elevated heat flow are concentrated along discrete zones. Most of these zones of elevated heat flow lie along or near the margins of the Earth's tectonic plates, as do most volcanoes and earthquakes. Tectonic plates are large slabs of the lithosphere (crust and uppermost mantle) that move continually (although movement is mainly fitful on human time scales), reflecting the huge amount of work done by Earth's internal heat energy. For example, consider the energy required to move a section of the Earth's crust about 10 m along a distance of about 1300 km. That is what happened in the 9.1-magnitude Sumatra–Andaman earthquake in Indonesia in 2004. The energy released at the surface, which is an indication of seismic potential for damage, amounted to about 20×10^{17} joules or 2.0 petajoules (PJ) (USGS, 2014a). That amount of work equals approximately 5.6×10^9 MWh. For comparison, total electricity generation in the United States from 2011 to 2014 averaged a little less than 4.1×10^9 MWh (EIA, 2016). The 2004 Sumatra earthquake released enough energy in a matter of minutes to power the current electrical needs of the United States for about 1.4 years. Understanding the distribution of geothermal energy across the surface of our planet requires examining plate tectonics. Also, as we began to explore in Chapter 3, the roles of earth materials (thermal and other physical properties of rocks and minerals) are critical in characterizing whether or not a geothermal resource is viable for development. Finally, forces imparted on rocks, mainly through the interaction of moving tectonic plates, produce structures, such as faults, that typically affect the flow of hydrothermal fluids.

PLATE TECTONICS

The theory of plate tectonics was developed in the late 1950s and 1960s and had become established by the mid-1970s. Its development is a classic example of inductive reasoning in which discrete lines of evidence, including the nature of seafloor topography, ages and magnetic patterns of ocean floor rocks, and ocean floor heat flow studies, were pooled to generate an overarching explanation of many earth processes and phenomena. These include the distribution of Earth's major mountain belts, earthquakes, volcanoes, many mineral resources, and geothermal energy. According to plate tectonic theory, the Earth's outer rigid layer or lithosphere, consisting of the crust and uppermost mantle, is broken into a dozen or so major tectonic plates (Figure 4.1). These plates are continually moving in response to forces including convection in the underlying asthenosphere, a zone whose rock has the ability to flow due to thermal weakening.

Within a tectonic plate, geologic activity is commonly minor, typified by few or small earthquakes, subdued topography, and little or no volcanism. Important exceptions to this generalization do occur, such as Hawaii being located near the middle of the Pacific tectonic plate. Hawaii and other places of intraplate volcanism reflect localized deep-seated internal processes (see discussion on intraplate settings later in the chapter). Most regions of geologic unrest (and geothermal energy potential) do lie along or near plate boundaries. The nature of that activity is dictated in good part by how adjoining plates are moving along their boundaries. There are three main types of boundaries based on the nature of plate movement (Figure 4.2). Plates can separate or diverge from each other (divergent boundary), they can collide or converge (convergent boundary), or they can slide past each other (transform boundary).

DIVERGENT PLATE BOUNDARIES

Divergent boundaries form where adjoining plates move away from each other in response to rising material in the underlying asthenosphere. As this rising material approaches the surface, the flow bifurcates, with material moving in opposite directions horizontally. The overlying lithosphere responds by extending and ultimately breaking into two plates that separate in opposite directions. As the underlying asthenosphere upwells below the boundary of separation, it also begins to partially melt due to the reduction in pressure. Thus, divergent boundaries are characterized by active volcanism. A good example of a divergent boundary exposed on land is Iceland (Figure 4.3). Iceland sits atop the Mid-Atlantic Ridge—a sub-ocean mountain chain that runs along the middle of the floor of the Atlantic Ocean and rises about 3 km above the adjacent ocean basins on either side of the ridge. Iceland, as a result, contains numerous active volcanoes and prodigious geothermal resources; in

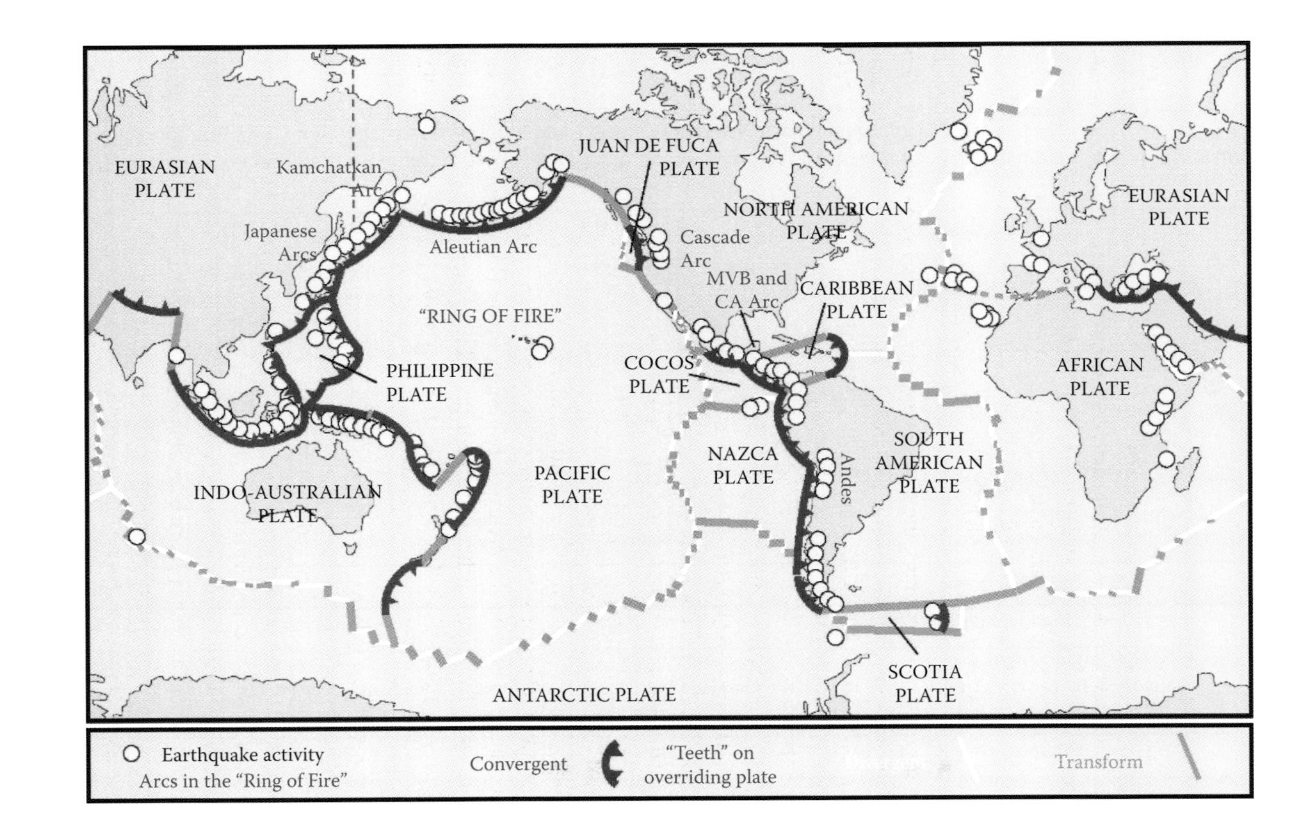

FIGURE 4.1 Map of the world showing the major tectonic plates. Yellow lines denote transform boundaries, red lines divergent boundaries, and black sawtooth lines convergent boundaries. See text for details for types of tectonic plate boundaries. (Illustration from http://www.nature.nps.gov/geology/education/images/GRAPHICS/Lillie_2005_Plate_Tectonic_Map-01.jpg.)

Type of Margin	Divergent	Convergent	Transform
Motion	Spreading	Subduction	Lateral sliding
Effect	Constructive (oceanic lithosphere created)	Destructive (oceanic lithosphere destroyed)	Conservative (lithosphere neither created or destroyed)
Topography	Ridge/Rift	Trench	No major effect
Volcanic activity?	Yes	Yes	No (or limited)

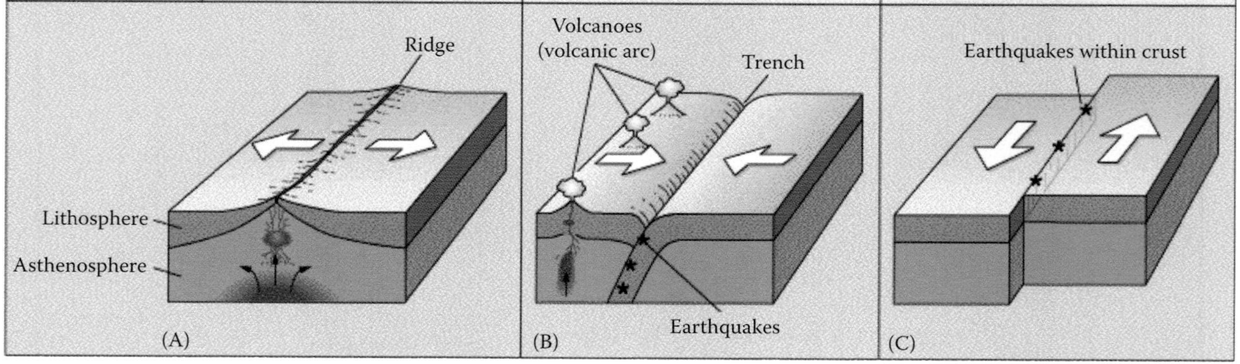

FIGURE 4.2 Illustration showing the three main types of plate tectonic boundaries: (A) divergent, (B) convergent, and (C) transform. See text for details. (Illustration from http://www.age-of-the-sage.org/tectonic_plates/volcanoes_earthquakes.gif.)

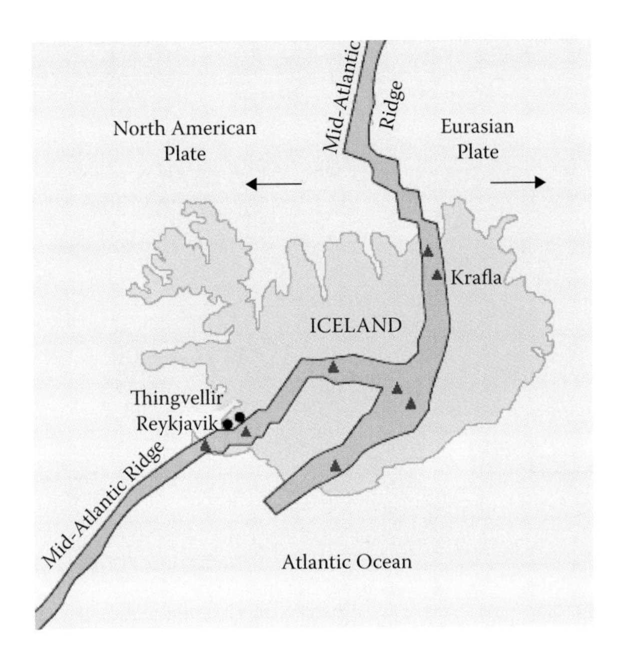

FIGURE 4.3 Map of Iceland showing the divergent Mid-Atlantic Ridge that separates the westward-moving North American plate and eastward moving Eurasian plate. Red triangles denote active volcanoes along the divergent ridge boundary. (Illustration from http://pubs.usgs.gov/gip/dynamic/graphics/Fig16.gif.)

fact, 30% of its power is provided by geothermal fluids, and over 90% of all houses in the country are heated by geothermal energy. The Mid-Atlantic Ridge is part of a 70,000-km-long mountain chain on the ocean floor stretching from the Atlantic, below the Indian Ocean, and extending below the western and eastern Pacific Ocean (Figure 4.4). This mid-ocean ridge system (divergent boundary) is thus the world's longest mountain chain and remains largely unexposed.

Where plates begin to separate on land, continental rifts form. An example is the East African Rift zone extending through Ethiopia, Kenya, Uganda, and Tanzania (Figure 4.5). Here, the continental crust is being stretched and thinned, resulting in partial melting of the upper mantle and development of numerous active volcanoes. As discussed further in Chapter 8, these countries are aggressively exploring and developing geothermal resources. Other examples of continental rifts or nascent divergent boundaries include the Basin and Range Province of Nevada, western Utah, southern Idaho, and southeastern Oregon and the Rio Grande rift of New Mexico and Colorado.

Convergent Plate Boundaries

If plates are separating along divergent boundaries and our Earth is not expanding, then material must be recycled. This occurs along convergent boundaries where old, cold, and dense lithosphere is returned to the mantle in subduction zones. Subduction

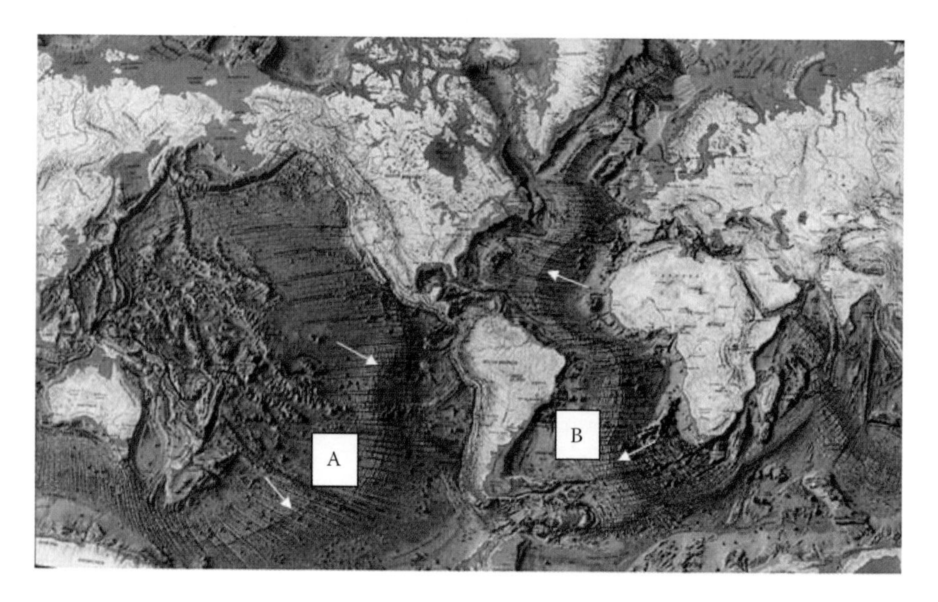

FIGURE 4.4 Physiographic map of the world including bathymetry of ocean floor and mid-ocean ridge system denoted locally by tips of white arrows. (A) East Pacific Rise, and (B) the Mid-Atlantic Ridge. (Base map modified from the World Ocean Floor Map created by Bruce Heezen and Marie Tharp and published by the Office of Naval Research, 1977.)

zones rim much of the Pacific Ocean basin. Where a lithospheric slab begins its downward decent, a deep linear to arcuate trough, called a *trench*, forms on the ocean floor. Such a trench lies off the western coast of South America, where the oceanic Nazca plate dives below the South American plate. Note that where these two plates meet, the more dense oceanic Nazca plate sinks below the less dense continental lithosphere of the South American plate. The deepest point on the ocean floor occurs in the Marianas Trench in the western Pacific where two oceanic plates converge. In this case, it is the older, colder, and more dense Pacific plate that subducts beneath the younger and less dense lithosphere of the Philippine plate.

Note that most of the world's on-land volcanoes are associated with subduction zones and occur in the overriding plate above the subduction zone. This is because as the downgoing plate enters the mantle, the added heat and pressure release water and other volatiles from the downgoing slab. The addition of volatiles lowers the melting point of the overlying mantle and leads to partial melting and formation of magma. Because magma is less dense than the surrounding rock, it rises. Some of it gets close enough to the surface to erupt and form volcanoes. The Andes Mountains of South America, the Cascades of the Pacific Northwest, and the volcanic islands of the western Pacific, such as Japan, are examples of subduction-related volcanism. Proximal to these volcanoes, the magma underlying the volcanoes can serve as a potent local heat source to heat deeply circulating groundwater, potentially producing hydrothermal convection cells and accessible geothermal reservoirs.

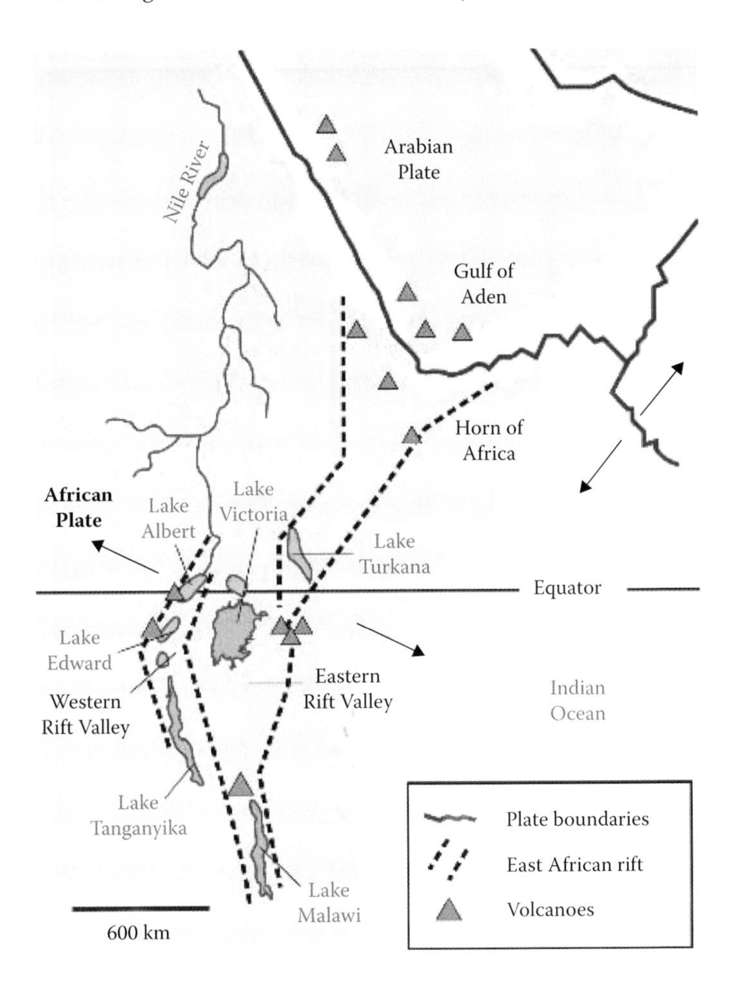

FIGURE 4.5 Map showing the East African Rift zone. To the north, an active divergent boundary occurs in the Red Sea and Gulf of Aden separating the African and Arabian plates. A southern prong of the divergent boundary extends southward onto land making up the rift zone. The rift is characterized by active normal faults and active volcanoes. Large bold arrows denote directions of crustal spreading. (Illustration from http://pubs.usgs.gov/gip/dynamic/East_Africa.html.)

Two types of subduction-related convergent boundaries occur, both resulting in volcanism and the potential for development of geothermal energy. Where two oceanic plates collide or converge, a chain of volcanic islands forms, such as the Aleutian Islands of Alaska or the western Pacific islands of Japan, Philippines, Marianas, and Tonga. A chain of islands created this way is called a *volcanic island arc*. Where an oceanic and a continental plate collide, on the other hand, a continental volcanic arc forms above the down-going oceanic slab to produce a series of volcanoes, as occurs along the Andes and Cascades mountain chains (Figure 4.6).

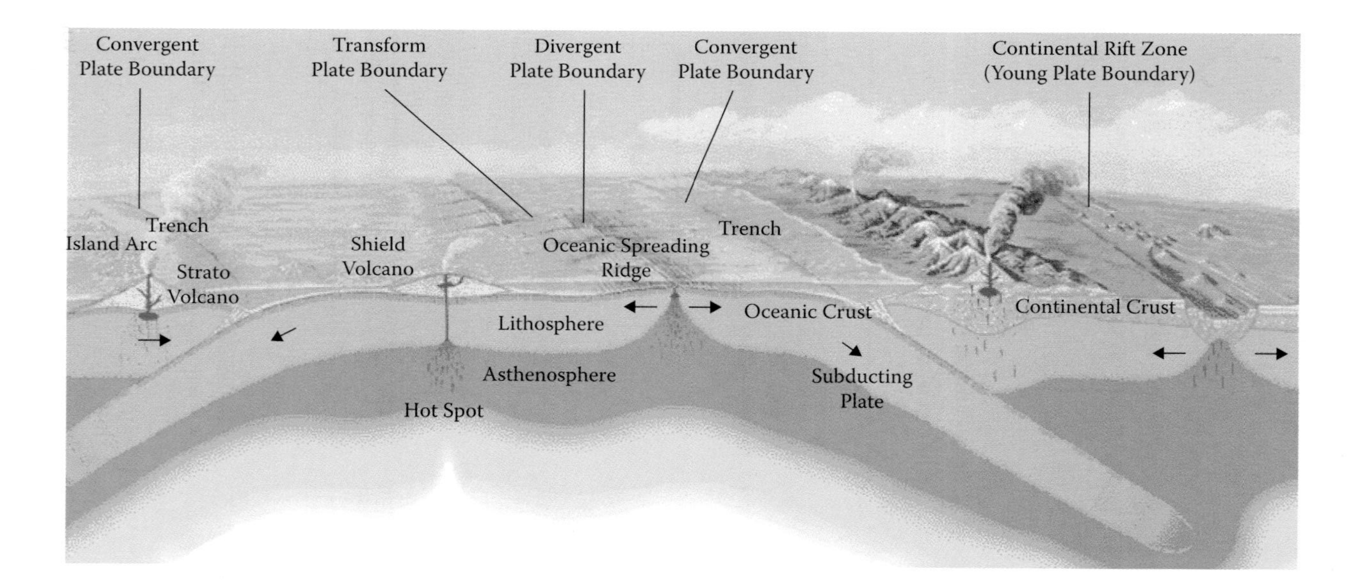

FIGURE 4.6 Cross-sectional view showing oceanic–continental collision on the right side and oceanic–oceanic collision on the left. The former produces a continental volcanic arc, such as the Andes of South America, and the latter produces a volcanic island arc, such as Japan. Hot spot volcanism and transform plate boundaries are discussed later in this chapter. (Adapted from Kious, W.J. and Tilling, R.I., *The Dynamic Earth: The Story of Plate Tectonics*, U.S. Government Printing Service, Washington, DC, 1996, http://pubs.usgs.gov/gip/dynamic/dynamic.html.)

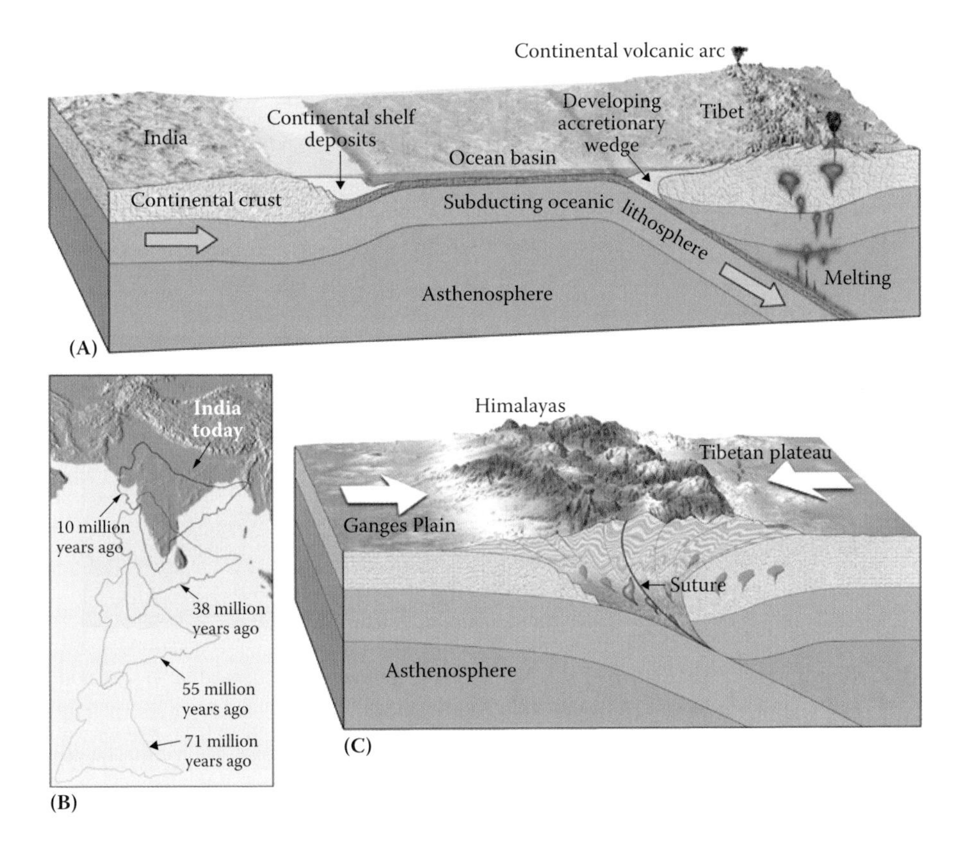

FIGURE 4.7 Sequential illustration of collision of the Indian subcontinent with Eurasia. (A) Cross-sectional representation of approximately 40 million years ago showing India approaching Eurasia and a shrinking marginal ocean basin being subducted below Eurasia. (B) Map showing approximate position of the Indian subcontinent with time. (C) Current cross-sectional view with India continuing to drive itself into Eurasia, pushing the Himalayas upward as the Indian continental crust cannot subduct because of its low density. This continental collision has been ongoing for approximately the last 30 million years. (From Tarbuck, E.J. and Lutgens, F.K., *Earth: An Introduction to Physical Geology*, Prentice Hall, Upper Saddle River, NJ, 1999.)

A third type of convergent boundary involves the collision of two continental plates. This occurred when the Indian subcontinent collided with Asia beginning around 25 million years ago. Because the collision involves only continental lithosphere, no subduction zone forms; the continental lithosphere is insufficiently dense to be subducted into the denser mantle. Instead, the continental plates are sutured and thrust upward to form high mountains, such as the great Himalayan chain (Figure 4.7). Little or no volcanism occurs because there is no subduction, although some deep crustal melting can occur due to frictional heating and thickened continental crust coming into contact with hot mantle. As the continental crust is thickened and its lower part is heated and thermally weakened, portions of the crust may spread laterally or perpendicular

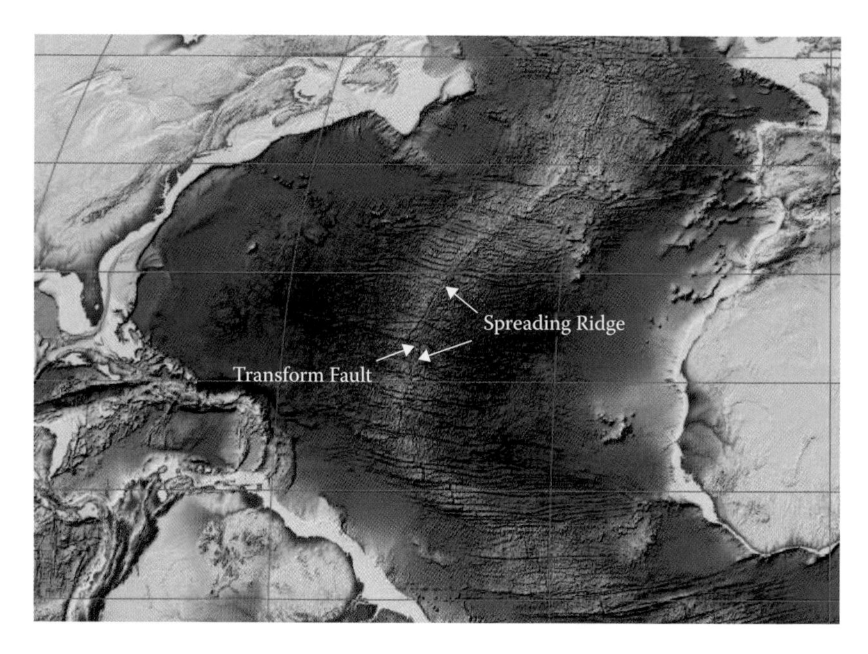

FIGURE 4.8 Bathymetric map of the North Atlantic showing the central divergent mid-ocean ridge boundary that is cut by numerous transform faults that run perpendicular to ridge segments. Two ridge segments and a transform fault are designated. Shown grid represents 10° of latitude and 40° of longitude. (Adapted from Amante, C. and Eakins, B.W., *ETOPO1 1 Arc-Minute Global Relief Model: Procedures, Data Sources, and Analysis*, NOAA Technical Memorandum NESDIS NGDC-24, National Oceanic and Atmospheric Administration, National Centers for Environmental Information, Washington, DC, 2009.)

to the direction of convergence to produce local zones of rifting or extension. Such a process can result in deeply penetrating faults. These faults can channel fluids to deep depths where they can be heated and buoyantly rise to produce local geothermal reservoirs, such as at Yangbajing, Tibet (discussed in Chapter 8).

Transform Plate Boundaries

A transform boundary occurs where plates neither diverge nor converge but instead slide past each horizontally. A transform boundary is also referred to as a *conservative boundary* because crust is neither being formed nor recycled as occurs along divergent and convergent boundaries, respectively. Most transform boundaries or faults occur on the ocean floor and are associated with the mid-ocean ridge divergent boundary system (Figure 4.8). These faults develop to accommodate different rates of spreading that occur along the mid-ocean ridge system (Figure 4.9). Some transform faults, however, slice through land, such as the well-known San Andreas fault of western California (Figure 4.10). Other major on-land transform faults include the North Anatolian fault of northern Turkey and the Alpine fault on the South Island of New Zealand.

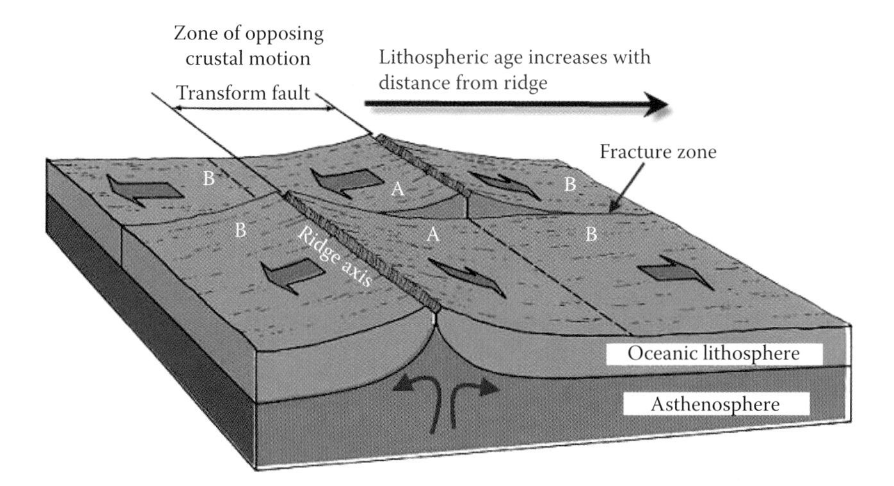

FIGURE 4.9 Close-up view of a transform fault segmenting a divergent mid-ocean ridge boundary. The sense of motion along the transform fault in this case is left lateral because when viewed across the fault the opposite plate is moving to the left (see text). Note that the left lateral motion is opposite to the apparent right-lateral offset of the ridge segments, indicating that the transform faults did not form after the ridges developed but formed in response to different rates of spreading along the spreading ridge. Also note that the transform fault is only in the region between the ridge segments (labeled A), and the areas beyond the ridge segments (labeled B) are called fracture zones because the plates move in the same direction. (Adapted from IMSA, *Earthquakes and Volcanoes: A Global Perspective*, Illinois Mathematics and Science Academy, Aurora, 2002.)

Movement along these transform faults can be of two types. In a left-lateral fault, the rock moves to the left when viewed looking across the fault. In a right-lateral fault, the rock moves to the right of an observer looking across the fault. Because the motion is sideways or near horizontal, transform faults are also called left- and right-lateral strike-slip faults depending on the sense of motion (strike is discussed later in this chapter). The San Andreas transform fault is a right-lateral transform plate boundary separating the Pacific and North American tectonic plates. San Francisco lies on the North American plate, while Los Angeles lies on the Pacific plate. Because of the right-lateral motion, San Francisco and Los Angeles are moving closer to each other at an average rate of about 4.6 cm/year, such that they will be juxtaposed in about 15 million years. Because of friction, motion along transform boundaries is not continuous, however, and the plates lurch past each other perhaps every 50 to 200 years, causing an earthquake that releases stress accumulated since the last earthquake.

Although volcanism is normally absent or minor along transform fault boundaries, unlike along divergent or convergent (subduction-related) boundaries, some important geothermal resources can occur. For example, two major geothermal systems occur within the San Andreas fault zone—The Geysers in northern California and Imperial Valley in southeastern California. Over most of the extent of the San Andreas fault, and along other major transform boundaries as well, volcanism and elevated crustal heat flow are commonly lacking.

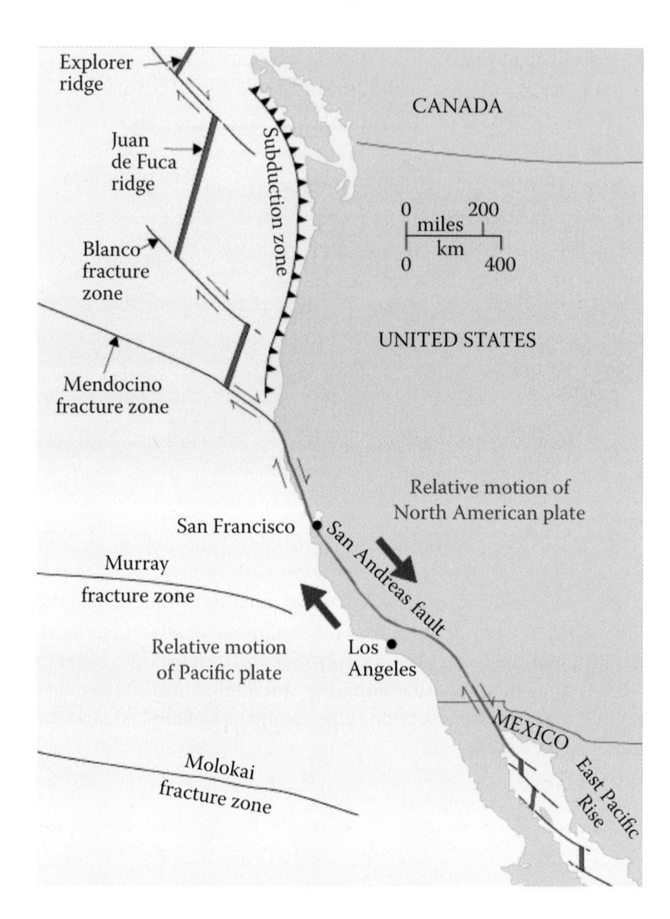

FIGURE 4.10 Map showing the San Andreas transform fault that slices through western California. This fault is a long transform that connects divergent boundaries in the Gulf of California with a divergent boundary and subduction zone off the northern coast of California. The relative sense of motion on the San Andreas fault is right lateral. Note that San Francisco and Los Angeles lie on opposite sides of the San Andreas fault. (From USGS, *Understanding Plate Motions*, U.S. Geological Survey, Reston, VA, 2014.)

Development of the geothermal systems for The Geysers and Imperial Valley reflects different local geologic conditions along the San Andreas fault zone. The Geysers field is temporally and spatially associated with the young and probably still active Clear Lake Volcanic Field. The origin of the Clear Lake Volcanic Field is related to the complex transition from an older subduction-related convergent boundary off the coast of California to the present transform boundary. This transition began upon subduction of an older divergent ridge–transform system of the now consumed Farallon plate, which shut down subduction. Because no plate was being further subducted, hot mantle material welled up in the wake of the downgoing Farallon plate. The upwelled mantle partially melted from reduced pressure, resulting in volcanism and development of the Clear Lake Volcanic Field and the eventual development of The Geysers geothermal field.

The Imperial Valley geothermal systems developed at the south end of the San Andreas fault zone, close to divergent spreading in the Gulf of California with its associated high heat flow. Moreover, this area of the San Andreas fault is also characterized by crustal extension (transtensional) which thins the crust, placing hot mantle rocks closer to the surface. The superposition of high heat flow and crustal extension makes for an ideal geothermal setting in which fluids are heated at shallow depths and extensional faults (discussed later in this chapter) direct the heated fluids toward the surface. Further details on the geologic settings of The Geysers and the Imperial Valley geothermal systems are discussed in Chapter 7.

Intraplate Settings

For the most part, the interiors of tectonic plates are geologically quiet, and areas of elevated crustal heat flow are restricted to some specific regions, including some deep sedimentary basins underlain by radiogenic granitic rocks. Overlying sedimentary rocks act like a thermal blanket, allowing heat to build in deep sedimentary basins. Decay of radiogenic isotopes, mainly uranium, thorium, and potassium, of underlying granitic rocks generates added heat. Such a situation exists in the Cooper Basin of east-central Australia, which was being explored for its potential as an engineered geothermal system. A 1-MWe pilot geothermal plant that was commissioned in 2013 operated for 160 days and produced fluids at 19 kg/s at a well-head temperature of 215°C and bottom-hole temperature of 242°C at a depth of 4200 m (Geodynamics, 2014).

The most attractive intraplate geothermal settings are related to isolated mantle upwellings called *hot spots* or *mantle plumes*. Perhaps the best established and known is the Hawaiian hot spot near the middle of the Pacific Ocean and tectonic plate. Here, a rising plume of mantle material has been operating for at least 75 million years and has resulted in a linear chain of volcanically produced islands and seamounts (submerged volcanic islands) that extend several thousands of kilometers (Figure 4.11). The chain of islands formed as the Pacific plate moved northwestward over the relatively stationary rising mantle plume. As mantle material rose, the pressure was reduced, causing the rock to partially melt to form magma that erupted onto the seafloor. The erupted lava eventually built up to breach the ocean surface to form an island. As the plate continued to move, the island was eventually carried away from the hot spot and island volcanism waned, but a new island began to form in its wake. Currently, the Big Island of Hawaii is comprised of five coalescing volcanoes. The two most southeastern volcanoes, Mauna Loa and Kilauea, are the most active because they are closest to lying directly above the hot spot (Figure 4.12). Moreover, offshore and to the southeast, a new seamount (Loihi) is growing and rises about 3000 m above the surrounding seafloor to within about 970 m of the sea surface (Figure 4.12). Applying the average growth rate of other active Hawaiian volcanoes (about 30 cm/year) to Loihi, Loihi could emerge as a new island in about 30,000 years (Malahoff, 1987).

Iceland is above sea level because it marks a hot spot coincident with the mid-ocean ridge divergent boundary. The combined action of both tectonic processes has resulted in voluminous volcanism allowing the volcanic rocks to pile up above sea level. Without the hot spot, Iceland would be like the rest of the Mid-Atlantic Ridge—under water.

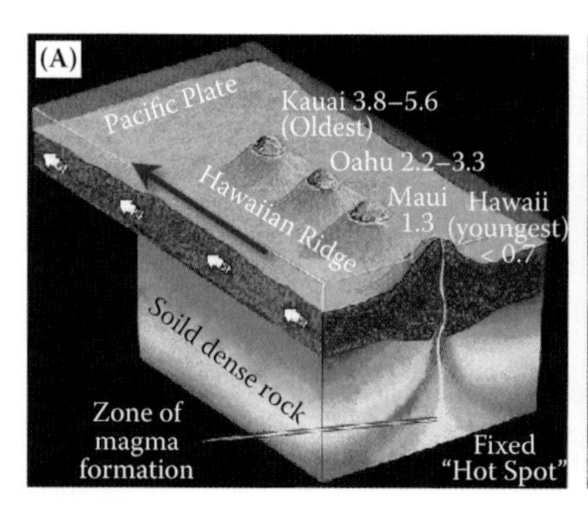

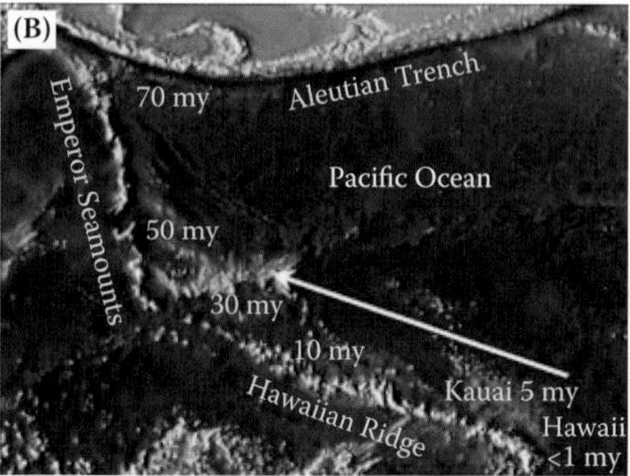

FIGURE 4.11 (A) Map showing Hawaiian island–seamount chain formed as the Pacific plate moves over a stationary plume of upwelling mantle material. Note the systematic increase in ages (numbers in millions of years next to name of island) of the islands to the northwest. (From USGS, *"Hotspots": Mantle Thermal Plumes*, U.S. Geological Survey, Reston, VA, 1999.) (B) Image of a portion of the Pacific Ocean showing the Hawaiian Ridge and Emperor Seamount chain. Note the progressive increase in ages of islands and seamounts (submerged islands) to the northwest, reflecting the direction of movement of the Pacific tectonic plate. The kink in the chain reflects a change in direction of plate motion about 40 million years ago. White arrow shows the direction of plate movement for the last 30 to 35 million years. (Illustration from http://ijolite.geology.uiuc.edu/02sprgClass/geo117/Ocean%20images/hotspot.html.)

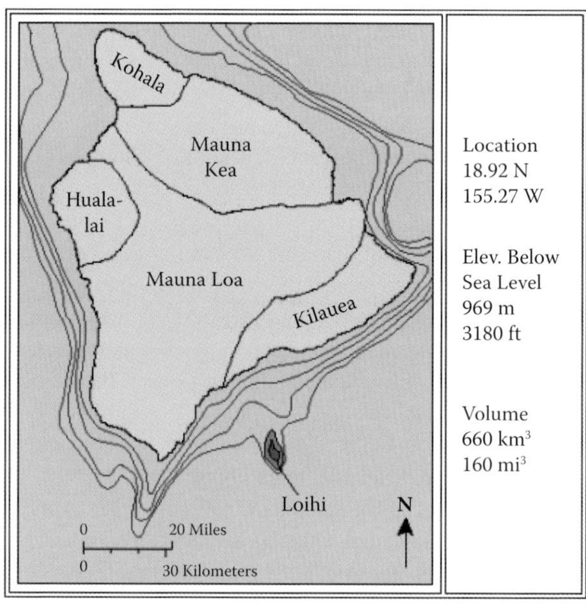

FIGURE 4.12 Map showing the five coalescing volcanoes that comprise the Big Island of Hawaii. A submerged sixth volcano, Loihi, is actively erupting and growing south of Kilauea. (From USGS, *Loihi Seamount: Hawaii's Youngest Submarine Volcano*, U.S. Geological Survey, Reston, VA, 2015.)

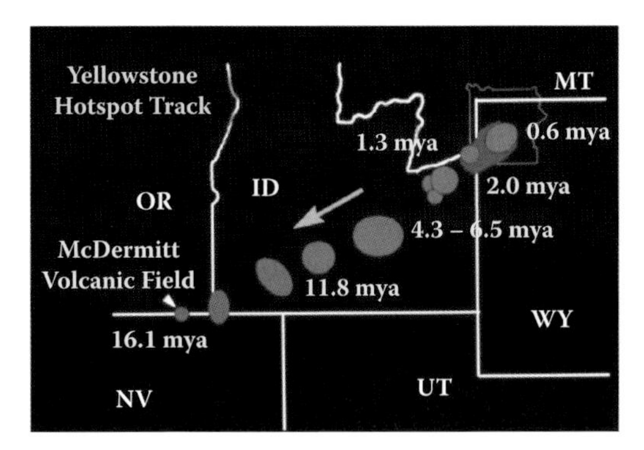

FIGURE 4.13 Map showing locations of volcanic centers (orange ellipses) marking the Yellowstone hotspot track. The abbreviation "mya" stands for millions of years ago. Note the systematic decrease in ages moving from northern Nevada to Yellowstone National Park in northwest Wyoming. The remarkable geothermal activity in Yellowstone is due to its location above the hot spot and presence of shallow (5 to 10 km deep) magma reservoir. (Illustration from http://www.nps.gov/features/yell/slidefile/graphics/diagrams/Images/15899.jpg.)

Because of Hawaii's hot spot setting, the two southeastern-most islands—Maui and the Big Island—have good geothermal power potential. Currently, the Puna geothermal plant, located on the southeastern flank of Kilauea volcano, has an installed capacity of 38 MWe. (The Puna geothermal facility is discussed further in Chapter 7 and in Chapter 12 as a flexible power provider.) The Puna plant is being evaluated for expansion, and geothermal exploration has taken place in Maui. Intraplate hot spots also occur on land. Yellowstone in northwest Wyoming is a good example. Yellowstone currently resides above the hot spot, but a trail of volcanic centers extends to the southwest into northwest Nevada and reflects the southwest movement of the overlying North American plate (Figure 4.13). Yellowstone hosts the largest geyser field (about half of the planet's geysers occur here), some of the most active hydrothermal features, and some of the largest volcanic eruptions on the planet. All are sustained by an underlying and focused upwelling of mantle material. The 1970 Geothermal Steam Act protects Yellowstone from development, but still much can be learned from its volcanic field about developing geothermal potential in other hot spot settings. If Yellowstone were developed, it would likely be able to provide several hundred to perhaps a few thousand megawatts of geothermal electrical power. Intraplate hot spots mark favorable sites for potential geothermal energy development as further explored in Chapter 7.

EARTH MATERIALS

Rocks are an important component for evaluating geothermal resources because the type of rock will influence geothermal reservoir characteristics. For example, compared to granite, a quartz-rich sandstone will have higher thermal conductivity and intrinsic permeability (ability to transmit fluids, as discussed in Chapter 5). Yet, a granite could have overall higher permeability if affected by secondary forces resulting in numerous fractures. Thus, some fundamental information on rocks and how forces affect them is required to help understand geothermal energy resources.

Rocks are classified by how they form and the minerals that compose them. A mineral is any inorganically formed substance that has an ordered atomic structure (crystalline) and a definite chemical composition. Window glass and quartz have the same chemical composition, SiO_2, but window glass is not a mineral because it lacks an organized arrangement of atoms as found in quartz. Rocks are commonly an aggregate of minerals, although some rocks may be just one mineral, such as limestone comprised mainly of the mineral calcite (calcium carbonate). A few rocks, such as obsidian, have no minerals (obsidian is volcanic glass). The three main groups of rocks are igneous, metamorphic, and sedimentary. Igneous and metamorphic rocks comprise the bulk of Earth's crust and mantle. Sedimentary rocks, on the other hand, are the most commonly exposed group at the Earth's surface; however, they are volumetrically minor as they generally form a relatively thin veneer overlying igneous and metamorphic rocks, which make up the bulk of the Earth's crust.

The formational relationships among the three rock groups are illustrated in the rock cycle (Figure 4.14). Although rocks may go through each part of the cycle, it is not required. In other words, the cycle can be short-circuited such that a metamorphic rock, for example, need not be melted to yield an igneous rock but could be uplifted, weathered, and eroded to form sediments that turn into sedimentary rock after burial and compaction.

Igneous Rocks

Igneous rocks form from the cooling and crystallization of molten rock or magma. There are two main groups of igneous rocks based on where they cool and solidify. The first type includes intrusive or plutonic igneous rocks that cool slowly at depth, and the second type includes extrusive or volcanic rocks that cool rapidly at or near the Earth's surface. Because plutonic igneous rocks cool slowly, crystals can grow and are easily visible, typically a few millimeters to more than a centimeter in size. Volcanic rocks, on the other hand, cool quickly, which limits the size of the formed crystals to less than about a millimeter. The environment in which an igneous rock forms, whether plutonic or volcanic, can be inferred by a rock's texture. The texture of a rock reflects the grain sizes and shapes of included minerals and how minerals are disposed to each other. Fine-grained volcanic rocks have an *aphanitic* texture (Figure 4.15A), whereas most medium- to coarse-grained plutonic rocks have *phaneritic* texture (Figure 4.15B). Very rapidly cooled rocks have a glassy texture, such as obsidian. In this case, the rate of cooling is so fast that few if any crystals have a chance to form. It is not uncommon for many volcanic rocks to have a mixed texture, consisting of a fine-grained or aphanitic groundmass along with larger, easily visible crystals called *phenocrysts*. This mixed texture is called *porphyritic* and reflects a two-stage cooling process (Figure 4.15C). The magma started to crystallize slowly at depth to form the phenocrysts and then erupted and cooled rapidly to form the aphanitic groundmass. Thus, the environment of an igneous rock is based on where it last cooled, so a porphyritic texture would be associated with a volcanic, not plutonic, environment even though the magma from which the rock formed began cooling slowly underground.

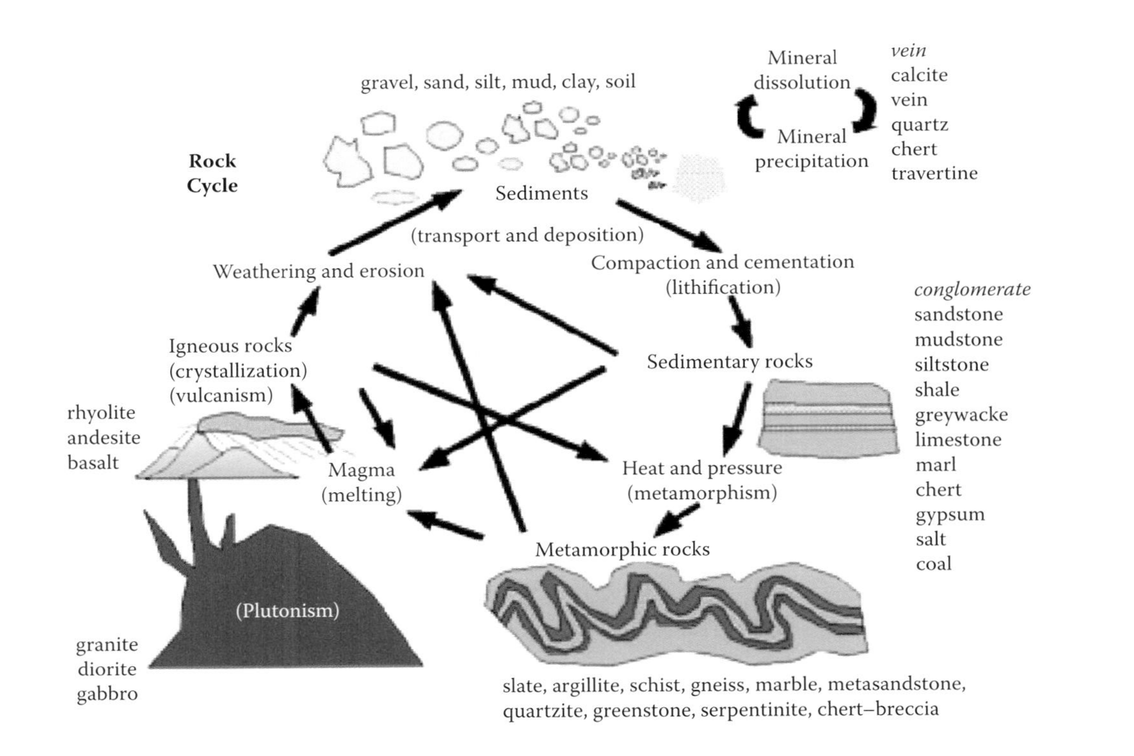

FIGURE 4.14 Rock cycle. Note the processes leading from one rock group to the next, and that the cycle can be short-circuited as indicated by the arrows that go between the outer loop. (From USGS, *Overview of Geologic Fundamentals*, U.S. Geological Survey, Reston, VA, 2015.)

FIGURE 4.15 Textures of igneous rocks. (A) Fine-grained or aphanitic texture of the volcanic rock basalt. (Image from https://en.wikipedia.org/wiki/Igneous_rock.) (B) Coarse-grained or phaneritic texture of the plutonic rock granite. (Image from https://www.flickr.com/photos/jsjgeology/8475800221.) (C) Porphyritic texture with conspicuous phenocrysts of plagioclase set in reddish-gray aphanitic groundmass; the rock is a porphyritic andesite. (Image from https://www.flickr.com/photos/jsjgeology/15661069958.)

As with all rocks, igneous rocks are also classified on their chemical and mineralogical makeup. The most important chemical constituent is silica (SiO_2), which ranges from a low end of about 40 wt% SiO_2 to a high end of about 78 wt% SiO_2. The amount of silica governs the overall color of the rock and its assemblage of minerals. Low-silica igneous rocks are typically dark colored, reflecting minerals rich in iron, magnesium, and calcium. High-silica igneous rocks are typically light colored because they are comprised of light-colored minerals, such as feldspar and quartz. These minerals are rich in silica, sodium, and potassium but low in iron, magnesium, and calcium.

Combining texture and composition yields a classification scheme as shown in the Figure 4.16. Basalt is a dark-colored volcanic rock, having low silica content but rich in iron, magnesium, and calcium, that erupts along mid-ocean ridge divergent boundaries and, therefore, underlies most of the ocean floor. Its intrusive counterpart

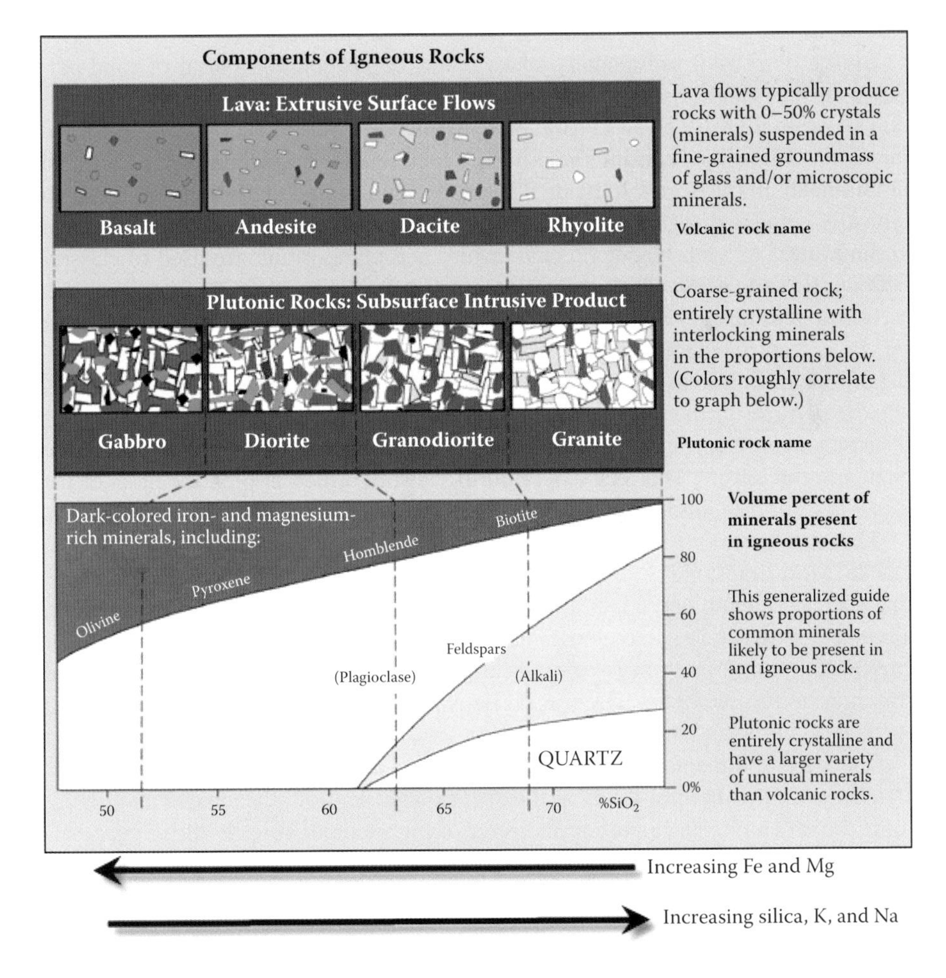

FIGURE 4.16 Simplified igneous rock classification chart. (From Johnson, J., *Glossary: Igneous*, U.S. Geological Survey, Reston, VA, 2015.)

is gabbro. On the other end of the spectrum is rhyolite, a common rock erupted from continental volcanoes. Its intrusive counterpart is granite. Intermediate compositions of silica form andesite, the main rock that makes up the Andes Mountains of South America. Its intrusive counterpart is diorite.

Sedimentary Rocks

Sedimentary rocks form at or near the surface of the Earth; therefore, they provide the best record of Earth's past life forms and climate regimes. The weathering and erosion of pre-existing rock form the components of sedimentary rocks. These components can be solid particles, such as sand, gravel, or clay. They can also be dissolved mineral constituents that precipitate from a solution either organically, such as coral, or inorganically, such as salt on the bottom of a dried-up lake bed. Thus, similar to igneous rocks, sedimentary rocks can be divided into two divisions. In this case, however, the divisions are clastic (detrital) and chemical.

Clastic or detrital sedimentary rocks reflect deposition and burial of solid particles, such as sand or silt, whereas chemical sedimentary rocks form by precipitation of originally dissolved constituents in solution. Clastic sediments are turned into stone following deposition and burial, such as in a floodplain or lake on land or ultimately in the ocean. Individual grains are compacted under pressure and also typically cemented together by silica or calcium carbonate that precipitates from groundwater. As with igneous rocks, texture and composition are used to classify both clastic and chemical sedimentary rocks (Figure 4.17).

Clastic sedimentary rocks are classified on the basis of texture and composition. In this case, texture reflects the size, shape, degree of rounding or angularity, and sorting of rock and mineral particles. Consolidated clay particles form shale or claystone, silt-size particles form siltstone, and sand-size particles form sandstone. Larger grains form either conglomerate or breccia, depending on whether the grains are rounded or angular, respectively. The rock can be further distinguished also on the basis of the mineralogy of the grains, such as a quartzose sandstone or chert–pebble conglomerate.

The most common chemical sedimentary rock is limestone, comprised largely of calcite. Most limestone is formed organically from biogenic processes of organisms that secrete calcium carbonate to form shells. These shells collect on the seafloor. The skeletal structure of coral reefs also turns into limestone upon burial. Another carbonate mineral that forms inorganically is dolomite, which can replace original limestone to form dolostone. Hence, limestone and dolostone are commonly referred together as carbonate rocks.

Chemical sedimentary rocks also form as a result of geothermal processes. Discharge of geothermal fluids at the surface can deposit siliceous sinter (Figure 4.18), which can be an important indicator of high-temperature (≥~180°C) geothermal fluids at depth (Fournier and Rowe, 1966). Calcium carbonate can also be deposited around hot springs to form travertine, which is an indicator of lower temperature fluids at depth (Figure 4.19). If springs are discharging at the bottom of a lake, tufa (another form of calcium carbonate) can be deposited (Figure 4.20). The subject of chemical geothermal minerals and chemical sedimentary rocks is discussed in further detail in Chapters 6 and 8.

Clastic Sedimentary Rocks

Texture (grain size)	Sediment Name	Rock Name
Coarse (over 2 mm)	Gravel (rounded fragments)	Conglomerate
	Gravel (angular fragments)	Breccia
Medium (1/16 to 2 mm)	Sand	Sandstone
Fine (1/16 to 1/256 mm)	Mud	Siltstone
Very Fine (less than 1/256)	Mud	Shale

Chemical Sedimentary Rocks

Composition	Texture (grain size)	Rock Name	
Calcite	Fine to coarse crystalline	Crystalline Limestone	
		Travertine	
	Shells and cemented shell fragments	Coquina	Biochemical Limestone
	Shells and shell fragments cemented with calcite cement	Fossiliferous Limestone	
	Microscopic shells and clay	Chalk	
Quartz	Very fine crystalline	Chert (light color) Flint (dark color)	
Gypsum	Fine to coarse crystalline	Rock Gypsum	
Halite	Fine to coarse crystalline	Rock Salt	
Altered plant fragments	Fine-grained organic matter	Bituminous Coal	

FIGURE 4.17 Simplified classification chart for sedimentary rocks. (Illustration from https://en.wikipedia.org/wiki/Clastic_rock#/media/File:Sedimentary_Rock_Chart.png.)

FIGURE 4.18 Geologist sampling Growler Hot Spring in northeastern California. The spring is actively boiling and white residue around spring is silica sinter deposited as the solution cools. (U.S. Geological Survey photograph.)

FIGURE 4.19 Mammoth Hot Springs in Yellowstone National Park. The impressive terraces are made of travertine (calcite) formed as CO_2 bubbles out of solution when fluids discharge at the surface. (Photograph by Jon Sullivan, http://www.public-domain-photos.com/travel/yellowstone/mammoth-hot-springs-free-stock-photo-4.htm.)

FIGURE 4.20 Tufa towers and fountaining geothermal well at the north end of Pyramid Lake, Nevada. The tufa is made of calcite and was formed when springs discharged at the bottom of ancestral Lake Lahontan of which Pyramid Lake is a remnant. The tufa towers are 60 to nearly 100 m high. (Photograph by author.)

Metamorphic Rocks

Metamorphic rocks also form from preexisting rocks, but unlike sedimentary rocks they form at depth due to increasing pressure, temperature, and the chemical action of fluids. These rocks are also classified on the basis of texture and mineral composition. In this case, the texture or grain size mainly reflects the grade (temperature and pressure) of the metamorphism, such that a fine grain texture reflects mainly low temperature and pressure (low grade), and a coarse grain reflects high temperature and pressure (high grade). All changes in texture and mineralogy occur in the solid state.

Similar to the other two rock groups, metamorphic rocks can be classified into two types: foliated and nonfoliated (Figure 4.21). In foliated metamorphic rocks, minerals have a preferred orientation. Generally, that orientation is perpendicular to the direction of greatest applied stress. A good example of a foliated metamorphic series is the progressive metamorphism of shale. With increasing temperature and pressure (increasing grade), a shale will change from slate, to phyllite, to schist, and finally to gneiss at highest temperatures and pressure, but prior to melting (Figure 4.21). In nonfoliated metamorphic rocks, a preferred orientation is lacking because either pressure was equally applied in all directions, as can occur in deep sedimentary basins, or the parent rock is mainly monomineralic. Monomineralic rock, such as a quartz-rich sandstone or limestone, turns into quartzite and marble, respectively.

A variety of metamorphic rock applicable to geothermal environments is the type formed at relatively low temperatures and pressures (typically, <300°C and <1.5 kilobars), but forms under the chemical action of fluids. Such metamorphism is referred to as *hydrothermal alteration*, and hydrothermally altered rocks can be an important indicator or tool for assessing the quality of a geothermal resource (discussed further in Chapter 6).

TECTONIC SETTINGS OF ROCK GROUPS

Igneous and metamorphic rocks are found together in subduction-related convergent boundaries. In active convergent settings, both volcanic and plutonic igneous rocks are commonly exposed. Metamorphic rocks may or may not be exposed, depending on the location and depth of erosion. In more deeply eroded island and continental arcs, metamorphic and plutonic igneous rocks are common. Close to the trench and subduction zone, metamorphic rocks are typified by mineral assemblages formed under high pressure but relatively low temperature, or what is referred to as *blue-schist metamorphism*. The name stems from the occurrence of a diagnostic bluish mineral (glaucophane) that forms under conditions of high pressure and low temperatures characteristic of subduction zones. Closer to the volcanic arc and farther from the trench, metamorphic rocks are characterized by high-temperature but low-pressure minerals as rocks are baked by rising bodies of magma. It is also in this region of high-temperature and low-pressure metamorphism where hydrothermal metamorphism (or hydrothermal alteration) occurs in association with geothermal systems.

Metamorphic Rock Classification

Original Rock	Texture	Rock Name	Metamorphic Process	Metamorphic Grade	Comments
Mudstone	Foliated	Slate	Regional	Lower	Breaks into plates (slaty cleavage)
Mudstone	Foliated	Phyllite	Regional	Moderate	More shiny and crenulated than slate
Mudstone	Foliated	Schist	Regional	Mod-high	Different schists recognized on the basis of mineral content
Mudstone Granite	Foliated	Gneiss	Regional	High	Well-developed light and dark banding
Quartz sandstone	Non-foliated	Quartzite	Contact	Low-high	Sugary texture composed of interlocking quartz grains; relatively hard; won't fizz with acid
Limestone	Non-foliated	Marble	Contact	Low-high	Sugary texture composed of interlocking calcite grains; relatively soft; may fizz with acid
Basalt	Non-foliated	Metabasalt	Contact	Low	Greenish color due to chlorite

FIGURE 4.21 Simplified metamorphic rock classification chart. (Illustration from https://www.mesacc.edu/sites/default/files/pages/section/academic-departments/physical-science/geology/images/mmrkid1.jpg.)

In convergent margins without subduction zones (continental collisional zones, such as the Himalaya Mountains), rocks consist mainly of variably metamorphosed sedimentary rocks along with local igneous intrusions and metamorphic rocks. Metamorphic rocks are commonly medium- to high-grade schists and gneisses reflecting high temperatures and pressures as rocks are squeezed and heated. Due to the considerable thickness of continental crust and lack of subduction zones to promote volcanism in continental collision zones, areas of elevated heat flow and geothermal energy potential are generally localized to specific regions. Examples are areas affected by a high degree of faults or penetrative fractures that can channel fluids to depth to be heated and then back to the surface.

In divergent tectonic settings, igneous rocks predominate along with basin-fill sedimentary rocks. Along the mid-ocean ridges, basalt is erupted. In divergent or rift environments on land, a wide compositional range of volcanic rocks is erupted from basalt to rhyolite with their intrusive counterparts occurring at depth. Because of the abundance of young volcanic rocks, elevated heat flow is fairly widespread in divergent tectonic settings and where such settings occur on land, such as in Iceland and the East African Rift zone. There, the potential for developing geothermal energy is quite high. Sedimentary rocks filling fault-bounded basins can serve as potentially permeable geothermal reservoirs in which heated water can freely circulate.

All three groups of rock types can be found in transform boundary settings. Young igneous rocks that could reflect localized areas of high heat flow are generally restricted to specific segments of transform faults, such as at the north and south ends of the San Andreas fault in California. In certain regions of a transform fault, crustal extension in addition to crustal shear (transtension) can occur, as is the case for the Imperial Valley of southeastern California—a major geothermal power-producing region. The Geysers, Salton Sea region, and Cerro Prieto in Baja Mexico are power-producing geothermal systems related to the San Andreas transform fault and are discussed in Chapter 7.

EARTH FORCES AND GEOLOGIC STRUCTURES

Forces in the Earth cause rocks to deform. These forces are typically concentrated along tectonic plate boundaries. Deformation of rocks occurs in two main ways: (1) brittle failure, which causes rocks to break or fracture, and (2) ductile failure, where rocks bend without breaking. Both types of rock structures can serve as pathways, traps, or barriers to geothermal fluid flow.

Stress vs. Strain

Stress is a force that is applied to rocks. How the rocks respond to that force or stress is called *strain* or *deformation*. Strain is what is recorded in the rocks, and the type of strain recorded reflects the type of stress applied. The three main types of stress are compressional, tensional, and shear. Compressional stress results in crustal shortening and thickening, whereas tensional stress elongates and thins the crust.

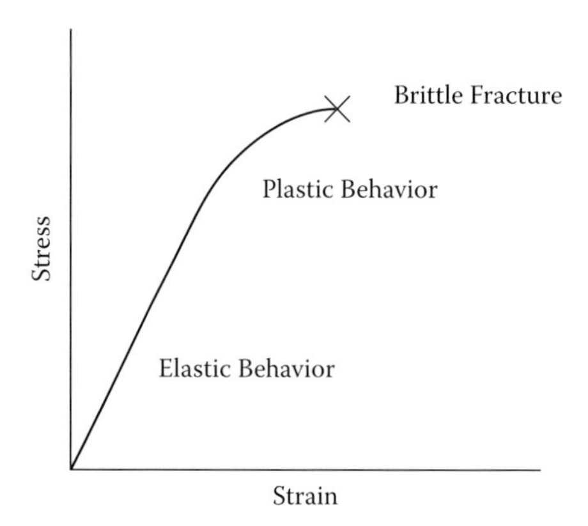

FIGURE 4.22 Graph of stress vs. strain and resulting styles of deformation.

Shear stress bends and distorts the crust. Convergent boundaries sustain largely compressive stresses, divergent boundaries produce tensional stresses, and shear stresses are localized along transform boundaries.

Strain or deformation of rocks can be brittle (breaking and fracturing), ductile (bending without breaking or folding), or elastic (a spring-like response to applied stress that occurs during earthquakes). The relationship between stress and strain can be shown on a graph, where low amounts of stress and strain result in elastic deformation or nonpermanent deformation (Figure 4.22). When enough stress is applied and the rock begins to fail (the yield point), it will typically bend, undergoing ductile or plastic deformation. If further stress is applied, the rock will break or undergo brittle deformation. The type of deformation is also a function of external parameters. If stress is rapidly applied, elastic or brittle, rather than ductile, deformation typically forms. On the other hand, if stress is applied under high temperatures, such as can be found in the lower and middle crust, rocks will tend to bend rather than break. Furthermore, the intrinsic strength of different rocks impacts the style of deformation. Shale is a relatively weak rock and can bend before breaking, reflecting the bendable nature of clay particles that make up shale. A quartz-rich sandstone or quartzite, on the other hand, consists of strong quartz grains that resist bending and will typically break before bending much.

Ductile Structures

The permanent bending of rocks without breaking is termed *folding*. Most folding occurs under a compressive stress regime at high temperature or in relatively weak rocks, such as shale. As rocks are squeezed they bend into a series of folds that effectively shorten and thicken the crust. A rather complex classification of folds, based on their geometry, exists, but for purposes of this book two main types of folds are considered—synclines and anticlines. A *syncline* is where rock layers bow downward, and an *anticline* is where rock layers bow upward (Figure 4.23). Because of this geometry, the older rocks are exposed in the core of an anticline and younger

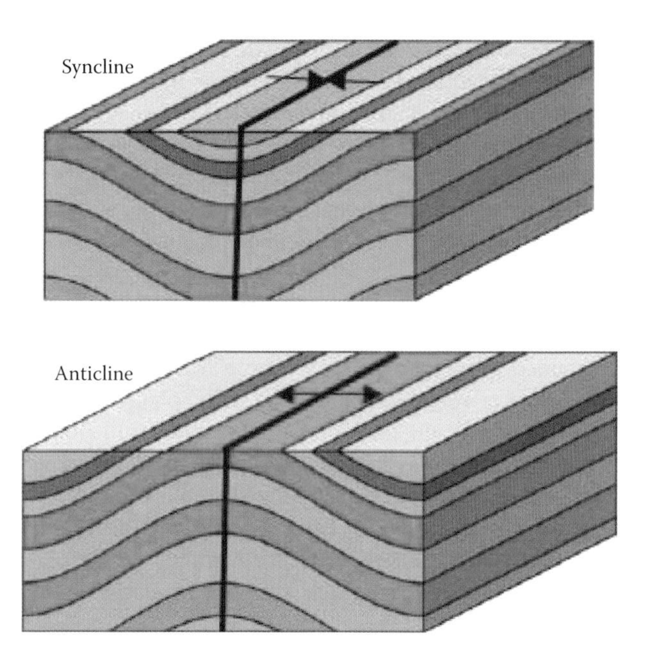

FIGURE 4.23 Block diagrams showing bowed-down layers (concave downward) of a syncline and bowed-up layers (convex upward) of an anticline. Note that younger rocks are exposed in the cores of synclines and older rocks in the cores of anticlines. The bold black lines in both diagrams mark the trace of the fold axis, which divides the folds in half. The rocks are inclined toward the fold axis in a syncline and away from the fold axis in an anticline, as shown by the black arrows. (Illustration adapted from http://www.slideshare.net/GP10/anticlines-and-synclines.)

rocks are exposed in the core of a syncline (Figure 4.23). Anticlines can serve as traps for low-density oil and natural gas if the apical portion (*hinge zone*) includes rock layers that do not transmit fluids easily. The same holds true for a hot geothermal fluid, which is more buoyant than cold water, and it too can be localized near the hinge zone of an anticline, especially if the anticline formed in response to an underlying intrusion of magma that arched the overlying rocks and heated deeply circulating groundwater.

Brittle Structures

When rocks break, fractures form. There are two types of fractures: (1) those that have no relative movement across the fracture, called *joints*, and (2) those where rocks on either side of the fractures have moved, called *faults*. Joints commonly form from cooling and contraction of lavas and are aptly called *cooling joints*. They can also form during tectonic uplift of rock when the rock expands due to unloading forming joints that can both parallel the rock surface or cut perpendicularly to the rock surface and angle steeply downward. Faults are special fractures in which rocks on either side of the fracture have moved relative to each other. These rocks can move past each other in three main ways: side to side, up and down, or a combination of the two styles. Faults are classified based on this relative sense of motion, which in turn reflects the type of stress applied. The two main classes of faults are dip-slip faults (up–down movement) and strike-slip faults (side-to-side motion). Dip-slip faults are further classified as normal or reverse faults and strike-slip faults as either right- or left-lateral faults. To understand this classification, some additional terms must be introduced that geologists use to define the spatial orientation of rock layers and faults.

Both rock layers and faults define planes. To describe the spatial orientation of a plane, geologists measure the orientation of two intersecting and mutually perpendicular lines contained in the plane. One of these lines is the *strike*, which is the compass direction of a horizontal line contained in the plane. The second line is the *dip*, which is the amount of inclination of the plane measured from horizontal and direction of inclination measured perpendicular to the strike line. The strike and dip of inclined rock layers are illustrated in Figure 4.24. Where a fault plane intersects the topographic surface (another plane), a line forms called the *fault trace*. In Figure 4.25, two fault planes (stippled patterns) are shown having different strikes and dips. Note that the compass direction of dip must be given, because there are two possibilities that are perpendicular to a given strike.

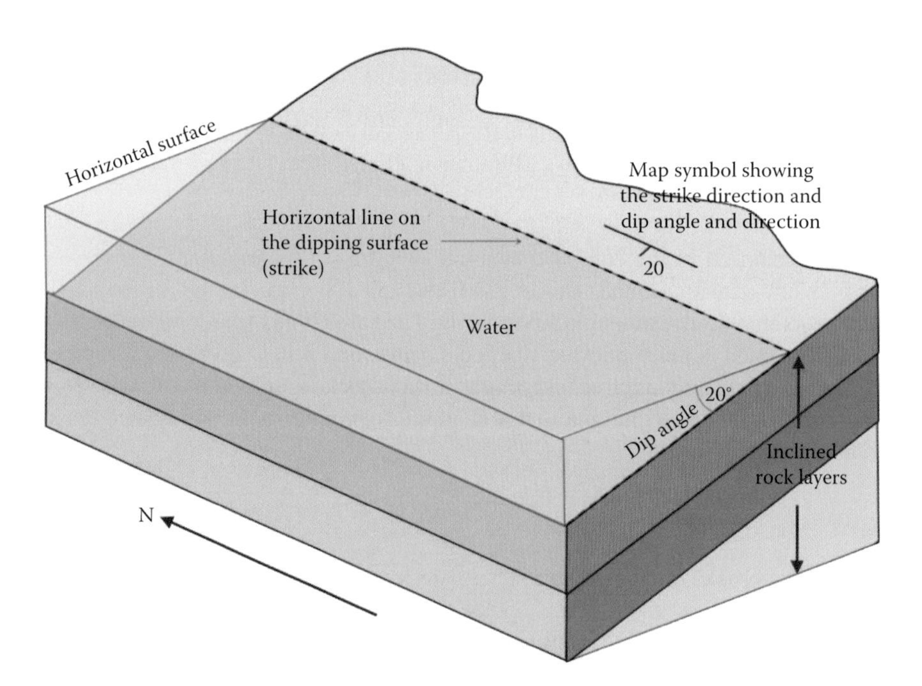

FIGURE 4.24 Block diagram illustrating the strike and dip of dipping rock layers. In this example, the strike is north and the dip direction is due west with rock layers dipping (inclined) 20° below the horizontal (water) surface. (Adapted from Earle, S., *Physical Geology*, Section 12.4, BCcampus, Victoria, BC, 2015, https://opentextbc.ca/geology/chapter/12-4-measuring-geological-structures/.)

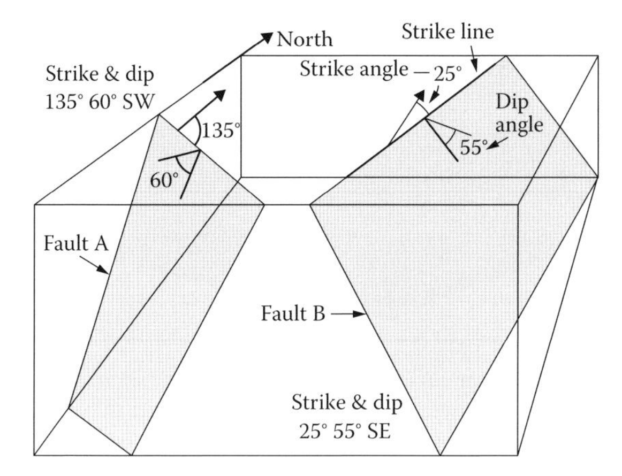

FIGURE 4.25 Block diagram showing two fault planes as indicated by the gray areas. Note that the dip direction must be specified to avoid ambiguity. (Illustration adapted from http://maps.unomaha.edu/maher/eurtoadstoolproject/toadstoolprojectinfo/strikedip.jpeg.)

The block of rock that lies above a dipping fault plane is the *hanging wall* and the block of rock below the fault plane is the *footwall*. Dip-slip faults occur where the hanging and footwall blocks move parallel to dip. In a normal dip-slip fault, the hanging wall moves down relative to the footwall block (Figure 4.26). Normal dip-slip faults form under extensional stress; consequently, the crust is thinned by movement along normal dip-slip faults. Movement along normal faults creates a characteristic topography in which the uplifted footwall forms mountain ranges called *horsts* and the downdropped hanging wall forms valleys called *grabens*. This horst and graben topography is exemplified in Nevada, where it forms the core of the Basin and Range Province and has resulted from ongoing crustal extension. Mountains continue to rise and valleys continue to drop with each earthquake.

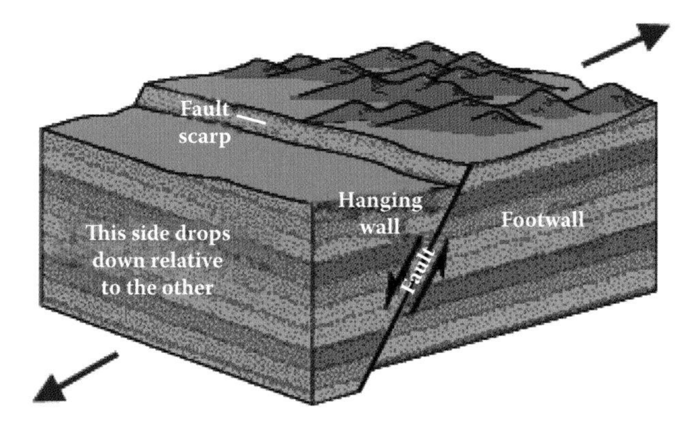

FIGURE 4.26 Block model showing a normal dip-slip fault with the hanging wall block moving down relative to the footwall. In addition to the up–down offset, note that normal faults form by crustal extension or pulling on the rocks as noted by the bold black arrows. (Adapted from USGS, *Visual Glossary: Normal Fault*, U.S. Geological Survey, Reston, VA, 2014.)

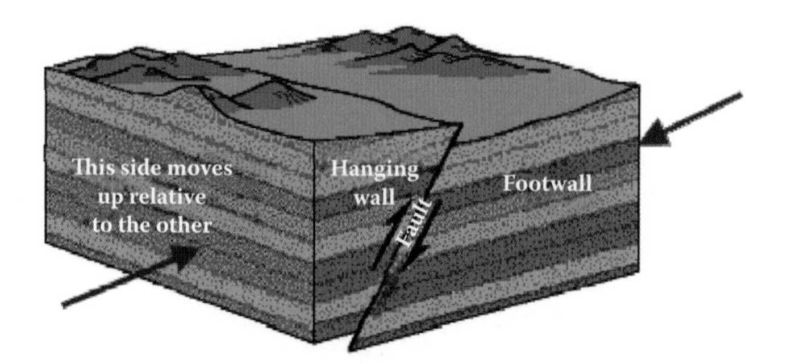

FIGURE 4.27 Block model of a reverse dip-slip fault. Note that in addition to uplift of the hanging wall block, rock layers are partially overlapped, which thickens the crust. Reverse faults form by crustal compression as noted by the bold arrows. (Adapted from USGS, *Visual Glossary: Reverse Fault*, U.S. Geological Survey, Reston, VA, 2014.)

A reverse dip-slip fault is where the hanging wall block moves upward relative to the footwall. Reverse dip-slip faults form during compressional forces that heave the hanging wall upward, and the footwall block is shoved underneath. Rock layers are overlapped or partially stacked over each other and as a result the crust is thickened (Figure 4.27).

In strike-slip faults, movement is more or less parallel to strike—that is, horizontal or side-to-side. As noted previously in the discussion on transform tectonic boundaries, motion along strike-slip faults is either right or left lateral (Figure 4.28). In a right-lateral strike-slip fault, the rock on the opposite side of the fault from an observer moves to the right, and in a left-lateral strike-slip fault, the rock on the other side of the fault moves to the left. A final type of a fault is an oblique-slip fault that combines both dip-slip and strike-slip movement (Figure 4.29). In this particular example, the shown fault has left-lateral, normal oblique-slip sense of movement.

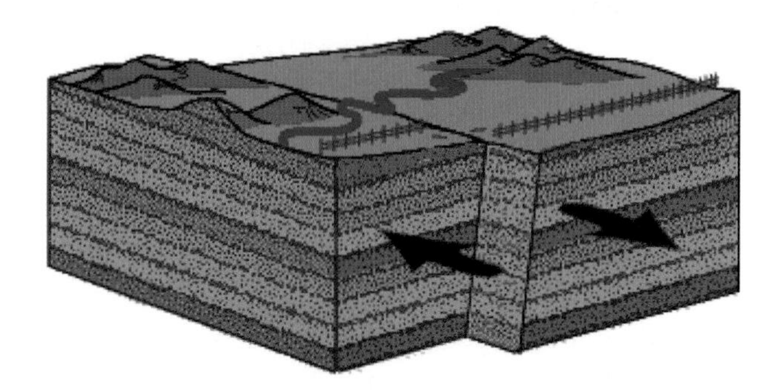

FIGURE 4.28 Block model illustrating a strike-slip fault in which rocks move parallel to the strike of the fault or horizontally. Can you correctly identify what type of strike-slip fault is shown? (Adapted from USGS, *Visual Glossary: Fault, Normal Faults, Reverse Faults, Strike-Slip Fault, Fault Scarp*, U.S. Geological Survey, Reston, VA, 2014.).

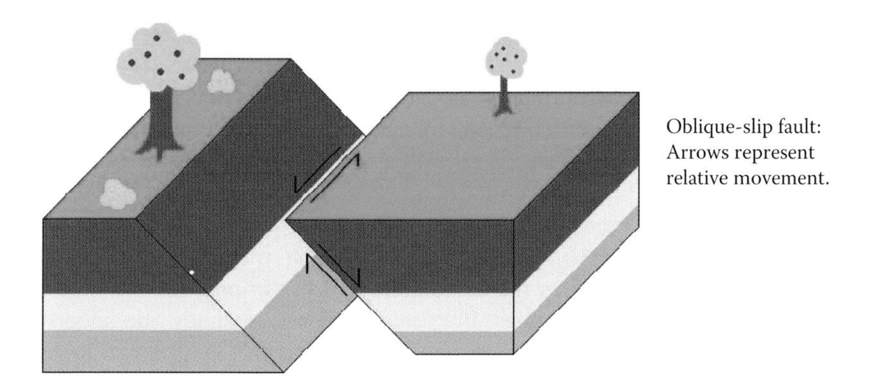

FIGURE 4.29 Block model of an oblique-slip fault having both strike-slip and dip-slip components of movement. In this case, the fault would be described as a normal left-lateral oblique-slip fault. (Illustration from https://upload.wikimedia.org/wikipedia/commons/0/0a/Oblique_slip_fault.jpg.)

In summary, the structures recorded in rocks, folds and faults, reflect the nature of stresses that produced them. From a geothermal standpoint, correctly identifying the type of structures in a region is very important because it reflects the forces that produced them. This in turn can impact assessing the potential for development of geothermal energy in a given region; for example, consider how subsurface fluid flow might be affected by faults produced by extensional forces (normal faults) compared to compressional forces (reverse faults) (see Question 3 at the end of the chapter).

SUMMARY

Although the flow of heat from the Earth's interior to its surface is very large (almost 50 terrawatts) (Davies and Davies, 2010), it is not uniformly distributed. Much of this heat flow is localized along discrete zones or localized regions associated with the boundaries of Earth's tectonic plates or isolated areas of focused upwelling mantle material called hot spots. It is in these regions where heat flow is elevated to allow potential capture of geothermal energy for electrical or thermal power. Boundaries of tectonic plates consist of three types: divergent, convergent, and transform. Along divergent boundaries, the tectonic plates are separating and magma wells upward to occupy the space formed as the plates separate. Most divergent boundaries underlie the ocean basins to form the approximately 70,000-km-long mid-ocean ridge system—the site of the most active or continual volcanism on the planet. A good example of a divergent boundary on land is Iceland, through which the Mid-Atlantic Ridge passes. Because of Iceland's geologic setting, about 90% of all buildings are heated geothermally and about 30% of its power comes from geothermal energy.

Another region of elevated heat flow occurs along convergent tectonic boundaries, of which there are three types based on the nature of the tectonic plates converging: (1) oceanic–continental convergence, (2) oceanic–oceanic convergence, and (3) continent–continent convergence. The first two types are associated with subduction zones in which the more dense plate dives below the less dense plate. The subducting plate is eventually recycled in the mantle, but before doing so the increasing heat and pressure drive off water, which reduces the melting point of hot overlying mantle rocks and leads to partial melting and formation of magma. Being less dense than the surrounding rock, the magma rises and may eventually erupt to form volcanoes. Proximal to these volcanoes, the underlying magma can heat deeply circulating groundwater to support geothermal systems. In continent–continent convergent boundaries, such as the Himalaya Mountains, a subduction zone does not exist as continental lithosphere is not dense enough to sink. Consequently, in this tectonic setting there is little or no volcanism, and geothermal systems are less well developed but can still occur if water-channeling faults cut deeply enough to intersect hot rocks at depth.

Transform boundaries occur where plates slide past each other horizontally, such as the San Andreas fault in California. Heat from volcanism is minor here, but local areas along the fault may have potential for geothermal development if there is extension causing the crust to thin, bringing hot rocks at depth closer to the surface. An example of such a setting is the Salton Sea region of southeastern California. The other major geothermal system associated with the San Andreas fault is The Geysers, near the northern end of the fault system. However, most of the San Andreas fault between The Geysers and the Salton Sea region has few or no geothermal systems present. Other plate-bounding transform faults, such as the Alpine fault in New Zealand or North Anatolian fault in Turkey, also have limited geothermal potential for most of their extents.

Within the interior of tectonic plates, the most attractive zones for hosting geothermal systems are mantle plumes (hot spots), such as Hawaii or Yellowstone. Although such hot spots provide a focused source of heat energy to support robust geothermal systems, they are widely spaced and localized to specific regions, such as the Big Island of Hawaii or Yellowstone. A potentially attractive intraplate setting for geothermal systems, unrelated to volcanism and which has yet to be developed, are deep sedimentary basins located in areas of high regional heat flow. The geothermal potential of these hot, deep sedimentary basins is explored in Chapter 11.

Earth materials of rocks and minerals also play a role in helping characterize geothermal systems. This is because rocks and minerals have variable thermal parameters, such as conductivity, and physical properties, such as permeability (discussed in detail in the next chapter. Also, the minerals that make up rocks have different mechanical properties, allowing some rocks to bend more easily rather than break, such as a shale compared to a granite. The three types of rocks that can make up a geothermal reservoir are igneous (volcanic and plutonic), sedimentary (clastic and chemical), and metamorphic (foliated and unfoliated). Igneous and metamorphic rocks generally have limited intrinsic permeability and require secondary fractures to allow fluids to flow. By contrast, some sedimentary rocks, such as sandstone or limestone, can have significant primary permeability to serve as potential geothermal reservoirs.

Plate tectonics produces forces that strain rocks to either break or bend, or both. Whether a rock breaks or bends depends on several factors, including temperature, pressure, and rate at which force is applied. Also, the nature of the rock itself will influence how it deforms. Bending or folding of rocks is ductile deformation, whereas breaking rocks, such as from faults, reflects brittle deformation. Folds typically form when rock is squeezed (compression), whereas faults can form from compression, extension (pulling apart), or shearing. Normal faults form under tensional forces, which thin the crust and elevate heat flow. Also, tension creates open space that improves permeability, fostering fluid circulation. Thus, regions undergoing extension and normal faulting, such as along a developing divergent tectonic boundary, can make attractive regions for hosting geothermal systems, such as Kenya in the East African Rift zone.

Throughout geologic time, geothermal systems have developed and died. Some fossil geothermal systems include important ore deposits, and some active geothermal systems are actively depositing ore-grade minerals. Minerals such as gold and silver and important industrial minerals such as lithium are deposited. The relationships between active geothermal systems and fossil geothermal systems and mineral deposits are reviewed in Chapter 10. Radiometric dating of minerals developed in geothermal systems indicates that their lifespans range from a few tens of thousands of years to a few million years. A waning geothermal system, however, can be attractive still for potential development because the geologic time frame is much longer than the time frame required for human use, provided the exploited system is managed in a sustainable manner as discussed in Chapter 12.

SUGGESTED PROBLEMS

1. On the world map below, the X (noted at tip of arrow and outlined by red ellipse) marks an area selected for geothermal exploration. Describe the tectonic setting and likely types of rocks and geologic structures that would be found. Justify your conclusions.

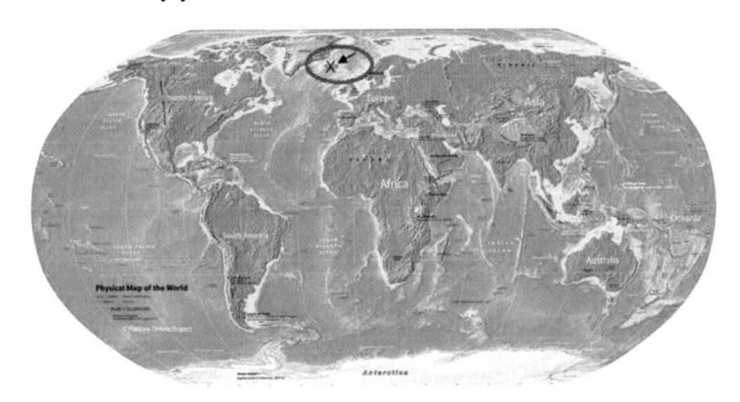

Landsat (aerial) view of fault referenced in Question 2. (NASA Earth Observatory image by R. Sigmon and J. Allen using Landsat data from USGS *Explorer*, https://upload.wikimedia.org/wikipedia/commons/4/48/Piqiang_Fault,_China_detail.jpg.)

2. Above is a satellite image of a portion of the Tien Shan Mountains in China. An important structure is evident in the image. What type of structure is it? Be as specific as possible. What is the approximate strike of the structure? Approximately how much displacement is indicated?
3. You are a geologist mapping a geothermal prospect and have mapped numerous normal faults. A colleague of yours is mapping another prospect and finds numerous reverse faults. Considering that other factors are equal, discuss whether you or your colleague has the more promising prospect. Your discussion should consider how the different fault types affect the ability of potential geothermal fluids to flow.

REFERENCES AND RECOMMENDED READING

Amante, C. and Eakins, B.W. (2009). *ETOPO1 1 Arc-Minute Global Relief Model: Procedures, Data Sources, and Analysis*, NOAA Technical Memorandum NESDIS NGDC-24. Washington, DC: National Oceanic and Atmospheric Administration, National Centers for Environmental Information (http://www.ngdc.noaa.gov/mgg/global/global.html).

Davies, J.H. and Davies, D.R. (2010). Earth's surface heat flux. *Solid Earth*, 1(1): 5–24.

Earle, S. (2015). *Physical Geology*, Section 12.4. Victoria, BC: BCcampus (https://opentextbc.ca/geology/chapter/12-4-measuring-geological-structures/).

EIA. (2016). *Electricity*. Washington, DC: U.S. Energy Information Administration (http://www.eia.gov/electricity/data.cfm; click on "Electricity data browser").

Fournier, R.O. and Rowe, J.J. (1966). Estimation of underground temperatures from the silica content of water from hot springs and wet-steam wells. *American Journal of Science*, 264(9): 685–697.

Geodynamics. (2014). *Quarterly Report Period Ending 30 September 2014: Operations*. Queensland, Australia: Geodynamics, Ltd. (http://www.geodynamics.com.au/Our-Projects/Innamincka-Deeps.aspx).

IMSA. (2002). *Earthquakes and Volcanoes: A Global Perspective*. Aurora: Illinois Mathematics and Science Academy (http://staff.imsa.edu/science/si/horrell/materials/Earthquakes/quakes55.html).

Johnson, J. (2015). *Glossary: Igneous*. Reston, VA: U.S. Geological Survey (https://volcanoes.usgs.gov/images/pglossary/VolRocks.php).

Kious, W.J. and Tilling, R.I. (1996). *This Dynamic Earth: The Story of Plate Tectonics*. Reston, VA: U.S. Geological Survey.

Malahoff, A. (1987). Geology and volcanism of the summit of Loihi submarine volcano. In: *Volcanism in Hawaii*, USGS Professional Paper 1350 (Decker, R.W., Wright, T.L., and Stauffer, P.H., Eds.), pp. 133–144. Reston, VA: U.S. Geological Survey.

Mattioli, G.S. (2008). *Geologic Structures* (lecture). Fayetteville: University of Arkansas (http://www.uta.edu/faculty/mattioli/geol_1113/pdf/lect_18_folds_faults_07.pdf).

Stoffer, P. (2002). *Rocks and Geology in the San Francisco Bay Region*, Bulletin 2195. Reston, VA: U.S. Geological Survey (http://pubs.usgs.gov/bul/2195/b2195.pdf). Provides a good description of the three main rock groups and pictures that illustrate rock names; also offers a concise discussion on the formation of the San Andreas fault.

Tarbuck, E.J. and Lutgens, F.K. (1999). *Earth: An Introduction to Physical Geology*. Upper Saddle River, NJ: Prentice Hall.

USGS. (1999). *"Hotspots": Mantle Thermal Plumes*. Reston, VA: U.S. Geological Survey (http://pubs.usgs.gov/gip/dynamic/hotspots.html).

USGS. (2014a). *FAQ—Everything Else You Want to Know About This Earthquake & Tsunami: Magnitude 9.1 Sumatra-Andaman Islands Earthquake FAQ*. Reston, VA: U.S. Geological Survey (http://earthquake.usgs.gov/earthquakes/eqinthenews/2004/us2004slav/faq.php).

USGS. (2014b). *Understanding Plate Motions*. Reston, VA: U.S. Geological Survey (http://pubs.usgs.gov/gip/dynamic/understanding.html).

USGS. (2014c). *Visual Glossary: Normal Fault*. Reston, VA: U.S. Geological Survey (http://geomaps.wr.usgs.gov/parks/deform/gnormal.html).

USGS. (2014d). *Visual Glossary: Reverse Fault*. Reston, VA: U.S. Geological Survey (http://geomaps.wr.usgs.gov/parks/deform/greverse.html).

USGS. (2014e). *Visual Glossary: Fault, Normal Faults, Reverse Faults, Strike-Slip Fault, Fault Scarp*. Reston, VA: U.S. Geological Survey (http://geomaps.wr.usgs.gov/parks/deform/gfaults.html).

USGS. (2015a). *Loihi Seamount: Hawaii's Youngest Submarine Volcano*. Reston, VA: U.S. Geological Survey (http://hvo.wr.usgs.gov/volcanoes/loihi/).

USGS. (2015b). *Overview of Geologic Fundamentals*. Reston, VA: U.S. Geological Survey (http://3dparks.wr.usgs.gov/nyc/common/geologicbasics.htm).

5 Subsurface Flow of Geothermal Fluids

KEY CHAPTER OBJECTIVES

- Explain how porosity and permeability are different and why they need to be considered when evaluating a geothermal system.
- Describe matrix porosity and permeability and fracture porosity and permeability and how they relate to each other.
- Apply Darcy's law to calculate various parameters of subsurface flow of fluids, including permeability, discharge, or total volume of flow through rock.
- Explain how fracture spacing and width of fractures affect porosity and permeability.
- Evaluate the change in porosity with depth compared to the change in permeability with depth, and explain their difference in behavior considering fracture transmissivity.

The flow of fluids, whether steam or liquid, is fundamental for the extraction of heat from the Earth that can be used by society. If no fluids are present or the fluids are prevented from flowing, obtaining geothermal energy, regardless of how hot the rocks are, becomes complicated. It requires expensive technical intervention to introduce fluids or make fluids flow or both; this is the realm of engineered or enhanced geothermal systems (EGSs). To understand how fluids move, we need to discuss the roles of porosity and permeability of rock materials.

PRIMARY MATRIX POROSITY AND PERMEABILITY

Porosity and permeability are not synonymous. Porosity is a ratio of the volume of voids to solid rock and is measured as a percentage. Permeability, on the other hand, measures the ability of a material to transmit a fluid and is commonly measured in darcys (d) or millidarcys (md) for geologic materials. A graphic illustration of the difference in porosity and permeability is shown in Figure 5.1. Both parts A and B have the same porosity ($\phi = 40\%$), ratio of open space to solid material, but in part A the pores are only partially connected, whereas in part B there is more pore connectivity, allowing for added fluid movement. Thus, interconnected porosity makes for superior permeability.

The SI unit for permeability is square meters (m^2) (see Equation 5.1, below); however, permeability is often expressed in darcies, named after Henry Darcy, who described fluid flow in porous media in the mid-1800s. A darcy is defined as the volumetric flow rate of a fluid of 1 cm^3/sec having a viscosity (shear strength or resistance to flow) of

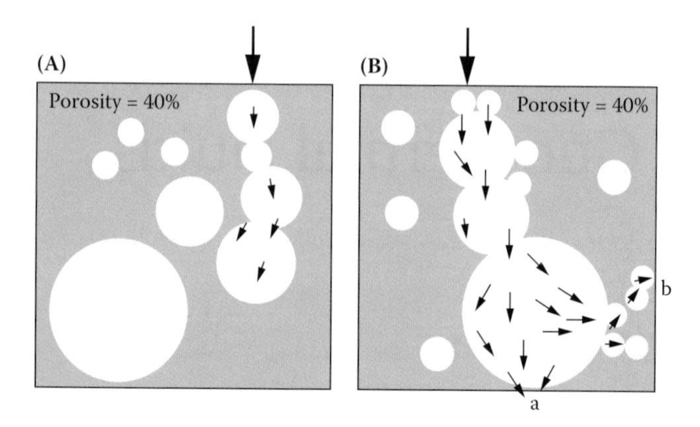

FIGURE 5.1 Illustration showing porosity (white circles) and permeability (small arrows). Large arrows indicate direction of hydraulic gradient. In part (B), which flow path, a or b, will have greater permeability? See text for detail. (Adapted from Glassley, W.E., *Geothermal Energy: Renewable Energy and the Environment*, 2nd ed., CRC Press, Boca Raton, FL, 2015.)

1 centipoise (which is typical for water at about 20°C) over a cross-sectional area of 1 cm^2 under a pressure gradient of 1 atmosphere per centimeter. The conversion is that 1 darcy equals 9.87×10^{-13} m^2 or 0.987×10^{-6} μm^2 (μ stands for micron or 1 millionth, 1×10^{-6}). Thus, 1 darcy equals approximately 1×10^{-6} μm^2. The darcy or milliidarcy is typically applied by geologists and groundwater hydrologists because the permeability of most geologic materials is relatively low and positive whole units are easier to use than fractions or negative exponents. Finally, because a unit of area does not intuitively translate into fluid transmissibility or flow rate, the permeability of soil or rock can also be expressed as the flux of water under a hydrostatic pressure or head of ~0.1 bar/m at 20°C. In this case, 1 darcy equals 0.831 m/day, a value not unusual for the flow rate of groundwater through *very* permeable material such as weakly cemented gravel. For reference, good permeability is generally considered to be greater than 10^{-14} m^2, which correlates to greater than about 10 md. In general, convection of hydrothermal fluids, due to heating and buoyancy effects, requires a minimum permeability of about 5 md. Results of porosity simulation models indicate a fracture permeability from 50 to 500 md (5×10^{-14} to 5×10^{-13} m^2) to match observed well flow rates (averaging about 200 kg/s per well) from producing geothermal wells.

Two types of porosity and permeability characterize geologic materials—one is matrix porosity and permeability and the other is fracture porosity and permeability, which will be discussed in the next section. Matrix porosity and permeability are primary or intrinsic qualities of the soil or rock. It is for this situation that Darcy developed his formula for fluid flow in porous media:

$$Q = -\frac{\kappa}{\mu} \times A \times \left(\frac{P_b - P_a}{L} \right) \tag{5.1}$$

where

Q = Discharge (m^3/s).
κ = Permeability (m^2).

μ = Dynamic viscosity (Pa·s or kg/(m·s)).
A = Cross-sectional area (m^2).
P_b = Final pressure (Pa).
P_a = Initial pressure (Pa).
L = Length over which pressure drop takes place (m).

The negative sign is necessary because fluid flows from high pressure to low pressure, so $P_b - P_a$ is negative, which then makes the flow in the positive x direction. If we divide both sides of Equation 5.1 by A, we get a simplified result:

$$q = -(\kappa/\mu) \times \nabla P \tag{5.2}$$

where
q = Darcy flux, discharge per unit area, Q/A (m/s).
κ = Permeability (m^2).
μ = Dynamic viscosity (Pa·s or kg/m·s).
∇P = Pressure gradient, $(P_b - P_a)/L$.

Note that the Darcy flux (q), despite its units of m/s, is not the velocity of fluid traveling through the pores. The actual velocity is related to the Darcy flux multiplied by porosity percentage, or

$$v = q \times \phi \tag{5.3}$$

where
v = Fluid velocity.
q = Darcy flux, discharge per unit area, Q/A (m/s).
ϕ = Porosity percentage.

The key implications of Darcy's law are as follows:

1. If there is no pressure gradient, fluids will not flow.
2. If there is a pressure gradient, flow will be from high to low pressure (hence the negative sign).
3. The greater the pressure gradient, the greater the discharge or flux.
4. For a given pressure gradient, the discharge rate will vary with different geologic materials.

Darcy's law is valid for only slow, viscous flow, which applies to groundwater in most cases. The law does not apply, for example, under high flow conditions as can occur in geothermal wells. In the latter case, flow rates are dictated mainly by buoyancy forces and the extent of hydrostatic head.

We can see that the flux in Figure 5.1A will be zero, which requires that permeability (κ) must also be zero. In Figure 5.1B, two flow paths are noted, a and b. Will both paths have equal permeability or flux? In path a, there is a greater pressure gradient (∇P), and the diameters of the connected pores are larger, resulting in greater permeability (κ). Thus, the same material can have anisotropy in permeability, in which flow can be favored in certain directions, depending on the geometry and distribution of pores. Other important factors affecting permeability besides pore connectivity include the following:

1. Tortuosity of fluid paths
2. Pore size
3. Surface tension affects

As we will learn more about in Chapter 6, water is a polar molecule; thus, water molecules tend to stick to one another, a form of hydrogen bonding. Small pores have high surface area-to-volume ratios compared to large pores, meaning that there is more surface area per given volume for water to adhere. A material having many small connected pores can have a lower permeability than another material with possibly lower connected porosity but a larger pore size. As Table 5.1 shows, well-sorted sand or gravel has a much higher permeability compared to very fine sand. Although the pore size may vary by a magnitude, such as between coarse and fine sand, the reduction in permeability varies by several orders of magnitude, mainly reflecting enhanced surface tension of small pores on fluid flow.

Another parameter worth noting is hydraulic conductivity (K), which measures a material's ability to permit fluid flow. Hydraulic conductivity is given by the κ/μ term in Darcy's law (Equation 5.1) multiplied by the specific weight of the fluid (density of material times acceleration of gravity):

$$K = -[\rho \times (g/\mu)] \times \kappa \tag{5.4}$$

where

K = Hydraulic conductivity (m/s).
ρ = Density of the fluid.
g = Acceleration of gravity.
μ = Dynamic viscosity (Pa·s or kg/m·s).
κ = Permeability (m^2).

TABLE 5.1
Permeablilities for Some Representative Geological Materials

	Permeability (κ)			
	Highly Fractured Rock	Well-Sorted Sand, Gravel	Very Fine Sand and Sandstone	Fresh Granite
κ (cm^2)	10^{-3}–10^{-6}	10^{-5}–10^{-7}	10^{-8}–10^{-11}	10^{-14}–10^{-15}
κ (millidarcy)	10^{-8}–10^{-5}	10^{-6}–10^{-4}	10^{-3}–1	10^{-3}–10^{-4}

Source: Glassley, W.E., *Geothermal Energy: Renewable Energy and the Environment*, 2nd ed., CRC Press, Boca Raton, FL, 2015.

Just as with permeability, hydraulic conductivity can vary over several orders of magnitude for a variety of geologic materials. Also like permeability, hydraulic conductivity can vary with direction and the distance over which it is considered. The permeability and hydraulic conductivity are key factors when assessing the prospective level of a geothermal reservoir.

FRACTURE POROSITY AND PERMEABILITY

This is a secondary process imposed on the rocks due to Earth forces causing rocks to break. These fractures could be joints or faults. Faults certainly form from tectonic forces, and different orientations of faults reflect a changing stress history with time. Joints can also reflect tectonic forces, but they can also form from unloading or heating and cooling of rock. As noted in Table 5.1, fracture permeability can be an order or more in magnitude greater than intrinsic matrix permeability. Indeed, secondary fracture permeability can make a rock with low matrix permeability, such as a granitic igneous rock, potentially permeable for fluid flow. Fracture porosity, similar to matrix porosity, is a ratio, but in this case it is the amount of open space in fractures to the volume of solid rock. Depending on the spacing and size of fractures, fracture porosity can be many times greater than matrix porosity. Furthermore, similar to matrix porosity, high fracture porosity does not necessarily imply high fracture permeability if the fractures fail to connect (e.g., fracture C in Figure 5.2).

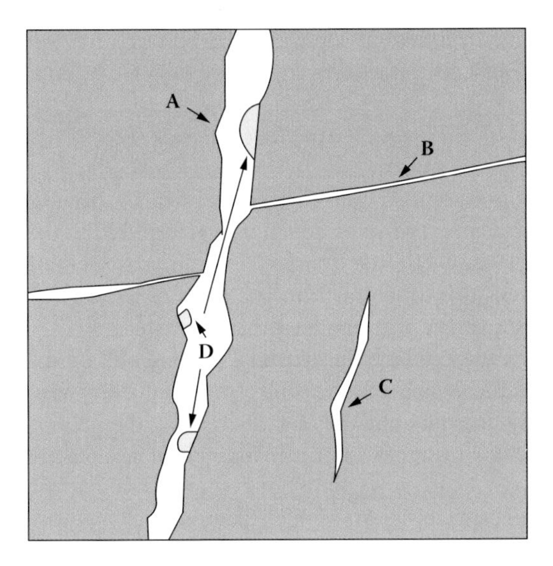

FIGURE 5.2 Diagrammatic view across a rock face showing three types of fractures. White areas denote open or pore space. This rock face has high matrix porosity but little matrix permeability, which is restricted to fractures A and B. Note that fracture walls of A are rough and include mineral precipitates, whereas fracture B has smooth walls. Fracture C, although porous, has no permeability. (Adapted from Glassley, W.E., *Geothermal Energy: Renewable Energy and the Environment*, 2nd ed., CRC Press, Boca Raton, FL, 2015.)

Hydraulic Fracture Conductivity and Permeability

Important controls on fracture permeability are the aperture size (width) of the fractures, the spacing of the fractures, and the orientation of fractures with relation to the hydraulic gradient. Similar to matrix permeability, the hydraulic fracture conductivity reflects the ability of a fluid to flow along a fracture and is defined as

$$K_{fr} = \rho \times (g/\mu) \times (a^2 \times 12) \quad (5.5)$$

where

K_{fr} = Fracture hydraulic conductivity (m/s).
ρ = Density of the fluid.
g = Acceleration of gravity.
μ = Dynamic viscosity (Pa·s or kg/m·s).
a = Aperture of fracture.

Substituting Equation 5.4 into Equation 5.5 for K_{fr} allows us to define the fracture permeability (κ_{fr}) as follows:

$$\kappa_{fr} = a^2/12 \quad (5.6)$$

Fracture Transmissivity

The actual discharge of a fluid at some velocity across a fracture of a given width or aperture size (a) reflects a fracture's transmissivity (T_{fr}) or flux and is defined by

$$T_{fr} = K_{fr} \times a = [\rho \times (g/\mu) \times (a^3/12)] \quad (5.7)$$

This tells us that the fracture's transmissivity or flux is strongly dependent on the aperture size of a fracture; thus, this equation is referred to as the "cube law." So, if a fracture is doubled in width, the transmissivity increases eightfold. The aperture, however, is quite sensitive to the surface roughness of the fracture and possible tortuosity of fractures and therefore can be difficult to interpret.

Flux and permeability change linearly on a log–log plot for a given pressure gradient (Figure 5.3). Increasing the pressure gradient for a given permeability will increase the flux by an equal amount (e.g., increasing the pressure gradient by four times increases the flux by about four times for a given permeability). Also plotted in Figure 5.3 are the ranges in permeability of common geologic materials from Table 5.1. Notice that the permeability for highly fractured rock is several orders of magnitude greater than that of relatively permeable well-sorted fine sand. The importance of fractures with regard to fluid permeability can be clearly shown by comparing fractured and unfractured granite, in which the permeability of the former is about ten orders of magnitude greater than that of the latter.

Figure 5.4 records changes in permeability and porosity as a function of fracture spacing and fracture width. It is clear from this graph that fracture width is a more important control on permeability than fracture spacing. For example, changing fracture spacing from 1000 cm to 100 cm increases permeability about an order

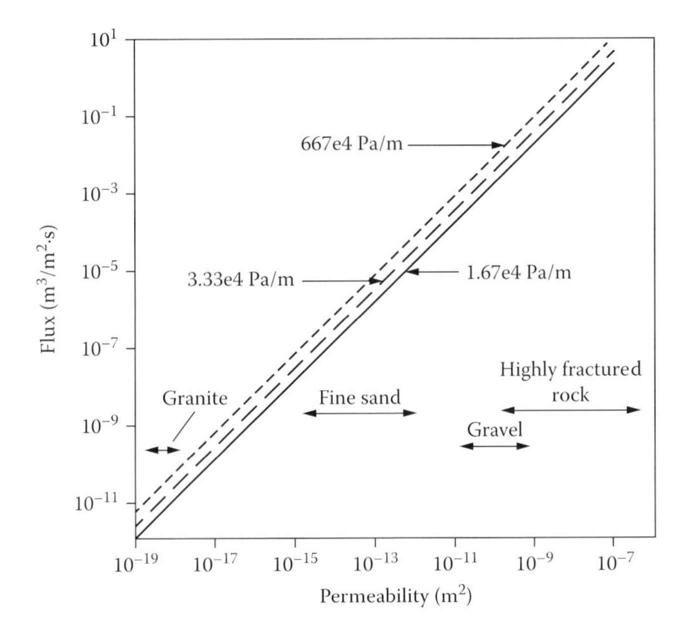

FIGURE 5.3 Log–log plot of flux and permeability as a function of different pressure gradients. A change in pressure gradient results in a comparable change in flux for a given permeability. Also shown are the permeabilities of common geologic materials. Note that the permeability for highly fractured rock is still much greater than porous and permeable loosely consolidated gravel. (Adapted from Glassley, W.E., *Geothermal Energy: Renewable Energy and the Environment*, 2nd ed., CRC Press, Boca Raton, FL, 2015.)

of magnitude, about the same as the reduction in spacing. On the other hand, for a given spacing, the permeability changes by about three orders of magnitude for every tenfold change in width (Figure 5.4). Thus, as Glassley (2015) stated, "A rigorous understanding of fracture characteristics is critical for evaluating whether flow rates are adequate for geothermal power production."

FLOW RATES AND POWER OUTPUT

To give a practical sense of what a flux (measured in $m^3/m^2 \cdot s$) means, let's convert a mass flow rate for a typical geothermal well in rocks having moderate flux of 10^{-5} $m^3/m^2 \cdot s$ and see how that relates to power output. Looking at Figure 5.3, a mid-range flux rate of 10^{-5} $m^3/m^2 \cdot s$ multiplied by the density of water (1000 kg/m^3) gives 10^{-2} $kg/m^2 \cdot s$, or the mass flow rate per unit of area. If a geothermal well has a diameter of 30 cm and intersects a geothermal reservoir for a depth of 30 m, the surface area of well pipe exposed to reservoir rock is about 28 m^2. Assuming that a slotted liner reduces the exposed surface area of the wellpipe and flow rate through the intersected rock by about 40%, the exposed surface area of wellpipe is about 17 m^2. Multiplying this result by 10^{-2} $kg/m^2 \cdot s$ yields a mass flow rate of 0.17 kg/s. As the fluid rises in the pipe, pressure is reduced, causing the fluid to start flashing to steam.

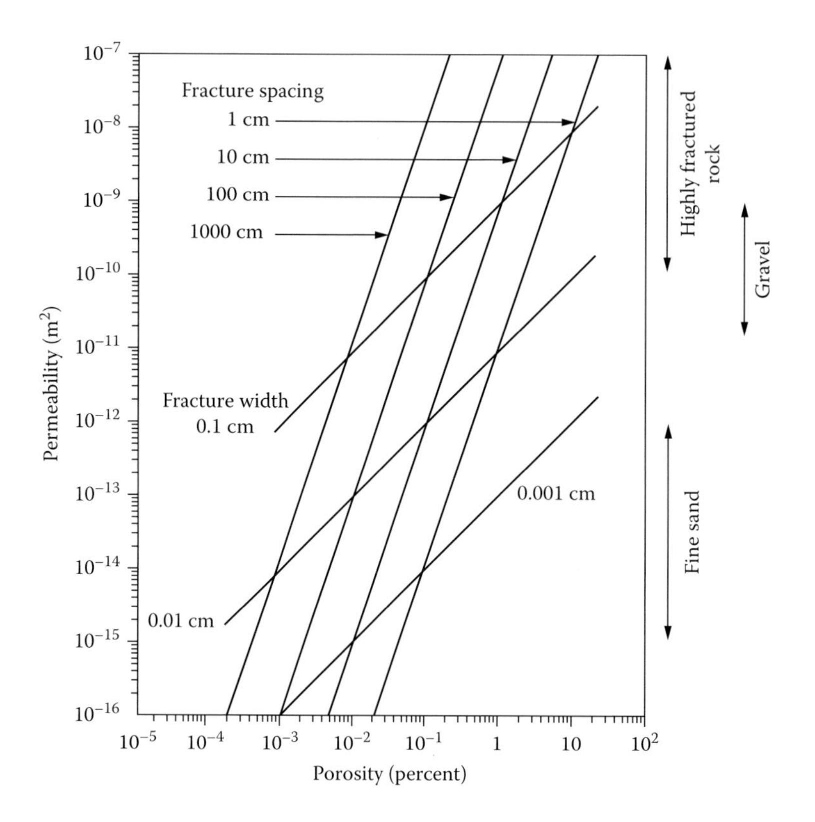

FIGURE 5.4 Graph plotting fracture permeability and fracture porosity as a function of fracture spacing and fracture width. Note that fracture width is a much more important control on permeability compared to fracture spacing. See text for discussion. (Adapted from Glassley, W.E., *Geothermal Energy: Renewable Energy and the Environment*, 2nd ed., CRC Press, Boca Raton, FL, 2015.)

Assuming that about 30% of the fluid flashes to steam at the well head, the mass flow rate of steam to the turbine is about 0.051 kg/s (0.17 kg/s × 0.30). Power is related to mass flow rate by the following equation:

$$P = (Q \times \rho) \times C_p \times \Delta T \tag{5.8}$$

where

P = Power (kJ/s).
$(Q \times \rho)$ = Mass flow rate (volume discharge rate times density).
C_p = Heat capacity of steam (about 2.0 kJ/kg·K).
ΔT = Temperature drop of steam entering and leaving turbine (K).

A common temperature drop of incoming and outgoing steam is about 50 K. Inserting these values yields the following:

$$\text{Power output} = 0.051\ \text{kg/s} \times 2.0\ \text{kJ/kg·K} \times 50\ \text{K} = 5.1\ \text{kJ/s} = 0.0051\ \text{MWe}$$

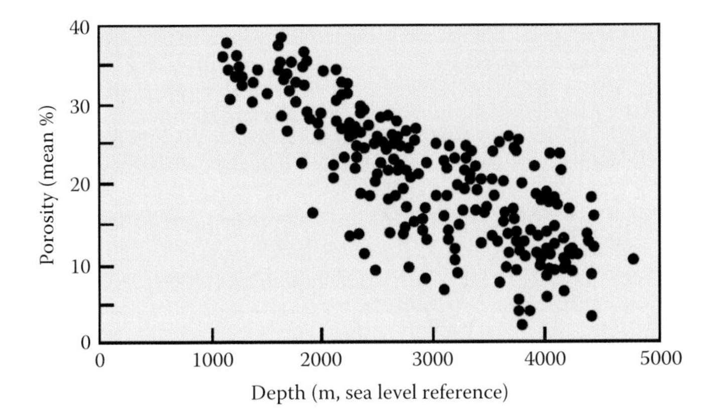

FIGURE 5.5 Graph showing change in porosity with depth of sediments for the Norwegian continental shelf. (Adapted from Glassley, W.E., *Geothermal Energy: Renewable Energy and the Environment*, 2nd ed., CRC Press, Boca Raton, FL, 2015.)

A typical geothermal well produces about 3 to 10 MWe. This means that the flux rate for the above example would have to increase by a factor of about 1000 or $10^{-2}/m^3 \cdot s$ (mass flow rate of 51 kg/s), yielding a power output of 5.1 MWe.

CHANGES IN POROSITY AND PERMEABILITY WITH DEPTH

As might be expected, porosity decreases with depth in response to increasing lithostatic pressure (weight of the overlying rock column). Both primary matrix and secondary fracture porosity is reduced as higher pressure squeezes or compresses grains and walls of fractures closer together. The extent of compression depends on rock strength. Loosely consolidated sediments will compress more at a given depth than a strong crystalline rock such as granite. The spread in the data shown in Figure 5.5 probably reflects different types of sediments encountered at depth in which coarser sandy layers will compress less for a given depth than clay-rich layers. Permeability is also reduced at depth. As we saw above, permeability is very sensitive to aperture size. A small reduction in pore size or fracture width, due to compression, can have a large impact on permeability. Whereas porosity decreases more or less linearly with depth (Figure 5.5), permeability decreases exponentially, with the greatest reduction in the first 5 km of depth (Figure 5.6). Note also that the permeabilities in the first 5 km vary by about 6 orders of magnitude, making predictive models difficult for geothermal wells (Figure 5.6). This first 5 km, however, is the main target zone for developing current and future geothermal energy resources.

POROSITY AND PERMEABILITY OF PRODUCING GEOTHERMAL RESERVOIRS

In a survey of existing geothermal reservoirs, those that are dominated by matrix porosity must have higher porosity to achieve a given permeability than that of fracture-dominated systems (Figure 5.7). Furthermore, low fracture porosity can

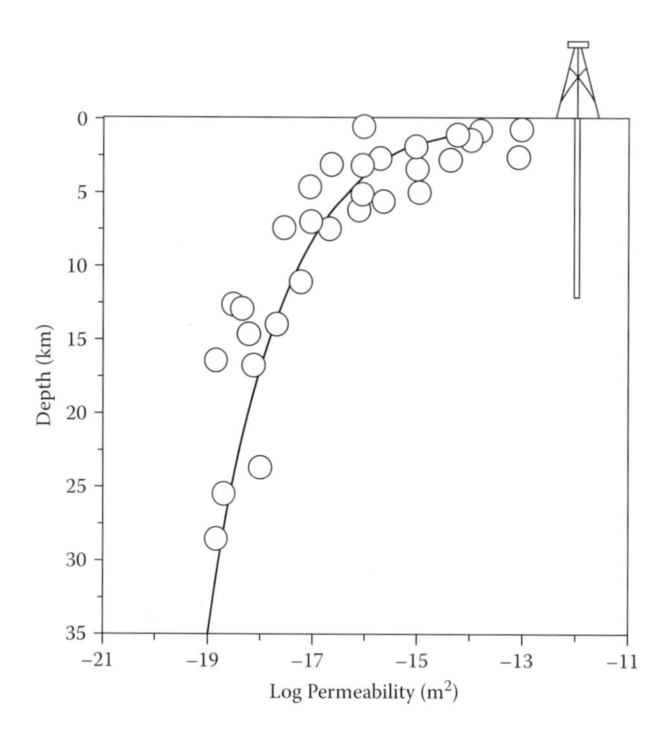

FIGURE 5.6 Change in permeability with depth. Note the exponential change compared to the linear change for porosity with depth and that the greatest change occurs in the first 5 km of depth. The deepest drill hole (about 12 km depth) is shown for reference in the upper right part of the diagram. See text for other details. (Adapted from Glassley, W.E., *Geothermal Energy: Renewable Energy and the Environment*, 2nd ed., CRC Press, Boca Raton, FL, 2015.)

still yield satisfactory permeability (10 to 100 md) to extract heat from the subsurface for geothermal energy development assuming fluids are present. This latter point illustrates the importance again of fractures in generating permeability so fluids can flow and move heat from the subsurface for geothermal direct use or production of electricity.

GEOLOGIC EXAMPLES OF MATRIX POROSITY

A developed geothermal system that makes significant use of matrix porosity is the Wairakei geothermal field in New Zealand (Figure 5.8). The Wairakei geothermal field is currently developed by two power stations, Wairakei and Poihipi, having a combined installed power output of about 235 MWe. A third power station, Te Mihi, taps a vapor-dominated zone at the north end of the Wairakei geothermal field and has an installed capacity of 166 MWe. Production from Te Mihi will eventually replace the aging 55-year-old Wairakei power station, whose operation will be phased out by or before 2025 (T. Montegue, Contact Energy, pers. comm., 2013).

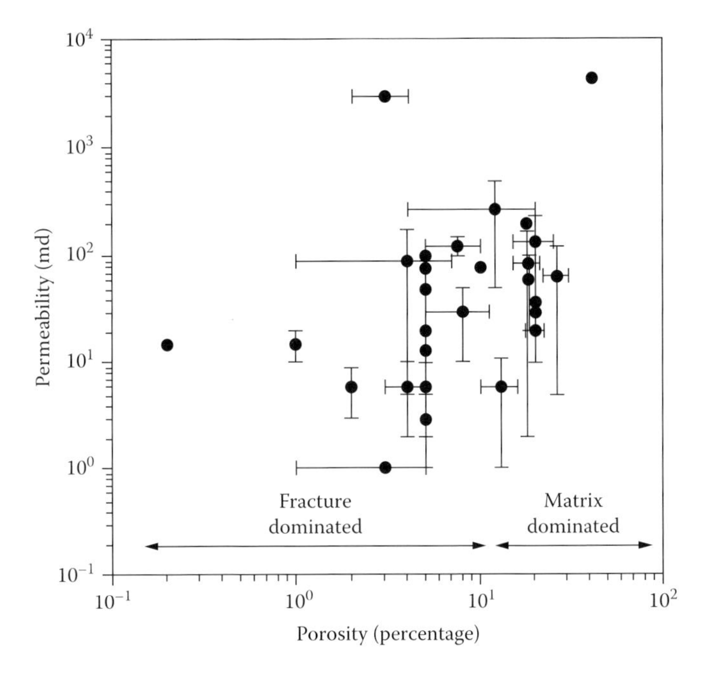

FIGURE 5.7 Graph plotting change in permeability and porosity for a variety of geothermal systems. See text for implications of shown results. (Adapted from Glassley, W.E., *Geothermal Energy: Renewable Energy and the Environment*, 2nd ed., CRC Press, Boca Raton, FL, 2015.)

The Wairakei geothermal reservoir is largely developed in the Waiora Formation, which consists of an interlayered mixture of pumice-bearing sandstones, pumiceous breccias, and ash-flow tuff units (Rosenberg et al., 2009). The pumiceous breccia units are particularly permeable and behave like sponges containing circulating ≥200°C geothermal fluids. Production depths are mainly between 500 and 1500 m. To maximize production flow rates newer wells have been drilled horizontally within permeable horizons of the reservoir. These horizontal wells intercept a larger volume of rock within the reservoir than vertical wells (e.g., well WK305 in Figure 5.9).

Nonetheless, faults associated with the Kaiapo fault zone along the western side of the field serve to channel geothermal fluids upward where they bleed out laterally in transected permeable rock layers of the Waiora Formation (Figures 5.8 and 5.9). Notice that most of the production wells occur outside of the fault zone, tapping into the bulk (matrix) permeable Waiora Formation at depth. However, as fluid is withdrawn from the reservoir and is only partially replaced, reservoir pressure has decreased, causing the reservoir to partially compress and the land surface to subside. Land subsidence can be an impact of geothermal production, especially where the injection of spent geothermal fluid is minimal. Potential land subsidence is discussed in Chapter 9.

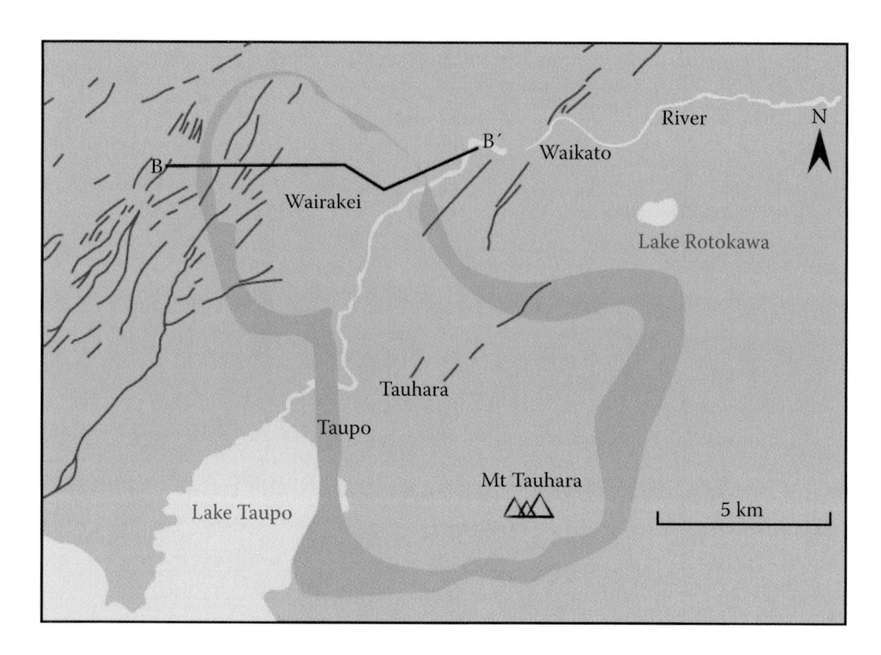

FIGURE 5.8 Map showing the Wairakei and contiguous Tauhara geothermal fields. Red lines denote active faults. Most of the geothermal wells in the Wairakei field lie to the east of the Kaiopo fault zone that transects the western part of the Wairakei geothermal field. The light blue–gray swath marks the edge of the geophysical negative resistivity anomaly that commonly indicates the subsurface extents of active geothermal fields due to the high electrical conductivity of circulating geothermal fluids. Line of section B–B′ is shown in Figure 5.9. (Adapted from Rosenberg, M.D. et al., *Geothermics*, 38(1), 72–84, 2009.)

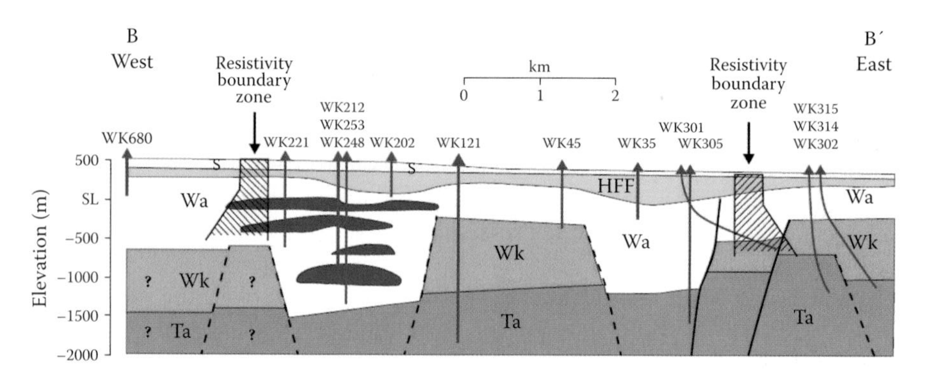

FIGURE 5.9 Line of section B–B′ whose location is shown in Figure 5.8. Wa stands for the Waiora Formation, the principal geothermal reservoir for the Wairakei geothermal field. Red lines are geothermal wells, some of which approach horizontal at depth (e.g., WK305), to maximize flow from permeable horizons. Red lenses denote rhyolite lavas that serve as aquitards (rock units having limited fluid flow) bounding underlying geothermal aquifers. Question marks denote regions of questionable stratigraphic affiliation (Adapted from Rosenberg, M.D. et al., *Geothermics*, 38(1), 72–84, 2009.)

A paleoanalog of bulk or matrix permeable rock is the Type II disseminated gold ore mined at Round Mountain, Nevada. The gold was deposited 26 million years ago by geothermal fluids that flowed through the lower, poorly welded zone of the tuff of Round Mountain (Henry et al., 1997; Sander and Einaudi, 1990). Indeed, most of the 13 million ounces of gold mined to date have come from matrix or bulk permeable rock. The matrix-permeable, poorly welded zone consists of pumiceous rhyolitic tuff that underwent vapor-phase alteration shortly after its deposition and prior to formation of the geothermal system. The vapor-phase alteration formed in response to rising vapors produced as the hot ash boiled surface and near-surface groundwater. This transformed the initial glassy ash and pumice particles to a fine-grained mixture of quartz and alkali feldspar (adularia), leaving interconnected open space as the glassy particles recrystallized. Upon later development of the geothermal system, the fluids could circulate through the recrystallized tuff, depositing gold during the cooling and ending stages of the geothermal system (Sander and Einaudi, 1990). Without the earlier vapor-phase alteration, the later geothermal fluids would have converted the glassy, poorly welded tuff to mainly impermeable clay. This would have made the lower tuff of Round Mountain an aquiclude (a barrier to fluid flow), forcing fluids to move along any open fractures rather than the geothermal aquifer that it became, serving as the principal host for gold ore (Figure 5.10).

FRACTURE PERMEABILITY AND CRUSTAL EXTENSION

Geothermal systems in the Great Basin of the western United States are mainly controlled by fracture permeability due to extensional or normal faulting. These faults control upwelling geothermal fluids that can locally flow out laterally adjacent to the faults if matrix-permeable horizons are encountered. An example of this is the Long Valley caldera and the Casa Diablo geothermal field in California. For the most part, however, the geothermal reservoirs of producing geothermal fields in the Great Basin are developed in strongly fractured zones of rock. Most of the fracturing results from crustal extension that breaks and pulls the rocks apart, creating open space and permeable zones for fluids to flow. Jim Faulds and coworkers from the University of Nevada, Reno, have examined over 200 geothermal systems in the Great Basin and have identified five critically stressed environments. These include fault step-overs, fault terminations, fault intersections, pull-apart zones, and fault-dip reversal zones (accommodation zones), associated with extensional normal and transtensional strike-slip fault zones (Faulds et al., 2011; Jolie et al., 2015). These critically stressed environments keep fractures open, promoting fluid circulation. Structurally or fault-controlled geothermal systems will be discussed in more detail in Chapter 8.

A good example of a geothermal system involving a step-over of a normal fault zone is the Desert Peak system, which has an installed capacity of about 25 MWe. Successful stimulation of a previously dry well boosted production by 38%. Wells to the north of the productive field were hot but largely dry or tight, even though they were drilled in the hanging wall of major fault zones. The step-over region, however, is characterized by a dense spacing of smaller faults and fractures that collectively make up a permeable reservoir (Figure 5.11). Thus, a high density of smaller

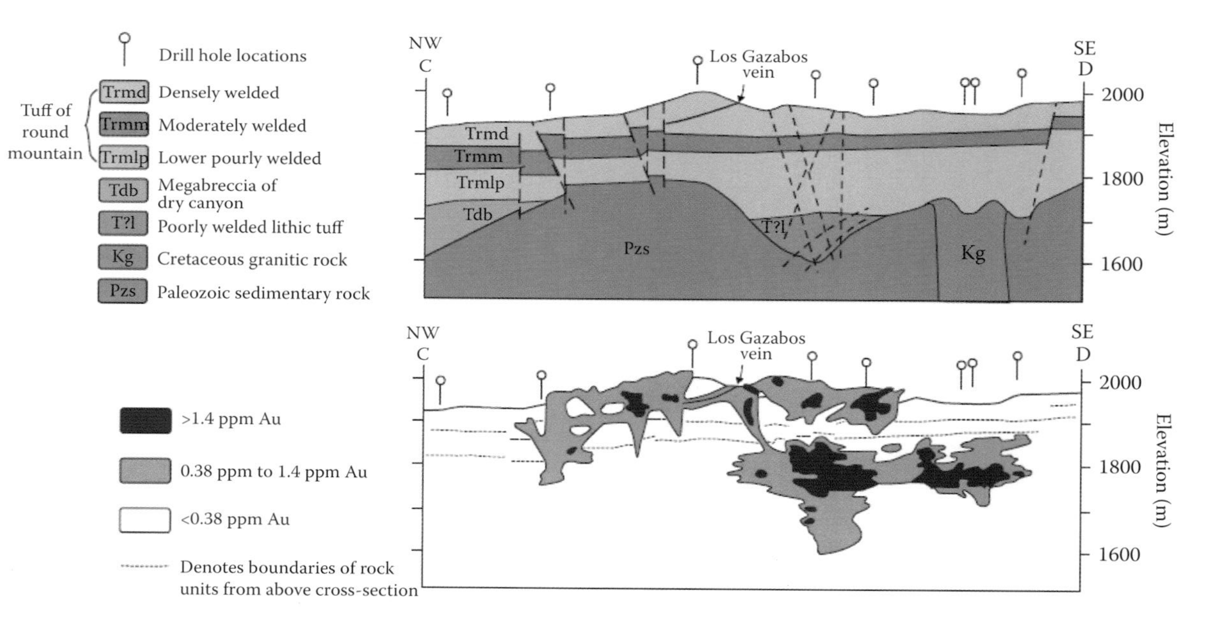

FIGURE 5.10 Cross-sections through the Round Mountain gold mine, Nevada. Upper cross-section shows geology in which the lower poorly welded and matrix permeable zone is the lower member of the tuff of Round Mountain. Dashed bold lines denote faults. The bottom cross-section shows the distribution of gold ore in which the red zones contain >1.4 ppm Au and the orange-colored zones contain between 0.38 and 1.4 ppm gold. Note the wide distribution of gold mineralization in the Trmlp layer and that most occurs in areas without faults, reflecting the high matrix permeability. However, upwelling gold-bearing geothermal fluids do appear to have been controlled in part by faults (based on narrow, steeply inclined zones of gold mineralization) that provided access to the matrix permeable lower tuff of Round Mountain. (Adapted from Sander, M.V. and Einaudi, M.T., *Economic Geology*, 85(2), 285–311, 1990.)

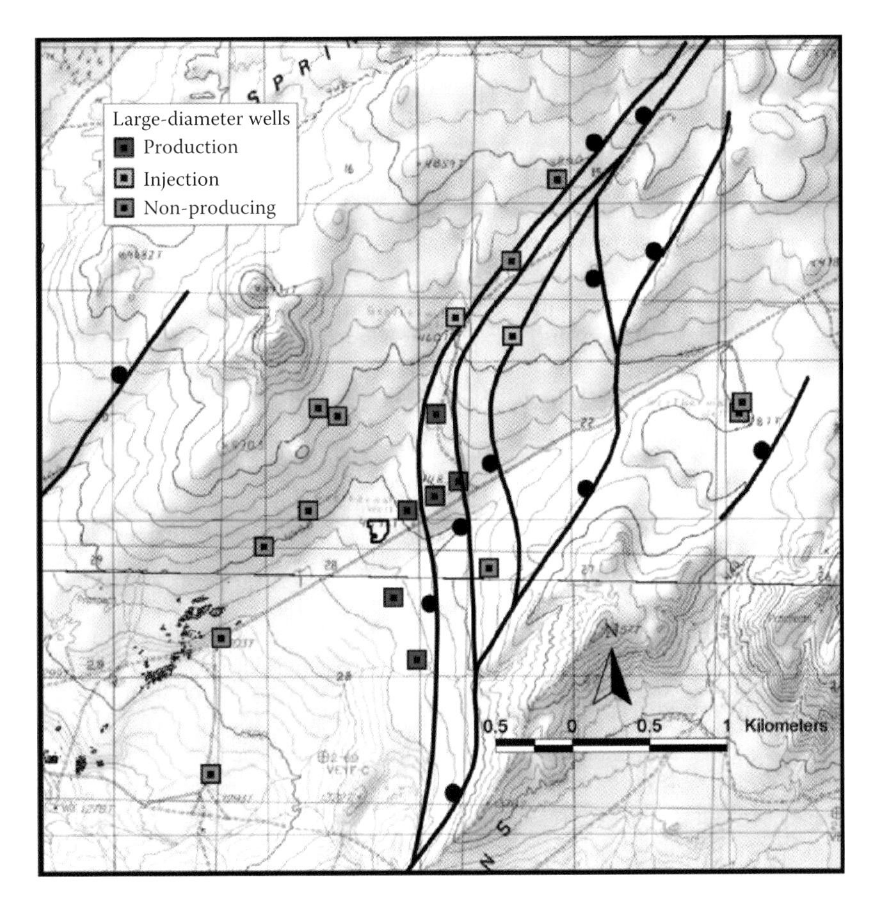

FIGURE 5.11 Map showing well locations for Desert Peak geothermal system. Note that the producing wells occur in the step-over region between the two major northeast-striking normal faults. See text for details. (From Faulds, J.E. et al., in *Great Basin Evolution and Metallogeny*, Steininger, R. and Pennell, B., Eds., DEStech Publications, Lancaster, PA, 2011, pp. 361–372. Copyright © Geological Society of Nevada.)

faults and fractures appears more important than one or two large faults having large displacement for developing the permeability necessary for forming a viable geothermal reservoir. Structurally developed secondary porosity and permeability are discussed further in Chapter 8.

SUMMARY

Porosity and permeability are key intrinsic properties affecting the viability of developing a geothermal system. Porosity reflects the open space in rocks and is measured in percent, such that 30% porosity indicates that about a third of the rock has open space that can be filled with liquid or gas. Unless those pores are connected, however, the fluid in the pores cannot flow. Permeability measures the flow of fluid through porous media and has SI units of square meters (m^2). Permeability is also expressed as millidarcy (md); 1 md equals about 1×10^{-15} m^2. Good rock permeability is generally considered to be greater than about 10 md or 1×10^{-14} m^2. The flow of groundwater generally follows Darcy's equation, which involves rock permeability, hydrostatic head (pressure drop), and dynamic viscosity of the fluid in question.

Two types of porosity and permeability are primary matrix permeability and secondary or fracture-controlled permeability. Factors affecting primary permeability include pore connectivity, tortuosity, and pore size and surface tension effects. Secondary or fracture-controlled permeability varies as a function of fracture spacing and width, with the latter exerting the stronger control (a tenfold change in fracture width results in a 1000-fold change in permeability). Matrix-dominated permeability requires high porosity to yield comparable permeability as provided by fracture-controlled permeability. For example, a strongly fractured but otherwise weakly porous rock, such as granite, can have permeability two to three orders of magnitude greater than porous but unfractured sandstone.

Power output depends strongly on mass flow rates, which in turn are governed by permeability. Although both porosity and permeability decrease with depth, porosity reduction is basically linear, whereas permeability decreases exponentially within approximately the first 5 km of the crust. This reduction is largely due to compression and cementation (deposition of minerals in pore spaces by circulating fluids) and more strongly affects permeability because, as open space of fluid pathways decreases, surface tension and frictional effects greatly increase, such that if the fracture width is halved from mineral deposition (cementation), fluid transmissivity decreases by a factor of eight (the so-called cube law). The reduction in porosity and permeability with depth has important implications for engineered geothermal systems, as discussed in Chapter 11.

A good example of a producing geothermal system making use of primary matrix porosity and permeability is the Wairakei geothermal field in New Zealand. The geothermal reservoir consists in part of particularly porous and permeable pumiceous breccias, which have responded to the withdrawal of fluid by partially compressing and causing the overlying land to subside. The Round Mountain gold deposit in central Nevada serves as a paleoanalog of primary matrix porosity and permeability, as gold mineralization is largely disseminated in a poorly welded ash-flow tuff layer.

Most developed geothermal systems, however, have mainly secondarily developed porosity and permeability produced by fractures and faults. This is particularly true of developed geothermal systems in the Basin and Range Province of the western United States, where crustal extension has faulted and broken the crust. This has allowed groundwater to circulate deeply, pick up heat from hot rocks at depth, and then buoyantly rise toward the surface, where relatively shallow, structurally controlled reservoirs can form.

SUGGESTED ACTIVITIES

1. Check out the YouTube video on porosity and permeability at http://www.youtube.com/watch?v=pKHI6GFIVHs. The video is only about 5 minutes but provides good visualization of the factors affecting both porosity and permeability.

   a. Explain why porosity is increased with improved sorting of sediment.
   b. Why does improved rounding of grains improve porosity?
   c. Explain why a sample with higher porosity may have lower permeability than a sample with lower porosity?
2. Go to the Exploring Earth website entitled "How Does Water Move through the Ground?" (http://www.classzone.com/books/earth_science/terc/content/investigations/es1401/es1401page01.cfm). Complete steps 1 to 5 and answer the questions posed for each step.

SUGGESTED PROBLEMS

1. Using Darcy's law, calculate the total volume of water at 80°C that would move through an open fracture having an aperture of 1.0 cm and length of 1 m over a 1-hour period and under a pressure gradient of 1 kPa per meter. (*Hint:* You will need to determine the dynamic viscosity of water at 80°C; go to the Engineering ToolBox website at www.engineeringtoolbox.com). Show your work.
2. What mass flux rate (in kg/s) is necessary in a steam well to yield 8 MWe of power in which the inlet temperature to the turbine is 165°C and the outlet temperature is 115°C? Show your work. (*Hint:* You will need to determine the specific heat of steam at 165°C; go to the Engineering ToolBox website at www.engineeringtoolbox.com).
3. Go to the Exploring Earth website entitled "How Does Water Move through the Ground?" (http://www.classzone.com/books/earth_science/terc/content/investigations/es1401/es1401page01.cfm).
   a. Complete steps 1 to 5.
   b. Answer the questions posed for each step.
   c. For step 5, take a screen shot of the produced graph to include with your answers to questions from the previous steps.
4. Looking at the figure below (same as Figure 3.7) of temperature profiles of wells drilled in the Casa Diablo geothermal field, pose two hypotheses that could account for the temperature profile encountered for well M-1. How do your hypotheses compare to those of your classmates? Which hypothesis do you consider the more likely and why?

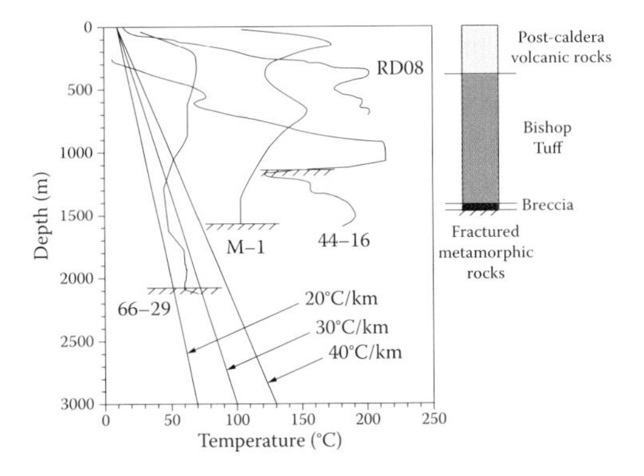

REFERENCES AND RECOMMENDED READING

Faulds, J.E., Hinz, N.H., and Coolbaugh, M.F. (2011). Structural investigations of Great Basin geothermal fields: applications and implications. In: *Great Basin Evolution and Metallogeny* (Steininger, R. and Pennell, B., Eds.), pp. 361–372. Lancaster, PA: DEStech Publications. Copyright © Geological Society of Nevada.

Glassley, W.E. (2015). *Geothermal Energy: Renewable Energy and the Environment*, 2nd ed. Boca Raton, FL: CRC Press.

Henry, C.D., Elson, H.B., McIntosh, W.C., Heizler, M.T., and Castor, S.B. (1997). Brief duration of hydrothermal activity at Round Mountain, Nevada, determined from Ar (super 40)/Ar (super 39) geochronology. *Economic Geology*, 92(7–8): 807–826.

Jolie, E., Moeck, I., and Faulds, J.E. (2015). Quantitative structural geological exploration of fault-controlled geothermal systems: a case study from the Basin-and-Range Province, Nevada (USA). *Geothermics*, 54: 54–67.

Rosenberg, M.D., Bignall, G., and Rae, A.J. (2009). The geological framework of the Wairakei–Tauhara geothermal system, New Zealand. *Geothermics*, 38(1): 72–84.

Sander, M.V. and Einaudi, M.T. (1990). Epithermal deposition of gold during transition from propylitic to potassic alteration at Round Mountain, Nevada. *Economic Geology*, 85(2): 285–311.

6 Physical and Chemical Characteristics of Geothermal Systems

KEY CHAPTER OBJECTIVES

- Relate the characteristics of the phase-change diagrams of water to its chemical characteristics, including its polar nature and hydrogen bonds.
- Explain why water is considered the universal solvent and how that applies to hydrothermal alteration of rocks in the vicinity of active and inactive geothermal systems.
- Using the pressure–enthalpy diagram for water, explain why dry steam resources are favored over other types of geothermal resources for electrical power generation.
- Given the temperature of a boiling fluid, determine the depth at which boiling begins.
- Compare the different styles of hydrothermal alteration and discuss their implications for fluid chemistry and operation of a geothermal power plant.

To successfully find and develop geothermal systems, an understanding of their physical and chemical attributes is crucial. This chapter begins with a discussion of the thermodynamic considerations of water that govern how geothermal systems behave and how they are ultimately utilized. Important characteristics of liquid- and vapor-dominated geothermal systems, in particular thermal manifestations and alteration of wallrocks, are then explored.

THERMODYNAMIC CHARACTERISTICS OF WATER

Although widespread on our planet and therefore commonly taken for granted, water is an extraordinary chemical compound. An important distinctive characteristic is that the solid phase of water (ice) is less dense than its liquid phase. This is important for geothermal resources because the high pressures found in the Earth would otherwise tend to favor the solid rather than liquid phase. Because solids cannot flow or flow much more slowly than liquids, solids are unable to transport sufficient heat at depth to the surface to run a geothermal power plant.

Heat Capacity and Specific Heat

Water has high heat capacity. *Heat capacity* is the ratio of the heat absorbed per associated temperature rise. In other words, it takes significant energy, either absorbed or released, to change the temperature of water. For example, if it takes 10 calories to raise the temperature of a cup of water by 2°C, then the heat capacity is 5 calories per degree C. *Specific heat* is the heat capacity per unit of mass. For water, it takes 1 calorie of heat to raise the temperature of 1 gram of water by 1°C, so the specific heat of water is 1 cal/g·°C or 4186 J/kg·°C. The high specific heat of water means that water can store considerable energy, compared to air or granite, for example, each of which has a much lower specific heat (about 1000 J/kg·°C and 800 J/kg·°C, respectively). As a very tangible example of water's high heat capacity and ability to store energy, consider that spent fuel rods at nuclear power plants are stored in large pools of water to prevent the rods from overheating and the potential escape of radioactive material. Groundwater heated by hot rocks at depth has much stored energy and therefore has the ability, when brought to the surface, to do work, such as spin a turbine connected to an electrical generator.

Polar Nature

Water is a polar molecule. Each molecule of H_2O has one end with a positive residual charge and one end with a negative residual charge. Although the oxygen and hydrogen atoms are covalently bonded (share electrons), this charge distribution arises because the electrons are not shared equally. Electrons spend most of their time with the oxygen, because of its high attraction for electrons (electronegativity), resulting in an overall negative charge on the oxygen end and a positive charge at the hydrogen side of the molecule. The polarity arises because the water molecule is bent, so that the central oxygen is not simply sandwiched between two hydrogen atoms, as is the case for the central carbon atom in carbon dioxide, which is linear and also nonpolar (Figure 6.1). Water is a bent molecule because the repulsive force between the unshared pair of electrons about the central oxygen atom "pushes" on the oxygen–hydrogen bond. The combination of unequal sharing of electrons and bent nature of the water molecule results in water's polar nature.

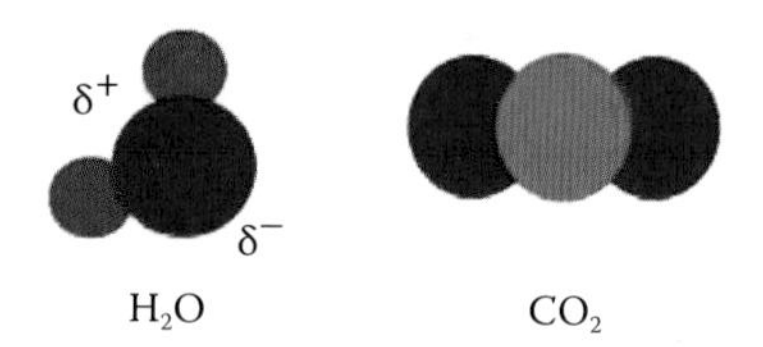

FIGURE 6.1 Models showing the shapes of water (H_2O) and carbon dioxide (CO_2) molecules. The blue circles are oxygen atoms, the red circles are hydrogen atoms, and the green ball is a carbon atom. Note how water is a bent molecule whereas the carbon dioxide molecule is linear. The delta (δ) symbols on the water molecule show how the oxygen end is negative and the hydrogen side is positive.

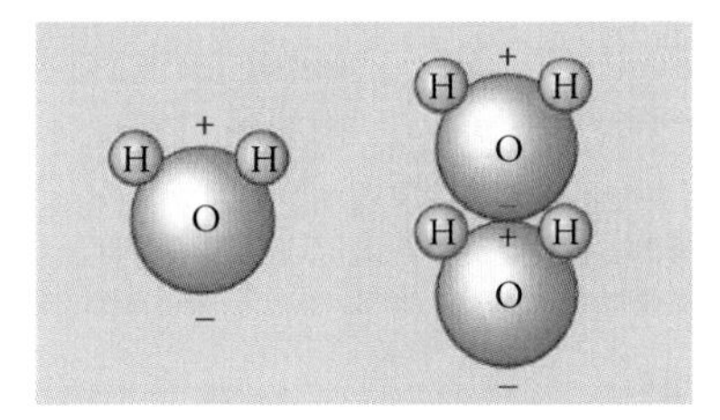

FIGURE 6.2 Schematic illustration showing how the polar and bent natures of water molecules interact to form weak hydrogen bonds. The hydrogen bond forms because of the polar nature and is bent because the positive end of a water molecule (H end) is attracted to the negative side (O end) of another water molecule. The formation of hydrogen bonds helps explains some of water's unique characteristics, including surface tension and high boiling point for a compound with low molecular mass. (Illustration from http://www.k12science.org/media/live/curriculum/waterproj/images/watermolecule.jpg.)

So what does this mean from a geothermal perspective?

1. It explains why water sticks to itself, resulting in surface tension. In other words, weak bonds (hydrogen bonds) form between adjacent water molecules so that the oxygen end of one molecule is attracted to the hydrogen end of another molecule (Figure 6.2). This helps explain why all the ground or geothermal water present underground cannot be fully accessed due to the effects of surface tension.
2. Another important aspect of the polar nature of water that pertains to geothermal systems is that many substances, particularly ionic substances such as salt, dissolve in water (Figure 6.3). This is especially true for rocks, as the majority of minerals are composed of ionically bonded elements. This solvent action can be accentuated at high temperatures for many substances, resulting in the alteration of rock (as discussed later in the chapter) in the vicinity of hot geothermal fluids. Geothermal fluids at high temperature, say, >150°C, can dissolve certain minerals, such as fine-grained silicate minerals, but can also precipitate other minerals, such as calcite ($CaCO_3$) or anhydrite ($CaSO_4$).
3. The bent nature of the water molecule and the development of hydrogen bonds between the hydrogen and oxygen atoms of adjacent water molecules cause water to expand by almost 10% when it freezes (Figure 6.4). As a result, increasing pressure at constant temperature will cause ice to melt or will help liquid water remain in liquid form. This is part of the reason why permafrost at high latitudes begins to melt at depth even if the temperature is still below freezing. For geothermal resources, melting with increased pressure means that water will be liquid phase in most cases at depth in the Earth with the ability to transfer heat and do work in a geothermal power facility.
4. The intermittently forming and disbanding nature of hydrogen bonds of liquid water molecules also helps explain water's high specific heat. Temperature is a measure of the speed of molecules, and as heat is added

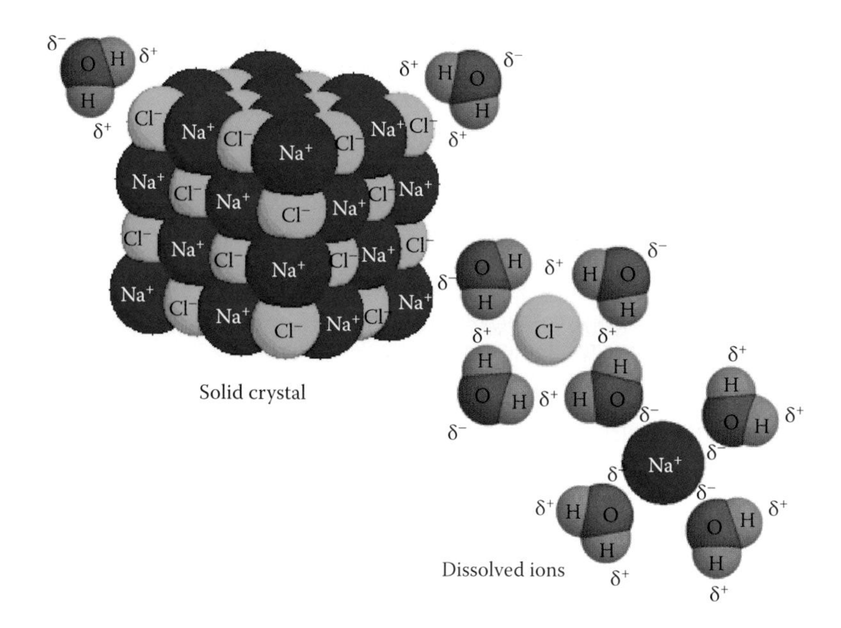

FIGURE 6.3 The dissolution of salt in water. The negative O ends of the water molecules surround the Na^+ ions, while the positive H ends of water molecules surround the Cl^- ions. The polar nature of water helps explain water's universal solvent capabilities. (Illustration from https://www.chem.wisc.edu/deptfiles/genchem/sstutorial/Text7/Tx75/tx75p3.GIF.)

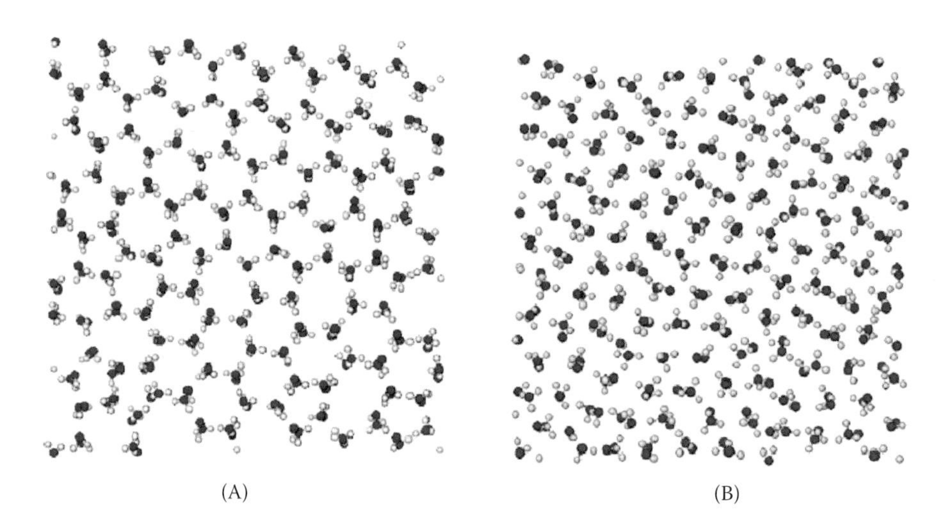

FIGURE 6.4 (A) Molecules for frozen water, and (B) H-bonded water molecules for liquid water. Note the more open (less dense) nature of the bonded frozen water molecules compared to unbonded liquid water molecules. (Illustration from http://www.nature.com/nphys/journal/v6/n9/images_article/nphys1708-f3.jpg.)

the speed of molecules increases. In the case of water, the molecules are sticking together, at least temporarily due to the hydrogen bonds, so more heat is needed to get molecules to move faster, which gives water a high thermal mass. This high stored heat in water means it has high capability to do work in a geothermal facility.

Water Phase Relationships and Critical Point

Before discussing the phases of water, which depend in part on temperature and heat (thermal energy) available, a short discussion on the two values is worthwhile to avoid potential confusion. Temperature measures how hot or cold an object is. The warmer the object, the faster its molecules vibrate; therefore, the object will contain higher internal heat or thermal energy. Temperature is measured in degrees on various scales, such as Celsius or Kelvin, but heat or thermal energy is measured in joules or calories.* To illustrate the difference between temperature and heat, consider a cup of coffee at 90°C and water in a hot tub at 40°C. The cup of coffee is hotter than the water in the hot tub, but the hot tub water contains much more heat or thermal energy because of its greater mass. In other words, if the water in the hot tub and water in a cup of coffee started at the same temperature of 20°C, for example, much more heat energy would be needed to raise the temperature of the water in the hot tub to 90°C than for the cup of coffee because of the greater mass of the water in the hot tub.

Water has three main phases—ice (solid), liquid, and vapor (a "fourth" phase, supercritical, is discussed in the next section). Of the three main phases of water, each phase has its own specific heat, with liquid water having the highest (the specific heat of supercritical water is variable depending on the conditions of temperature and pressure). Specific heat is the heat required to change the temperature of a unit mass of a single phase by a unit mass of temperature, such as 1°C or 1 K; therefore, units for specific heat are energy per unit mass per degree.† The specific heat for ice is about 2.1 J/g·K,‡ meaning that it takes about 2.1 joules of energy per gram of ice to increase the temperature by 1 K. For liquid water, the specific heat is about double that for ice, or about 4.2 J/g·K, and for steam the value is similar to that of ice at about 2.1 J/g·K. The different specific heat capacities of each phase are reflected by the slope in the change in temperature of the single phase in Figure 6.5. Note the gentler slope for liquid water compared to ice and steam, meaning that it takes more heat to change the temperature of liquid water than for ice and steam.

---

* *Joule* (J) is the SI measure of energy, whereas *calorie* is a metric measurement of energy, where 1 calorie measures the heat required to raise the temperature of 1 gram of liquid water by 1°C (or 1 K). One calorie equals approximately 4.2 J.

† A related term is *heat capacity*, which reflects the specific heat multiplied by the mass in question. For example, a gallon of water has a higher heat capacity than a quart of water, but both have the same specific heat.

‡ The abbreviation J/g·K indicates joule per gram per degree Kelvin. Because the specific heat for water is about 4.2 J/g·K, 1 calorie equals approximately 4.2 J, as noted in the previous footnote.

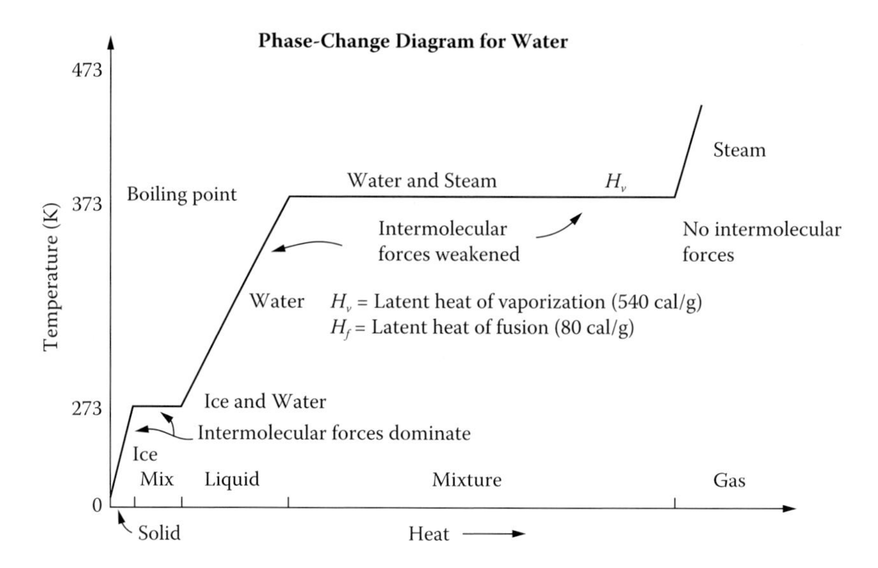

FIGURE 6.5 Phase-change graph for water illustrating changes in temperature as a function of added heat (energy)—as heat is added, temperature increases. Where two phases are present, the temperature remains constant until only one phase remains. Note that the higher specific heat for liquid water is reflected by the gentler slope compared to that of steam and ice. Steam contains a high amount of latent heat energy that becomes partly available to do work (spin a turbine connected to an electrical generator) when condensed back to liquid water. See text for discussion.

During the transition between one phase and the next, temperature stays constant,* as the added heat goes to driving the phase change (Figure 6.5). The heat required to drive a phase change is called *latent heat* and varies depending on the type of phase change. For example, the latent heat of fusion (melting) of ice is 80 calories per gram (or 336 joules per gram), whereas the latent heat of vaporization (liquid water to steam) is a considerable 540 calories per gram (or about 2268 joules per gram) (Figure 6.5). In order for water to move from a low energy state, such as ice, to a high energy state, such as steam, it must absorb energy (latent heat) from its environment. Likewise, to go from higher energy steam to lower energy liquid water, energy must be given up to the environment. From a geothermal standpoint, this means that some of that energy becomes available to do work as steam is condensed to liquid water. This work is used to spin turbine blades and drive an electrical generator.

* Constant temperature during a phase change is an outcome of the Gibbs phase rule, which is given by the equation $F = 2 + C - P$, where F is the degree of freedoms, C represents the chemical components, and P represents the phases present. For the water system in Figure 6.5, when water is boiling to steam, two phases are present for a one-component (water) system. Therefore, $F = 2 + 1 - 2 = 1$, which means temperature must be constant and hence the slope of the curve is zero (horizontal). If only one phase is present, both temperature and heat change because $F = 2$, and the slope of the curve is no longer zero but is equal to the specific heat for that particular phase of water.

Figure 6.5 illustrates changes in temperature of the different phases of water at constant pressure, but for geothermal purpose it is also necessary to look at water phase changes as a function of both temperature and pressure (Figure 6.6). The negative slope of the ice–liquid boundary in Figure 6.6 reflects the volume expansion that occurs upon freezing. In other words, at constant temperature ice will melt with an increase in pressure. This results in the liquid field expanding at higher pressures. The other point of interest is the critical point, which occurs at about 374°C and 22 MPa or 220 bars (218 atmospheres). At temperatures and pressures above the critical point, a distinction between liquid and vapor ceases to exist, and water has properties of both liquid and vapor, such as the ability to diffuse like a vapor for increased permeability but also the ability to dissolve minerals like a liquid (in other words, potentially being corrosive or hostile to equipment). Some particularly hot geothermal systems, such as those in Iceland, probably contain supercritical fluids in their deeper portions (depths > ~3 km); if these were accessed and developed, as discussed in Chapter 11, their power output would be 5 to 10 times that of a well tapping a subcritical reservoir (as much as 50 MWe per well) (e.g., Fridleifsson and Elders, 2005; Fridleifsson et al., 2014).

Development of supercritical geothermal fluids is currently being studied in the Iceland Deep Drilling Project (IDDP). A second experimental well (IDDP-2) is planned to begin drilling in the near future at the Reykjanes geothermal field in southwest Iceland; the first test well (IDDP-1) was abandoned when it hit magma

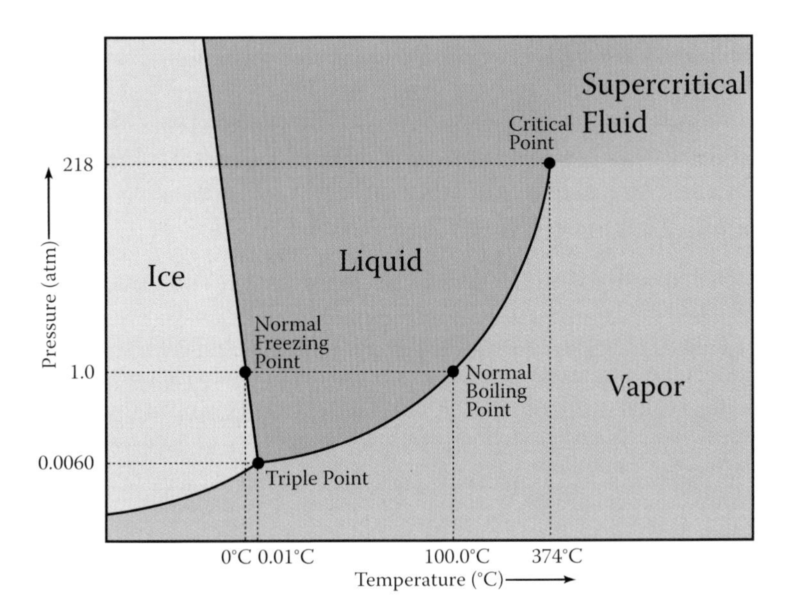

FIGURE 6.6 Phase diagram for water. Note the negative slope of the ice–liquid water boundary which favors the liquid phase at higher pressure and the region for supercritical water above 374°C and 221 bars.

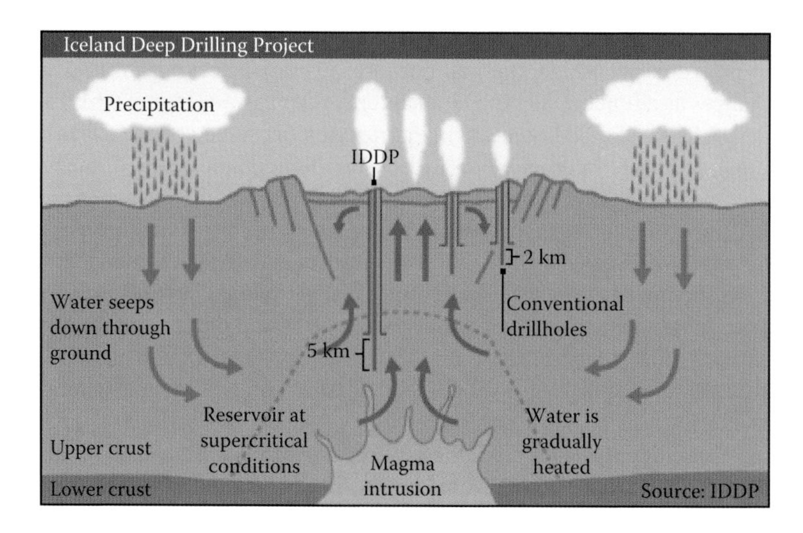

FIGURE 6.7 Schematic geologic cross-section of IDDP hole #1 drilled in the Krafla caldera in northern Iceland showing the intended target of supercritical fluids at depths of 4 to 5 km. The hole was abandoned at about 2 km depth after intersecting a dike (narrow intrusion) of magma. The hole now represents the world's first potential magma-engineered geothermal system. (Illustration courtesy of Iceland Deep Drilling Project.)

before reaching its intended target depth (Figure 6.7). The abandoned IDDP-1 well was nonetheless a partial success, as a 2-year-long flow test produced superheated (or dry) steam* at a wellhead temperature of 450°C and pressures of 40 to 140 bars, capable of generating about 36 MWe of power (Fridleifsson et al., 2015). During that time, it was the hottest geothermal well on the planet.

Pressure and Enthalpy (Heat) Relationships

A way to represent the work potential of heated water is to graph changes in enthalpy (a thermodynamic term for heat energy) as a function of pressure (Figure 6.8). The dome-shaped curve in the middle outlines the region of co-existing liquid and vapor. The region to the left of the dome-shaped boundary is the liquid-only field, and the region to the right is the vapor-only field. The top of the dome-shaped curve marks the critical point. Temperature contours are plotted and are horizontal in the region of coexisting liquid and vapor under the dome-shaped curve. Temperature contours are horizontal there because two phases coexist (liquid and vapor); therefore, added heat converts liquid water to steam and the temperature remains constant (as shown earlier in Figure 6.5). Also plotted in the coexisting liquid–vapor region is the steam percentage, which increases toward the right as more water is boiled to steam with increasing enthalpy. To relate Figure 6.8 to the

* Superheated or dry steam is steam that is at a temperature above its vaporization point for a given pressure and no liquid water is present. In contrast, wet steam is at its vaporization point and both liquid water and steam coexist.

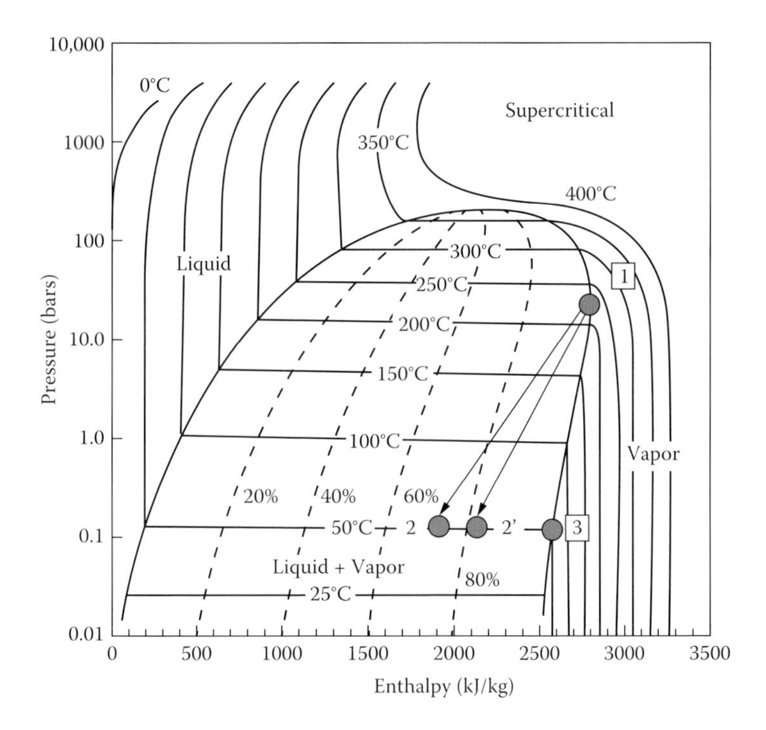

FIGURE 6.8 Pressure–enthalpy diagram for water with isotherms plotted. The dome-shaped region in the middle of the graph is the region of coexisting liquid and vapor. The dashed lines denote contours of percent steam. Can you plot the critical point on this diagram? (From Glassley, W.E., *Geothermal Energy: Renewable Energy and the Environment*, 2nd ed., CRC Press, Boca Raton, FL, 2015.)

latent heat of vaporization of water, discussed above, at one atmosphere of pressure (slightly more than 1 bar), consider the 100°C isotherm where it intersects the dome-shaped curve at both the liquid and vapor ends of the diagram. At 100°C, the enthalpy of liquid water is about 420 kJ/kg and the enthalpy of vapor is about 2688 kJ/kg. The difference in these two values is the latent heat of vaporization, which equals 2268 kJ/kg, as noted previously, and represents the energy stored in steam that can be used to do work (spin a turbine).

Let's look at an example of how to use Figure 6.8, such as for a dry steam (no liquid fraction) geothermal reservoir at 235°C and a pressure of 30 bars (point 1 in Figure 6.8), to determine the potential work that can be done by a system with these characteristics. The work that can be done is given by Equation 6.1:

$$W = H_i - H_f \tag{6.1}$$

where

W = Work

H_i = Initial enthalpy of steam entering the turbine chamber.

H_f = Final enthalpy of steam leaving the turbine chamber.

Using Figure 6.8, the initial enthalpy is about 2800 kJ/kg. In the ideal case where efficiency is 100%, the final enthalpy is about 1980 kJ/g if cooled to 50°C (point 2 in Figure 6.8), and the work done is 820 kJ/kg. However, efficiency is never 100%, as some energy is lost to friction and unrecoverable heat (about 2150 kJ/kg at point 2′ in Figure 6.8), so the amount of work done is typically 15 to 20% less than the ideal case. The ratio of work (actual work/ideal work) is a measure of turbine efficiency, so in the example here that would be 2800 – 2150 (actual work)/2800 – 1980 (ideal work) = 0.79, or 79%. An additional element that can affect steam turbine efficiency is the presence of any liquid water with the steam. Empirical relationships indicate that, for every percentage increase of liquid in the steam, steam turbine efficiency decreases by 0.5% (Glassley, 2015). This is because rapidly moving droplets of liquid water can damage turbine blades and slow down blade rotation.

The power output from a well can be determined if the flow rate and work done (using Figure 6.8) are known:

$$\text{Power } (P) = \text{Work done on turbine} \times \text{Flow rate} \tag{6.2}$$

For example, if the flow rate from a well is 8 kg/s, then the power output, using a realized work value of 640 kJ/kg (enthalpy value at point 1 minus enthalpy value at point 2′ in Figure 6.8), would be

$$P = 8 \text{ kg/s} \times 640 \text{ kJ/kg} = 2560 \text{ kJ/s} \times 1 \text{ MWe/1000 kJ/s} = 2.56 \text{ MWe}$$

The benefit of a dry-steam or vapor-dominated system compared to a liquid-dominated geothermal system is illustrated in Figure 6.9. Many producing geothermal reservoirs have conditions shown by the shaded region in the upper left part of the pressure–enthalpy diagram or on the low enthalpy side of the critical point, simplified to point 1 for the sake of discussion. At the conditions represented by point 1, the fluid has about 1000 kJ/kg of enthalpy. As pressure is decreased, the fluid begins to flash at about 30 bars and 235°C. Only about a third of the fluid goes to steam if the temperature is reduced to 50°C or if pressure is reduced to 0.15 bar as indicated by point 3 in Figure 6.9. *Therefore, a power plant tapping a liquid-dominated geothermal system will require about three times the mass flow rate to achieve the same power output of a power plant accessing a dry-steam reservoir at the same conditions of temperature and pressure.* Figure 6.9 illustrates that steam content is maximized when pressure and temperature are reduced as much as possible. For example, engineers working at the Wairakei geothermal field in New Zealand discovered empirically that for every 10-millibar increase in pressure in the condenser, the power output goes down by about 1 MWe (T. Montegue, pers. comm., 2013).

LIQUID-DOMINATED GEOTHERMAL SYSTEMS

Liquid-dominated systems are the most common types of hydrothermal systems developed for power generation and direct use. The fluid is in the liquid state because the fluid remains below the boiling point curve with depth. The boiling point

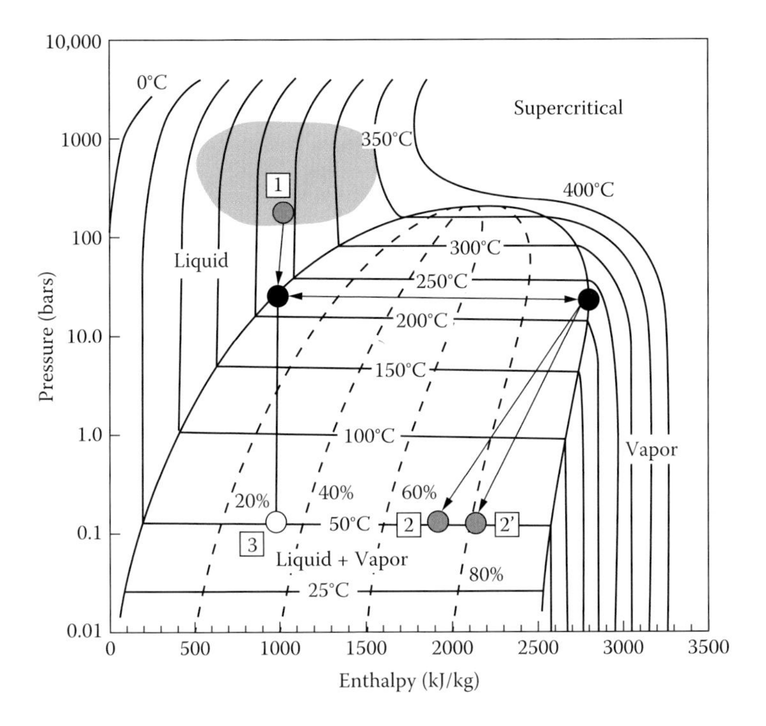

FIGURE 6.9 Pressure–enthalpy graph of water illustrating that in a liquid-dominated system only about 30% of liquid is flashed to steam to power a turbine generator, meaning that, mass for mass, liquid systems provide only about a third of the power potential as dry steam reservoirs. See text for discussion. (From Glassley, W.E., *Geothermal Energy: Renewable Energy and the Environment*, 2nd ed., CRC Press, Boca Raton, FL, 2015.)

increases due to increased pressure, as illustrated in Figure 6.10. In such systems, the fluid does not begin to boil or flash until it rises to lower pressure and intersects the curve, at which point the fluid cools along the curve with further reduction in pressure or depth. In addition to pressure, the boiling point of geothermal fluids is also influenced by dissolved solids and gases. Increasing amounts of dissolved solids will increase the boiling point, whereas increasing dissolved concentrations of non-condensable gases, such as CO_2, will lower the boiling point (Figure 6.10).

Temperature Range of Fluids

The temperature of liquids ranges from <100°C to about 350°C in some of the hottest brine systems in the Imperial Valley of southeastern California. For power production, fluids are typically greater than about 125°C but can be as low as about 80°C such as at Chena Hot Springs, Alaska. As previously discussed in Chapter 2, binary geothermal power plants are typically in the range of 125 to 175°C, single-flash plants from about 175 to ~210°C, and double- or triple-flash plants above ~210°C. Direct-use systems are typically in the range of 50 to 100°C.

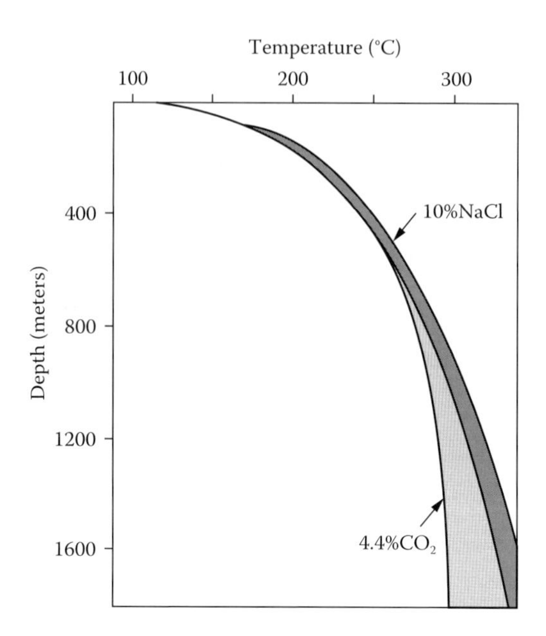

FIGURE 6.10 Changes in temperature of boiling point of water with depth. Dissolved solids, such as NaCl, will increase the boiling point temperature, whereas dissolved gases, such as CO_2, will lower the boiling point. (Adapted from Haas, Jr., J.L., *Economic Geology*, 66(6), 940–946, 1971.)

Fluid Compositions

Fluid compositions can be grouped into three main categories: (1) neutral chloride, (2) acid-sulfate, and (3) brine. Neutral-chloride systems are some of the most commonly exploited; they have near-neutral pH (weakly acidic to weakly alkaline, pH ~6 to ~8.5). Total dissolved solids (TDS) range from <1000 ppm to about 10,000 ppm. Because of the benign chemical make-up of the fluids in these systems, they are some of the easiest to develop. The fluids are noncorrosive to equipment, and precipitates of silica or carbonate scale, which can reduce flow in wells and pipelines, are fairly easy to manage. Essentially all of the developed geothermal fields in Nevada, including the Steamboat Springs, Beowawe, Tuscarora, and recently commissioned McGinness Hills power plants, utilize geothermal fluids of this type.

Acid-sulfate fluids, as the name implies, are acidic (pH commonly <4 and sometimes <1) and are rich in dissolved sulfate. Such fluids commonly occur near the vents of active volcanoes where magmatic vapors rich in SO_2 and HCl react to produce acidic conditions. Specifically, SO_2 gas disproportionates (undergoes a redox reaction in which SO_2 is simultaneously oxidized and reduced) to form sulfuric acid (H_2SO_4-oxidized) and hydrogen sulfide (H_2S-reduced) as illustrated by the following reaction:

$$2SO_2 + 2H_2O \rightarrow H_2SO_4 + H_2S + O_2 \tag{6.3}$$

The H_2S can further react with near-surface oxygen to also produce sulfuric acid:

$$H_2S + 2O_2 \rightarrow H_2SO_4 \tag{6.4}$$

As such, these systems can be very corrosive to equipment. Even though temperatures and enthalpy can be quite high (>220°C), they are either not developed or require acid neutralization procedures. For example, the high-temperature portion of the Miravalles geothermal well field in Costa Rica, which is located closest to the flanks of an active andesitic volcano, requires injection of dilute sodium hydroxide (NaOH) to help neutralize the pH and reduce corrosion of well casings and surface equipment (DiPippo, 2012). A good example of an undeveloped high-temperature (>300°C), acid-sulfate hydrothermal system is White Island in the Bay of Plenty on the North Island of New Zealand, which lies in the throat of an active volcano (Wood, 1994). Other examples of undeveloped active acid-sulfate geothermal systems include the steaming crater lakes at Mt. Ruapehu, New Zealand (Deely and Sheppard, 1996) and Kawah geothermal systems in Java, Indonesia (Purnomo, 2014).

Brine geothermal systems are those that have more than about 3.5% (or 35,000 ppm) total dissolved salts (which is comparable to that of seawater). Some geothermal brines are hypersaline, such as those in the Salton Trough (a transtensional tectonic rift basin that is discussed in Chapter 7) of southeastern California (Figure 6.11). The Salton Trough includes the Salton Sea, Brawley, Heber, and East Mesa geothermal fields, whose fluids contain as much as 26 wt% total dissolved solids (TDS) with temperatures as high as 365°C at depths of 2 to 3 km (McKibben et al., 1988a). The high brine content results from the geothermal fluids encountering and dissolving evaporite rock layers (consisting mainly of gypsum with halite). The neighboring Cerro Prieto geothermal field to the south in Mexico is also part of the Salton Trough (Figure 6.11). The geothermal fluids there, however, are not hypersaline because beds of evaporite are absent or minimally developed in the rock section.

The brine fluids are also modestly acidic because the dissolution of gypsum forms sulfuric acid and a base calcium hydroxide as shown by the reaction below:

$$H_2O + CaSO_4{\cdot}H_2O \rightarrow H_2SO_4 + Ca(OH)_2 \tag{6.5}$$

$$\text{Water} + \text{Gypsum} \rightarrow (\text{Strong acid}) + (\text{Moderate base})$$

Because sulfuric acid is a strong acid and calcium hydroxide is a moderate base, the resulting solution is not pH neutral but acidic. As a result, these brines can result in both considerable corrosion of equipment and clogging due to the buildup of scale in wells and pipelines. The recently opened 49.9-MWe Hudson Ranch I geothermal plant accesses 350°C hypersaline brine containing over 200,000 ppm TDS, including high concentrations of lithium (Li), manganese (Mn), and zinc (Zn). To prevent scaling of equipment from the hypersaline brine at the Hudson Ranch facility (now renamed the John L. Featherstone geothermal plant), seed material is added to the residual fluid after flashing. The supersaturated solids of the hypersaline brine then precipitate on the seed material rather than on the surfaces of the flash chamber and pipelines. The resulting slurry is sent to a chemical processing plant where the

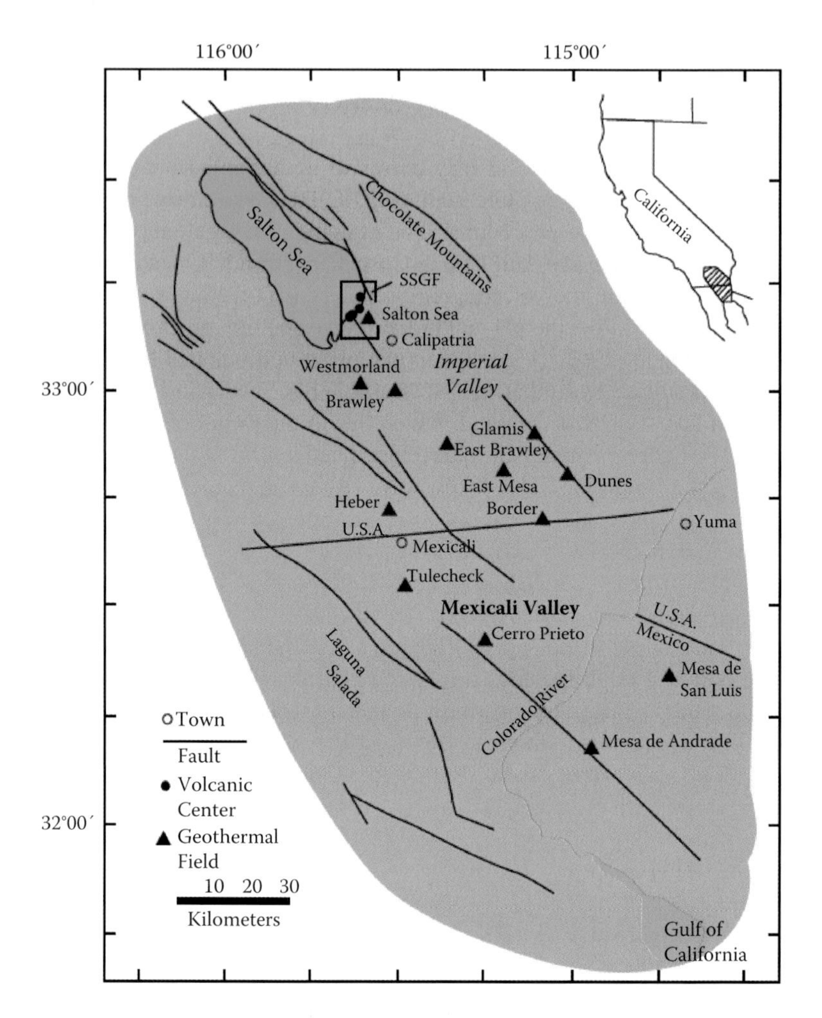

FIGURE 6.11 Simplified geologic map of southeastern California and Baja California showing the Salton Trough and associated geothermal fields. SSGF stands for the Salton Sea geothermal field. (Adapted from McKibben, M.A. et al., *Geochimica et Cosmochimica Acta*, 52(5), 1047–1056, 1988.)

dissolved solids are largely removed prior to reinjection to avoid clogging of injection wells and reservoir permeability. The plan is to eventually recover Li, Mn, and Zn from the separated dissolved solids for ready markets in battery and steel manufacturing. An additional remarkable feature of the Hudson Ranch I power plant is that its 49.9-MWe output is basically provided by just two of three production wells. In fact, only one well is needed at a time, making these some of the most powerful producing geothermal wells on the planet.*

* An informative video about the Hudson Ranch I geothermal plant can be viewed at http://www.youtube.com/watch?v=dS9p4tfR1RE. Although the video does a good job of discussing the power generation process, the mineral recovery process is not addressed.

Wallrock Alteration

Due to water's polar nature and strong ability to dissolve minerals, rocks in the vicinity of geothermal systems become transformed or altered from their original state. The type and degree of alteration vary as a function of temperature and the chemical nature of the hydrothermal fluids, including near-neutral-pH alkali-chloride fluids, acid-sulfate-rich solutions, or brine-rich fluids as discussed above. Also affecting wallrock alteration are the processes of boiling and mixing of different fluids. Typically, wallrock alteration is zoned, with the most intense alteration occurring where temperatures and fluid flow are highest and becoming progressively less strong moving away to lower temperatures and less fluid flow. From a practical standpoint, the significance of hydrothermal alteration provides information on the following:

1. The geophysical properties of the rocks, including their densities, porosities, permeabilities, and electrical properties (discussed further in Chapter 8)
2. Temperature distributions of geothermal fluids both laterally and with depth
3. Identifying the zones of fluid upflow and highest temperatures (the geothermal targets) and recharge
4. The chemical composition of the geothermal fluids
5. Locating the zones of enhanced permeability as well as barriers for fluid flow
6. Geotechnical properties that influence drilling and plant operations

Figure 6.12 shows common hydrothermal alteration minerals and their associated range of temperatures of formation and stability based on studies of active geothermal systems. Some minerals such as quartz and calcite are stable over a wide range of temperatures, whereas other minerals are more restrictive, such as epidote, which forms generally above 200°C, and actinolite above 300°C. Therefore, the presence of common hydrothermal minerals such as quartz and calcite is not particularly informative with regard to temperature conditions because of their wide range of temperature stability. The presence of minerals such as epidote or actinolite, however, does provide useful constraints on temperature because of their more limited range of temperature stability.

Furthermore, select minerals identified from drill samples can also provide information on whether a system may be waxing or waning in temperature. A geologist can observe whether certain mineral assemblages are stable or are replacing each other on the basis of textural relationships observed under a microscope from drill cuttings. For example, a geologist might observe that epidote is being replaced by prehnite. From calculated stability phase relationships (Figure 6.13), such a mineralogical replacement suggests that the system could be cooling as prehnite is stable at a lower temperature.

Hydrothermal mineral assemblages also reflect fluid compositions as reflected by activity–activity plots (Figure 6.14). The generation of these plots is determined from calculations using experimentally determined equilibrium constants and observed field relationships of mineral stabilities. Figure 6.14 was calculated at a temperature of 250°C and shows the relative stability fields of different hydrothermal minerals

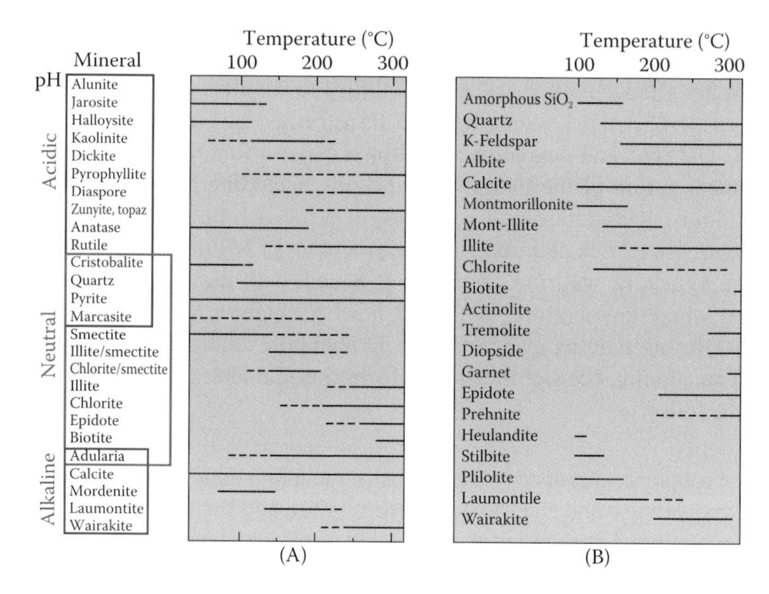

FIGURE 6.12 Temperature stability ranges, as determined from borehole temperature observations, of common hydrothermal alteration minerals. (A) Minerals are grouped according to the pH of the geothermal fluids, from acidic to alkaline. (Adapted from Hedenquist, J.W. et al., *Reviews in Economic Geology*, 13, 245–277, 2000.) (B) Chart showing temperature stability ranges of common alteration minerals. (Adapted from Henley, R.W. and Ellis, A.J., *Earth-Science Reviews*, 19(1), 1–50, 1983.)

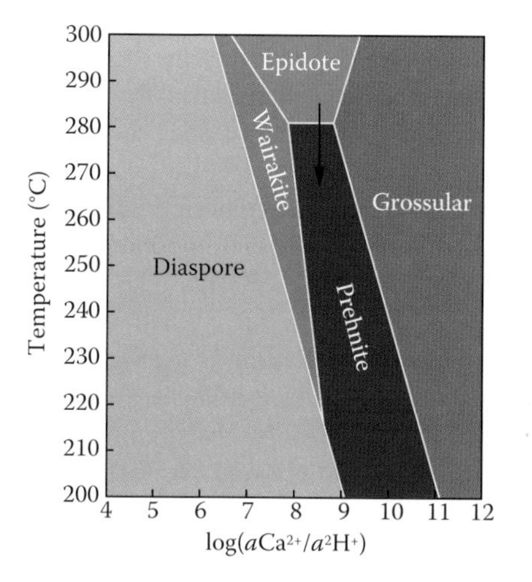

FIGURE 6.13 Temperature–activity plot showing stability fields of select calcium-bearing silicate minerals found in geothermal systems. The *x*-axis simply reflects the concentrations of Ca^{2+} ions relative to H^+ ions. As shown by the arrow, as temperature decreases, epidote will be replaced by prehnite below about 280°C for these modeled conditions. (Graph taken from a presentation by J. Moore at a workshop on geochemistry of geothermal fluids at University of California, Davis, in 2012. Adapted from Bruton, 1996.)

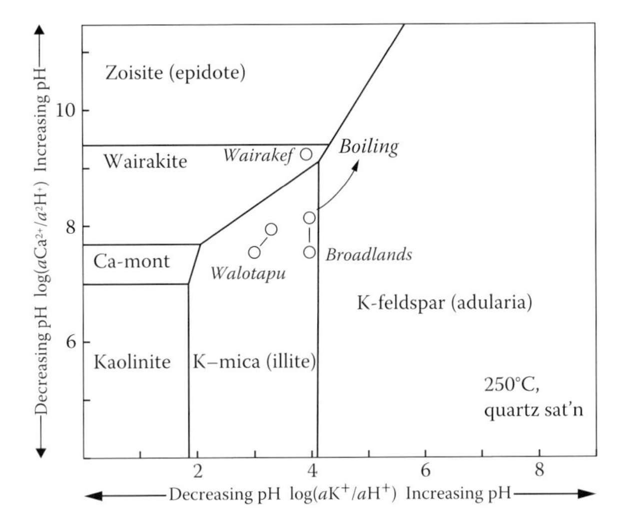

FIGURE 6.14 Activity–activity diagram showing the stability fields of some important hydrothermal alteration minerals as a function of Ca and K ion activity per H ion activity at 250°C and for fluids saturated with quartz. Acidity decreases toward the right and toward the top. Thus, kaolinite is stable under the most acid conditions and lowest concentrations of Ca and K ions. Some fluid compositions for known geothermal systems in New Zealand are plotted (e.g., Wairakei). As discussed in the steam-heated section below, boiling drives solutions at the Broadlands geothermal field to higher pH as acidic gases are preferentially partitioned into the vapor phase. (Adapted from figures presented in Browne, 1978; Browne and Ellis, 1970; Simmons and Browne, 2000.)

as a function of K^+ and Ca^{2+} concentrations (technically, activities*) per H^+ ion concentration. We see that adularia is stable at high K^+ concentration and low H^+ ion concentration (or neutral to mildly alkaline pH). As H^+ ion concentration (acidity) increases, however, adularia will be replaced by illite and eventually by kaolinite.

For our purposes, three main styles of wallrock alteration in geothermal systems will be considered: (1) low sulfidation formed by near-neutral-pH alkali-chloride geothermal fluids; (2) high sulfidation formed by acidic solutions involving degassing of subjacent magmatic intrusions; and (3) steam-heated, high-level acidic alteration formed by boiling of near-neutral-pH chloride waters boiling at depth and liberating gases such as H_2S and CO_2 that combine with water vapor and near-surface groundwater to produce acidic conditions near the surface.

The term *sulfidation* reflects the oxidation state and abundance of aqueous sulfur species of hydrothermal solutions. Alteration of low-sulfidation types results from fluids that are generally reduced, have lower amounts of dissolved aqueous sulfur, and are mainly of groundwater origin (meteoric). Alteration of high-sulfidation types results from oxidized and acidic fluids rich in both sulfate and sulfide species due to the disproportionation of magmatically derived SO_2 as discussed earlier in this chapter.

* Chemical activity is the effective concentration of elemental species and is related to concentration by the activity coefficient that takes into account ionic interactions or deviations from ideal behavior in a multicomponent solution.

Low-Sulfidation Alteration

Due to the near-neutral-pH and alkali-chloride composition of the fluids, the style of alteration (type of minerals formed) in wallrocks is governed mainly by temperature, rather than the more extensive physical and chemical changes associated with alteration of the high-sulfidation type. Geothermal upflow zones denote permeable areas where hot fluids are actively rising and are productive targets for producing geothermal wells with temperatures ranging from about 200°C to about 350°C. Actual fluid pathways, such as fractures and faults, commonly contain minerals of quartz, adularia, and calcite. Wallrocks adjoining the fluid pathways are replaced by variable mixtures of minerals that are stable depending on conditions of temperature and pressure, including epidote, actinolite, chlorite, albite, K-feldspar (adularia), and calcite. At the upper temperature end (>~300°C), minor garnet, biotite, and diopside can also occur. Altered wallrocks in this region typically have a greenish hue reflecting the green color of chlorite, epidote, and possible actinolite (at the upper temperature range), which typically replace primary mafic minerals and plagioclase. For the economic geologist looking for deposits of gold and silver in fossil geothermal systems, this type of wallrock alteration is referred to as *propylitic*. For modern or active geothermal systems, the propylitic zone typically marks the geothermal sweet spot or reservoir region for systems dominated by near-neutral-pH alkali-chloride fluids.

At shallower depths and cooler temperatures, propylitic alteration commonly gives ways to variably adularized and silicified rock, typically expressed as very fine-grained adularia and microcrystalline silica that flood and precipitate in the pores of the rock. This flooding thus can destroy primary permeability, but at the same time makes the rock hard and brittle, which can promote secondary or fracture permeability when subjected to thermal or tectonic stresses.

If fluids rising along fractures reach the surface, hot spring travertine or silica sinter can form, depending on the temperature of fluids rising from depth. Where the geothermal reservoir is relatively cool, at temperatures less than about 150°C, the mineral travertine (a banded form of calcite) is typically deposited. Travertine-type hot springs may be cool enough to soak in (but not always!). If the fluids are hotter, however, the travertine would be deposited at depth because calcite becomes less soluble with increasing temperature, due to its *retrograde solubility*. The formation of calcite can be expressed by the following reaction:

$$Ca^{2+} + 2HCO_3^- \rightarrow \underset{\text{(Calcite)}}{CaCO_3} + H_2O + CO_2 \tag{6.6}$$

The retrograde solubility of calcite occurs because CO_2 is less soluble in high-temperature solutions than in low-temperature solutions so it tends to escape. This drives the above reaction to the right, as predicted by Le Chatlier's principle,* promoting the precipitation of calcite.

* Le Chatlier's principle indicates that a system perturbed from equilibrium will work to maintain equilibrium. In Reaction 6.4, for example, if CO_2 is being lost or escaping from the system, the reaction is driven toward the right to make more CO_2 (and $CaCO_3$) so the amount of reactants and products remains about the same.

FIGURE 6.15 (A) View of silica sinter terrace at Steamboat Hills near Reno, Nevada, as it looked in 2014. Water-vapor-emitting fissure strikes north and issued sinter-depositing hot water as recently as the mid-1980s. Note clipboard and hammer for scale. (Photograph by author.) (B) Close-up view of Steamboat silica sinter as it appeared in the mid-1980s when the terrace was still active and precipitating sinter. The red arrow points to a nickel for scale. Note the frothy texture of the silica sinter, reflecting deposition of silica from agitated, boiling fluids. (Photograph courtesy of D. Hudson.)

If fluid reservoir temperatures exceed 180°C, then silica sinter* typically deposited as hot water is discharged at the spring. This is because silica has prograde solubility, meaning that as temperature increases more silica can be dissolved in solution (Fournier, 1985). As the temperature cools, which occurs most dramatically at or near the surface, the dissolved silica precipitates. The white bluffs visible on the west side of U.S. Highway 395, about 15 km south of Reno, Nevada, are terraces of silica sinter of the power-producing Steamboat Springs geothermal field (Figure 6.15) (discussed further in Chapter 10).

* *Sinter* refers to silica deposited around the margins of high-temperature (boiling) hot springs.

Where temperatures are cooler, such as on the margins of the upflow zones or in recharge regions of downflowing fluids, wallrocks can be altered to a mixture of clays and very fine-grained micas, such as illite and sericite. This kind of altered rock, unlike propylitic and silicic altered zones, is relatively soft and has an overall whitish color that makes the rock conspicuous and thus an important exploration tool (discussed further in Chapter 8). The development of clays, such as kaolinite, or fine-grained micas is mainly due to hydrolytic alteration where hydrogen ions of the geothermal fluid replace cations of silicate minerals to produce clays or fine-grained micas. This is illustrated in the chemical reaction below, where K-feldspar is converted to kaolinite:

$$2KAlSi_3O_8(s) + 3H_2O(aq) \rightarrow Al_2Si_2O_5(OH)_4(s) + 4SiO_2(aq) + 2KOH(aq) \quad (6.7)$$

K-feldspar + Geothermal fluid → Kaolinite (clay) + Silica + Potassium hydroxide

At slightly higher temperatures (>~180°C), fine-grained micas, such as illite or sericite, would form instead of kaolinite.

High-Sulfidation Alteration

The high-sulfidation type of alteration is characterized by minerals stable at high temperatures (200° to 350°C) and under acidic conditions (pH typically <3, sometimes <1). Typical nonsulfide minerals present include alunite (hydrated potassium aluminum sulfate, $KAl_3(SO_4)_2(OH)_6$) and aluminous-rich clays, such as kaolinite and dickite, anhydrite (calcium sulfate, $CaSO_4$), and locally native sulfur.* Sulfide minerals can include pyrite and sulfur-rich, copper-bearing minerals such as enargite. As noted above, such zones are localized proximal to vent areas of active volcanoes, where magmatic gases, such as SO_2 and HCl, are high. As sulfur dioxide encounters near-surface groundwater, it disproportionates:

$$4SO_2 + 4H_2O \rightarrow 3H_2SO_4 \text{ (sulfuric acid)} + H_2S \quad (6.8)$$

The H_2S can further react with oxygen in the atmosphere or at shallow depths to also produce sulfuric acid:

$$H_2S + 2O_2 \rightarrow H_2SO_4 \quad (6.9)$$

Due to the acidic conditions, surrounding wallrock is commonly leached of cation components (K, Na, Ca, and Mg), leaving a silica residue commonly referred to as *vuggy quartz*. Unlike low-sulfidation alteration, where silica can be added to the rock, making for a hard and potentially impermeable rock (pore space filled with silica), vuggy quartz can be quite porous and permeable, leading to good fluid circulation.

* This assemblage of minerals formed under conditions of low pH and temperatures of 200°C to about 300°C is also referred to as *advanced argillic alteration*, as opposed to *argillic alteration*, which also includes clays but occurs under lower temperature and less acidic conditions.

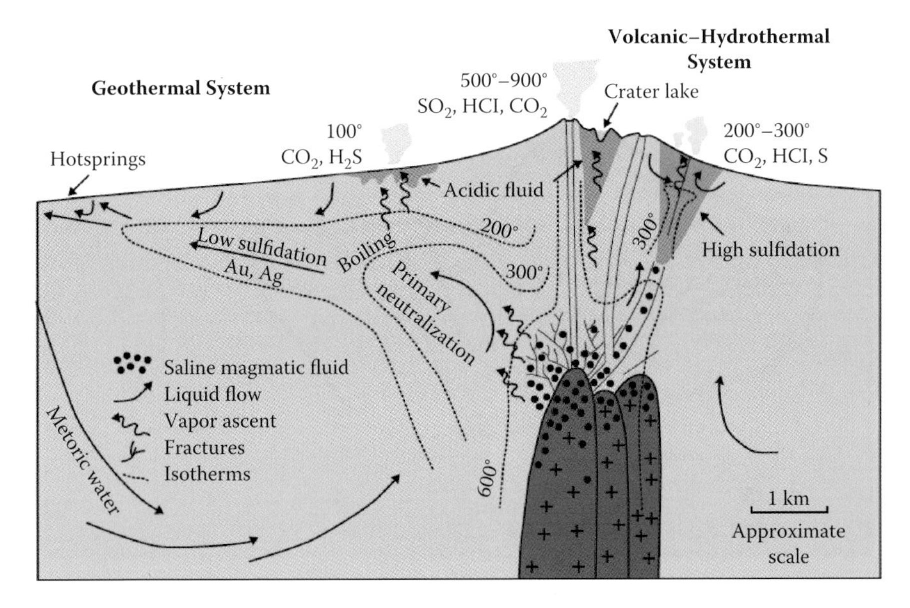

FIGURE 6.16 Schematic cross-section illustrating the lateral transition from high sulfidation alteration near the vent region of an active volcano (volcanic–hydrothermal systems) to low sulfidation alteration on the flanks of the volcano (geothermal system). The finger-shaped areas below the volcano vent/crater lake represent igneous intrusions that are the heat source for both the geothermal and volcanic–hydrothermal systems. Also depicted are regions of different styles of mineralization that form at different levels of depth in addition to the different forms of hydrothermal alteration, high sulfidation and low sulfidation, found at and near the surface. (Adapted from Hedenquist, J.W. et al., *Reviews in Economic Geology*, 13, 245–277, 2000.)

Although fluids associated with high-sulfidation alteration typically have high temperatures and enthalpy contents, they are not normally developed due to their corrosive nature. Nonetheless, the margins of such systems can transition into areas characterized by near-neutral-pH alkali-chloride solution and associated alteration (Figure 6.16). A good example of lateral transition from high-sulfidation alteration near the vent area of a volcano to marginal low-sulfidation alteration, where the bulk of geothermal production occurs, is the Tiwi geothermal field in the Philippines (Figure 6.17) (Moore et al., 2000).

Steam-Heated Acid-Sulfate and Bicarbonate Alteration

Areas of acid-sulfate and bicarbonate alteration form at and near the surface above zones of low-sulfidation propylitic and silicic alteration at deeper levels in response to boiling of near-neutral-pH alkali-chloride solutions. The rising steam condenses upon encountering near-surface groundwater that in turn is heated as the steam condenses (latent heat of condensation). To best understand this process, we need to examine briefly the nature of boiling.

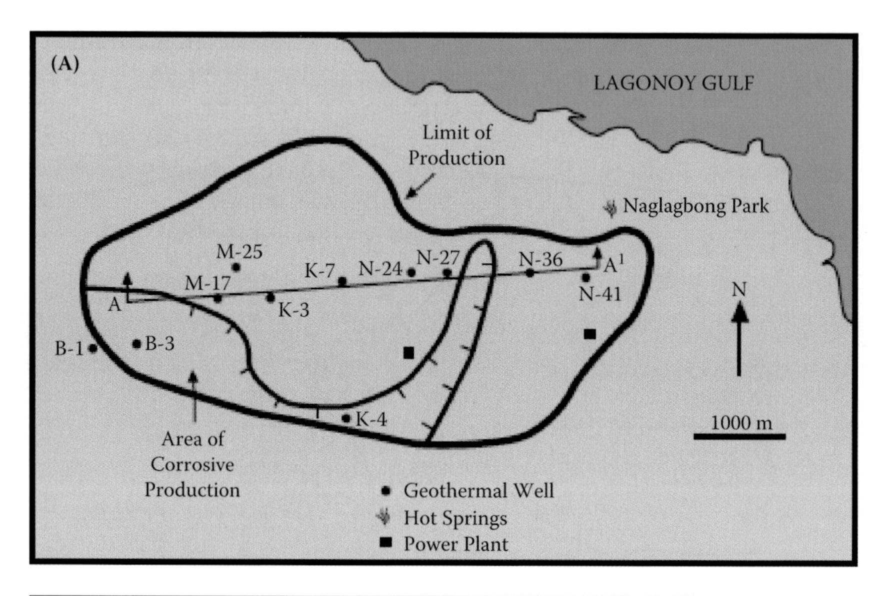

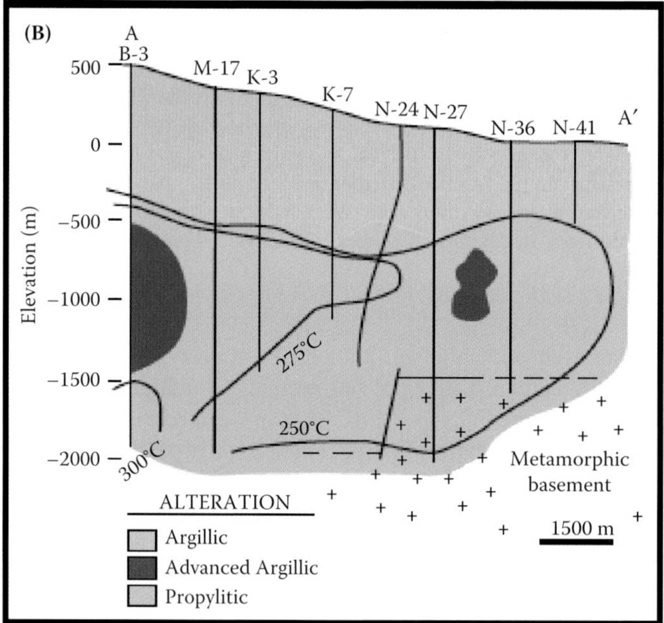

FIGURE 6.17 Part A is a plan map of the Tiwi geothermal field showing locations of geothermal wells and line of cross section A–A′ shown in B. Note that most of the wells are drilled in the low sulfidation region of propylitic alteration. The areas of high sulfidation alteration (denoted as advanced argillic) and highest geothermal temperatures lie mainly to the west of the outlined production field. Argillic stands for mainly clay alteration possibly developed as a steam-heated zone above boiling solutions in the propyltic zone (see discussion in text). (Adapted from Moore, J.N. et al., *Economic Geology*, 95(5), 1001–1023, 2000.)

As noted earlier, the boiling point of geothermal fluids is mainly a function of temperature and pressure (depth), which is illustrated in the boiling point-to-depth curve (Figure 6.10). Other factors that control the boiling point include the total dissolved solids (TDS) content and dissolved gases. TDS raise the temperature of the boiling point, while dissolved gases such as CO_2 and H_2S lower the boiling point (Figure 6.10). As fluids rise in a convecting geothermal reservoir, temperature changes little with depth (approximately isothermal) due to thermal mixing from circulation. When they rise high enough (and pressure is low enough) to intersect the boiling point-to-depth curve, the liquid will begin to boil. With further rise, their cooling follows the boiling point-to-depth curve (in this case, the temperature changes with two phases present because pressure is not fixed as shown in Figure 6.5). Because of this behavior, the depth of boiling can be determined if the temperature is known from an independent means.

An effective way to find the depth of boiling is to look for fluid inclusions trapped in minerals that were growing in the presence of the boiling solutions. Fluid inclusions are tiny pockets of fluid (1 to 100 microns or 1/1000 to 1/100 of a millimeter) trapped by the mineral as it was growing. If fluids are boiling, the liquid-to-vapor ratio of inclusions will be variable, as some inclusions will trap more liquid and other inclusions will trap more vapor. Because the liquid contracts upon cooling, the vapor volume of any given inclusion is larger than at the time of trapping or formation. If those inclusions are reheated in the laboratory, the liquid will expand until the temperature of original trapping is reached. This is called the *homogenization temperature*, and it reflects the actual temperature of the boiling fluids. Once the homogenization temperature is determined, the depth can be determined by reference to the boiling point-to-depth curve. Note that, if there is no evidence for boiling from the fluid inclusions or other lines of evidence (such as geologic reconstructions from cross-sections), then the pressure of formation, or depth, cannot be determined. By analyzing fluid inclusions from different minerals that formed at different times,[*] the temperature history of geothermal system can be disclosed to see if temperature has been stable, waxing, or waning.

If the boiling fluids contain a high CO_2/H_2S ratio, then the condensed steam produces bicarbonate (HCO_3^-)-rich springs that are mildly acidic (pH of ~5) due to the presence of weak carbonic acid (H_2CO_3). On the other hand, boiling fluids with higher H_2S content will yield acidic, sulfate-rich springs or mudpots when the water table is low. This is because H_2S oxidizes in air to produce sulfuric acid, resulting in fluids having a pH of ~2 or even lower. This acidic condensed steam typically alters the rock to soft clay, which serves as an impermeable cap to the underlying geothermal reservoir containing near-neutral-pH alkali-chloride waters. Indeed, the high-level, steam-heated acid sulfate alteration can resemble the high-temperature, high-sulfidation alteration produced by volcanic degassing. In the latter, however, the alteration mineralogy is consistent with a lower pH (<1 to 2) and higher temperatures, and includes clay such as dickite (a polymorph of kaolinite), hydrated

[*] Relative age relationships are determined from geologic observations, including replacement textures or different veins of fluid inclusion-bearing minerals cross-cutting each other.

aluminum oxide minerals such as diaspore, and visible or coarser grained sulfate minerals such as alunite, anhydrite (calcium sulfate, $CaSO_4$), or barite (barium sulfate, $BaSO_4$).

Alteration Associated with Hypersaline Brine Systems

Alteration associated with hypersaline brine systems, such as the geothermal fields in the Salton Trough in southern California, is of the low-sulfidation type. Rocks in the geothermal reservoirs there are propylitically altered and contain widespread epidote and chlorite (McKibben et al., 1988a). In this system, veins of epidote (along with quartz and pyrite) change mineralogy as a function of depth and temperature. Calcite occurs with epidote above depths of 2000 m, alkali feldspar (adularia) at depths between 1700 and 2745 m, and actinolite below 2890 m (Caruso et al., 1988). At the deepest levels, depths of 3000 to 3180 m, hydrothermal pyroxene occurs (Cho et al., 1988). Partial dissolution of evaporite beds (gypsum plus halite) at depths of 1 to 3 km has created local solution collapse of interbedded shale layers (McKibben et al., 1988b). Such volume contraction can lead to increased permeability, at least locally, and improved fluid circulation. Most hypersaline brine systems have limited or no surface expression in part due to the dense, saline-rich nature of the geothermal fluid, making it difficult for them to reach the surface. In the Salton Trough region, the geothermal reservoirs are also overlain by impermeable cap rocks consisting of shales and evaporite deposits and deltaic sandstones. The precipitation of anhydrite and calcite from downward-percolating water clogged pores in the deltaic sandstone and reduced permeability (Moore and Adams, 1988). Where fluids do reach the surface, through localized fractures in the otherwise impermeable cap rocks, clay-rich mudpots form locally, reflecting both the low volume of flow and acidic nature of high-level geothermal fluids. The acidic character of these high-level geothermal fluids is probably due to boiling of solutions at depth giving rise to ascending H_2S, which oxidizes to form sulfuric acid (steam-heated alteration) as described by Reaction 6.8 above.

VAPOR-DOMINATED GEOTHERMAL SYSTEMS

Vapor-dominated geothermal systems are the gems of the geothermal world because the enthalpy or heat energy used to produce electricity is not partitioned between liquid and vapor; all fluid mass is in the form of dry steam and goes to the turbine. Unfortunately, the systems are the rarest geologically because they require very special geologic conditions to develop (as described in the section below). The two largest developed vapor-dominated systems on the planet are The Geysers in northern California and Larderello in Italy. These two systems together produce about 1700 MWe, or about 14% of the world's geothermal power capacity as of 2015. About 10 years ago, these two vapor-dominated systems comprised about 25% of the world's geothermal electrical power capacity, but the percentage has declined mainly due to new flash and binary systems coming online and the relatively static but sustainable production at The Geysers and Larderello.

Formation and Rarity

Large vapor-dominated systems, such as The Geysers, are geologically rare for several reasons. First, these systems typically begin as liquid-dominated systems and evolve into vapor-dominated systems. For this to happen, a potent heat source and low rates of recharge are required; as the liquid boils, a steam cap develops over time (on the order of tens of thousands of years). Second, in order for the steam cap to grow, some leakage of steam to the surface must occur. Without leakage, the system would pressurize to the point that boiling would stop, and a steady-state condition would develop consisting of a stagnant steam cap overlying a nonboiling liquid zone. With some surface leakage, however, the steam cap grows as the boiling liquid zone lowers due to evaporation. Most of the upflow steam probably condenses on the margins of the steam reservoir and moves downward, allowing for the coexistence of minor liquid with vapor in the reservoir. At The Geysers, some wells are almost 4 km deep and have yet to encounter the inferred underlying boiling liquid zone from which the steam cap formed over time (Donnelly-Nolan et al., 1993; M. Walters, pers. comm., 2015).

Exploration and development of these systems have revealed a relatively uniform temperature of about 240°C and a pressure of about 35 kg/cm^2, or about 34 bars. As a result, for systems occurring at depths greater than 300 m, the steam reservoir pressure is less than hydrostatic and becomes increasingly more so with depth. For this to occur, the steam reservoir must be basically isolated from surrounding water-saturated rock by a more or less impermeable zone. Otherwise, water would flow into the reservoir under the hydrostatic gradient at a rate exceeding the rate of steam discharge and effectively collapse the underpressured (vapor static or less than hydrostatic) steam reservoir. Formation of large vapor-dominated reservoirs therefore requires a special orchestration of geological processes, including the necessary imbalance between rates of discharge and recharge, more or less hydrologic reservoir isolation, and a potent underlying heat source for a steam zone to grow over time; such constraints are rarely achieved geologically.

Surface Wallrock Alteration

Surface rock alteration above a large vapor-dominated geothermal reservoir is similar to that found in local shallow steam-heated zones above a boiling liquid-dominated reservoir consisting of near-neutral-pH alkali-chloride water. In both cases, fumaroles (gas vents), mudpots, mud volcanoes, collapse craters, turbid pools, and acid-leached ground are found. Springs are typically low in Cl content because Cl is not a particularly volatile element and stays with the brine, similar to evaporation of seawater. However, sulfate contents are relatively high and springs are generally acidic, reflecting the oxidation of H_2S to produce sulfuric acid (as shown in Reaction 6.8). Some mildly acidic springs (pH of ~5 to 6 due to carbonic acid) may be found in areas with abundant escaping CO_2 producing bicarbonate-rich springs. The difference in distinguishing between hydrothermally altered ground above a vapor-dominated zone and high-level acidic steam-heated alteration above

FIGURE 6.18 View of mudpot (in foreground) and surrounding steaming acid-leached ground in a portion of The Geysers geothermal field, California. Mudpot measures about 1.5 m across and reflects local or perched steam-heated groundwater made acidic by rising H_2S that oxidizes near the surface to form sulfuric acid. As a result, the rock is turned to clay and mud. Unlike in liquid-dominated geothermal systems, Cl-bearing springs and siliceous sinter are largely absent above The Geysers vapor-dominated reservoir. Condensed water vapor plume in the distance comes from cooling towers, hidden by the trees, of a geothermal power plant. (Photograph by author.)

a boiling liquid-dominated reservoir would include the areal coverage of the two types of systems. In the former, the acid-leached ground and associated Cl-poor, sulfate-rich springs or mudpots and steaming collapse pits would be quite widespread. Areas of Cl-rich pools or siliceous sinter or silicified rock, however, which require deposition from liquids, are absent or minor (Figure 6.18). Above liquid-dominated systems, on the other hand, acidic steam-heated zones are more localized, and Cl-bearing hot springs and siliceous sinter or travertine will be more widespread.

Artificially Produced Vapor-Dominated Systems

In some, hot (>200°C), electricity-producing liquid-dominated geothermal systems, withdrawal of the liquid has exceeded recharge and the geothermal water table has dropped. As a result, hydrostatic head is reduced, which lowers hydrostatic pressure, fostering boiling and production of steam in thermally robust systems. In such cases, a steam cap can form as production continues, and wells may transition with time from liquid, to liquid-plus-vapor, to eventually dry steam. Two examples of this phenomenon are Matsukawa, Japan, and Wairakei, New Zealand. As noted previously, the latter represents the world's first commercial development, beginning in 1958, of a liquid-dominated system for geothermal electrical power production. As might be expected, reservoir conditions have changed due to the more than 50 years of production. Some of the shallow wells (<500 m deep) serving the aging Wairakei power station now provide dry steam from the steam cap. The neighboring 55-MWe Poihipi power station is largely fed by dry steam wells tapping the steam cap of the overall liquid-dominated system. A new power facility at Wairakei, Te Mihi, was commissioned in 2014, and is discussed below.

Matsukawa Geothermal System, Japan

The Matsukawa geothermal power plant (23.5 MWe) began production in 1966 to provide power for a metal fabricating facility. Since 1991, when metal fabricating ceased, electrical power has been sold to the local utility company (Hanano, 2003). For the first 6 months of production, wells produced liquid or a liquid–vapor mixture. After initial production and drawdown of the geothermal water table, some of the wells produced dry steam as the steam cap expanded to greater depth. The growth of the vapor cap occurred because of an existing impermeable clay cap formed by steam-heated acid-sulfate alteration of shallow rocks overlying the boiling liquid-dominated reservoir (Figure 6.19). Low-permeability barriers, probably consisting of intrusive dikes, also exist along the sides of the reservoir, restricting liquid recharge as production proceeded. At the onset of production, the reservoir had a pressure of about 85 bars at 1000 m, which dropped quickly to about 25 to 30 bars as the liquid was replaced by steam. Since the early drawdown, the decline in pressure has been very slow, but the steam production per well has dropped, requiring more production wells to be drilled to maintain the current power output of about 23 MWe. In the early and mid-1970s, six wells accounted for about 23 MWe. From 2000 to 2005, however, eight new production wells were added in order to maintain the same output. During that 25- to 30-year period, the average steam production per well decreased by about half, from 38 tonnes/hr (10.6 kg/s) to about 18 tonnes/hr (5 kg/s). Reinjection of condensate and river water did not begin until 1988, but it has helped stabilize production and steam pressure. The steam reservoir after 40 years of production has expanded from a base depth of about 700 to 900 m to more than about 2000 m (Hanano, 2003).

Te Mihi Geothermal Field, Wairakei, New Zealand

Te Mihi is the newest geothermal power plant (166 MWe operating capacity) in the long history of electrical power production (since 1958) at the Wairakei geothermal field (Figure 6.20). The Te Mihi power plant began operation in 2014. It is planned that over about a 12-year period power from Te Mihi will supplant production from the aging 157-MWe Wairakei geothermal power plant, located about 5 km to the east of Te Mihi. The new power plant uses dual-flash technology and produces about 25% more power for the same amount of fluid as achieved at the aging Wairakei power station (Peltier, 2013).

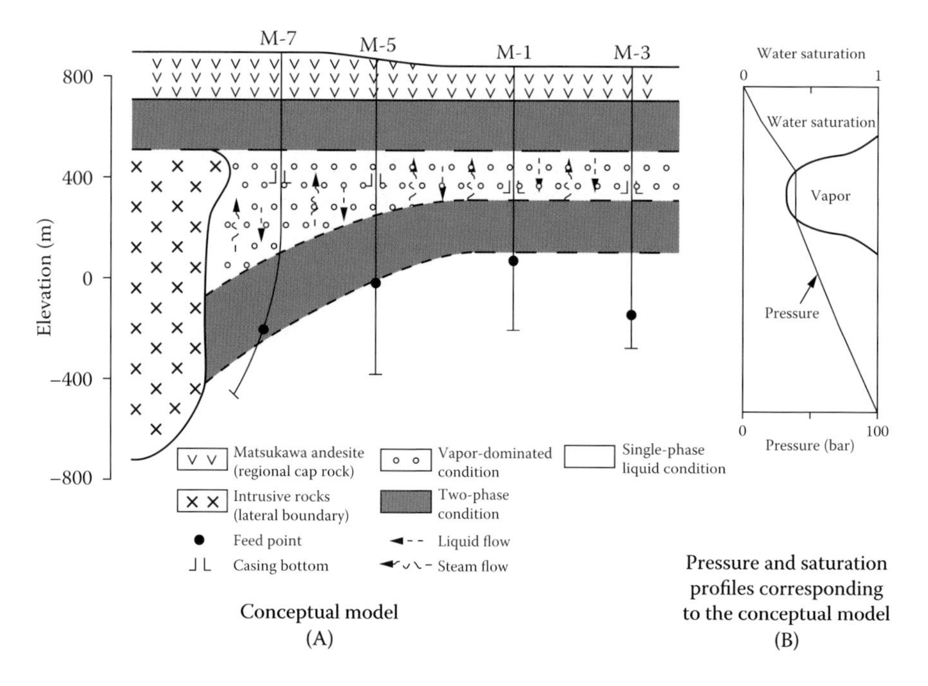

FIGURE 6.19 (A) Conceptual model of initial subsurface conditions of the Matsukawa geothermal field. Upper two-phase region corresponds to the steam condensate zone, where both vapor and liquid coexist below the clay-altered andesite cap rock. The lower two-phase region represents the boiling zone of the underlying single-phase liquid region. The vapor-dominated zone develops and expands as the low-density steam collects below the impermeable cap and steam condensate zone. The feed point marks the approximate bottom of the reservoir, where significant vapor and/or liquid enter the wellbore at depth. (B) Graph of the change in pressure and water saturation with depth with respect to the conceptual model. (From Hanano, M., *Geothermics*, 32(3), 311–324, 2003.)

Te Mihi derives its power from two geothermal fields. About 40% of the power comes from the western borefield, consisting of 30 wells that tap a liquid-dominated reservoir at an average depth of about 600 m. The remaining 60% comes from the hybrid or stacked borefield that taps a shallow vapor-dominated zone at depths of 300 to 500 m, with a pressure of about 10 bars, and a deeper liquid-dominated zone at depths of as much as 2500 m. Temperatures for both zones are about 240° to 250°C, but recall that, although the temperature is the same, the enthalpy from the steam is much higher, as we saw in Figures 6.8 and 6.9.

Wells originally drilled in the Te Mihi area during the 1960s were primarily liquid dominated, but when drilling recommenced in the mid-1980s a shallow vapor-dominated zone had developed due to a drop in reservoir pressure from fluid withdrawal of earlier drilled production wells. The shallow steam zone is being developed first and then will be augmented by production from the deeper liquid-dominated zone that will provide two-phase fluid resulting in long-term sustained production. As at Matsukawa, the upper steam zone developed below an impermeable cap of clay-altered volcanic rock formed from high-level, steam-heated, acid-sulfate alteration. Unlike at

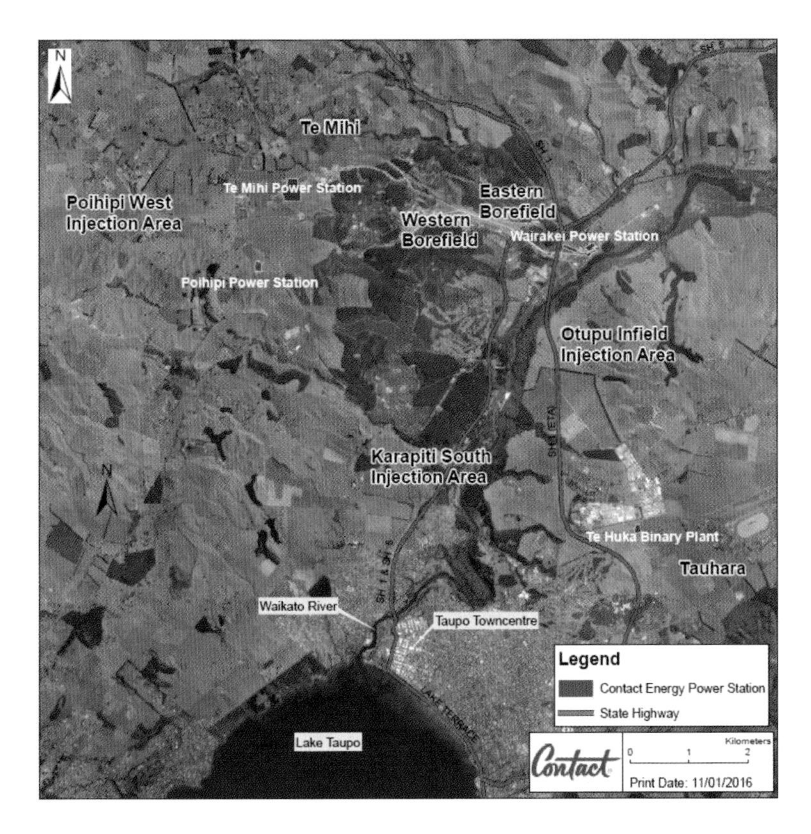

FIGURE 6.20 Aerial view of the Wairakei and neighboring Tauhara geothermal fields, New Zealand. The locations of geothermal power facilities in both fields are shown. (Annotated image courtesy of T. Montegue, Contact Energy, Wellington, New Zealand.)

The Geysers, however, where the inferred deep boiling brine has yet to be intercepted by drilling, Te Mihi will be utilizing the deep brine to compensate for anticipated pressure declines during the course of production in the upper vapor-dominated zone.

SUMMARY

The unique chemical properties of water are a major driving force behind harnessing geothermal energy. These facets include the following:

1. Its mobility as a liquid or gas allows the transfer of energy from hot rocks to the surface.
2. Its high heat capacity or enthalpy to do work means it can spin a turbine to generate electricity.
3. Its polar nature yields a high boiling point for its molecular weight. Without this property, water would occur mainly as a gas and, because of the higher mobility of a gas, most of it would escape to the atmosphere where it could not be heated by Earth's internal heat.

4. Its bent molecular shape causes water to expand when frozen, which means the dominant phase in the Earth is a liquid as pressure increases, not an immobile solid.

The two main types of producing geothermal systems are liquid and vapor dominated, with the former being the far more common. The liquid-dominated systems can be divided into three main types based on fluid chemistry:

1. Near-neutral-pH alkali-chloride solutions (low-sulfidation systems) typically located distally from volcanic vents or underlying magmatic intrusions
2. Acid-sulfate solutions (high-sulfidation systems) often located near volcanic vents due to degassing of acidic gases (HCl and SO_2) from magma at depth
3. Brine-rich systems, such as found in the Salton Sea region of southeastern California, where hot geothermal fluids have encountered and leached salt-rich evaporite deposits at depth

These different fluid types lead to characteristic styles of wallrock alteration that can help exploration geologists locate and distinguish the various liquid-dominated systems. The near-neutral-pH alkali-chloride solutions form alteration of adjacent wallrocks consistent with low-sulfidation conditions. In the main geothermal reservoir, chlorite, epidote, calcite, and possible adularia (propylitic mineral assemblage) are typical. If temperatures are >300°C, the propylitic assemblage may also include actinolite, diopside, biotite, and garnet. Illite, sericite, and clay (depending on temperature and pH) can form in the recharge zones and in the upper, capping parts of such geothermal systems. Silicification of wallrocks can form at the geothermal water table with silica sinter deposited in boiling hot springs. Low-sulfidation alteration can form on the distal flanks of a stratovolcano, such as Tiwi in the Philippines, or in extensional basins with or without magmatic sources of heat, such as Steamboat Springs, Nevada (recent rhyolitic volcanism) and Dixie Valley, Nevada (no recent shallow magmatism).

Acidic sulfate-rich fluids lead to high-sulfidation alteration, typified by aluminous-rich clays (kaolinite and dickite), residual silica or vuggy quartz due to acid attack and leaching of rock, and general absence of siliceous sinter at hot springs. Other minerals that can occur are alunite and anhydrite. Fumaroles or gas vents with native sulfur and clay-rich mudpots are not uncommon at the surface for these fluids. High-sulfidation alteration is similar to what can develop at the top of a boiling low-sulfidation system that forms an acid-steam-heated zone at or near the surface. In this case, however, H_2S is liberated from the boiling near-neutral-pH alkali-chloride fluids and oxidizes to form sulfuric acid that attacks overlying rocks. Steam-heated altered rock is typified by kaolinite, but usually without the higher temperature-formed dickite or coarse-grained alunite and possible anhydrite as commonly found in primary high-sulfidation alteration.

Alteration associated with hypersaline brine systems is similar to that of the low-sulfidation type dominated by propylitic alteration in the vicinity of the geothermal reservoir. These hypersaline brine systems have little or no surface expression because their dense, saline-rich nature can make them too heavy to reach the surface easily. Where fluids do reach the surface, clay-rich mudpots typically form. This reflects both the low volume of flow and modestly acidic nature of fluids due to the dissolution of gypsum, the principal mineral of evaporite deposits. Gypsum dissolves in water to form both sulfuric acid and a base calcium hydroxide in about equal concentrations, but because the sulfuric acid is a stronger acid than calcium hydroxide is a base, the resulting solution is acidic.

Large vapor-dominated systems, such as The Geysers in California and Larderello in Italy, are a geological rarity because special geologic conditions are required, including a long-lived and potent heat source and just the right conditions of permeability to allow some escape of steam at the top. Permeability must be limited also along the sides of the reservoir to restrict recharge to prevent collapse of the steam reservoir that is underpressurized compared to hydrostatic conditions of surrounding rocks. Some escape of steam above the reservoir is required so the steam zone can grow over time; otherwise, the pressure would build to stymie further boiling of the underlying liquid-dominated zone. Production-induced steam caps can form atop liquid-dominated systems where production is not balanced by reinjection as has occurred at Wairakei in New Zealand and Matsukawa, Japan. Wallrock alteration above vapor-dominated reservoirs is generally similar to steam-heated alteration above a boiling liquid-dominated reservoir. The exception is that, in the former, Cl-bearing pools or hot springs depositing siliceous sinter are generally lacking but can occur locally in the latter.

SUGGESTED PROBLEMS

1. How do the chemical properties of water make it the universal solvent and an important agent for hydrothermal alteration?
2. Looking at the figure on the next page (Lagat, 2009), explain the significance of garnet for interpreting the temperature history of the well at this depth? Does this same inferred temperature history apply for the upper part of the well? Why or why not?
3. A geothermal well is tapping a liquid-dominated reservoir at a temperature of 235°C at a depth of 800 m. Assuming a hydrostatic pressure gradient and dilute TDS:
   a. Is the fluid boiling at the bottom of the well? Why or why not? If not, at what pressure will it begin to boil or flash?
   b. Using Figure 6.8 and Equation 6.2, calculate the power output from the well if the turbine inlet temperature is 235°C, the exit temperature is 50°C, and the mass flow rate is 5 kg/s. Recall that power equals the change in enthalpy Assume an ideal expansion process across the turbine and a turbine efficiency of 85%.
4. Imagine you are monitoring the progress of a geothermal drill rig. At a depth of 300 m, you encounter a mixture of liquid and vapor. At a depth of 400 to 600 m, only steam is encountered. At depths greater than 600 m, a liquid-only zone is intercepted. How do you interpret such a transition, and what kinds of hydrothermal features and wallrock alteration might you find at the surface?
5. In developing a geothermal resource, is it advisable to draw down the water table to generate steam caps? Consider the advantages and disadvantages of doing such in your response.

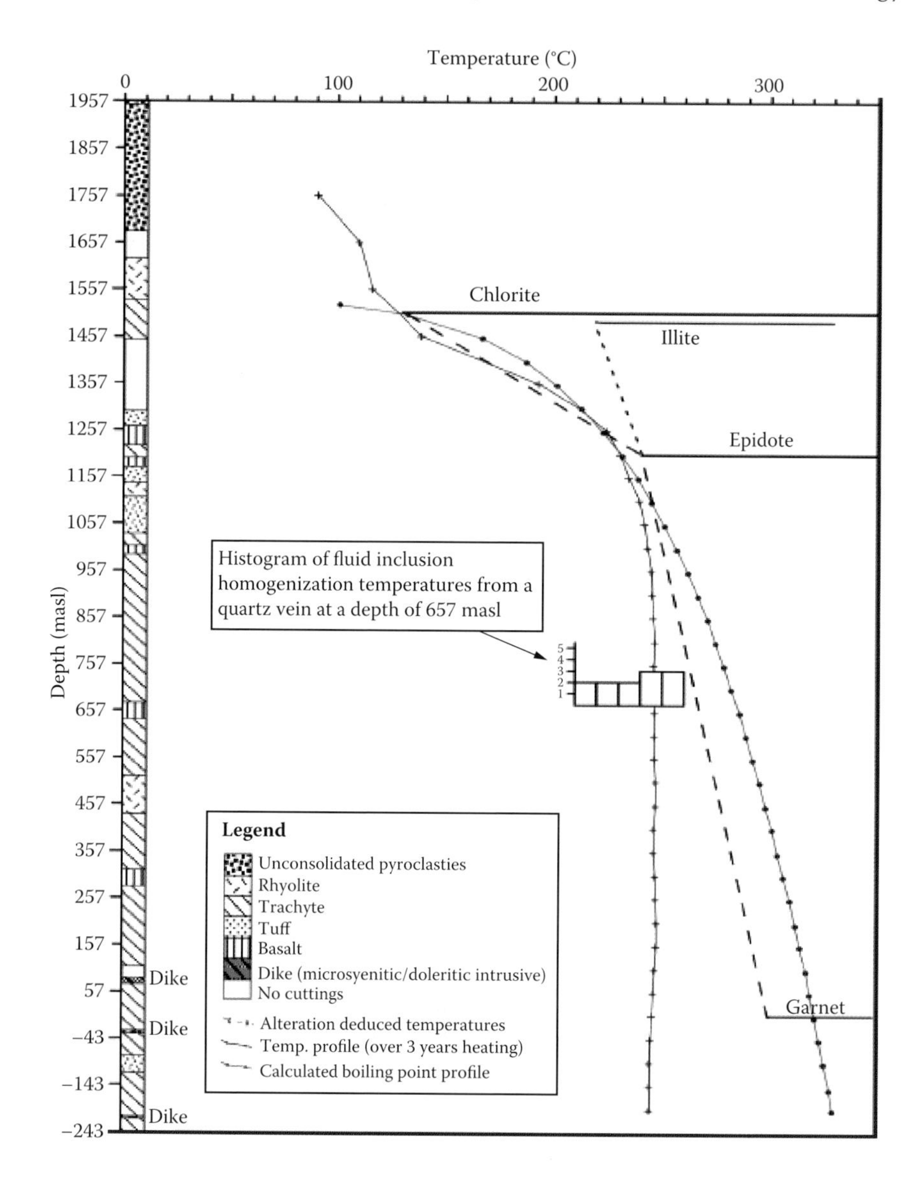

Figure for Problem 2.

REFERENCES AND RECOMMENDED READING

Browne, P.R.L. (1978). Hydrothermal alteration in active geothermal fields. *Annual Review of Earth and Planetary Sciences*, 6(1): 229–248.

Browne, P.R.L. and Ellis, A.J. (1970). The Ohaki-Broadlands hydrothermal area, New Zealand; mineralogy and related geochemistry. *American Journal of Science*, 269(2): 97–131.

Caruso, L.J., Bird, D.K., Cho, M., and Liou, J.G. (1988). Epidote-bearing veins in the State 2-14 drill hole: implications for hydrothermal fluid composition. *Journal of Geophysical Research*, 93(B11): 13123–13133.

Cho, M., Liou, J.G., and Bird, D.K. (1988). Prograde phase relations in the State 2-14 well metasandstones, Salton Sea geothermal field, California. *Journal of Geophysical Research*, 93(B11): 13,081–13,103.

Deely, J.M. and Sheppard, D.S. (1996). Whangaehu River, New Zealand; geochemistry of a river discharging from an active crater lake. *Applied Geochemistry*, 11(3): 447–460.

DiPippo, R. (2012). *Geothermal Power Plants: Principles, Applications, Case Studies, and Environmental Impacts*, 3rd ed. Waltham, MA: Butterworth-Heinemann.

Donnelly-Nolan, J.M., Burns, M.G., Goff, F.E., Peters, E.K., and Thompson., J.M. (1993). The Geysers–Clear Lake area, California: thermal waters, mineralization, volcanism, and geothermal potential. *Economic Geology*, 88(2): 301–316.

Fournier, R.O. (1985). The behavior of silica in hydrothermal solutions. *Reviews in Economic Geology*, 2: 45–61.

Fridleifsson, G.O. and Elders, W.A. (2005). The Iceland Deep Drilling Project: a search for deep unconventional geothermal resources. *Geothermics*, 34(3): 269–285.

Friethleifsson, G.O., Elders, W.A., and Albertsson, A. (2014). The concept of the Iceland Deep Drilling Project. *Geothermics*, 49: 2–8.

Fridleifsson, G.O., Palsson, B., Albertsson, A.L., Stefansson, B., Gunnlaugsson, E., Ketilsson, J., and Gislason, I. (2015). IDDP-1 Drilled into Magma—World's First Magma-EGS System Created, paper presented at World Geothermal Conference 2015, Melbourne, Australia, April 19–24 (http://www.geothermal-energy.org/pdf/IGAstandard/WGC/2015/37001.pdf).

Glassley, W.E. (2015). *Geothermal Energy: Renewable Energy and the Environment*, 2nd ed. Boca Raton, FL: CRC Press.

Haas, Jr., J.L. (1971). The effect of salinity on the maximum thermal gradient of a hydrothermal system at hydrostatic pressure. *Economic Geology*, 66(6): 940–946.

Hanano, M. (2003). Sustainable steam production in the Matsukawa geothermal field, Japan. *Geothermics*, 32(3): 311–324.

Hedenquist, J.W., Simmons, S.F., Giggenbach, W.F., and Eldridge, C.S. (1993). White Island, New Zealand, volcanic-hydrothermal system represents the geochemical environment of high-sulfidation Cu and Au ore deposition. *Geology*, 2 (8): 731–734.

Hedenquist, J.W., Arribas, A., and Gonzalez-Urien, E. (2000). Exploration for epithermal gold deposits. *Reviews in Economic Geology*, 13: 245–277.

Henley, R.W. and Ellis, A.J. (1983). Geothermal systems ancient and modern: a geochemical review. *Earth-Science Reviews*, 19(1): 1–50.

Lagat, J. (2009). Hydrothermal Alteration Mineralogy in Geothermal Fields with Case Examples from Olkaria Domes Geothermal Field, Kenya, paper presented at Short Course IV on Exploration for Geothermal Resources, Lake Naivasha, Kenya, November 1–22 (http://www.os.is/gogn/unu-gtp-sc/UNU-GTP-SC-10-0102.pdf).

McKibben, M.A., Andes, Jr., J.P., and Williams, A.E. (1988a). Active ore formation at a brine interface in metamorphosed deltaic lacustrine sediments; the Salton Sea geothermal system, California. *Economic Geology*, 83(3): 511–523.

McKibben, M.A., Williams, A.E., and Okubo, S. (1988b). Metamorphosed Plio-Pleistocene evaporites and the origins of hypersaline brines in the Salton Sea geothermal system, California: fluid inclusion evidence. *Geochimica et Cosmochimica Acta*, 52(5): 1047–1056.

Moore, J.N. and Adams, M.C. (1988). Evolution of the thermal cap in two wells from the Salton Sea geothermal system, California. *Geothermics*, 17(5–6): 695–710.

Moore, J.N., Powell, T.S., Heizler, M.T., and Norman, D.I. (2000). Mineralization and hydrothermal history of the Tiwi geothermal system, Philippines. *Economic Geology*, 95(5): 1001–1023.

Peltier, D. (2013). Contact Energy Ltd.'s Te Mihi power station harnesses sustainable geothermal energy. *Power Magazine*, 8: 38–42 (http://www.slthermal.com/3_Recognition/Te%20Mihi%20Power%20Magazine%20Marmaduke%20Award%20Aug%202013.pdf).

Purnomo, B.J. and Pichler, T. (2014). Geothermal systems on the Island of Java, Indonesia. *Journal of Volcanology and Geothermal Research*, 285: 47–59.

Simmons, S.F. and Browne, P.R.L. (2000). Hydrothermal minerals and precious metals in the Broadlands–Ohaaki geothermal system: implications for understanding low-sulfidation epithermal environments. *Economic Geology*, 95(5): 971–999.

Wood, C.P. (1994). Mineralogy at the magma-hydrothermal system interface in andesite volcanoes, New Zealand. *Geology (Boulder)*, 22(1): 75–78.

7 Geologic and Tectonic Settings of Select Geothermal Systems

KEY CHAPTER OBJECTIVES

- Explain the differences between magmatic and amagmatic geothermal systems and their implications for energy development.
- Infer the types of geothermal resources, associated power plants, and overall geothermal energy potential for different tectonic settings.

This chapter focuses on select geothermal systems and their relationships to the tectonic environment in which they formed. An overview of two main geologic types of geothermal systems—magmatic and amagmatic—and the geologic and geochemical characteristics of each begins the chapter. The tectonic settings and geologic types of geothermal systems exert a clear influence on the physicochemical characteristics of geothermal resources and on how those resources might ultimately be developed. The discussion here generally follows earlier ideas of occurrence models of geothermal systems proposed by previous workers (Erdlac et al., 2008; Muffler, 1976; Walker et al., 2005) and recently extended to characterizing geothermal resources as play concepts similar to the oil industry (Moeck, 2014). The effort here is to move toward a classification system that removes arbitrary boundaries, such as temperature or enthalpy to distinguish geothermal systems, and focuses on fundamental geological parameters that dictate the characteristics of geothermal resources and how those resources might best be found and developed.

MAGMATIC AND AMAGMATIC GEOTHERMAL SYSTEMS

Magmatic geothermal systems are heated directly, largely from conduction, by a reservoir of molten rock or magma that underlies the geothermal resource or reservoir. Amagmatic geothermal systems, on the other hand, have no obvious connections to a magmatic heat source and typically occur in regions where the crust is thinned from extension, which results in elevated geothermal gradients (>50°C/km) and high heat flow (80 to >120 mW/m^2). In amagmatic systems, deeply circulating fluids are heated and rise buoyantly to feed potential geothermal reservoirs at modest depths (generally <3 km) with modest temperatures (mainly 150° to 200°C but locally as high as 250°C, such as at Dixie Valley, Nevada).

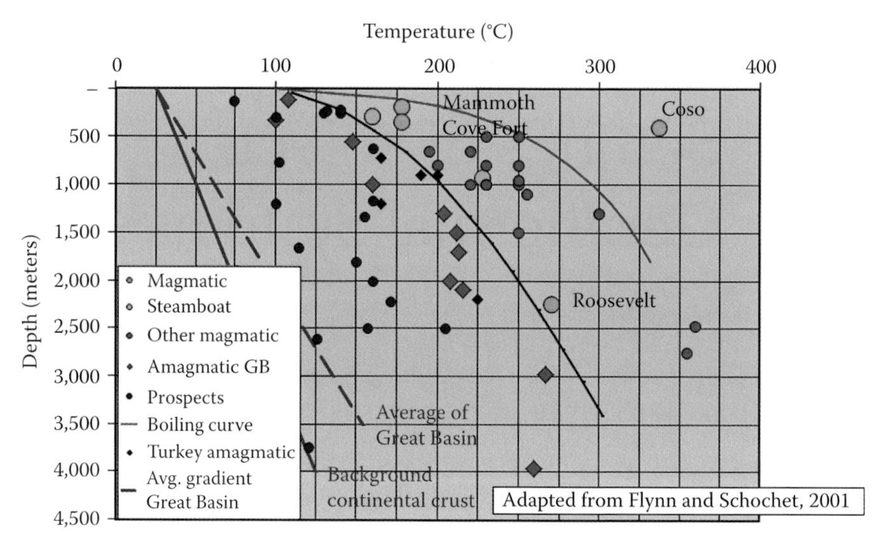

FIGURE 7.1 Temperature with depth graph of select geothermal reservoirs. The black curve in the middle of the graph is an empirical division between magmatic systems to the right and amagmatic systems to the left. Note that the amagmatic systems require drilling to a greater depth to get the same temperature of a magmatic system. The red curve is the boiling point with depth curve, indicating that the fluids in most reservoirs are not boiling, except for Coso. If boiling continues at Coso, a vapor-dominated system could develop if the correct balance between recharge and discharge occurs. (Figure courtesy of M. Coolbaugh.)

In most magmatically heated geothermal systems, fluids are hotter for a given depth compared to those in amagmatic geothermal systems, so generally wells need not be drilled as deep to obtain a given temperature (Figure 7.1). This obviously can save costs on drilling. Moreover, most magmatically heated geothermal systems support flash plants. These plants deliver more power per unit fluid mass, reflecting the typical higher temperature and enthalpy of tapped fluids, than can be accomplished from binary power plants. The majority of amagmatic systems are developed using binary geothermal plants due to their improved efficiency in generating power from geothermal fluids having lower enthalpy than those utilized in flash power plants.

Magmatic Geothermal Systems

Magmatically heated geothermal systems most commonly form in areas of active or young volcanism. Volcanic rocks in the vicinity of the geothermal system are less than ~1.5 million years old (and commonly <0.5 million years old) and are fairly widespread. Volcanic rocks are usually andesitic to rhyolitic in composition, but basalt can be a dominant rock type depending on the tectonic setting, as discussed below. At least portions of magmatic geothermal systems have unique geochemical signatures reflecting degassing of magmatic volatiles, including detectable sulfur dioxide (SO_2) or elevated bisulfate ions (HSO_4^-) in thermal waters and potentially significant concentrations of carbon dioxide (CO_2) that can kill or stunt trees. This has occurred at Steamboat Springs, near Reno, Nevada, and at Mammoth in east-central California. Furthermore, due to the presence of magmatic volatiles, including HF and HCl, portions of many magmatic systems can be quite acidic, making development in such affected areas difficult due to corrosion of equipment. For example, the northwest part of The Geysers steam field,* although hotter than the rest of the field (280° to 400°C compared to about 240°C), has thwarted conventional development, in part due to the corrosive, acidic nature of the steam in this area (Lutz et al., 2012; Rutqvist et al., 2013; Walters et al., 1992). Indeed, all known naturally produced vapor-dominated systems are magmatic systems because a potent, long-lived heat source is required to boil the water for a long enough time (probably at least several tens of thousands of years) to form an extensive steam cap (Hulen et al., 1997). Such a process may be occurring now at the Coso geothermal field in eastern California, if the cap rock can leak some steam and the side boundaries to the reservoir restrict recharge to sustain steam static conditions. This is because the temperature of the Coso reservoir is above the boiling point with depth curve (Figure 7.1).

Liquid-dominated systems, however, can also be magmatically heated. For example, the potent geothermal systems of the Taupo Volcanic Zone in New Zealand and volcanically active divergent rift zones of Iceland are liquid dominated and provide about 15% and 30% of their country's produced power, respectively. Another typical characteristic of magmatically heated geothermal systems is that their fluids have a high $^3He/^4He$ ratio (Kennedy and van Soest, 2007). The ^3He isotope is primordial, being left over from the formation of the Earth, and is stored largely in the mantle. The ^4He isotope, however, forms from the decay of uranium (U) and thorium (Th), both of which are enriched in the crust. As U and Th decay, the amount of produced radiogenic ^4He in the crust increases with time, whereas ^3He slowly decreases with time, mainly from magma degassing. Therefore, magmas that form by partial melting of mantle rocks will have a relatively high $^3He/^4He$ ratio, which is passed on to the geothermal fluids. In certain cases, however, if faults penetrate deep enough to interact directly or indirectly with the mantle, high $^3He/^4He$ ratios can occur without magmatism, as illustrated in deep oil wells located along a deeply seated fault zone in the Los Angeles Basin (Boles et al., 2015).

AMAGMATIC GEOTHERMAL SYSTEMS

Amagmatic systems form in areas where young volcanic rocks or active volcanism are lacking. As noted above, they are largely restricted to regions undergoing active crustal extension, which thins the crust and brings hot mantle rocks closer to the surface, resulting in a high geothermal gradient and elevated heat flow. Consequently, amagmatic systems are also referred to as *extensional geothermal systems*, which represent many, but not all, developed geothermal systems in Nevada. Moreover, rocks are weakest under tensional stress, which causes them to break and produce normal faults

* The Northwest Geysers is now part of an EGS demonstration project in which treated wastewater from the town of Santa Rosa is being reinjected into an existing well. As a result, noncondensable gas concentrations in steam from neighboring potential production wells have been reduced by about 90%, in turn reducing steam acidity.

that can form pathways for fluids to circulate deeply. This brings cool water to depth, where it is heated and then buoyantly rises toward the surface. Fluids associated with amagmatic systems are typically of the neutral-pH alkali-chloride type, unless boiled to produce high-level (near surface) acid-sulfate solutions. Fluids usually have low to modest total dissolved solids (TDS), commonly less than 5000 ppm. Because fluids of amagmatic systems are heated in and circulate through the crust, they typically have low $^{3}He/^{4}He$ ratios, reflecting the high radiogenic ^{4}He content in crustal rocks. All developed amagmatic geothermal systems are liquid dominated, generally reflecting the lower temperatures of these systems compared to magmatically heated systems.

Exploration and Production Implications of Magmatic and Amagmatic Systems

Magmatically heated geothermal systems commonly support flash- or dry-steam geothermal power plants. However, in places where production taps outflow rather than the main upwelling zones, binary geothermal plants are utilized such as at Casa Diablo, near Mammoth, California. At Steamboat Springs, Nevada, both flash and binary plants are employed. Wells for the Steamboat Hills flash plant tap higher temperature fluids (180° to 200°C) closer to the inferred zone of upwelling. Wells for the binary power plants at Steamboat, however, mainly tap cooler (about 150° to 160°C), shallow (<~300 m deep) outflow plumes but also include some mixing with the higher temperature wells (Figure 7.2) (Klein et al., 2007). Because magmatically heated reservoirs are generally less deep for a given temperature and fluid temperatures are typically higher compared to amagmatic or extensional geothermal systems, drilling costs are lower and power output is higher, resulting in a favorable economic scenario. Nonetheless, amagmatic systems are still profitable when well-studied and designed (and therefore not overbuilt) for the resource at hand. Furthermore, a few amagmatic systems are of high enthalpy with temperatures >200°C, such as Dixie Valley, east of Fallon, Nevada, and Beowawe in northeastern Nevada. Both of these systems are harnessed by flash power plants, along with bottoming binary plants, but productions wells are deep (~3 km at Dixie Valley and >1.5 to ~3 km at Beowawe) (Benoit, 2014). Both of these plants are discussed in more detail near the end of this chapter.

TECTONIC SETTINGS OF SELECT GEOTHERMAL SYSTEMS

Tectonic settings play a strong role in the thermal and chemical characteristics of geothermal systems. Geothermal systems in the following tectonic settings are examined:

1. Divergent boundaries as exemplified by systems in Iceland
2. Convergent settings, which are subdivided into (a) those with subduction (island and continental volcanic arcs) and (b) non-subduction continental collisional zones
3. Transform boundaries in regions of transtension, such as the Salton Sea system, or in the wake of a migrating triple junction, as at The Geysers in northern California

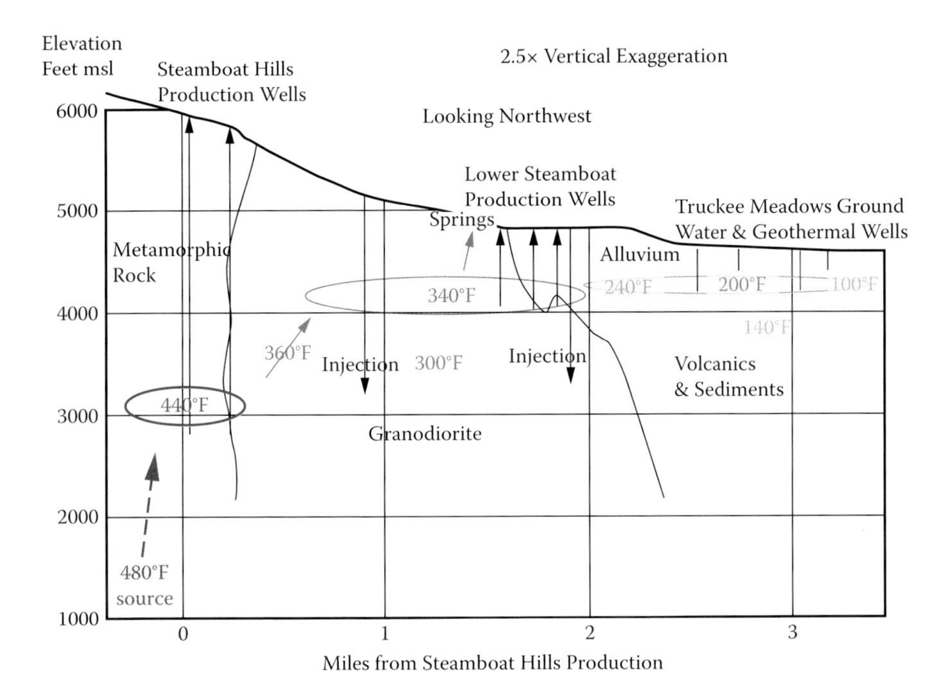

FIGURE 7.2 Southwest-to-northeast-oriented cross-sectional view of the Steamboat geothermal system. Temperatures of the inferred upflow zone below Steamboat Hills and the outflow zone of lower Steamboat are about 10° to 15°C (15° to 25°F) cooler than shown on this diagram due to the cooling effects of injection over time. (Adapted from Klein, C.W. et al., *Geothermal Resources Council Transactions*, 31, 179–186, 2007.)

4. Continental rifting as occurring in eastern Africa and the Basin and Range Province of the western United States
5. Hot spot zones such as Hawaii and the Azores
6. Radiogenic granitic rocks and deep sedimentary basins in stable cratonic interiors of continents, such as the Cooper Basin of Australia and Williston Basin in central North America.

Divergent Setting of Icelandic Geothermal Systems

Iceland, as we learned in Chapters 2 and 4, lies across the Mid-Atlantic Ridge (MAR), which is a divergent boundary marking the eastern edge of the North American tectonic plate and the western edge of the Eurasian plate (Figure 7.3A). Iceland is the focus of particularly high rates of volcanism along the ridge or boundary because of the convergence of seafloor spreading with a geologic hot spot or mantle plume. Indeed, researchers consider Iceland to be a hot spot coincident with a spreading ridge. This overlapping of tectonic processes has allowed the ridge to grow above sea level to produce an island (Arnorsson et al., 2008; Vogt, 1983). In

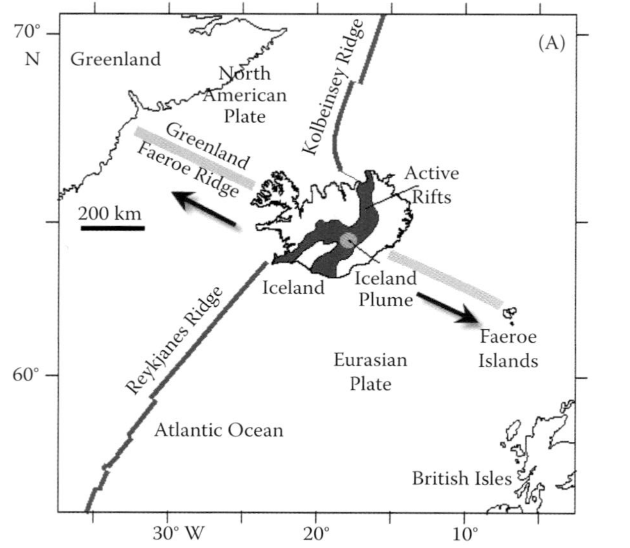

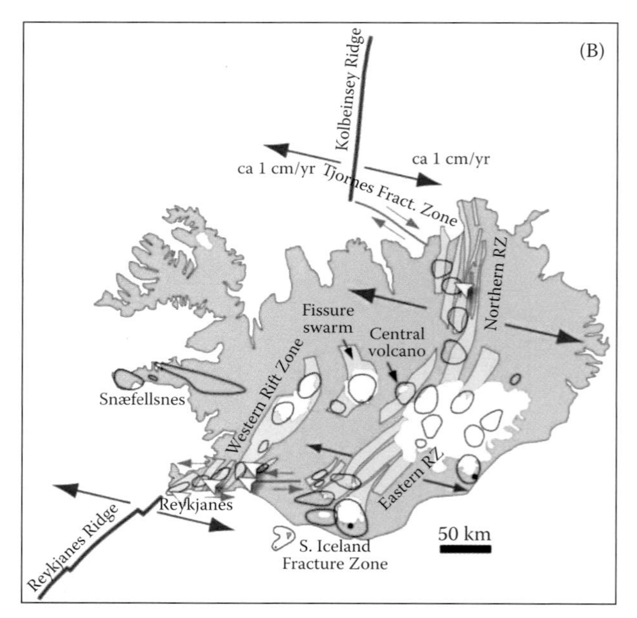

FIGURE 7.3 (A) Map showing tectonic setting of Iceland. The divergent mid-Atlantic Ridge occurs to the south (Reykjannes Ridge) and north (Kolbeinsey Ridge) of Iceland. On Iceland, the divergent boundary is expressed by active volcanic rift zones. The hot spot or Iceland plume is located near the eastern edge of the active rifts. As the plates separate above the Iceland plume a series of largely now submerged volcanic islands or seamounts, called a hot-spot track, extends to the northwest and southeast of Iceland (Greenland–Faeroe Ridge). The blue arrows denote the direction of separation of the North American and Eurasian plates, as also indicated by the orientation of the Greenland-Faeroe Ridge. (Illustration adapted from http://brennanjordan.org/IcelandNAtl.jpg.) (B) A more detailed view of the volcano–tectonic components of Iceland. Yellow zones denote active fissure swarms, and circular to elliptical lines outline main active central vent volcanoes. The arrow or cursor symbols denote the developed geothermal fields of Reykjannes, Svartsengi, Hellisheidi, Nesjavellir, and Krafla extending from southwest to northeast, respectively. (Illustration adapted from http://images.sciencedaily.com/2010/04/100416193630-large.jpg.)

southern Iceland, the MAR consists of two subparallel branches connected to a single ridge zone, extending through central and northern Iceland, by a transform fault (Figure 7.3B).

Developed geothermal systems in Iceland occur in three principal regions: the Reykjanes peninsula, the Hengil volcano, and at Krafla (Figure 7.4). The Reykjanes and Hengil systems lie within the western branch of the MAR, and Krafla is located within the central northern branch. The Reykjanes region has two developed geothermal power plants, one at Reykjanes and the other at Svartsengi. In the Hengil region, the Hellisheidi geothermal power plant is located on the south side of the Hengil volcano, and on the north side of the volcano is the Nesjavellir geothermal power facility. The Hellisheidi facility is the largest combined geothermal power and thermal plant

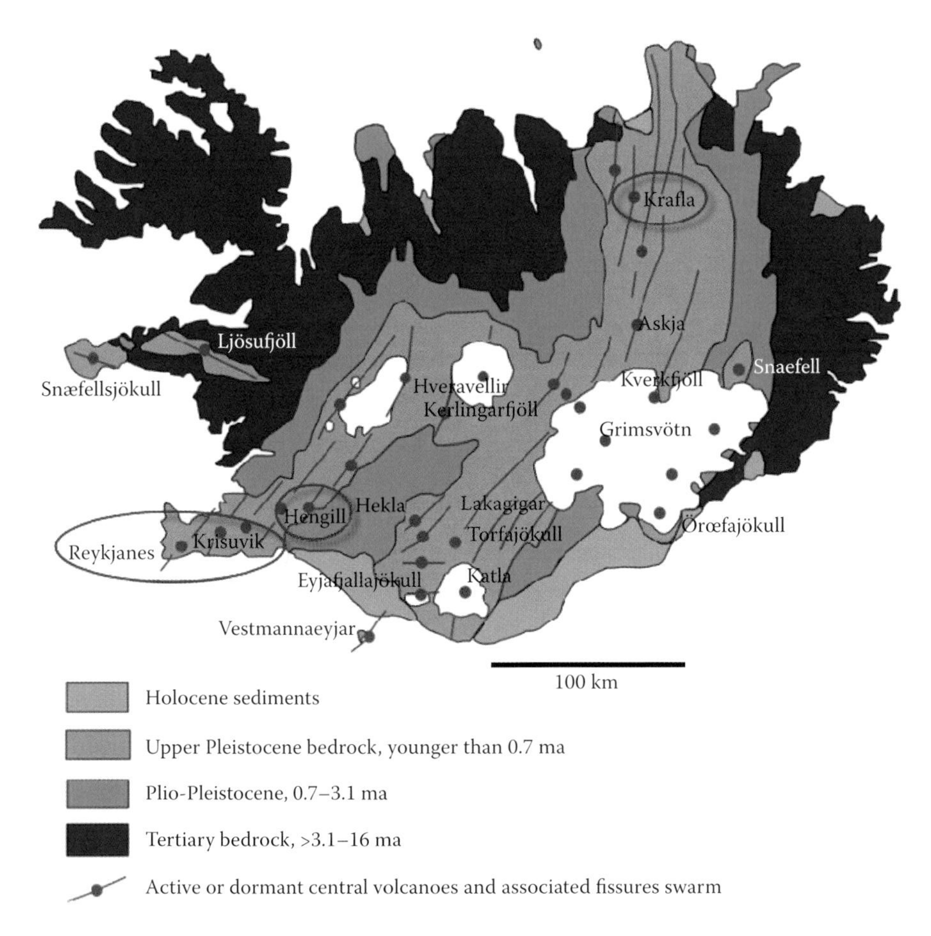

FIGURE 7.4 Simplified geologic map of Iceland showing the three main developed geothermal regions of Reykjanes, Hengil, and Krafla (outlined by red ellipses). Reykjanes regions supports the Reykjanes and Svartsengi geothermal power facilities, Hengil hosts the Hellisheidi and Nesjavellir geothermal power plants, and the Krafla geothermal power plant is named after the host volcanic center. (Adapted from Weisenberger, T., Zeolite Facies Mineralization in the Hvalfjördur Area, Iceland, diploma thesis, University of Freiburg, 2005.)

in the world. It has an electric power capacity of 303 MWe and a planned maximum thermal capacity of 400 MWt for direct use in nearby Reykjavik. Indeed, over 90% of buildings in Iceland (99% of buildings in Reykjavik) are heated using geothermal fluids. Moreover, geothermal power accounts for about 30% of Iceland's electrical power generation, tying it with El Salvador as the highest geothermal power percentage of any country (Orkustofnun, 2016). The balance of power comes from hydroelectric facilities; thus, 100% of Iceland's produced electrical power comes from renewable sources.

Because of its geologic setting of coincident tectonic divergence and hot spot volcanism, all geothermal systems in Iceland, both developed and undeveloped, are magmatically heated. Temperatures at production depths of about 1 km are consistently greater than 200°C, so the plants are single or double flash. In some cases, the temperatures are much higher. For example, the first Iceland Deep Drilling Project (IDDP-1) well was drilled at Krafla (2008–2009) to explore for supercritical fluids. The wellbore, however, stopped short of its target depth (~5 km) because it intersected a zone of magma at a depth of about 2.1 km. Since then, the well has been flow tested several times and was found to be capable of producing about 36 MWe (more than half of the nearby 60-MWe Krafla geothermal power plant's capacity), or about five times the output of a typical Icelandic geothermal well (Fridleifsson et al., 2015). Unfortunately, the 450°C super-heated steam condensate is quite acidic (pH ~2.6 from magmatic inputs of HCl and HF) and has proved to be very corrosive to piping and equipment. Scrubbing to lower the acidity, possibly using alkaline geothermal waters from other Krafla wells, would be required if the well were to be utilized commercially, and tests are continuing. If deemed feasible, the IDDP-1 well would be the world's first geothermal well heated directly by magma and could increase Krafla's power plant capacity by about 60%.

Convergent Continental and Island Volcanic Arcs

Subduction of oceanic lithosphere leads to partial melting of overlying mantle rocks (from the addition of volatiles that lower the melting point of rocks) and the formation of magmas that rise into the upper crust where they locally erupt to produce volcanoes. Where oceanic lithosphere subducts beneath continental lithosphere, a continental volcanic arc develops, such as the Andes of South America or the Cascades of North America. Island arcs, on the other hand, form where old and cold oceanic lithosphere subducts beneath young and relatively warm oceanic lithosphere, such as the many volcanic island chains in the western Pacific, including Japan, the Philippines, the Marianas, and Tonga-Kermadec.

Underlying the volcanoes, in continental and island volcanic arcs, are subjacent, shallow magma reservoirs that serve as the heat sources for locally developed overlying geothermal systems. Good examples of producing geothermal systems in continental volcanic arc settings include Los Azufres (Martinez, 2013) and Los Humeros (Elders et al., 2014) in Mexico, Miravalles in Costa Rica (Ruiz, 2013), San Jacinto Tizate in Nicaragua (Chin et al., 2013), and Berlin and Ahuachapan in El Salvador (Herrera et al., 2010). Interestingly, geothermal power production in South America has yet to occur, but several advanced exploration projects are underway in Chile. For example, the 48-MWe Cerro Pabellon geothermal plant, being built by Enel Green Power and the state-owned oil company ENAP, is scheduled to begin producing electricity in mid-2017. Why might western South America be lagging behind in geothermal power development, despite its position along an active continental volcanic arc? The reason is somewhat complicated, but more than heat is required to engender a viable geothermal system. Other key attributes are permeability, availability and chemical composition of fluids, and, in many cases, a cap rock to help constrain fluid and heat energy. Rock permeability has been shown to correlate positively with zones of increased structural (fault) complexity. In particular, many if not most producing geothermal systems occur in regions undergoing active, high (>0.5 mm/yr) rates of extensional strain (Faulds et al., 2012). In volcanic arcs, strain can range from mainly compressional, where convergence is head on, to mainly extensional, as in the case of subduction zone rollback (e.g., the Taupo Volcanic Zone), to variably transtensional to transpressional, where convergence is oblique (Hinz et al., 2015). For most of western South America, the intra-arc strain is mainly compressional, as expressed by arc-parallel active reverse and thrust faults along with local arc-orthogonal normal faults. As a result, extensional strain and related areas of improved permeability appear relatively limited. Moreover, potential cap rocks in many places may have been breached due to high rates of uplift in the Andes and ensuing erosion (Coolbaugh et al., 2015). Other non-geological factors for the current limited geothermal development in the Andes include the high cost and risk of test drilling (exacerbated by the remoteness of promising prospects) and the expense of constructing transmission lines in areas of challenging topography, high elevation, and harsh mountainous climates. In addition, much of the existing renewable power in Chile is largely derived from hydroelectric sources, whose technical development is well established and for which the risk is minimal (when a dam is constructed, power is ensured).

Although producing geothermal systems occur in both compressional and extensional regions of strain in volcanic arcs, Wilmarth and Stimac (2015) observed that those systems with higher power density and power output are generally associated with arcs having complex structural settings induced by oblique convergence, involving particularly transtension (e.g., Salak, Indonesia) or intra-arc rift-related extension (e.g., Wairakei, New Zealand) (Figure 7.5). Where convergence is oblique, strike-slip faults can form in the overlying plate. In areas of fault step-overs, zones of transtension can occur, forming possible pull-apart basins that can foster crustal dilation (improved permeability for convection of geothermal fluids) and the rise of magma into the upper crust (heat source).

High-level intrusions of magma can also thermally weaken overlying rocks, leading to gravitational collapse and dilation, generally orthogonal to the direction of plate convergence. As a result, a series of extensional basins or grabens can form that in association with heat from volcanism can help engender or support prospective or producing geothermal systems, such as Los Humeros and Los Azufres in the Trans-Mexican Volcanic Belt (Figure 7.6). More informaion on transtension and pull-apart basins is provided in the section on geothermal systems associated with transform boundaries later in the chapter.

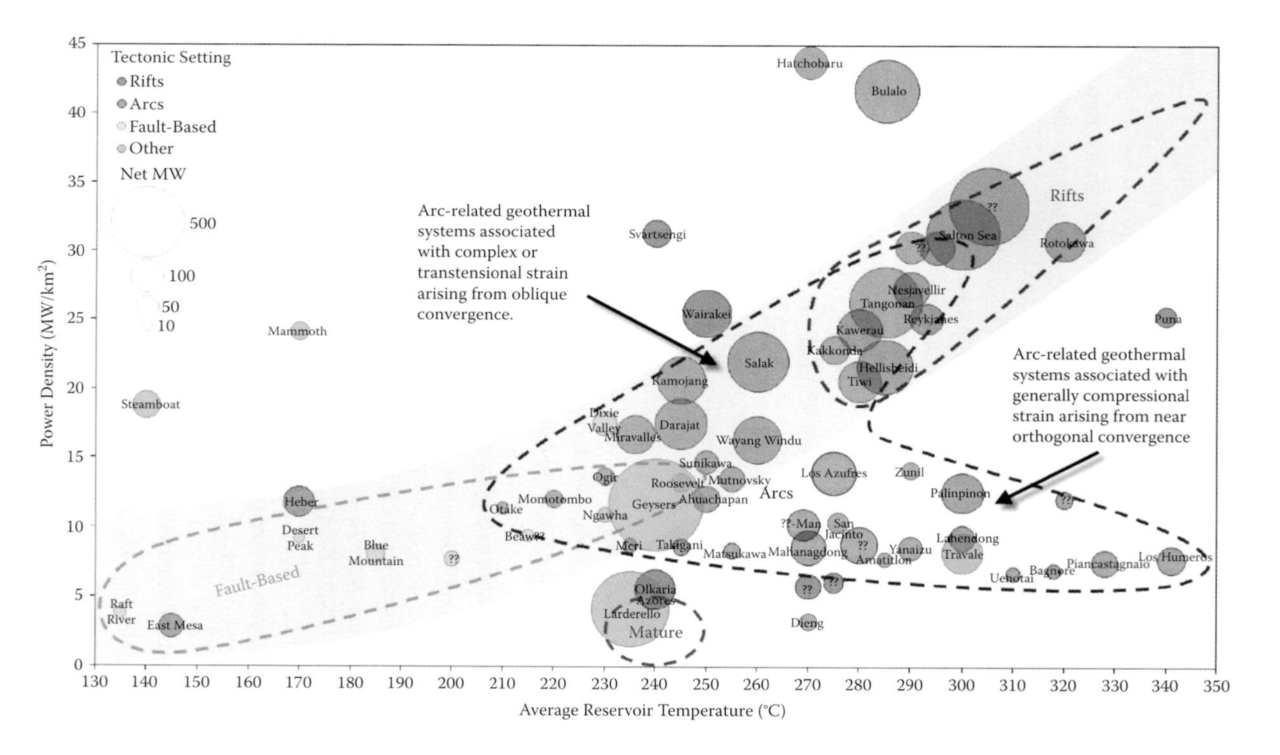

FIGURE 7.5 Power density as a function of geothermal reservoir temperatures. Plotted geothermal systems are color coded and grouped according to some geologic settings. Note that in the arc-related environment, geothermal systems associated with complex or transtensional strain have higher power density and power output compared to those where compressional strain is dominant. (Adapted from Wilmarth, M. and Stimac, J., in *Proceedings of World Geothermal Congress 2015*, Melbourne, Australia, April 19–24, 2015.)

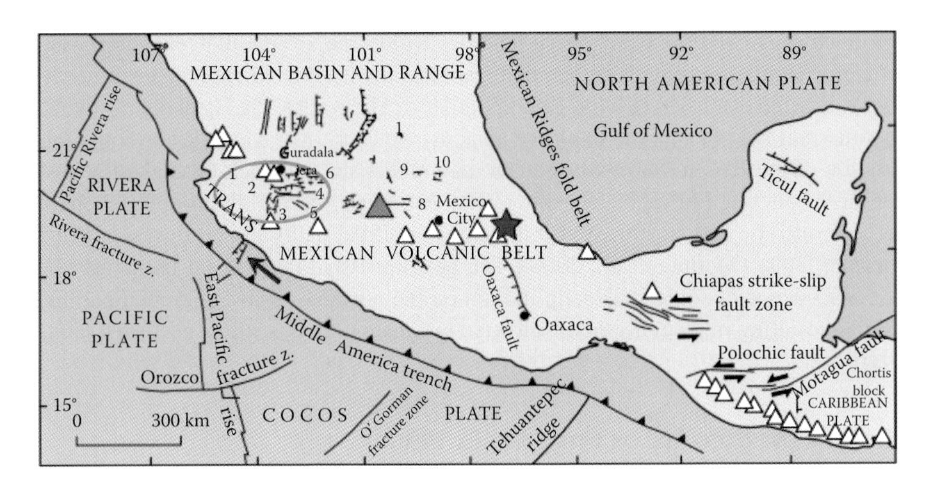

FIGURE 7.6 Tectonic map of central Mexico showing Trans-Mexican Volcanic Belt as continental volcanic arc. Short lines with hachures, such as the Oaxaca fault, denote extensional normal faults. Area enclosed by green ellipse shows both arc-parallel and arc-orthogonal normal faults. Grabens bounded by normal faults are indicated where hachures on fault lines point toward each other, such as the one noted by the red arrow south of the green ellipse. Red star and green triangle denote locations of Los Humeros and Los Azufres geothermal fields, respectively. (From Suter, M. et al., *Geological Society of America Bulletin*, 113(6), 693–703, 2001.)

Another example of a geothermal system that has developed in a local zone of extension (transtension) in a continental volcanic arc is the Miravalles geothermal field in Costa Rica (Figure 7.5). The Miravalles geothermal field is currently developed by three flash plants and a bottoming binary plant for a total capacity of 162.5 MWe (DiPippo, 2012). The producing geothermal field is located on the south flank of the Miravalles Quaternary* stratovolcano,† which last erupted about 7000 years ago. The production area covers about 16 km^2 and occurs within a north-northeast-trending graben on the southwest flank of the volcano. The north-northeast-trending graben may be related to possible left steps in west-northwest-striking left-lateral faults (transtension) arising from the oblique (left-lateral) convergence between the downgoing Cocos plate and overriding Caribbean plate. The bounding and internal faults of the graben have produced secondary fracture permeability of the volcanic rocks, facilitating convection of geothermal fluids and development of this field as a significant power producer.

An example of a developed geothermal system in a volcanic island arc setting is the Hatchobaru geothermal power plant (installed capacity of 112 MWe), located on the island of Kyushu in southwestern Japan. Hatchobaru is the largest geothermal power producer in Japan and is supplied by 26 production wells having an integrated mass flow rate of 560 kg/s (Tokita et al., 2000). A significant factor for the high

* A geological time period covering the most recent 2 million years.

† Stratovolcanoes are the classic conical-shaped volcanoes, such as Mt. Fuji in Japan, and are the most common type of volcano found in continental and island arc volcanic settings.

productivity is that the Hatchobaru field lies within the arc-parallel, east-northeast-trending Beppu–Shimbara graben that transects the island and reflects north-north-west-directed extension (Ehara, 1989). Similar to Miravalles, the Hatchobaru–Otake geothermal field lies on the flanks of an active volcano (Mt. Kuju). Extensional formation of the Beppu–Shimbara graben may reflect slab rollback (discussed below) of the subducting Philippine oceanic plate. Flow of geothermal fluids at Hatchobaru is controlled by both northeast-striking graben-parallel faults and northwest-striking normal faults (Momita et al., 2000). The northwest-striking normal faults may be a consequence of some strike-slip motion on the northeast-striking graben-parallel faults, resulting in local northeast-directed extension in areas where northeast faults step-over.

Convergent Back-Arc or Intra-Arc Extension

In addition to compression, associated with head-on convergence, or transtension and transpression, related to oblique convergence, a phenomenon known as *back-arc* or *intra-arc extension* (spreading) can also occur. In this case, extension is oriented perpendicular to the arc, resulting in elongated grabens that run parallel for a significant part of the arc. Such dilation is thus more extensive than the localized zones of extension related to transtension or gravitational collapse as noted in the previous section. The development of back- or intra-arc extension is not fully understood, and several models have been put forth to explain their formation. What is known is that development of back-arc or intra-arc spreading is more common in island arcs compared to continental arcs. Moreover, it occurs most commonly where subducted oceanic lithosphere is relatively old (>55 million years) and where the intermediate dip angle of the subducting slab is >30°. Some models advanced to help explain extension behind or within the arc include the following:

1. Trench-slab rollback occurs, where the downgoing slab steepens with time due to the vertical component of motion and pulls the arc with it due to hypothesized trench suction forces (Figure 7.7). This results in tension as the arc pulls away from the rest of the overriding plate.

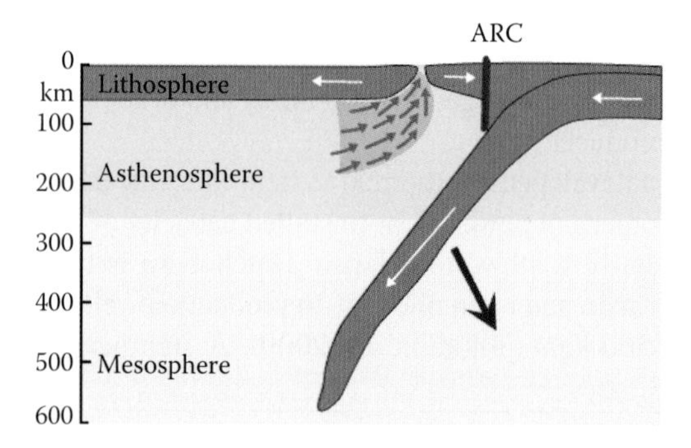

FIGURE 7.7 Illustration of back-arc spreading passively induced by slab roll back.

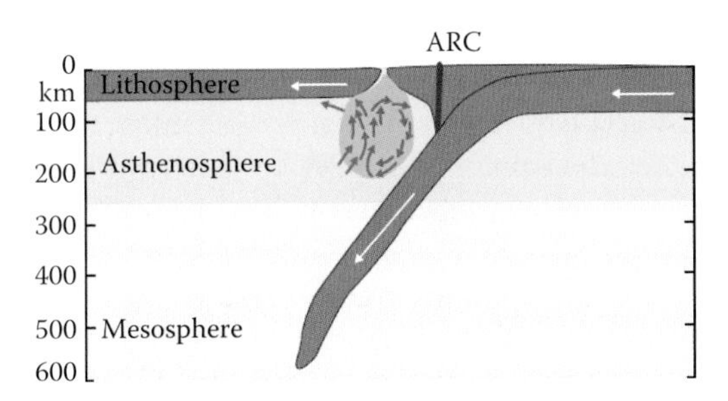

FIGURE 7.8 Illustration of back arc extension resulting from subsidiary convection induced by drag forces above the subducting slab and upwelling buoyant, volatile-charged mantle.

2. Subsidiary cells of convection develop above the downgoing slab due to downward drag forces adjacent to the slab and upward rise of buoyant mantle produced by the rise of volatiles (mainly water and CO_2) squeezed out of the downgoing slab (Figure 7.8). This might also explain why back-arc or intra-arc spreading is more common with island arcs than with continental arcs because of the comparative thinness of oceanic crust.

Whether extension and crustal rifting occur behind or within the arc probably reflects the rheological or mechanical characteristics of the overriding plate and steepness of the subduction zone. Where the subduction zone dips less steeply, back-arc spreading will be promoted as compressive stresses are distributed over a wider area in the overriding plate including the active volcanic arc. On the other hand, if the subduction zone dips more steeply, compressive stresses of convergence are more narrowly distributed to the area of the accretionary wedge and forearc region of the overriding plate. Under these conditions, spreading or extension is more likely developed within or closely behind the arc, where the crust has been rheologically weakened by the added heat from magma (Figure 7.9).

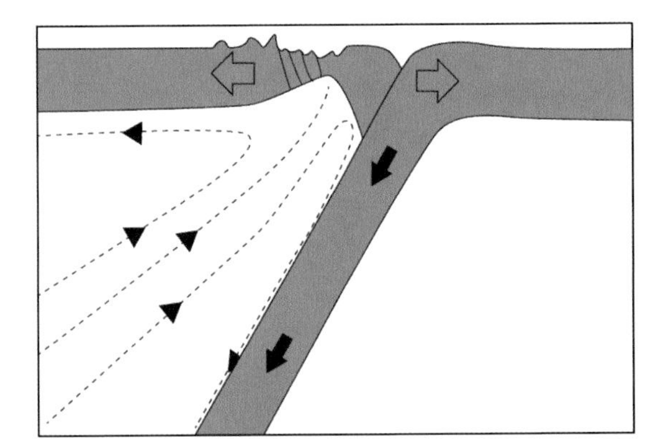

FIGURE 7.9 Cross-section of a steeply dipping subduction zone inducing extension within the volcanic arc. (From Martinez, F. and Taylor, B., *Geological Society Special Publications*, 219, 19–54, 2003.)

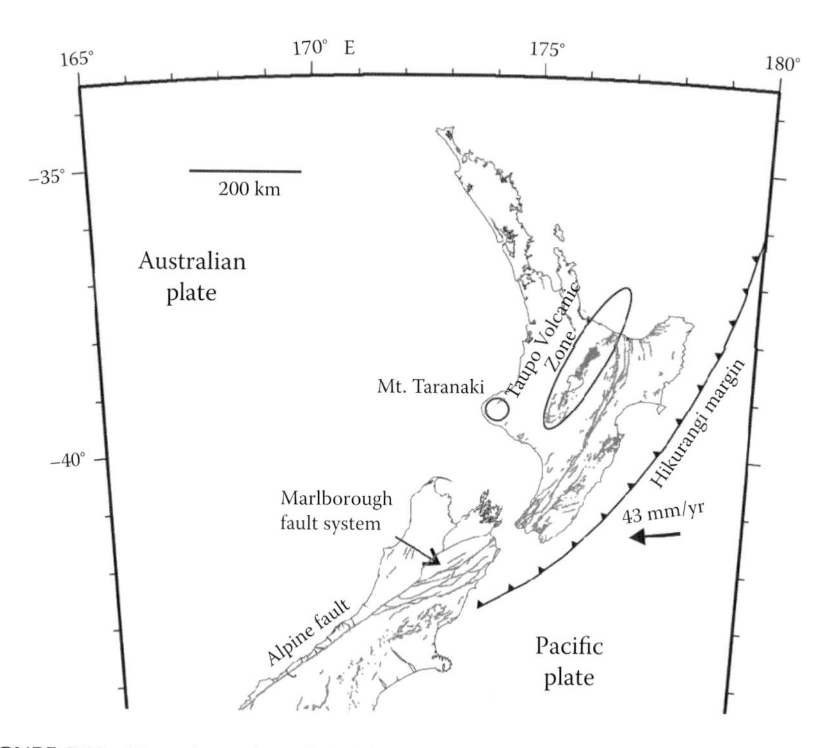

FIGURE 7.10 Tectonic setting of the Taupo Volcanic Zone (TVZ), which resides in an arc-parallel extensional graben. Rate of extension is about 10 mm/year within the TVZ and is directed west-northwest–east-southeast. Line with sawtooth pattern marks the trench of the subducting Pacific plate. (Adapted from Otago, *Tectonic Setting of New Zealand: Astride a Plate Boundary Which Includes the Alpine Fault*, University of Otago, Department of Geology, Dunedin, New Zealand, 2016.)

What does back-arc or intra-arc spreading have to do with geothermal systems? Because of the extensional forces, secondary rock permeability and crustal dilation are promoted, aiding intrusion of magma to high crustal levels that can serve as a heat source and development of overlying convecting and potentially exploitable geothermal systems. This prospective environment for geothermal development is demonstrated by the elevated power density and power output of some of the rift-related systems shown in Figure 7.5, such as Wairakei and Rotokawa in the Taupo Volcanic Zone, New Zealand. Two main types of back-arc or intra-arc extensional settings can be distinguished—magmatic and amagmatic.

Magmatic Intra-Arc Extensional Setting

Probably the best example of this environment is the Taupo Volcanic Zone (TVZ) on the North Island of New Zealand. The TVZ is largely constrained within a northeast-trending extensional graben that lies above a west-dipping subduction zones whose trench lies offshore to the east (Figure 7.10). The zone extends for about 200 km in length and is about 30 to 50 km wide. It is one of the most volcanically active regions on the planet and has produced, on average, 0.3 m^3/s (~26,000 m^3/day) of magma for the last 350,000 years (Wilson et al., 1995). Not surprisingly, the TVZ is also one of the most geothermally productive regions on the planet. Approximately 108 m^3 of 250°C fluid are discharged per year through the central rhyolitic portion of the TVZ (Bibby et al., 1995; Rowland and Simmons, 2012); this volume and temperature of fluid represent a heat power output of about 4000 ± 500 MW (Hochstein, 1995). All together, the TVZ supports 13 of the 14 geothermal power stations in New Zealand and has a combined installed capacity of 1005 MWe (Bertani, 2015), which represents about 15% of New Zealand's installed electrical capacity.

Geothermal power development began in 1958 with commissioning of the Wairakei geothermal plant located in the southern part of the TVZ (Figure 7.11). Wairakei was the planet's first geothermal facility to produce electricity from a liquid-dominated reservoir. Prior to that time, the only electricity-producing geothermal facility was at Larderello, Italy, which continues to exploit a vapor-dominated

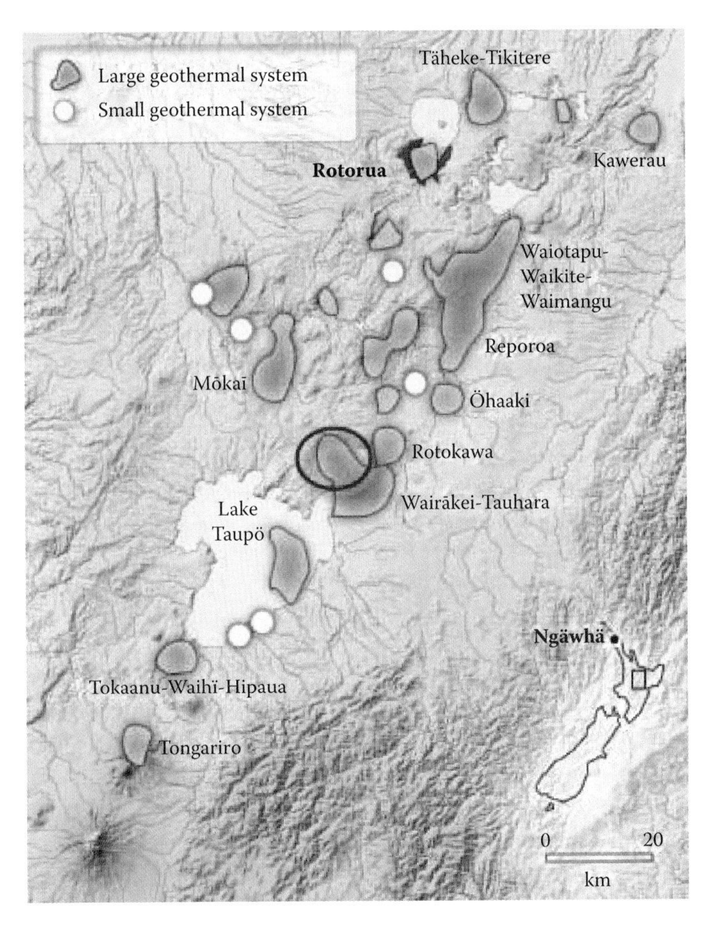

FIGURE 7.11 Map of Taupo Volcanic Zone showing locations of geothermal fields. The Wairakei geothermal field is enclosed by the red ellipse and is contiguous with the Tauhara geothermal field to the southeast. Collectively, the TVZ hosts about 20 geothermal systems. (Illustration adapted from http://www.teara.govt.nz/files/m-5418-enz.jpg.)

reservoir. The Wairakei power station has a current total installed capacity of 171 MWe, consisting of two flash plants (67 MWe and 90 MWe) and one bottoming-cycle binary plant with a capacity of 14 MWe; the net power output is 132 MWe.

Most production wells are between 500 and 1000 m deep and were initially drilled to tap permeable steeply dipping normal faults; however, more recent holes intersect rock layers having primary matrix permeability, resulting in some wells having as much as 2 km of horizontal throw to maximize wellbore surface area and flow rate within the reservoir. Some of these wells having significant horizontal extents produce as much as 20 MWe per well because of their high flow rates. For much of the production history of the Wairakei geothermal field, spent fluid was not reinjected partly because of cost-saving measures and because the benign chemical nature of the geothermal fluids (near-neutral pH and low TDS, <1500 ppm Cl) that could easily be disposed of in the nearby Waikato River without causing any significant degradation of water quality. Moreover, engineers were concerned that injection might adversely cool the reservoir and produced geothermal fluids, thereby reducing power output and revenue.

As a result of not reinjecting spent fluids, however, two issues arose. First, due to the pressure drop from not replenishing the system, boiling was induced which led to the formation of vapor-dominated zones that were formerly liquid dominated. Because of this, 22 geysers and numerous hot springs in nearby Geyser Valley, now renamed Wairakei Valley, disappeared, leaving remnant mounds of silica sinter as the only evidence of the former geothermal surface activity. The other main consequence of not reinjecting fluids was ground subsidence, which occurred at a rate of as much as 0.45 m/year. Total subsidence, over the past 50 years, has been about 15 to 20 m, but rates of subsidence have been reduced to 10 to 15 cm/year since the onset of reinjection began in the mid-1990s (Bromley et al., 2013). Some subsidence can also be attributed, however, to municipal groundwater withdrawal, as much of the current subsidence is due to compaction of a hydrothermally altered and weakened pumiceous breccia unit (groundwater aquifer) that overlies the geothermal reservoir. Further discussion of geothermally induced subsidence is explored in Chapter 9.

Amagmatic Back-Arc Extensional Setting

A good example of this tectonic setting occurs in southwestern Turkey in the Menderes graben. Most of Turkey's operating geothermal power plants occur in this region, the oldest of which is the Kizildere plant (commissioned in 1984), located at the eastern end of the graben (Figure 7.12). A second geothermal facility at Kizildere went into operation in 2013 and has 80 MWe of installed capacity. As of the end of 2014, Turkey had an installed geothermal power capacity of 400 MWe with an additional 165 MWe under construction (Mertoglu et al., 2015). The Menderes graben extends east-west for about 200 km and varies from 10 to 20 km in width. The overall north-south-directed extension probably results from slab rollback (steepening of the Hellenic subduction zone), as evidenced by the southwestward migration of volcanism from the Miocene arc to the current south Aegean volcanic arc. The Hellenic subduction zone marks where ocean crust associated with the African tectonic plate is slowly subducting beneath the overriding Eurasian plate that includes western Turkey (Figure 7.13). Although a volcanic arc exists, as exhibited by the volcanic Greek islands of Santorini

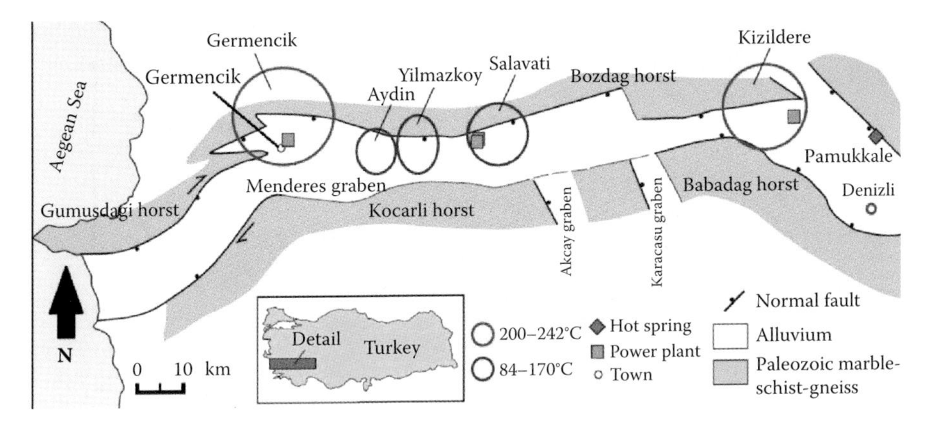

FIGURE 7.12 Map showing the location of the Menderes graben and the location of the geothermal power facilities, including Kizildere, which is the largest. The east-west north-striking normal faults that bound the graben formed due to north-south-directed extension. (From Faulds, J. et al., in *Proceedings of World Geothermal Congress 2010*, Bali, Indonesia, April 25–30, 2010.)

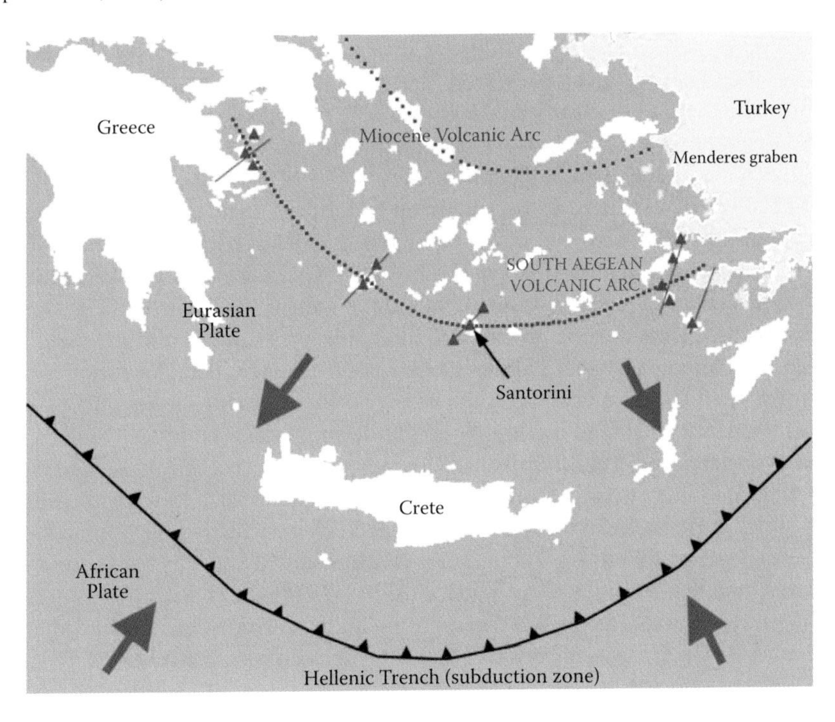

FIGURE 7.13 Map showing tectonic setting of the Menderes graben in southwestern Turkey as a near back-arc rift system behind the currently active south Aegean volcanic arc. Slab rollback of the Hellenic subduction zone is indicated by southward migration of volcanism from the Miocene (20 to 5 million years ago) to the current volcanic arc shown by the red dotted line. Red triangles mark active volcanoes. (Adapted from Vassilakis, E. et al., *Earth and Planetary Science Letters*, 303(1–2), 108–120, 2011.)

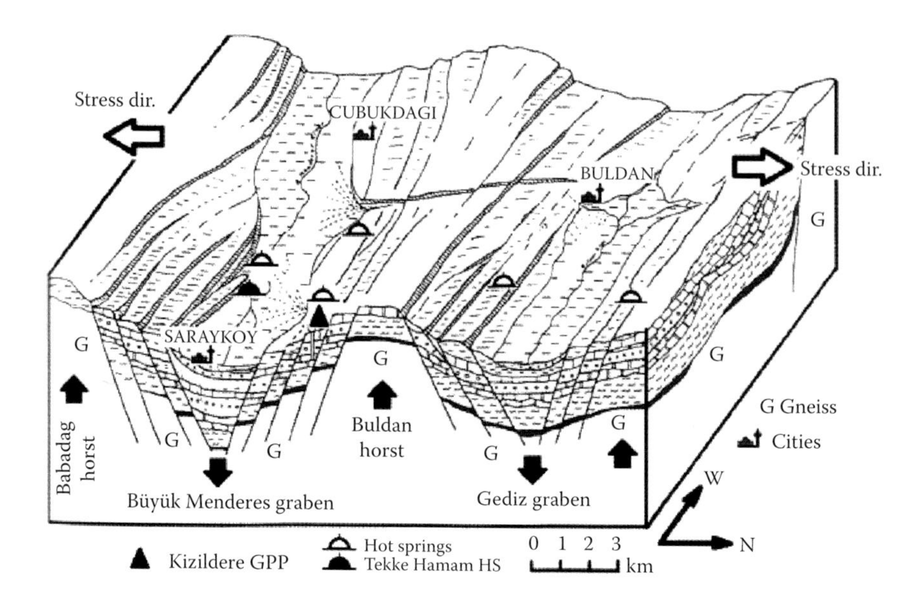

FIGURE 7.14 Perspective block view, looking west, of the Menderes–Gediz graben system. Note that the location of the Kizildere geothermal power plant is indicated by the solid triangle; its location is centered on numerous normal faults along the northern margin of the Menderes graben. (From DiPippo, R., *Geothermal Power Plants: Principles, Applications, Case Studies, and Environmental Impacts*, 3rd ed., Butterworth-Heinemann, Waltham, MA, 2012.)

and Milos, the dearth of back-arc volcanism is not well understood. It could reflect the slow rate of subduction, possibly limiting the amount of back-arc aesthenospheric upwelling and partial melting to form magma. Back-arc extensional stresses, nonetheless, are effectively transmitted from depth to the surface, resulting in the east-west-trending Menderes and Gediz grabens (Figure 7.14).

The Kizildere geothermal system is located at the eastern end of a major easterly striking normal fault zone bounding the northern side of the Menderes graben. In the area of Kizildere, the fault zone breaks up into several splays, creating a high density of fractures (and permeability) that serve to channel the geothermal fluids (Figure 7.15). Indeed, this is also the structural setting of the double-flash Germencik geothermal facility at the western end of the Menderes graben, which is the second largest geothermal power facility in Turkey (69.9-MWe installed capacity), reflecting high reservoir fluid temperatures of as much as 232°C (Mertoglu et al., 2015). Moreover, the structural setting of geothermal systems in Turkey is similar to many geothermal systems in the Great Basin of Nevada, such as Gerlach hot springs, reflecting the overall extensional amagmatic settings of both regions (Faulds et al., 2010).

The Kizildere geothermal power facility recently consisted of a 15-MWe flash plant and a 6.8-MWe binary plant. Production wells access three reservoirs, the shallowest at about 400 to 600 m and the deepest at 2000 to 3000 m. All reservoirs are liquid dominated, with fluid temperature generally increasing with depth from about 200°C to about 240°C. Although fluids have low to modest TDS (~2500 to 3200 ppm), they contain high concentrations of noncondensable gases (NCGs),

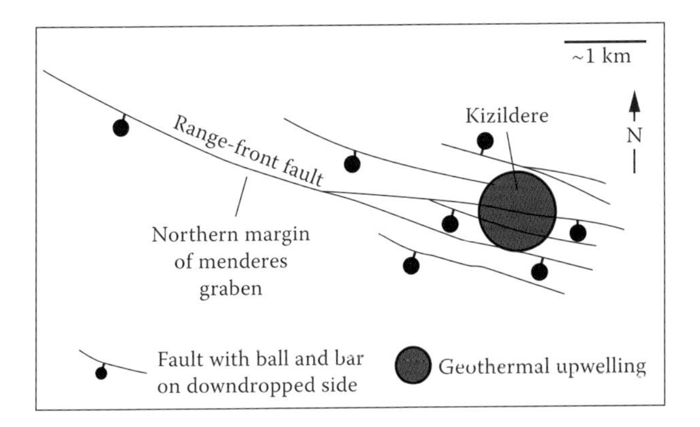

FIGURE 7.15 Structural setting of Kizildere geothermal system with heavy black lines denoting normal faults with filled circle on the upthrown side of fault. Splaying of a single large fault into several smaller faults results in improved permeability for upwelling geothermal fluids. (Adapted from Faulds, J. et al., in *Proceedings of World Geothermal Congress 2010*, Bali, Indonesia, April 25–30, 2010.)

averaging 13% by weight of steam at the turbine inlet. Carbon dioxide makes up 96 to 99 wt% of the NCGs, with the balance consisting of hydrogen sulfide. The high concentration of CO_2 arises from the reservoir rocks consisting mainly of limestone (Haizlip et al., 2013). The escape of CO_2 as fluids rise and pressure drops in production wells creates a significant scaling problem of calcite in production wells (Figure 7.16), necessitating the application of chemical downhole scale inhibitors to

FIGURE 7.16 Buildup of calcite scale near wellhead of one of the production wells in the Kizildere geothermal field, Turkey. (From Haizlip, J.R. et al., in *Proceedings of the 37th Workshop on Geothermal Engineering*, Stanford, CA, January 30–February 1, 2012, http://www.geothermal-energy.org/pdf/IGAstandard/SGW/2012/Haizlip.pdf.)

maintain satisfactory flow rates. The reaction governing the precipitation of calcite is shown below:

$$2HCO_3^- + Ca^{2+} \rightarrow CaCO_3 + H_2O + CO_2 \quad (7.1)$$

(Bicarbonate) + (Dissolved Ca ion) → (Calcite) + (Water) + (Carbon dioxide)

As CO_2 escapes due to the drop in pressure as fluids rise in the well, the reaction is driven toward the right, promoting the precipitation of calcite.

A new 80-MWe flash plant at Kizildere was completed in 2013 that involved drilling several new production wells and deepening select existing wells. Moreover, the binary plant has been refurbished and expanded to 15-MWe installed capacity, for a total combined installed capacity of 95 MWe. These new plants also recover the abundant CO_2 to make dry ice, and spent fluid is used directly for heating greenhouses and 2500 households in the nearby community.

Continental Convergent Setting

Of all tectonic settings, regions of continental convergence, such as the Himalaya, Hindu Kush, and Zagros mountain regions in southern central Asia, are not particularly conducive for engendering accessible geothermal systems. Due to the continental convergence, the crust is over-thickened, which lowers the geothermal gradient because the increase in temperature with depth is spread over a greater distance. Furthermore, the overall compressional stresses in this tectonic setting tend to limit dilation associated with fractures and faults and thereby reduce permeability for any hydrothermal fluids. Finally, because colliding continental crust cannot subduct, an important vehicle for magma generation found in convergent continental and island arcs, which can serve as a heat source to drive upper crustal geothermal systems, is lacking. However, the high temperature and pressure metamorphism associated with continental collision zones can induce partial melting of rocks in the lower and middle crust, resulting in magma that can serve as potential sources of heat. Other sources of heat can be buried radiogenic granitic and high-grade metamorphic rocks that would heat deeply circulating groundwater reflective of an amagmatic geothermal environment. With that said, geothermal systems do occur and are associated mainly with local zones of extension within an otherwise compressional stress regime.

As the crust is thickened during continental collision, extensional forces begin to operate orthogonal to the main direction of compression as the uplifted, thickened crust begins to collapse or spread under its own weight. Depending on the type and strength of rocks, extensional normal faults can develop in local zones and bound horsts and grabens that generally trend perpendicular to the main trend of the mountain range. The extensional normal faults and associated dilation foster improved permeability, helping groundwater to circulate to significant depths where it is heated and then returns to the surface, along other normal faults, to form hot springs and relatively shallow geothermal reservoirs. Good examples of geothermal systems formed in localized extensional settings include Yangbajing and the newly developed Yangyi geothermal fields in Tibet.

Yangbajing, Tibet

Yangbajing is located about 90 km northwest of Lhasa (Figure 7.17). At an altitude of 4300 m, it is the highest geothermal power facility in the world. Production began in 1976 at 1 MWe and was increased to an installed capacity of 25 MWe in 1991. Produced electricity supplies part of the electrical demand of Lhasa. The plant is served by wells tapping a shallow reservoir (~50 to 150 m deep in the southern part and ~100 to 300 m deep in the northern part of the field) containing fluids at temperatures of 150 to 175°C, which are generally insufficient to operate the plant at installed capacity. Hydrologic studies and results of drilling indicate that the currently exploited shallow reservoir is an outflow plume from a higher temperature upwelling zone at the north end of the field (Ji and Ping, 2000; Xiaoping, 2002). A 2000-m deep well encountered a temperature of 329°C, but unstable flow, due to limited permeability, made the well unusable. Another deep well (ZK 4001) (Figure 7.18), 1459 m deep, was drilled in 1996 and had well-head conditions of 200°C, 15 bars, and a flow rate of 84 kg/s, which would be enough to generate about 12 MWe of electricity; it is unclear, however, if this well has been connected to the plant. The power plant is a double-flash design, which seems overbuilt, considering produced fluid temperatures are generally <200°C.

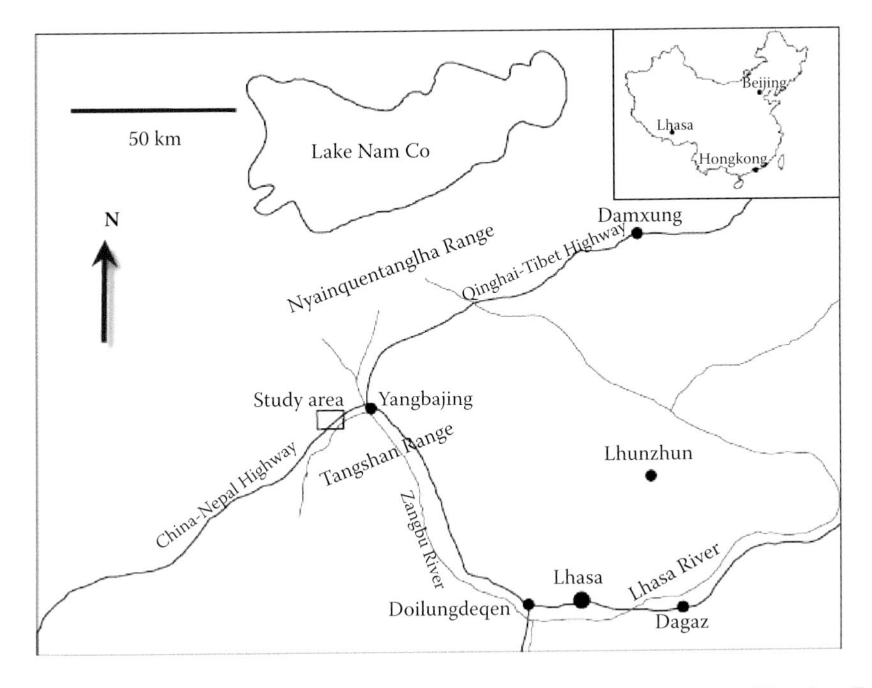

FIGURE 7.17 Map showing location of Yangbajing geothermal field. (From Xiaoping, F., *Conceptual Model and Assessment of the Yangbajing Geothermal Field, Tibet, China*, The United Nations University, Geothermal Training Programme, Reykjavik, Iceland, 2002.)

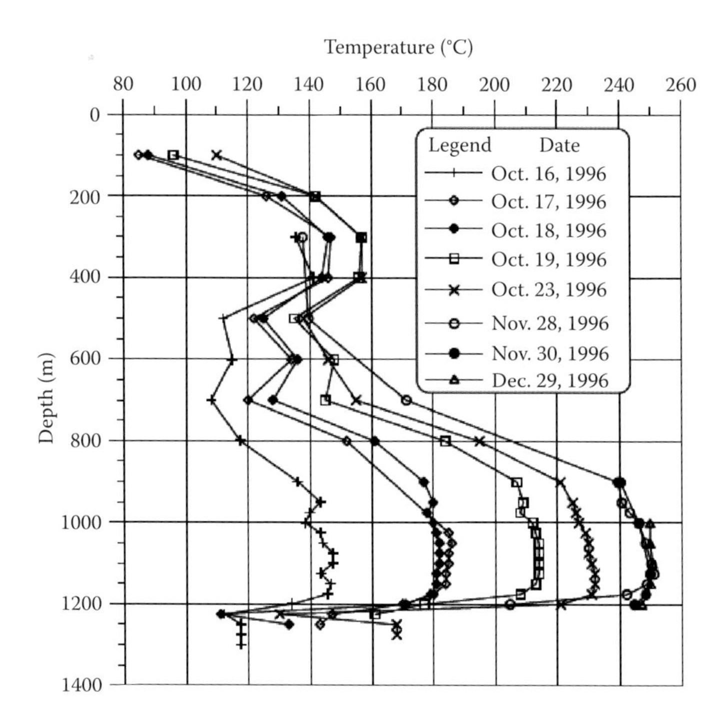

FIGURE 7.18 Temperature–depth profile for well ZK 4001, drilled in the northern part of the field, which encountered a geothermal reservoir between 1000 and 2000 m having an equilibrated temperature of ~250°C. (From Xiaoping, F., *Conceptual Model and Assessment of the Yangbajing Geothermal Field, Tibet, China*, The United Nations University, Geothermal Training Programme, Reykjavik, Iceland, 2002.)

The Yangbajing geothermal field lies in a northeast-southwest-trending graben bounded by normal faults. The higher temperature (160° to 170°C) northern part of the shallow (200 to 400 m deep) geothermal field lies in the strongly faulted, northwestern margin of the graben and appears to lie above the probable upflow region that feeds the shallow, cooler (120° to 160°C) outflow plume to the southeast (Figures 7.19 and 7.20). Recharge for the geothermal system appears to be infiltration of melt water from glaciers capping the mountain range on the northwest side of the graben in which the geothermal facility is located. Faults channel the meteoric fluids (groundwater) to depth, where they are heated by radiogenic granitic and metamorphic rocks and then rise back toward the surface along faults forming the inferred upwelling zone. The fluids are near-neutral pH and have low TDS content ranging from 2800 ppm in the deeper upwelling zone to 1500 ppm in the shallow reservoir.

A total of 54 wells have been drilled into the Yangbajing geothermal system; it is not clear from the literature reviewed to date, however, if any drilled holes are used for reinjection. What is known is that from 1983 to 1993 a total of 360 mm of subsidence has occurred in the southern part of the field due to fluid withdrawal for power generation and pore volume collapse (Figure 7.21). Concurrent with this, pressure

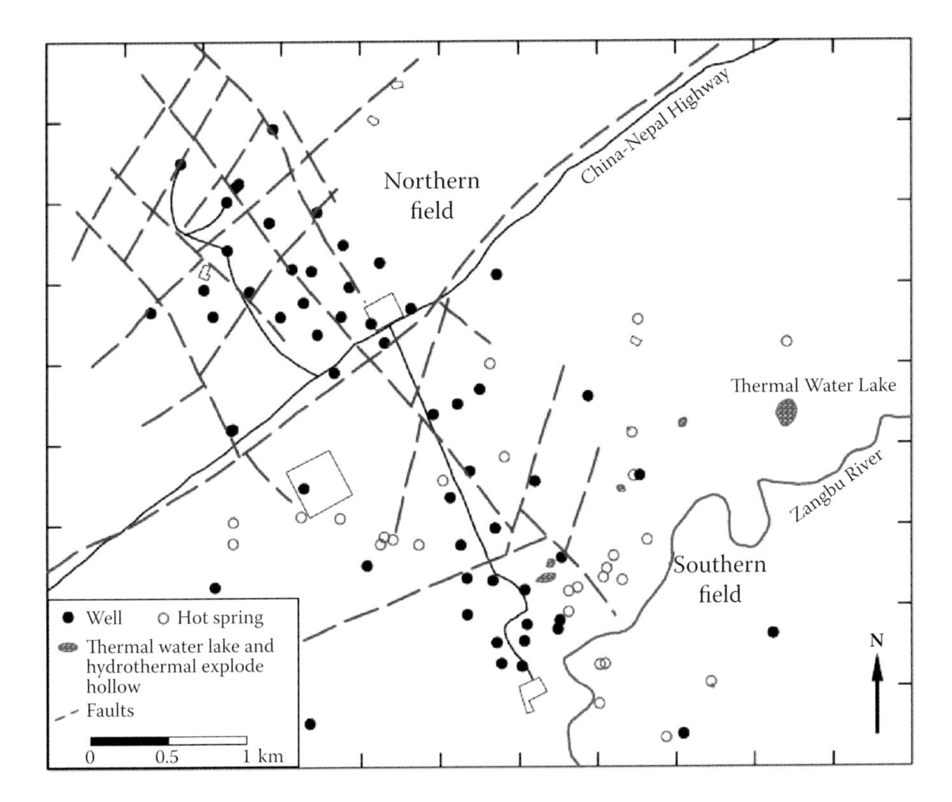

FIGURE 7.19 Map showing locations of wells of the northern and southern parts of the Yangbajing geothermal field and faults. (From Xiaoping, F., *Conceptual Model and Assessment of the Yangbajing Geothermal Field, Tibet, China*, The United Nations University, Geothermal Training Programme, Reykjavik, Iceland, 2002.)

and temperature have dropped, resulting in declining power generation. At the same time, however, temperature and pressure have increased slightly in the northern part of the field, stimulated likely from an enhanced pressure gradient between the upflow zone and the southern part of the outflow plume (Figure 7.22). Indeed, while the southern part of the field has subsided, the northern part of the field has actually risen (Figure 7.21).

Yangyi, Tibet

The Yangyi geothermal field is located 45 km south-southwest of Yangbajing and 75 km west of Lhasa. It is currently being developed for power generation with 40-MWe installed capacity (Zheng et al., 2013). The type of plant was not specified. It appears to have a similar extensional setting as Yangbajing, characterized by north-south-striking graben-bounding faults, whose intersection with northeast-striking faults localize geothermal activity. Reservoir fluid temperatures are about 180° to 210°C. Geothermal fluids at Yangyi are chemically similar to those at Yangbajing and are of the alkali-chloride-bicarbonate type with low TDS (1400 to 1800 ppm) and near-neutral to modestly alkaline pH (7.5 to 9.5). Bicarbonate (HCO_3^-) is the

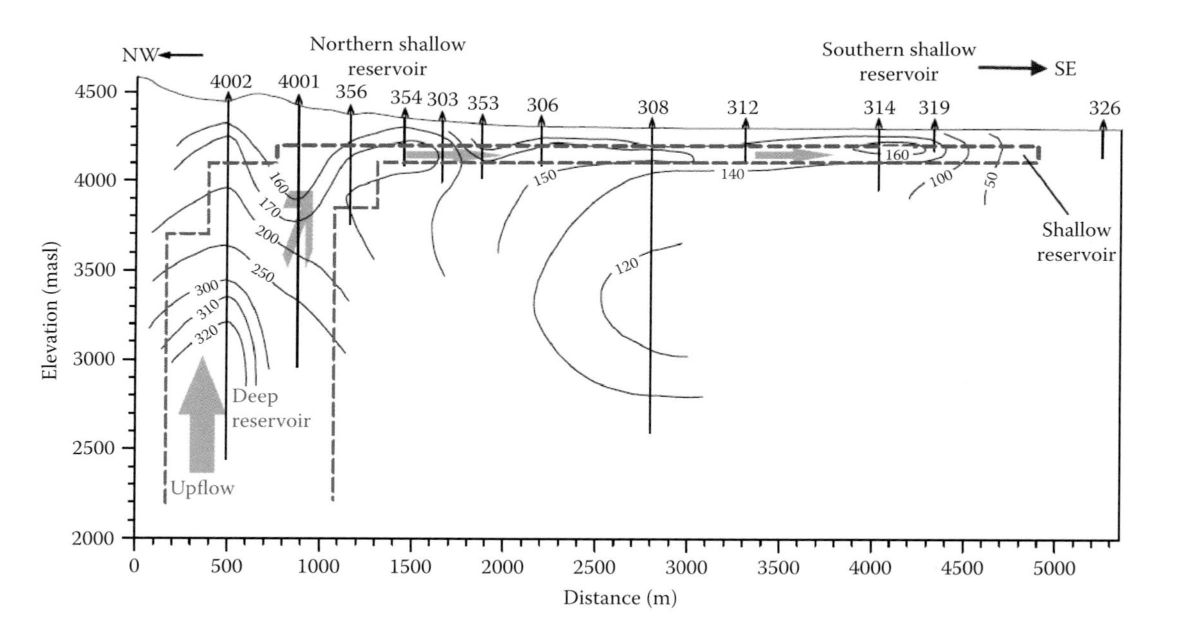

FIGURE 7.20 Northwest–southeast cross showing the location of select wells (shown by numbers) and their depths along with plotted isotherms (in °C). Diagram shows the inferred upwelling zone that bleeds off to the southeast to supply the north and south shallow reservoir zones. Faults, shown in Figure 7.19, have been omitted from cross-section. (From Xiaoping, F., *Conceptual Model and Assessment of the Yangbajing Geothermal Field, Tibet, China*, The United Nations University, Geothermal Training Programme, Reykjavik, Iceland, 2002.)

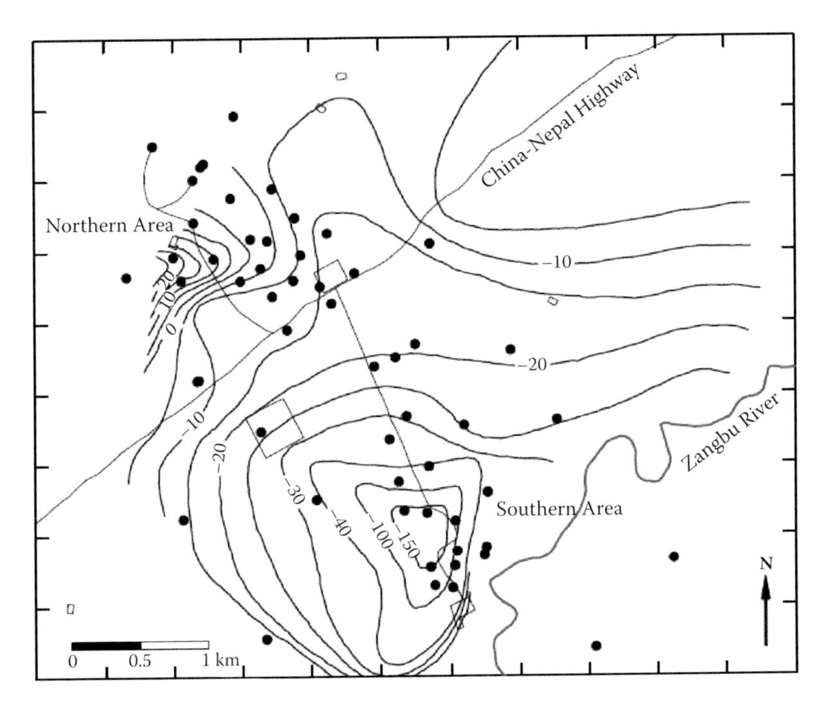

FIGURE 7.21 Map showing contours (in mm) showing changes in surface elevation during the period from 1989 to 1993. Note that, while the southern area has subsided by more than 150 mm, the northern area has risen as much as 25 mm during the same period. (From Xiaoping, F., *Conceptual Model and Assessment of the Yangbajing Geothermal Field, Tibet, China*, The United Nations University, Geothermal Training Programme, Reykjavik, Iceland, 2002.)

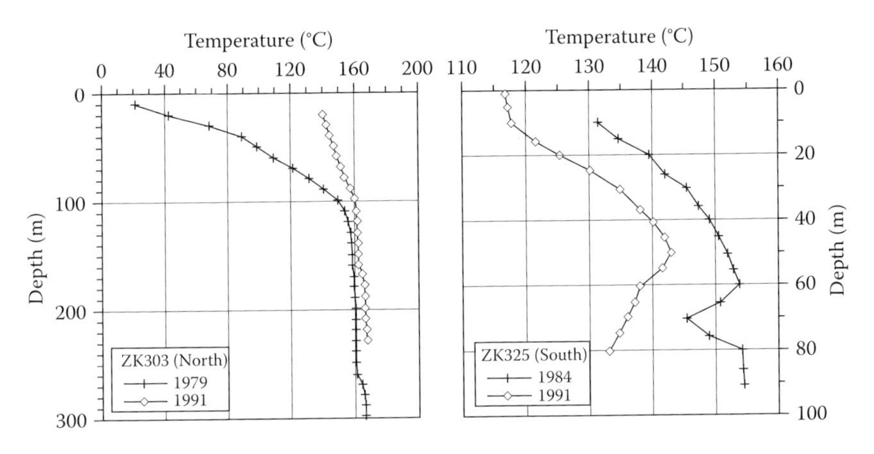

FIGURE 7.22 Temperature with depth profiles for wells in the northern and southern part of the Yangbajing geothermal reservoir as a function of time. Note that the temperature in the northern part of reservoir has increased slightly with time, whereas that of the well in the southern part of the field has decreased notably with time. (From Xiaoping, F., *Conceptual Model and Assessment of the Yangbajing Geothermal Field, Tibet, China*, The United Nations University, Geothermal Training Programme, Reykjavik, Iceland, 2002.)

most common anion (60% of total anion content), and sodium (Na^+) is the most common cation (85% of total cation content). Although CO_2 is the main gas component, the low amount of calcium suggests that calcium carbonate (calcite) scale may not be much of a problem because sodium carbonate or sodium bicarbonate (the chemical name for baking soda) is much more soluble than calcite.

Transform Boundary Settings

As discussed in Chapter 4, most transform boundaries are found in the ocean floor associated with mid-ocean ridge spreading zones. However, a few notable ones occur on land, including the Anatolian fault of northern Turkey, the Alpine fault on the South Island of New Zealand, and the San Andreas fault extending from northwestern Baja Mexico to northwestern California. Inboard of the San Andreas fault is a zone of distributed right-lateral shear reflected by the Eastern California Shear Zone and Walker Lane of western Nevada. Geologic findings and global positioning system (GPS) studies indicate that about 20% of the right-lateral motion between the Pacific and North American tectonic plates is taken up by discontinuous north- to northwest-striking right-lateral faults making up the Walker Lane and Eastern California Shear Zone (Faulds et al., 2005). Geothermal systems associated with transform boundaries are generally limited. However, some important exceptions occur in places where extension and/or magmatism occur in association with the transform motion.

San Andreas Fault System

As with many transform fault settings, geothermal activity is limited along most of the San Andreas fault zone. The two important exceptions are near the fault zone's two ends—the Salton Sea/Cerro Prieto fields at the south end and The Geysers at the north end (Figure 7.23). As was introduced in Chapter 4, the south end of the San Andreas fault system is a zone of transtension produced by a series of right-stepping, northwest-striking, right-lateral faults. This right-stepping creates localized areas of extension for right-lateral faults to form pull-apart basins that fill with sedimentary and volcanic rocks, producing potentially permeable, productive geothermal reservoirs and relatively impermeable cap rocks. Heat flow is elevated in these localized zones of extension due to crustal thinning and decompression that lower the melting point of hot rocks at depth. Furthermore, it is a region where seafloor spreading and the opening of the Sea of Cortez transition into transform faulting as narrow spreading ridge segments die out northward (Figure 7.23). In places, volcanism is localized along the strike-slip faults ("leaky transforms") or along northeast-striking normal faults that serve as links between right-stepping, northwest-striking, right-lateral faults (Figure 7.24). Thus, the southern region of the San Andreas fault has two important attributes favorable for promoting geothermal systems. First, it is an area of localized transtension forming northeast-striking normal faults that provide open space and improved permeability for the rise and circulation of heated fluids (Bennett, 2011). Second, potent, shallow sources of heat from magma are available because of the buried spreading ridge segments and thinning of the crust from extension and associated lower pressure to induce melting of hot rock.

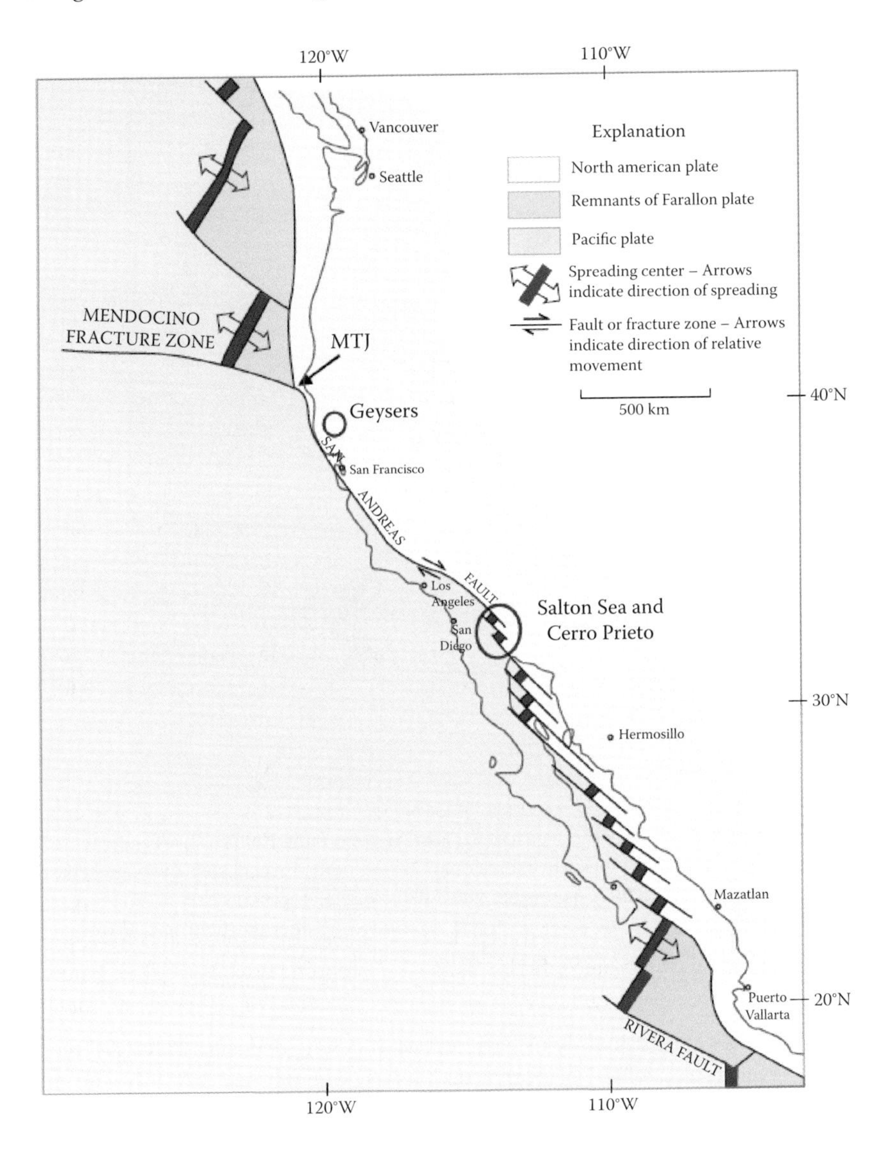

FIGURE 7.23 Map of San Andreas fault showing locations of The Geysers and Salton Sea/Cerro Prieto geothermal fields. Also shown is the current location of the Mendocino triple junction (MTJ) which continues to migrate northward with time.

For these reasons, Cerro Prieto is the largest geothermal field in Mexico. It has an over 800-MWe installed power capacity and taps fluids with temperatures as high as 350° to 370°C. The nearby geothermal fields in the Salton Sea and Imperial Valley region to the north tap fluids of similarly high temperatures but contain very high total dissolved salt contents (as much as 30 wt% compared to 3 wt% at Cerro Prieto). The difference in the total dissolved salt content between the two regions stems from the different types of rocks making up the geothermal reservoirs. At Cerro

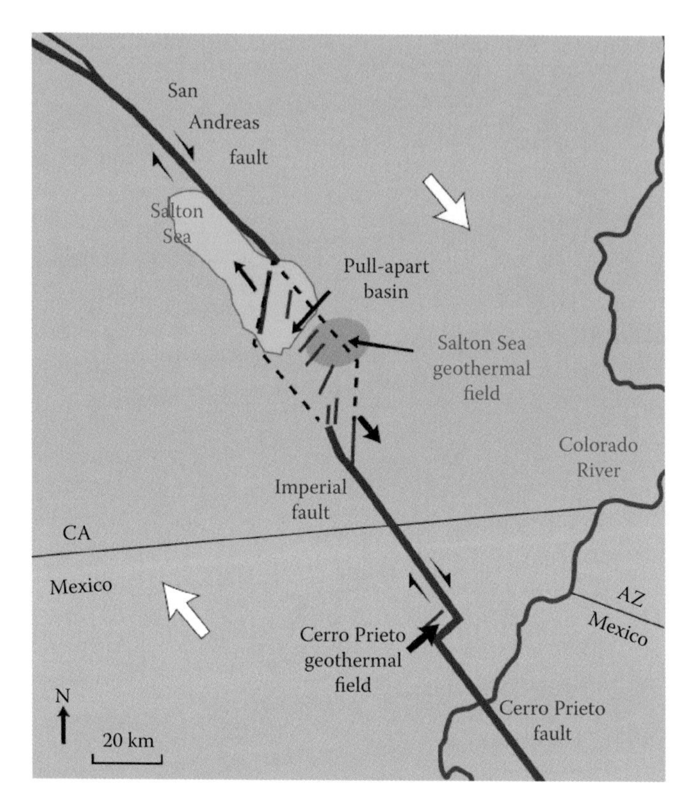

FIGURE 7.24 Map of the southern part of the San Andreas fault. Bold red lines indicate strike-slip faults, and thin red lines indicate extensional normal faults linking step-overs in strike-slip faults. Pull-apart basins localize extension, volcanism, and geothermal systems. The Salton Sea region (Imperial Valley) is the second largest developed geothermal field in California, and Cerro Prieto is the largest producing geothermal field in Mexico. (Illustration adapted from http://citizensjournal.us/wp-content/uploads/2014/12/SanAndreasSalton-835x1000.jpg.)

Prieto, the rocks consist mainly of buried coastal deltaic deposits. In the Salton Sea area, however, rocks consist of evaporites (deposits of salt and gypsum) and dried lake bed deposits, indicative of internal drainage in an arid environment. Because salt and gypsum are readily soluble, the geothermal fluids in the Salton Sea region acquired an extremely high total dissolved salt content. The geothermal fields in the Salton Sea regions have been slower to develop compared to Cerro Prieto due to the difficulty in building new transmission lines for environmental reasons ("not in my backyard" syndrome) and the extra costs of equipment to handle the high TDS content of geothermal fluids. However, with California's aggressive portfolio standards for renewable energy (33% of energy used in California must come from renewable sources by 2020 and a proposed 50% by 2030), added transmission lines will likely be built in the near future. Installed geothermal power capacity for the Salton Sea region (which includes the neighboring geothermal power facilities at East Mesa, Brawley, and Heber) is about 610 MWe and was expected to expand to about 660 MWe by 2015 with completion of the Hudson Ranch II. The developable geothermal power potential for the Salton Sea region is estimated to be as much as 1800 MWe by 2030 (Gagne et al., 2015), which means that in the next 5 to 10 years it could overtake The Geysers as the largest producing region of geothermal energy in California.

The Geysers, located near the north end of the San Andreas fault, is the world's foremost vapor-dominated geothermal resource. It has an installed capacity of about 2000 MWe from 22 power plants. Net power production from the field, however, averages about 800 MWe (M. Walters, pers. comm., 2015). The large difference between installed capacity and net power generation results from initial overproduction from the mid-1980s to early 1990s, which resulted in rapid pressure decline in the reservoir, as only about 25% of the steam mass removed for power production was reinjected (most of the mass being lost from the evaporative cooling towers). To arrest the decline, two pipelines carrying a combined 20 million gallons per day of treated municipal sewage from Clear Lake and Santa Rosa were completed in 1997 and 2003, respectively. Those piped fluids are now injected into the reservoir, which has increased the amount of injectate to about 80% of that removed for production. It has a stabilized power output of about 800 MWe. Although significant (enough to power the city of San Francisco), this is less than half of what The Geysers produced in the late 1980s and early 1990s and serves as a good illustration that geothermal systems must be carefully managed to remain sustainable (discussed further in Chapter 12).

Geologically, the source of heat at The Geysers is magmatic but unrelated to spreading or extension, as is the case at the southern end of the San Andreas fault. As noted in Chapter 4, magmatism at The Geysers is related to the transition from a previous subduction margin to the current transform margin. The San Andreas fault formed as a spreading ridge, and the transform margin of the now extinct Farallon plate was subducted (Figure 7.25). When this occurred, two triple junctions formed—(1) a ridge–transform–subduction junction (Rivera triple junction) that moved south with time, and (2) a transform–subduction–transform junction (Mendocino triple junction) that moved north with time. In the wake of the northward-moving Mendocino triple junction, hot mantle material welled into the "gap" formed behind the earlier subducted slab of the Farallon plate. As the mantle welled upward, it began to melt as pressure was reduced, forming pockets of magma that rose into the upper crust to form shallow magma chambers. Some of these chambers erupted to form the young Clear Lake Volcanic Field (noted as CL in Figure 7.25) that lies adjacent to The Geysers steam field and contains volcanic units as young as 10,000 years old (Donnelly-Nolan et al., 1981, 1993).

Underlying much of The Geysers steam field is a granitic intrusive complex that has been previously termed the "felsite" or Geysers plutonic complex. It consists of three texturally and mineralogically distinct units (Dalrymple et al., 1999; Hulen and Walters, 1993; Hulen et al., 1997). Radiometric dating of samples from the intrusive complex (U–Pb ages on zircons) indicates that the intrusive complex was episodically emplaced between ~1.1 and ~1.8 Ma (Schmitt et al., 2003). Modeling and heat-flow calculations indicate that a 1-Ma-old intrusive mass the size of that underlying The Geysers could no longer generate enough heat to sustain the modern geothermal system (Stimac et al., 2001). Instead, the heat source for The Geysers steam field appears to stem from numerous small shallow silicic intrusions emplaced over the entire region (about 750 km^2) of high heat flow. Most of these intrusions are now solidified or nearly so (Stimac et al., 2001).

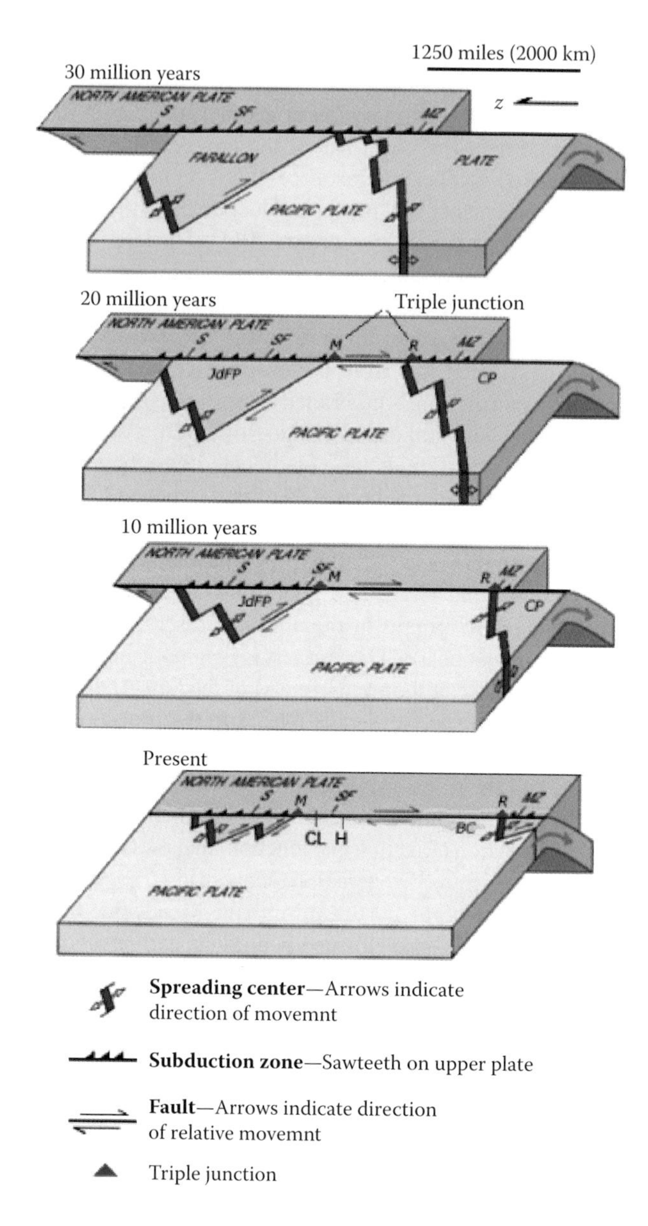

FIGURE 7.25 Formation of the San Andreas fault with time. The Geysers and the adjacent Clear Lake Volcanic Field (CL) formed in the wake of the northward migrating Mendocino triple junction (M). At the beginning of the formation of the San Andreas fault, the Farallon plate is split into the Juan de Fuca plate (JdFP) to the north and the Cocos plate (CP) to the south. The southward migrating Rivera triple junction (R) bounds the south end of the San Andreas fault. Other abbreviations are Seattle (S), San Francisco (SF), Hollister (H), MZ (Mazatlan), and BC (Baja California). (From USGS, *Geologic History of the San Andreas Fault System*, U.S. Geologic Survey, Washington, DC, 2006.)

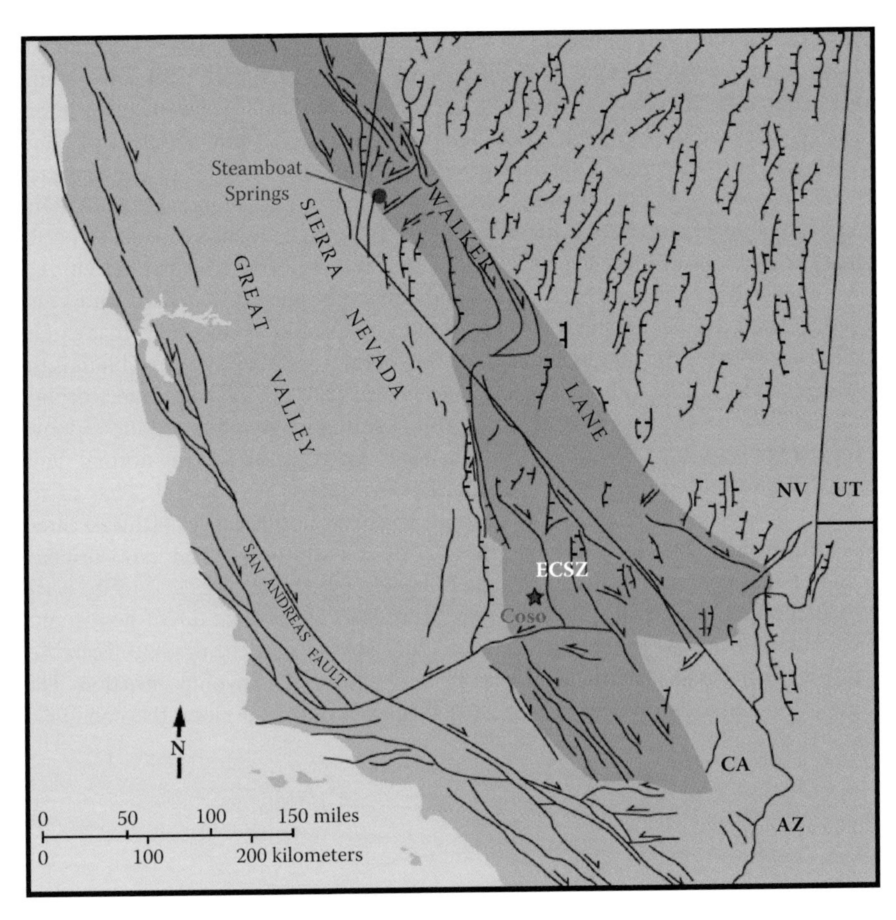

FIGURE 7.26 Map showing location of Steamboat Springs geothermal system. Shaded regions represent the Walker Lane and Eastern California Shear Zone (ECSZ), which consists of a region of discontinuous northwest-striking right-lateral faults and associated faults. Arrows along faults denote a large component of strike-slip movement, and faults with short hachures denote mainly normal movement with hachures on downdropped side. (Adapted from Henry, C.D. and Faulds, J.E., *Geological Society of America Special Papers*, 434, 59–79. 2007.)

Walker Lane and Eastern California Shear Zone

As introduced earlier, about 20% of the right-lateral motion between the Pacific and North American tectonic plates is taken up along a zone of diffuse right-lateral shear along the east side of the Sierra Nevada (Eastern California Shear Zone) and Walker Lane in western Nevada (Figure 7.26). Rather than a major through-going fault like the San Andreas, here motion is taken up along a series of north and northwest-striking right-lateral faults. Two good examples of geothermal systems include Steamboat Springs in the Walker Lane and Coso in the Eastern California Shear Zone.

Steamboat Springs is located just south of Reno, Nevada, and began producing geothermal power in the mid-1980s. It consists of one flash plant with a bottoming binary unit, and five other operating binary, air-cooled plants. Total installed power capacity is about 140 MWe with an annualized average net power output of about 90 MWe, reflecting parasitic loads and seasonal and daily changes of condensation efficiencies (higher output in winter, lower output in summer; higher output at night, lower output during the day). Although Steamboat is located within a zone of northwest-striking right-lateral shear, faults that largely control geothermal activity are generally north- to northeast-striking normal faults. The northeast-striking faults also have localized a series of small rhyolitic domes (<1.5 million years old) that extend northeast across the Steamboat Hills to the foot of the Virginia Range that bounds the east side of the valley. Although young, these domes are not the heat source driving the Steamboat geothermal system, but deeper residing magma that fed these domes is the likely heat source (White, 1985). The north- to northeast-striking normal faults result from extension associated with northwest-directed right-lateral shear of the Walker Lane. When this happens, a component of west-northwest to northwest direction extension occurs which forms north to northeast-striking normal faults oriented more or less perpendicular to the direction of extension (Figure 7.27).

Although the young rhyolite domes at Steamboat are not the direct heat source for geothermal fluids, the temperature of fluids, based on drilling data, increases to 220°C or more toward the southwest in the vicinity of a rhyolite intrusion. This relationship suggests that elevated heat flow may be focused along the conduit of

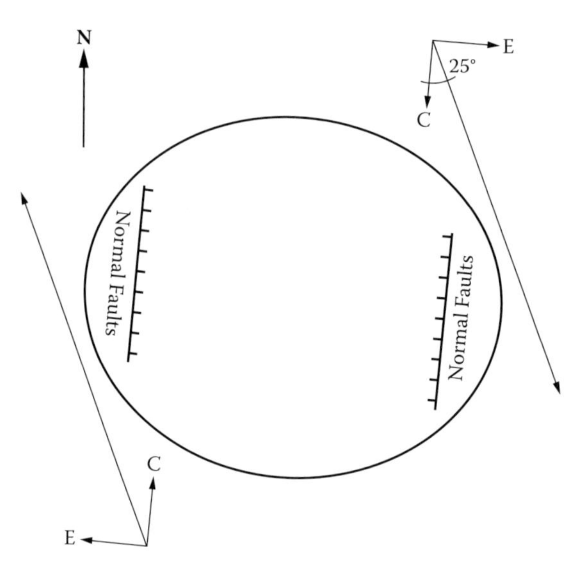

FIGURE 7.27 Strain ellipse for a northwest-directed right-lateral shear. If a circle is subjected to shear, such as right-lateral as shown in this case, it is deformed into an ellipse. The long axis is being pulled or extended (E) in a west-northwest direction, whereas the short axis is being compressed (C) in a north-northeast direction. As more shear is applied, the ellipse is further stretched and rotated clockwise to produce a range of north- to northeast-striking normal faults.

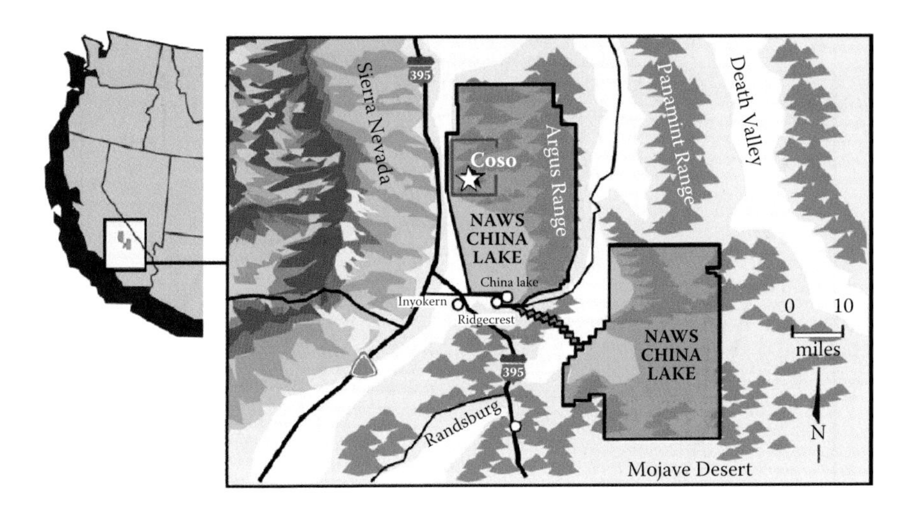

FIGURE 7.28 Map showing location of Coso geothermal field and major nearby physiographic features. Coso is entirely enclosed within the Naval Air Weapons Station (NAWS). (From Monastero, F.C., *Geothermal Resources Council Bulletin*, 31(5), 188–193, 2002.)

the rhyolite dome that leads to a deep, large residing body of magma. The higher temperature fluid in this region is utilized by the Steamboat Hills flash plant and the Galena 2 combined cycle power plant, in which separated steam goes to a back pressure steam turbine and the exhaust steam and separated brine are used to vaporize and preheat a secondary working fluid that powers a separate turbine. Most of the other binary plants tap the shallow (200 to 300 m deep) outflow plume that flows to the northeast and averages about 165°C.

The Coso geothermal system is located at the south end of the Eastern California Shear Zone (Figures 7.28 and 7.29). It is harnessed by four geothermal power plants having an installed capacity of 270 MWe. Net power output, however, has been averaging <200 MWe mainly due to a deficit of water for cooling the steam condensate. The Coso facility is in an arid environment and must share the groundwater with neighboring ranches. Temperatures of fluids in the reservoir range from 200° to 328°C, resulting in the use of double-flash technology for power generation. Fluids are moderately saline, as they contain 7000 to 18,000 ppm total dissolved solids. It is a liquid-dominated reservoir, although as a consequence of production local steam pockets have formed, not unlike Wairakei in New Zealand, indicating that the reservoir is liquid, but not heat, limited.

Coso, like Steamboat, is a magmatically heated geothermal system as evidenced by 38 rhyolite domes, which are as young as 40,000 years old. These domes are upper extensions of an underlying silicic magma reservoir that is supplying heat to the overlying geothermal system. Structurally, the Coso geothermal field lies in an extensional or releasing bend (step-over) of two right-lateral north-northwest-striking faults: the Little Lakes fault to the southwest and an unnamed fault to the northeast bounding the Argus Range (Figure 7.29). The releasing bend consists of a dense array of northeast-striking normal faults and the northwest-striking right-lateral

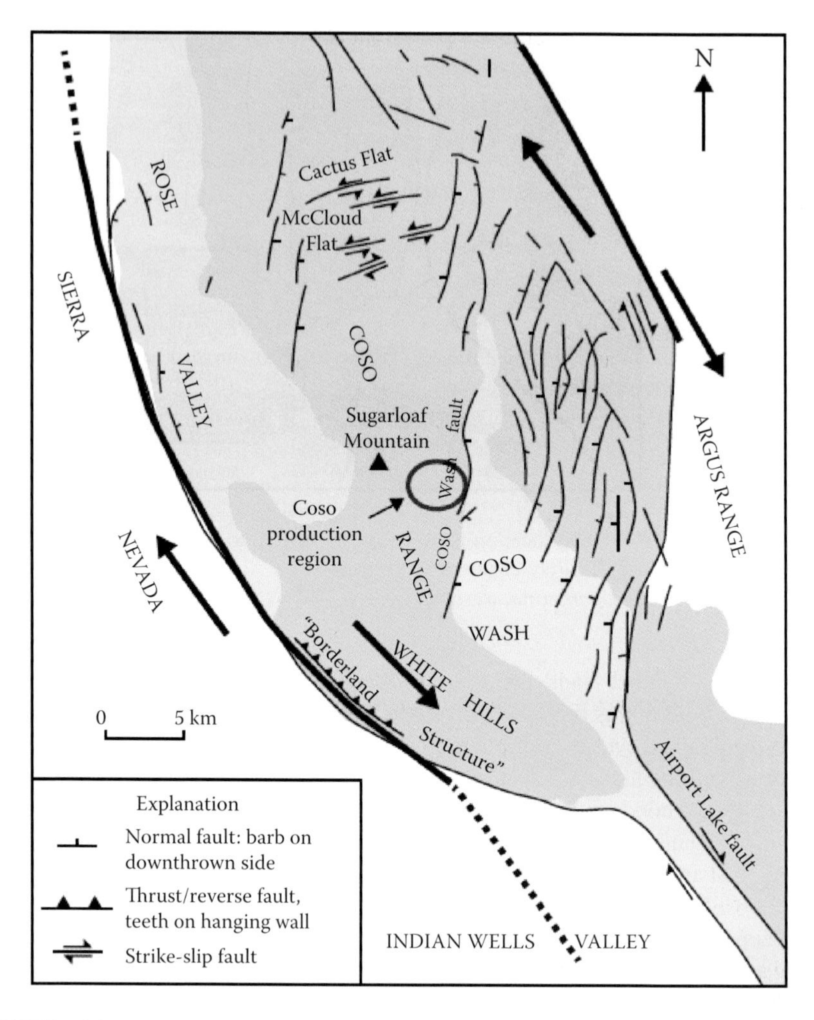

FIGURE 7.29 Map showing faults in the vicinity of the Coso geothermal field. The northeast-striking faults are normal faults linking a step-over (or releasing bend) in right-lateral strike-slip faults making up part of the Eastern California Shear Zone. (Adapted from Monastero, F.C., *Geothermal Resources Council Bulletin*, 31(5), 188–193, 2002.)

Airport Lake fault, which extends across the releasing bend and projects through the geothermal production area. The dilation provided by the northeast normal faults and cross-cutting northwest strike-slip faults forms a zone of high fracture permeability that likely fostered the rise of magma to form the rhyolite domes and promotes permeability and the circulation of hot water. Using the results of global positioning technology, researchers have determined that the Argus Range in the vicinity of Coso is moving 6.5 mm/year to the northwest relative to the Sierra Nevada Range to the west, reflecting the active right-lateral sense of shear of the region (Dixon et al., 2000; Unruh et al., 2003).

Coso is also under consideration as a candidate for the Frontier Observatory for Research in Geothermal Energy (FORGE) program sponsored by the U.S. Department of Energy (EERE, 2016). Scientists and engineers will explore and test new concepts at these sites with the goal of achieving breakthroughs that will lead to widespread development of enhanced geothermal systems (EGSs), which currently remain experimental and expensive; EGSs are discussed further in Chapters 11 and 12.

Continental Rifting and Geothermal Systems

Continental rifting consists of regions undergoing crustal extension. If allowed to go to completion a new ocean basin can begin to form, such as the Red Sea, which formed as the Arabian plate rifted from the African plate about 5 million years ago (although continental rifting in that region began about 30 million years ago; thus, about 25 million years of rifting occurred before the ocean basin formed). Two different continental rifts are reviewed here—the East African Rift and the Basin and Range Province lying mainly within Nevada and western Utah. The former largely involves magmatic heat sources, whereas the latter is mainly, but not entirely, amagmatic.

Magmatic East African Rift Zone

The East African Rift zone extends south-southwest from the junction of the south end of the Red Sea and the Gulf of Aden (Figure 7.30). Indeed, in a desolate region called the Afar Triangle, the Red Sea and Gulf of Aden have already begun invading the north end of the East African Rift. The rift extends mainly across the countries of Djibouti, Burundi, Ethiopia, Kenya, Uganda, and Tanzania, but it also affects Eritrea, Rwanda, Democratic Republic of the Congo, Zambia, Malawi, and Mozambique. Hydropower and diesel-powered generators are currently the main sources of electrical energy for these countries, but with climatic fluctuations, including extended periods of drought, and silting of reservoirs and high costs of diesel fuel, concerns have arisen regarding the reliability of such resources for the long term. Consequently, these countries, especially Kenya, are making use of their geologic setting and actively exploring and developing geothermal resources, as inherent fossil fuel resources are essentially absent.

Kenya is the current leader in developing geothermal resources in the region, with geothermal power production beginning there in the mid-1980s. Most of the current production comes from the Olkaria volcanic complex with a recently completed installed capacity of 540 MWe. This represents an increase of more than 100% from just 2 years earlier when installed capacity was 198 MWe. Altogether Kenya has 14 known geothermal regions within the rift having an estimated geothermal power potential of 3000 MWe, which would be sufficient to supply the power needs of Kenya for the next 20 years. Exploration underway at the Menengai geothermal field, located about 50 km north of Olkaria, indicates that it has the potential for developing an additional 400 MWe. By 2020, the geothermal power capacity of Kenya could approach 1000 MWe from the most promising geothermal fields at Olkaria, Menengai, Longonot, and Eburru (Figure 7.31), assuming additional financial capital can be secured.

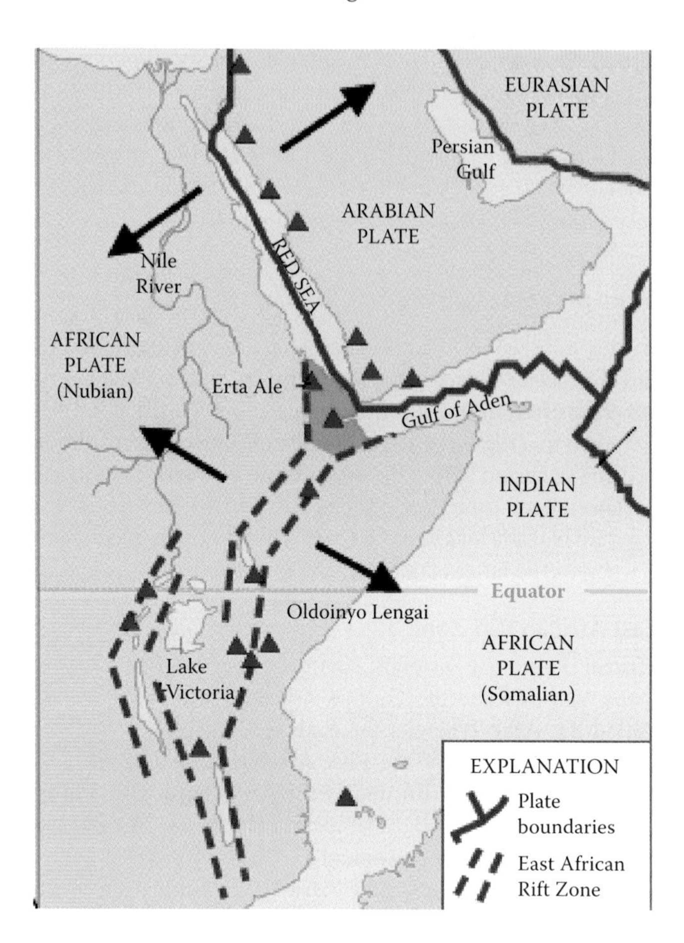

FIGURE 7.30 Map of east African rift zone outlined by bold dashed lines. Blue arrows denote the direction of plate movement. The eastern horn of Africa will eventually split away, just as the Arabian plate did about 5 million years ago to form the Red Sea. (Illustration adapted from https://upload.wikimedia.org/wikipedia/commons/0/0d/EAfrica.png.)

Geologically, Olkaria consists of a series of rhyolite lavas and domes, the youngest of which is only a few hundred years old. The geothermal system covers about 80 km^2 and appears localized along the margin of the volcanic center and normal faults associated with the west side of the East African Rift zone. Some normal faults provide fluid upflow zones for the hottest geothermal fluids, whereas others appear to serve as recharge zones, resulting in discrete zones of higher and lower fluid temperatures within the overall geothermal field. Geothermal power plants tap fluids generally in the 200° to 220°C range, although some temperatures of >300°C occur in parts of the field. Flash plants occur at Olkaria I and Olkaria II power facilities in the eastern and northern part of the field, respectively, whereas combined flash binary units have been developed at Olkaria III in the western part of the field. Alteration mineralogy consists of silica, calcite, and illite—consistent with near-neutral to modestly alkaline pH fluids. Well depths include an average depth of

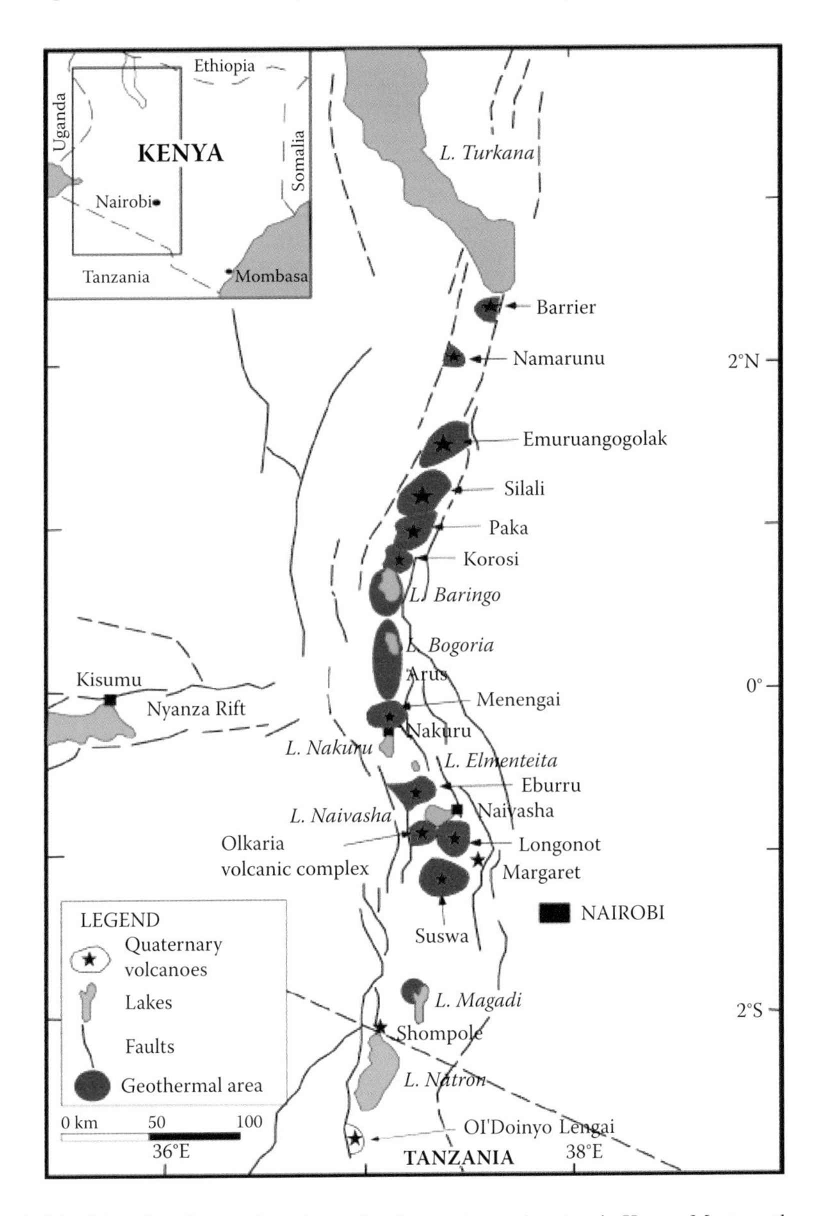

FIGURE 7.31 Map showing locations of major geothermal system in Kenya. Most geothermal fields are associated with young or active volcanoes indicating a magmatic heat source for driving the geothermal systems. (Illustration from http://www.renewableenergy.go.ke/asset_uplds/images/Geo%20sites.jpg.)

1200 m at Olkaria I and 2200 m at Olkaria II, with more recently drilled production wells averaging about 3000 m deep. For a magmatically heated system, these well depths are unusually deep, suggesting that perhaps the magmatic heat sources are rather small or widely scattered.

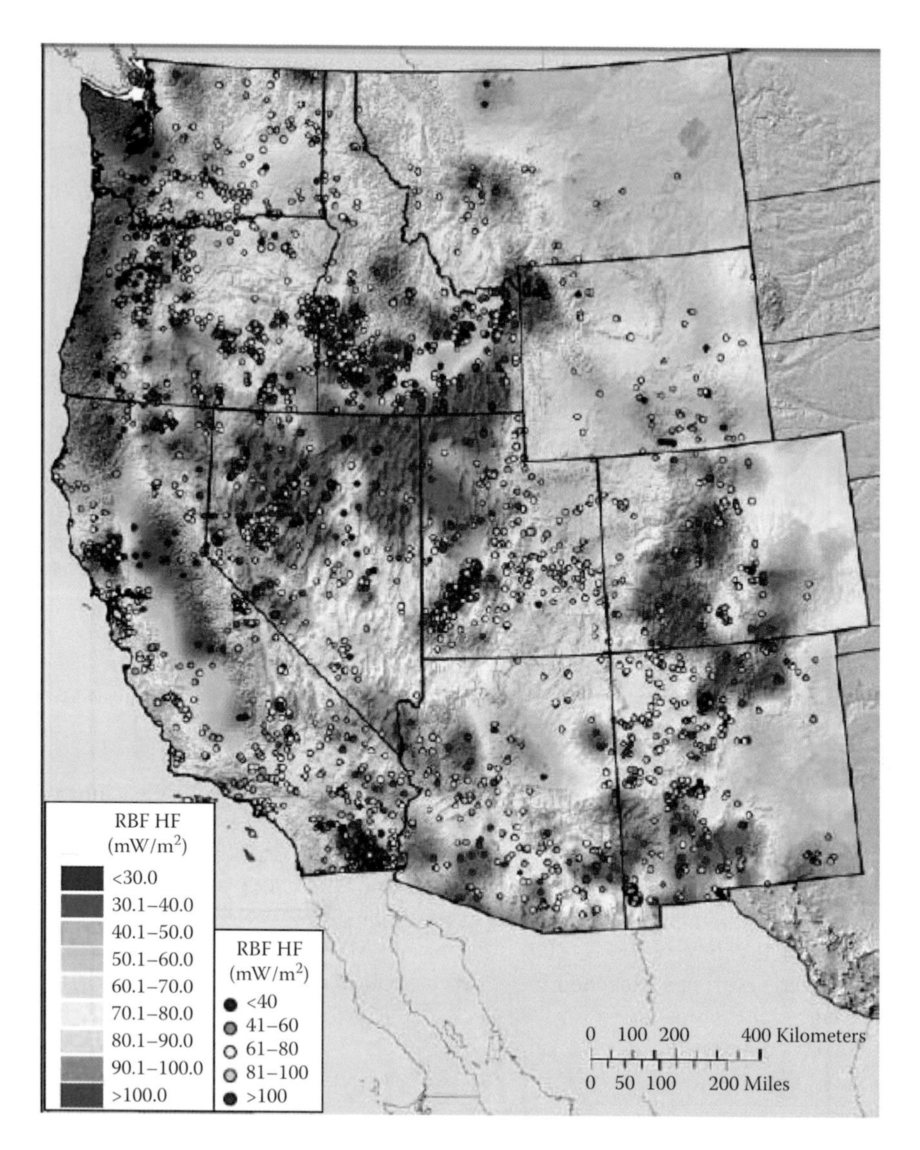

FIGURE 7.32 Heat flow map of the western United States compiled from measured heat flow data of wells. (From Williams, C.F. and DeAngelo, J., *Geothermal Resources Council Transactions*, 35, 1599–1605, 2011.)

Amagmatic Northern Basin and Range Province

This region is concentrated mainly in northern Nevada but also extends into eastern Oregon, southern Idaho, and western Utah (Figure 7.32). Geologically, the region is undergoing crustal extension on the order of several millimeters per year. As a result, the crust is thinned, which effectively brings hot mantle rocks closer to the surface, resulting in widespread elevated geothermal gradients (on average, 50° to 60°C/km, which is more than twice the crustal average) and high heat flow (in many places greater than 150 mW/m^2, again about twice the crustal average). Furthermore, due to the ongoing crustal extension, normal faults continue to form and move, providing channels for fluids to circulate to depth, where they are heated and then rise toward the surface along other normal faults to form potentially accessible geothermal reservoirs.

Despite the pressure reduction on underlying mantle rocks from ongoing extension and crustal thinning, magmatism or volcanism is largely lacking through much of the central Basin and Range Province. This dearth of volcanism is not well understood but could reflect ponding of magma in the lower and middle crust, at depths of 20 to 25 km, due to poor density contrasts between the magma and crustal rocks. Without strong density contrasts, buoyancy forces to drive magma upward into the upper crust are limited. Lower crustal rocks were transformed into less dense varieties from an earlier period of magmatism that affected much of the central and northern Basin and Range Province between 30 and 20 million years ago. As a result, without the presence of high-level crustal magma bodies, geothermal fluids are generally <200°C, with most in the range of 150° to 180°C for developed geothermal systems. Consequently, most geothermal plants in Nevada are of the binary type, having installed power capacities ranging mainly between 15 and 30 MWe (Figure 7.33). Notable exceptions are the dual-flash plant at Dixie Valley (67 MWe installed) and the new twin binary power plants at McGinness Hills, with 72 MWe installed capacity. Production fluids at Dixie Valley are in the 210° to 230°C range. The high fluid temperatures in part reflect that most production wells are quite deep, averaging around 3 km, compared to typically <2 km deep for most wells in the other producing geothermal fields in Nevada.

A second possible reason for the high temperatures of geothermal fluids at Dixie Valley is that geophysical data (magnetotelluric studies) suggest that a region of partially molten rock may underlie the Dixie Valley region at depths of 10 to 15 km, adding more heat than is found elsewhere in the province (see Figure 8.22 in Chapter 8) (Wannamaker, 2005). The new McGinness Hills geothermal facility consists of two binary power plants rated at 30 MWe each.* Actual power production, however, has exceeded installed capacity for both plants by about 10%. Although fluid temperatures are a modest 150° to 170°C, the added power production reflects high well flow rates, reported to be as much as 677 gpm, along with high productivity indexes† of several tens to several hundreds of gpm/psi (Nordquist and Delwiche, 2013). These high flow rates of geothermal fluids in wells from Phase I resulted in Ormat building the second geothermal power plant at McGinness Hills, making the two-power-plant facility the largest geothermal power producer (72 MWe) in the Great Basin after Coso. Most wells achieve bottom-hole temperatures of 150°C or more at depths between about 600 and 1000 m, with one well reaching a temperature of about 160°C at a depth of about 1600 m (Nordquist and Delwiche, 2013).

* Phase I of the facility, completed in 2012, is supported by five production wells and three injection wells. Phase II was completed in 2015 and is also supported by five production wells and three injection wells.

† The productivity index is given by the formula $Q/(P_s - P_f)$, where Q is the flow rate, P_s is static pressure at the bottom of the well while shut in, and P_f is flowing pressure at the bottom of the well. A small difference between P_s and P_f indicates very good rock permeability, as P_f is only slightly reduced during flow.

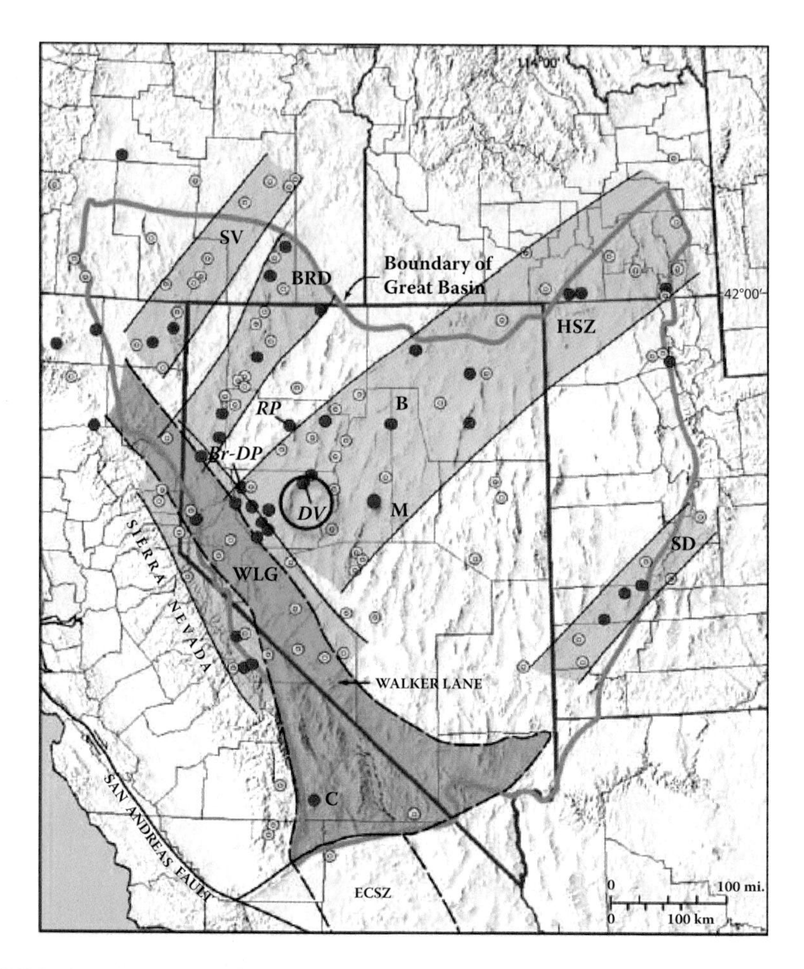

FIGURE 7.33 Map showing locations of Dixie Valley (DV) and Beowawe (B) flash power plants. Yellow circles denote geothermal systems with temperatures of 100° to 160°C, and red circles denote geothermal systems with temperatures greater than 160°C. Green zones denote geothermal belts. Abbreviations: SV, Surprise Valley; BRD, Black Rock Desert; HSZ, Humboldt structural zone; SD, Sevier desert; WLG, Walker Lane belt; ECSZ, Eastern California Shear Zone; RP, Rye Patch, Br-DP, Brady's and Desert Peak; C, Coso; M, McGinness Hills. (From Faulds, J.E. et al., in *Great Basin Evolution and Metallogeny*, Steininger, R. and Pennell, B., Eds., DEStech Publications, Lancaster, PA, 2011, pp. 361–372. Copyright © Geological Society of Nevada.)

Hot Spots and Associated Geothermal Systems

Hot spots are surface expressions of upwelling and relatively stationary plumes of mantle material. As the mantle material rises, pressure is reduced, which induces partial melting and the rise of magma to form volcanic centers. As the tectonic plate moves over the mantle plume, a series of volcanic centers or volcanic islands form. One of the best examples of hot spot volcanism is the Hawaiian Islands, which lie in the middle of the large Pacific tectonic plate. The Big Island of Hawaii is currently

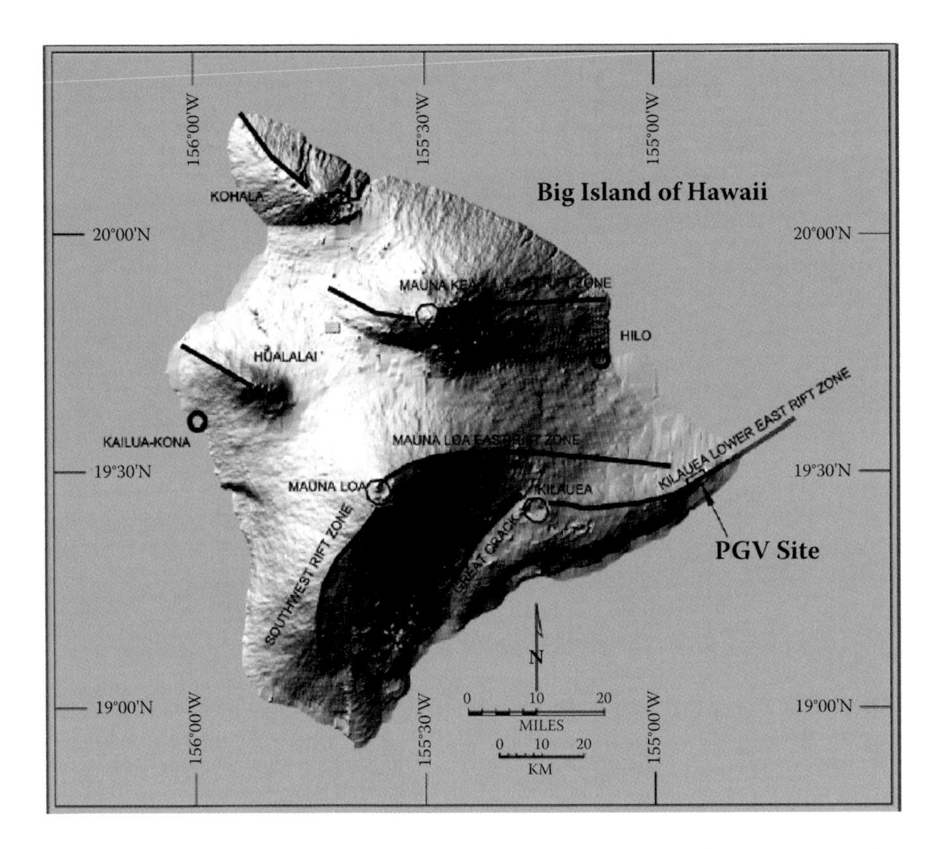

FIGURE 7.34 A labeled DEM image of the Big Island of Hawaii and location of the Puna Geothermal Venture (PGV) at the east end of the Kilauea east rift zone. The Big Island is made up of the five coalescing volcanoes of Kohala, Mauna Kea, Hualalai, Mauna Loa, and Kilauea (http://www.soest.hawaii.edu/coasts/images/hawaii_dem.jpg).

over the plume and is comprised of five volcanoes—the most active of which is Kilauea located at the southeast end of the island (Figure 7.34) and most directly above the plume. A tectonically more complicated example of geothermal development associated with mantle plume upwelling is the Azores, located about 1600 km west of Portugal.

Hawaii

On the Big Island of Hawaii, the Puna geothermal field and power plant facility are located near the east end of the East Rift zone of the Kilauea volcano (Figure 7.34). As testimony to the volcanic activity of the region, in 2005 one of the drill holes intersected magma at a depth of 2488 m. The temperature of the magma was 1050°C, hot enough to dislodge the carbide teeth of the drill bit and seize up the drill string. The Puna geothermal reservoir is hosted in basaltic flows and fragmental rocks, with main production zones associated with fractures and faults of the east Kilauea rift zone. The reservoir lies at depths between about 1500 and 2000 m and is liquid dominated, although steam pockets are intersected in some of the

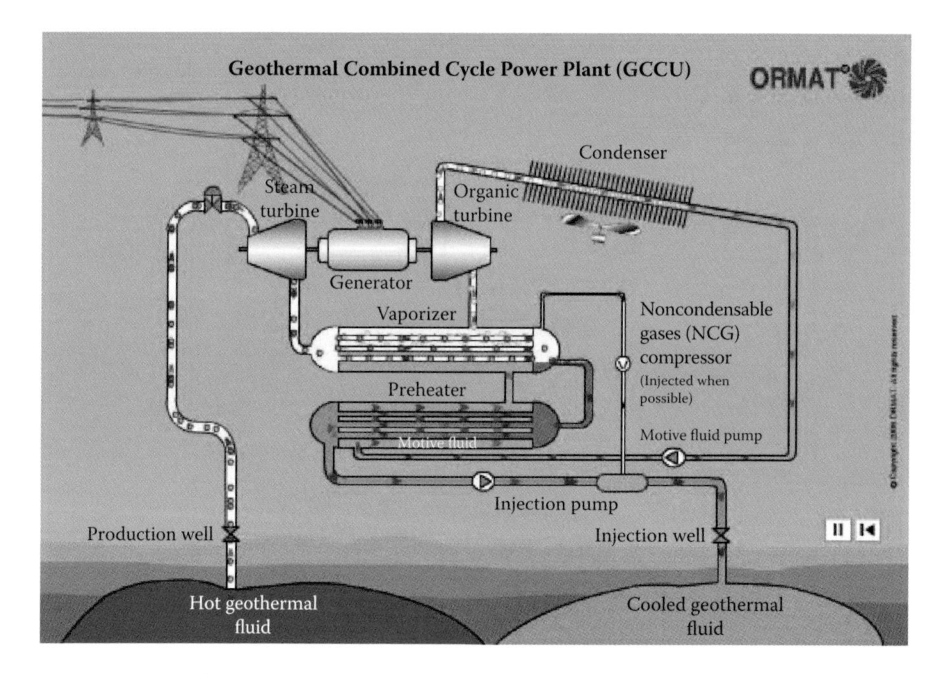

FIGURE 7.35 Diagram showing components and flow paths of geothermal fluid and working fluid in a combined cycle power plant as utilized at Puna. The separate 8-MWe binary cycle that uses brine from the wellhead separators is not shown in this diagram. (Illustration from http://www.ormat.com/solutions/Geothermal_Combined_Cycle_Units.)

wells. Temperatures of fluids are uniformly >260°C and in places exceed 300°C. The fluids are generally acidic with a pH of around 4.5, requiring some preventative measures to reduce corrosion of equipment. The relatively high acidity reflects gases escaping from nearby magma sources of mainly SO_2 (which disproportionates to H_2SO_4 and H_2S, as discussed in Chapter 6), HF, and HCl. The Puna geothermal system, because of its hot spot geologic setting, is one of the hottest geothermal systems developed, comparable to those in the Salton Sea region and in Iceland. The Puna geothermal power plant utilizes a combined cycle and binary system having an installed capacity of 38 MWe, which is enough to supply about 20% of the Big Island's power needs. A combined-cycle power plant (Figure 7.35) exploits geothermal systems having high steam pressure (as much as 70 bar at the wellheads of some Puna wells). The high-pressure steam is directed to a back-pressure turbine, which means the spent steam is not condensed in cooling towers to generate low pressure (hence back pressure) but rather goes to a vaporizer. In the vaporizer, the latent heat of condensation of the steam is used to vaporize a working fluid that powers a separate turbine.

This configuration originally produced 30 MWe beginning in 1993. A separate 8-MWe binary unit was installed in 2011 to make use of the 200°C brine at the wellhead separators to bring the current capacity to 38 MWe. It is currently the only geothermal facility in the state, and efforts to expand have been stymied by concerns about disturbing cultural values by using the land for power generation. Critics should be alerted to the fact that the Puna geothermal plant saves nearly 150,000 barrels of heavy fuel oil per year that otherwise would be imported to provide power. Also, residents could be informed of other benefits that the plant could provide, such as hot water for bathing and washing and perhaps even for prawn farming, that could economically benefit this rural, low-income region of Hawaii.

The Azores

The tectonic setting of the Azores is not as straightforward as that of Hawaii. The Azores, somewhat similar to Iceland, the Galapagos Islands, and Reunion Island, appear to have formed in response to the interaction of seafloor spreading, crustal rifting, and mantle plume upwelling. Because of the intersection of these overlapping processes, a simple linear age progression of volcanic islands, as is the case of the Hawaiian Islands, is not so evident with the islands of the Azores archipelago. Nonetheless, the existence of a mantle plume below the Azores is supported by geochemical and geophysical data (e.g., Adam et al., 2013; Madureira et al., 2011). The Azores currently contain two binary geothermal power plants, having a combined installed capacity of 29 MWe (23 MWe net) (Bertani, 2015). These two plants tap the Ribeira Grande geothermal field, a 240°C liquid-dominated reservoir on the island of Sao Miguel. Both plants provide about 42% of the electrical power for the large island of Sao Miguel and about 22% of the total power needs for all the islands of the Azores (Carvalho et al., 2015). Curiously, the high temperature of the geothermal reservoir could certainly support a flash or double-flash facility and the reason why this was not done is not clear. It could reflect in part fluid chemistry; if there is high dissolved silica, significant scaling in flash plants can occur, which can certainly cause problems with power production. In binary plants, however, geothermal fluids remain under pressure and do not flash, inhibiting dissolved constituents from precipitating. Also, the fluids may contain high noncondensable gas contents (H_2S or CO_2), which can result in emission problems in a flash plant. Another factor could be costs, in that binary plants can be assembled in modular form, reducing construction costs. Generation of geothermal power on the Azores costs, including depreciation and taxes, about $0.08/kWh, which is less than half the cost of power produced by diesel-fueled generators. An additional geothermal resource has been found on the neighboring island of Terceira. Well tests at Pico Alto indicate the capacity to support a 2.5- to 3.0-MWe pilot power plant, and additional wells are planned for the Ribeira Grande field to increase power by about an additional 5 MWe (Carvalho et al., 2015).

Stable Cratons

These are regions located away from tectonic plate boundaries in areas of continents that are geologically stable (few or no earthquakes and no active volcanism). A couple examples of this setting would be hot water associated with producing oil fields in deep sedimentary basins (e.g., Teapot Dome oil field in central Wyoming, Williston Basin centered in western North Dakota) and deeply buried old, but hot granitic rocks due to radiogenic decay of U and Th, such as in the Cooper Basin in east-central Australia. Both examples are mainly in the experimental phase—in the former case due to the low temperature (90 to 95°C) of coproduced water with oil and the latter due to permeability issues and power consumed for pumping fluids down injection wells and up production wells. A third example of an intracratonic setting utilized for direct geothermal use is the Paris Basin.

Intracratonic Oil-Bearing Sedimentary Basins

A consortium of the National Renewable Energy Laboratory (NREL), the Department of Energy's Geothermal Technology Office (GTO), and the Rocky Mountain Oil Testing Center (RMOTC) has been investigating harnessing low-temperature geothermal water coproduced with oil at the Teapot Dome field. Beginning in 2008, a 280-kW organic Rankine-cycle (binary) plant has utilized 92°C water separated from producing oil wells to produce electricity (net output of about 200 kW of electricity). The plant has been online 97% of the time and has produced 1918 MWh of energy from 10.9 million barrels of coproduced geothermal water. NREL is experimenting with a hybrid cooling system in which water is sprayed on the air-cooled condensers to improve cooling efficiency and power production during the warm temperatures of the day and summer (NREL, 2011). Although geothermal waters coproduced from oil wells in the Williston Basin range from 100°C to more than 150°C at depths of 4 km, the flow rates are generally too low to be used for coproduced geothermal power production (Gosnold et al., 2013). The low rate of pumping is to maximize the amount of oil produced. However, when wells are watered out, flow rates could be increased significantly, allowing for possible geothermal power production potentially on the order of hundreds of megawatts and perhaps more. Currently, the economics greatly favor oil and gas extraction over fluid heat extraction for power production. Even coproduction is not economic, as oil and natural gas go directly to market, whereas utilization of fluid heat energy requires building a geothermal plant. Nonetheless, the money from oil and gas could also be used to build low-temperature organic Rankine-cycle plants that would supply the power needs of the field and additional power to the grid and likely improve revenue flow in the long term.

Buried Radiogenic Granitic Rocks

Another potential source of geothermal energy in stable cratonic interiors consist of deeply buried (depths of 3 to 5 km), old radiogenic granitic rocks. These granites contain high concentrations of uranium, thorium, and potassium and yield radiogenic heat outputs that are 4 to 20 times the outputs of normal granites. Where these granitic rocks are buried by shales, having low thermal conductivity, the heat from radiogenic decay can lead to temperatures well over 200°C at drillable depths (3 to 5 km). In Australia, two energy companies have been actively exploring and trying to develop geothermal energy for power from such geologic environments. One is Geodynamics' Habenero project in the Cooper Basin of northeast South Australia and the other is Petratherm's Paralana project located about 250 km to the south of the Cooper Basin. Both projects are targeting deep (~4 km) fracture zones in radiogenic granites. Temperatures of 240°C have been encountered at depths of about 4.4 km and 280°C at a depth of almost 5 km in wells at the Habenero project. A pilot demonstration plant for an injection–production well couplet at Habenero, with the two wells spaced about 0.5 km apart, produced 1 MWe for 160 days at a well-head temperature of 215°C and a mass flow rate of 19 kg/s. Although this project has demonstrated the feasibility of tapping such resources, whether such projects can be scaled up to make them economical will strongly depend on the price of the traditional fossil fuels, financial governmental incentives, and policies governing greenhouse gas emissions. As of mid-2015, both Australian projects had been put on hold due mainly to competition from a plentiful supply of low-cost fossil fuels.

Paris Sedimentary Basin

Geothermal development of the sedimentary Paris Basin, in particular the Dogger aquifer, began in the 1970s. Wells tap fluids ranging in temperature from 65° to 85°C at depths of 1500 to 2000 m. The reservoir is accessed by over 30 doublet well pairs (one production and one injection well); an additional 10 triplet and 18 doublet pairs are projected to come online. An obvious concern of using well doublet technology is the possibility of thermal breakthrough into the production well from the injection of the cooler spent fluid back into the reservoir. Most of the doublet well sets have been in operation for more than 20 years, and thermal decline has been noted in only one pair. Currently, about 170,000 homes in Paris are heated by geothermal fluids provided by the well network. An additional 10,000 homes are expected to be heated from geothermal fluids by 2015 (Patel, 2014).

SUGGESTED PROBLEMS

1. On the next page is a tectonic map of the New Guinea region located in the southwestern Pacific and north of Australia. As an exploration geologist, you have discovered some interesting signs indicating geothermal activity, including hydrothermally altered rock and local hot springs, noted by the red X on the map. For that location (the red X), discuss (a) the tectonic setting, (b) whether the geothermal system is likely magmatic or amagmatic, (c) associated local stress regime(s) and likely types of faults, and (d) what type of geothermal power plants might be warranted. Justify your decisions and please be as specific as possible. Please note the labeling for trenches, spreading ridge segments, and transform faults on the map. Black arrows denote absolute motion of the different shown plates.
2. Using the graph shown on the next page, what can we infer about conditions in the Coso geothermal reservoir and what could happen to the reservoir with time? (*Hint:* Consider the position of the Coso field in relation to the plotted boiling point with depth curve—the red curve).
3. What did you learn from reading Moeck's paper (Moeck, 2014) that was different or in addition to what was covered in this chapter? For example, how is Moeck's classification of The Geysers and Olkaria similar to or different from what is presented in this chapter?

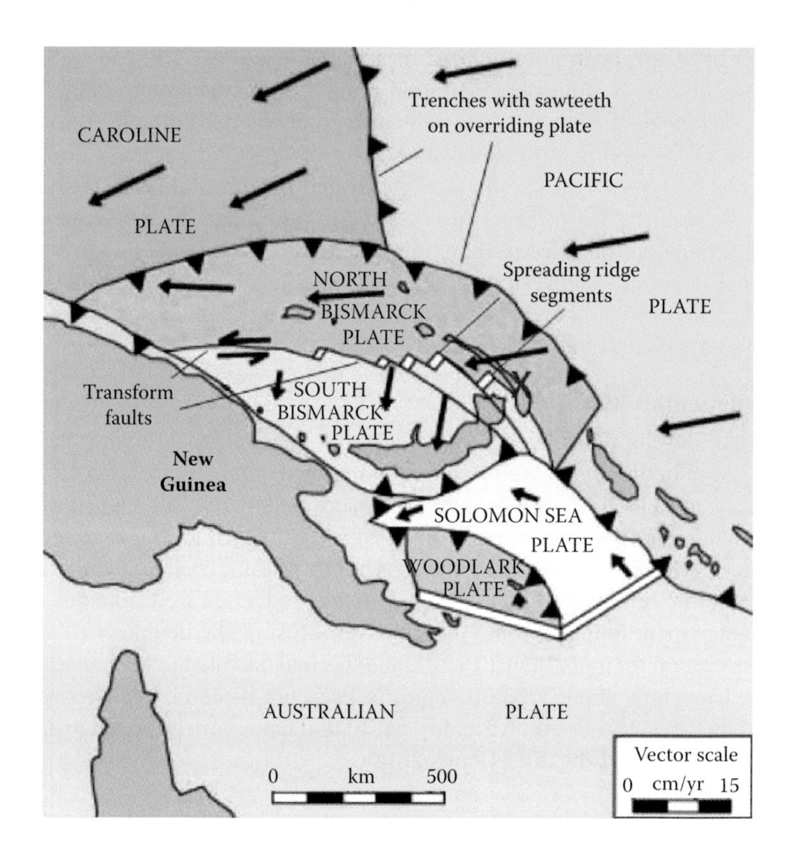

Illustration for Problem 1.

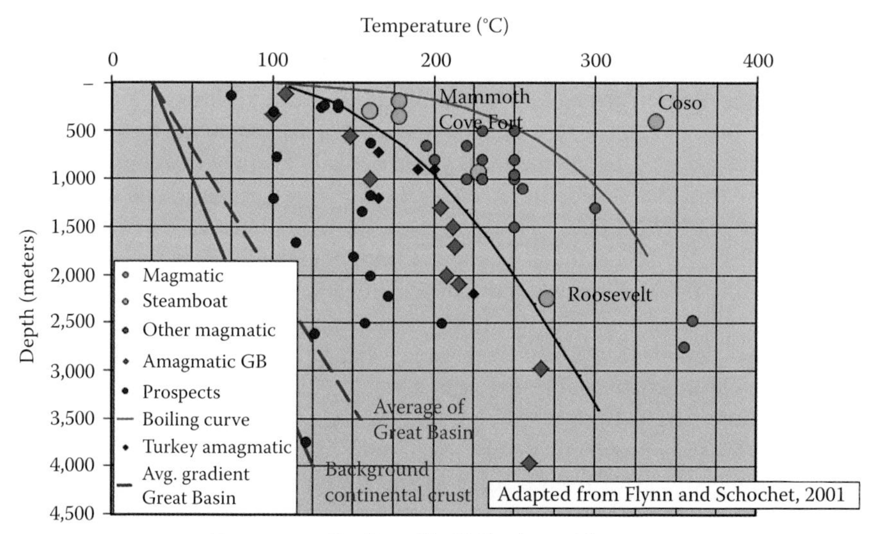

Illustration for Problem 2.

REFERENCES AND RECOMMENDED READING

Adam, C., Madureira, P., Miranda, J.M., Lourenco, N., Yoshida, M., and Fitzenz, D. (2013). Mantle dynamics and characteristics of the Azores Plateau. *Earth and Planetary Science Letters*, 362: 258–271.

Arnorsson, S., Axelsson, G., and Samundsson, K. (2008). Geothermal systems in Iceland. *Jokull*, 58: 269–302.

Bennett, S. (2011). Geothermal potential of transtensional plate boundaries. *Geothermal Resource Council Transactions*, 35(1): 703–707.

Benoit, D. (2014). The long-term performance of Nevada geothermal projects utilizing flash plant technology. *Geothermal Resources Council Transactions*, 38: 977–984.

Bertani, R. (2015). Geothermal power generation in the world—2010–2015 update report. In: *Proceedings of World Geothermal Congress 2015*, Melbourne, Australia, April 19–24 (http://www.geothermal-energy.org/pdf/IGAstandard/WGC/2015/01001.pdf).

Bibby, H.M., Caldwell, T.G., Davey, F.J., and Webb, T.H. (1995). Geophysical evidence on the structure of the Taupo Volcanic Zone and its hydrothermal circulation. *Journal of Volcanology and Geothermal Research*, 68(1–3): 29–58.

Boles, J.R., Garven, G., Camacho, H., and Lupton, J.E. (2015). Mantle helium along the Newport–Inglewood fault zone, Los Angeles Basin, California: a leaking paleo-subduction zone. *Geochemistry, Geophysics, Geosystems*, 16(7): 2364–2381.

Bromley, C., Brockbank, K., Glynn-Morris, T. et al. (2013). Geothermal subsidence study at Wairakei–Tauhara, New Zealand. *Proceedings of the Institution of Civil Engineers—Geotechnical Engineering*, 166(2): 211–223.

Carvalho, J.M., Coelho, L., Nunes, J.C. et al. (2015). Portugal country update (2015). In: *Proceedings of World Geothermal Congress 2015*, Melbourne, Australia, April 19–24 (http://www.geothermal-energy.org/pdf/IGAstandard/WGC/2015/01065.pdf).

Chin, C., Wallace, K., Harvey, W., Dalin, B., and Long, M. (2013). Big Iron in Nicaragua: a muscular new geothermal plant! The San Jacinto–Tizate geothermal project, *Geothermal Resources Council Transactions*, 37: 687–691.

Coolbaugh, M., Shevenell, L., Hinz, N.H., Stelling, P., Melosh, G., Cumming, W., Kreemer, C., and Wilmarth, M. (2015). Preliminary ranking of geothermal potential in the Cascade and Aleutian volcanic arcs. Part III. Regional data review and modeling. *Geothermal Resources Council Transactions*, 39: 677–690.

Dalrymple, G.B., Grove, M., Lovera, O.M., Harrison, T.M., Hulen, J.B., and Lanphere, M.A. (1999). Age and thermal history of The Geysers plutonic complex (felsite unit), Geysers geothermal field, California: a 40Ar/39Ar and U–Pb study. *Earth and Planetary Science Letters*, 173(3): 285–298.

DiPippo, R. (2012). *Geothermal Power Plants: Principles, Applications, Case Studies, and Environmental Impacts*, 3rd ed. Waltham, MA: Butterworth-Heinemann.

Dixon, T.H., Miller, M., Farina, F., Wang, H., and Johnson, D. (2000). Present-day motion of the Sierra Nevada block and some tectonic implications for the Basin and Range Province, North American Cordillera. *Tectonics*, 19(1): 1–24.

Donnelly-Nolan, J.M., Hearn, Jr., B.C., Curtis, G.H., and Drake, R.E. (1981). *Geochronology and Evolution of the Clear Lake Volcanics*, USGS Professional Paper 1141. Reston, VA: U.S. Geological Survey.

Donnelly-Nolan, J.M., Burns, M.G., Goff, F.E., Peters, E.K., and Thompson, J.M. (1993). The Geysers–Clear Lake area, California: thermal waters, mineralization, volcanism, and geothermal potential. *Bulletin of the Society of Economic Geologists*, 88(2): 301–316.

EERE. (2016). *FORGE*. Washington, DC: Office of Energy Efficiency & Renewable Energy, U.S. Department of Energy (http://energy.gov/eere/forge/forge-home).

Ehara, S. (1989). Thermal structure and seismic activity in central Kyushu, Japan. *Tectonophysics*, 159(3–4): 269–278.

Elders, W.A., Fridleifsson, G.O., and Albertsson, A. (2014). Drilling into magma and the implications of the Iceland Deep Drilling Project (IDDP) for high-temperature geothermal systems worldwide. *Geothermics*, 49: 111–118.

Erdlac, Jr., R.J., Gross, P., and McDonald, E. (2008). A proposed new geothermal power classification system. *Geothermal Resources Council Transactions*, 32: 322–327.

Faulds, J.E., Henry, C.D., Coolbaugh, M.F., Garside, L.J., and Castor, S.B. (2005). Late Cenozoic strain field and tectonic setting of the northwestern Great Basin, Western USA: implications for geothermal activity and mineralization. In: *Symposium 2005: Window to the World: Symposium Proceedings* (Rhoden, H.N., Steininger, R.C., and Vikre, P.G., Eds.), pp. 1091–104. Reno: Geological Society of Nevada.

Faulds, J.E., Bouchot, V., Moeck, I., and Orgiz, K. (2009). Structural controls on geothermal systems in western Turkey: a preliminary report. *Geothermal Resources Council Transactions*, 33 :375–381.

Faulds, J.E., Coolbaugh, M., Bouchot, V., Moeck, I., and Oguz, K. (2010). Characterizing structural controls of geothermal reservoirs in the Great Basin, USA, and western Turkey: developing successful exploration strategies in extended terranes. In: *Proceedings of World Geothermal Congress 2010*, Bali, Indonesia, April 25–30.

Faulds, J.E., Hinz, N.H., and Coolbaugh, M.F. (2011). Structural investigations of Great Basin geothermal fields: applications and implications. In: *Great Basin Evolution and Metallogeny* (Steininger, R. and Pennell, B., Eds.), pp. 361–372. Lancaster, PA: DEStech Publications. Copyright © Geological Society of Nevada.

Faulds, J.E., Hinz, N., Kreemer, C., and Coolbaugh, M. (2012). Regional patterns of geothermal activity in the Great Basin region, western USA: correlation with strain rates. *Geothermal Resources Council Transactions*, 36: 897–902.

Flynn, T. and Schochet, D.N. (2001). Commercial development of enhanced geothermal systems using the combined EGS–hydrothermal technologies approach. *Geothermal Resources Council Transactions*, 25: 9–13.

Fridleifsson, G.O., Palsson, B., Albertsson, A.L., Stefansson, B., Gunnlaugsson, E. et al. (2015). IDDP-1 Drilled into Magma—World's First Magma-EGS System Created, paper presented at World Geothermal Conference 2015, Melbourne, Australia, April 19–24 (http://www.geothermal-energy.org/pdf/IGAstandard/WGC/ 2015/37001.pdf).

Gosnold, W.D., Barse, K., Bubach, B. et al. (2013). Co-produced geothermal resources and EGS in the Williston Basin. *Geothermal Resources Council Transactions*, 37: 721–726.

Haizlip, J.R., Gunney, A., Tut Haklidir, F.S., and Garg, S.K. (2012). The impact of high noncondensible gas concentrations on well performance: Kizildere Geothermal Reservoir, Turkey. In: *Proceedings of the 37th Workshop on Geothermal Reservoir Engineering*, Stanford, CA, January 30–February 1 (http://www.geothermal-energy.org/pdf/IGA standard/SGW/2012/Haizlip.pdf).

Haizlip, J.R., Tut Haklidir, F., and Garg, S.K. (2013). Comparison of reservoir conditions in high noncondensible gas geothermal systems. In: *Proceedings of the 38th Workshop on Geothermal Reservoir Engineering*, Stanford, CA, February 11–13 (http://www.geothermal-energy.org/pdf/IGAstandard/SGW/2013/Haizlip.pdf).

Henry, C.D. and Faulds, J.E. (2007). Geometry and timing of strike-slip and normal faults in the northern Walker Lane, northwestern Nevada and northeastern California: strain partitioning or sequential extensional and strike-slip deformation? *Geological Society of America Special Papers*, 434: 59–79.

Herrera, R., Montalvo, F., and Herrera, A. (2010). El Salvador Country Update. In: *Proceedings of World Geothermal Congress 2010*, Bali, Indonesia, April 25–30 (http://www.geothermal-energy.org/pdf/IGAstandard/WGC/2010/0141.pdf).

Hinz, N.H., Coolbaugh, M., Shevenell, L. et al. (2015). Preliminary ranking of geothermal potential in the Cascade and Aleutian volcanic arcs. Part II. Structural–tectonic settings of the volcanic centers. *Geothermal Resources Council Transactions*, 39: 717–725.

Hochstein, M.P. (1995). Crustal heat transfer in the Taupo volcanic zone (New Zealand): comparison with other volcanic arcs and explanatory heat source models. *Journal of Volcanology and Geothermal Research*, 68(1–3): 117–151.

Hulen, J.B. and Walters, M.A. (1993). The Geysers felsite and associated geothermal systems, alteration, mineralization, and hydrocarbon occurrences. *Society of Economic Geologists Guidebook*, 16: 141–152.

Hulen, J.B., Heizler, M.T., Stimac, J.A., Moore, J.N., and Quick, J.C. (1997). New constraints on the timing of magmatism, volcanism, and the onset of vapor-dominated at The Geysers steam field, California. In: *Proceedings of the 22nd Workshop on Geothermal Reservoir Engineering*, Stanford, CA, January 27–29 (http://www.geothermal-energy.org/pdf/IGAstandard/SGW/1997/Hulen.pdf).

Ji, D. and Ping, Z. (2000). Characteristics and genesis of the Yangbajing geothermal field, Tibet. In: *Proceedings of World Geothermal Congress 2000*, Kyushu-Tohoku, Japan, May 28–June 10 (http://www.geothermal-energy.org/pdf/IGAstandard/WGC/2000/R0070.pdf).

Kennedy, B.M. and van Soest, M.C. (2007). Flow of mantle fluids through the ductile lower crust; helium isotope trends. *Science*, 318(5855): 1433–1436.

Klein, C.W., Johnson, S., and Spielman, P. (2007). Resource exploitation at Steamboat, Nevada: what it takes to document and understand the reservoir/groundwater/community interaction. *Geothermal Resources Council Transactions*, 31: 179–186.

Lutz, S.J., Walters, M., Pistone, S., and Moore, J.N. (2012). New insights into the high-temperature reservoir, Northwest Geysers. *Geothermal Resources Council Transactions*, 36: 907–916.

Madureira, P., Mata, J., Mattielli, N., Queiroz, G., and Silva, P. (2011). Mantle source heterogeneity, magma generation and magmatic evolution at Terceira Island (Azores archipelago): constraints from elemental and isotopic (Sr, Nd, Hf, and Pb) data. *Lithos*, 126(3–4): 402–418.

Martinez III, A.M. (2013). Case History of Los Azufres: Conceptual Modelling of a Mexican Geothermal Field, paper presented at Short Course V on Conceptual Modelling of Geothermal Systems, Santa Tecla, El Salvador, February 24–March 2 (http://www.os.is/gogn/unu-gtp-sc/UNU-GTP-SC-16-08.pdf).

Martinez, F. and Taylor, B. (2003). Controls on back-arc crustal accretion; insights from the Lau, Manus and Mariana Basins. *Geological Society Special Publications*, 219: 19–54.

Mertoglu, O., Simsek, S., and Basarir, N. (2015). Geothermal country update of Turkey (2010–2015). In: *Proceedings of World Geothermal Congress 2015*, Melbourne, Australia, April 19–24 (http://www.geothermal-energy.org/pdf/IGAstandard/WGC/2015/01046.pdf).

Moeck, I.S. (2014). Catalog of geothermal play types based on geologic controls. *Renewable and Sustainable Energy Reviews*, 37: 867–882. This article catalogs geothermal systems with respect to geologic and tectonic settings from a slightly different perspective than presented in this chapter (http://www.sciencedirect.com/science/article/pii/S1364032114003578).

Monastero, F.C. (2002). Model for success: an overview of industry–military cooperation in the development of power operations at the Coso geothermal field in Southern California. *Geothermal Resources Council Bulletin*, 31(5): 188–195.

Momita, M., Tokita, H., Matsudo, K., Takagi, H., Soeda, Y., Tosha, T., and Koide, K. (2000). Deep Geothermal Structure and the Hydrothermal System in the Otake–Hatchobaru Geothermal Field, paper presented at the 22nd New Zealand Geothermal Workshop, Auckland, November 8–10 (http://www.geothermal-energy.org/pdf/IGAstandard/NZGW/2000/Momita.pdf).

Muffler, L.J.P. (1976). Tectonic and hydrologic control of the nature and distribution of geothermal resources. In: *Proceedings of the Second United Nations Symposium on the Development and Use of Geothermal Resources*, San Francisco, CA, May 20–29, pp. 499–507.

Nordquist, J. and Delwiche, B. (2013). The McGinness Hills Geothermal Project, paper presented at Geothermal Resources Council 37th Annual Meeting, Las Vegas, NV, September 29–October 2.

Otago. (2016). *Tectonic Setting of New Zealand: Astride a Plate Boundary Which Includes the Alpine Fault*. Dunedin, New Zealand: University of Otago, Department of Geology (http://www.otago.ac.nz/geology/research/structural-geology/alpine-fault/nz-tectonics.html).

Rowland, J.V. and Simmons, S.F. (2012). Hydrologic, magmatic, and tectonic controls on hydrothermal flow, Taupo Volcanic Zone, New Zealand: implications for the formation of epithermal vein deposits. *Economic Geology*, 107(3): 427–457.

Ruiz, O.V. (2013). The Miravalles Geothermal System, Costa Rica, paper presented at Short Course V on Conceptual Modelling of Geothermal Systems, Santa Tecla, El Salvador, February 24–March 2 (http://www.os.is/gogn/unu-gtp-sc/UNU-GTP-SC-16-32.pdf).

Rutqvist, J., Dobson, P.F., Garcia, J. et al. (2013). The Northwest Geysers EGS Demonstration Project, California: pre-stimulation modeling and interpretation of the stimulation. *Mathematical Geosciences*, 47(1): 3–29.

Schmitt, A.K., Grove, M., Harrison, M.T., Lovera, O., Hulen, J., and Walters, M. (2003). The Geysers–Cobb Mountain Magma System, California. Part 2. Timescales of pluton emplacement and implications for its thermal history. *Geochimica et Cosmochimica Acta*, 67(18): 3443–3458.

Stimac, J.A., Goff, F., and Wohletz, K. (2001). Thermal modeling of the Clear Lake magmatic–hydrothermal system, California, USA. *Geothermics*, 30(2): 349–390.

Suter, M., Lopez-Martinez, M., Quintero-Legorreta, O., and Carrillo-Martinez, M. (2001). Quaternary intra-arc extension in the central Trans-Mexican Volcanic Belt. *Geological Society of America Bulletin*, 113(6): 693–703.

Tokita, H., Harauguchi, K., and Kamensono, H. (2000). Maintaining the related power output of the Hatchobaru geothermal field through integrated reservoir management. In: *Proceedings of World Geothermal Congress 2000*, Kyushu-Tohoku, Japan, May 28–June 10 (http://www.geothermal-energy.org/pdf/IGAstandard/WGC/2000/R0381.pdf).

Unruh, J., Humphrey, J., and Barron, A. (2003). Transtensional model for the Sierra Nevada frontal fault system, eastern California. *Geology*, 31(4): 327–330.

USGS. (2006). *Geologic History of the San Andreas Fault System*. Washington, DC: U.S. Geologic Survey (http://geomaps.wr.usgs.gov/archive/socal/geology/geologic_history/san_andreas_history.html).

Vassilakis, E., Royden, L., and Papanikolaou, D. (2011). Kinematic links between subduction along the Hellenic trench and extension in the Gulf of Corinth, Greece: a multidisciplinary analysis. *Earth and Planetary Science Letters*, 303(1–2): 108–120.

Vogt, P.R. (1983). The Iceland mantle plume: status of the hypothesis after a decade of new work. In: *Structure and Development of the Greenland-Scotland Ridge* (Bott, M.H.P. et al., Eds.), pp. 191–213. New York: Springer.

Walker, J.D., Sabin, A.E., Unruh, J.R., Combs, J., and Monastero, F.C. (2005). Development of genetic occurrence models for geothermal prospecting. *Geothermal Resources Council Transactions*, 29: 309–313.

Walters, M.A., Haizlip, J.R., Sternfield, J.N., Drenick, A.F., and Combs, J. (1992). A vapor dominated high-temperature reservoir at The Geysers, California. In: *Monograph on The Geysers Geothermal Field*, Special Report 17 (Stone, C., Ed.), pp. 77–87. Davis, CA: Geothermal Resources Council.

Wannamaker, P., Maris, V., Sainsbury, J., and Iovenitti, J. (2013). Intersecting fault trends and crustal-scale fluid pathways below the Dixie Valley geothermal area, Nevada, inferred from 3D magnetotelluric surveying. In: *Proceedings of the 38th Workshop on Geothermal Reservoir Engineering*, Stanford, CA, February 11–13.

Weisenberger, T. (2005). Zeolite Facies Mineralization in the Hvalfjördur Area, Iceland, diploma thesis, University of Freiburg.

Williams, C.F. and DeAngelo, J. (2011). Evaluation of approaches and associated uncertainties in the estimation of temperatures in the upper crust of the western United States. *Geothermal Resources Council Transactions*, 35: 1599–1605.

Wilmarth, M. and Stimac, J. (2015). Power density in geothermal fields. In: *Proceedings of World Geothermal Congress 2015*, Melbourne, Australia, April 19–24.

Wilson, C.J.N., Houghton, B.F., McWilliams, M.O., Lanphere, M.A., Weaver, S.D., and Briggs, R.M. (1995). Volcanic and structural evolution of Taupo volcanic zone, New Zealand: a review. *Journal of Volcanology and Geothermal Research*, 68(1–3): 1–28.

Xiaoping, F. (2002). *Conceptual Model and Assessment of the Yangbajing Geothermal Field, Tibet, China*. Reykjavik, Iceland: The United Nations University, Geothermal Training Programme (http://www.os.is/gogn/unu-gtp-report/UNU-GTP-2002-05.pdf).

Zheng, X., Duan, C., and Liu, H. (2013). The chemical properties of Yangyi high temperature geothermal field in Tibet, China. In: *Proceedings of the 38th Workshop on Geothermal Reservoir Engineering*, Stanford, CA, February 11–13 (http://www.geothermal-energy.org/pdf/IGAstandard/SGW/2013/Zheng.pdf).

8 Exploration and Discovery of Geothermal Systems

KEY CHAPTER OBJECTIVES

- Describe the major steps involved in the exploration for and development of geothermal systems.
- Explain how geothermal systems hosted in a stratovolcano setting and in a rhyolite dome field or caldera setting are different and how that might influence the type of geothermal facility constructed.
- Recognize the main structural settings of geothermal systems and explain why large young (active) faults, capable of producing significant earthquakes, are less attractive for hosting a potential geothermal system than numerous small faults.
- Estimate geothermal reservoir temperatures from chemical analysis of thermal spring water using geothermometers.
- Identify the application and limitations of geophysical studies used in the search for geothermal resources.

INTRODUCTION

During the early years of exploring for geothermal energy, drilling of areas displaying obvious geothermal surface manifestations, such as hot springs, mudpots, fumaroles, and even geysers, was the primary exploration tool. Some of the resulting wells, however, produced only for short periods—the actual source of the geothermal energy, in such circumstances, being largely missed. Moreover, the geothermal surface features that sparked the initial interest may have been compromised or even destroyed as a result of drilling. To avoid such occurrences, minimize the cost and amount of drilling, and maximize chances of success, an exploration strategy has developed that involves the following components and usually in the following order:

1. Review of existing information in the literature, much of which is now available online
2. Satellite remote sensing studies, including interferometric synthetic aperture radar (InSAR)
3. Airborne photographic and light detection and ranging (LiDAR) surveys
4. Ground-based geologic studies

5. Geochemical and hydrologic studies
6. Geophysical studies (gravity, magnetic, electrical resistivity, and temperature/heat flow)

Some of these steps are typically combined or take place concurrently for the sake of expediency and the synergy that results by workers from different disciplines sharing collected information as the work proceeds. Close collaboration is typical in the case for field geologic and geochemical/hydrologic studies. Results of geologic and geochemical studies are helpful to determine the type and location of geophysical techniques that might be most useful for helping delineate prospective geothermal targets.

As a result, the objectives of the exploration program, prior to deep drilling (1000 to 3000 m or more), are to

1. Locate areas underlain by hot rock.
2. Estimate the size and boundaries of a potential reservoir, including the type and permeability of rocks comprising the reservoir.
3. Determine the temperature and chemical nature of geothermal fluids.
4. Predict the nature of reservoir fluids (dry steam, liquid, steam cap above boiling liquid, or a broad range at boiling producing a two-phase condition).
5. Identify structural controls that can focus geothermal flow (such as recharge zones, upwelling zones, and outflow plumes).
6. Forecast power output for a minimum of 20 to 25 years.

In the final analysis, however, geothermal reservoirs are discovered by the results of drilling (from inexpensive shallow temperature-probe holes, to 100- to 150-m temperature gradient wells, to a full bore production test well if all other data support the expense of doing so). Drilling of temperature gradient wells or a possible production test well is the most expensive part of exploration but also provides the most definitive information on rock types and structures, permeability, temperature changes with depth, and fluid flow rates. But, the expense of drilling must be justified by the results of less expensive studies involving geology, geochemistry, and geophysics.

Clearly, a heat source must exist (hotter than normal ground) if there is a chance for developing geothermal energy for either electrical power or direct use (thermal power). Many of those geothermal systems having obvious surface manifestations, such as hot springs, have already been found. The next challenge is to identify hidden (or "blind") heat sources—those that are within drillable depths (generally less than about 4 km) but lack obvious surface expressions of hydrothermal activity.

A significant volume (several cubic kilometers or more) of permeable rock must also exist; otherwise, production will be small and short lived. Many of the producing geothermal reservoirs in Nevada are generally less than about 5 km^3 and produce on average about 20 to 25 MWe. By contrast, The Geysers geothermal vapor-dominated reservoir has a volume of about 100 km^3 with a sustained power output of approximately 800 MWe (M. Walters, pers. comm., 2015). The Geysers field has produced for over 50 years, whereas many of the small fields in Nevada have come into production within the last 10 to 15 years, although some, such as Steamboat Springs, Dixie Valley, and Beowawe, have been producing geothermal electrical power for about 30 years.

If surface features are present, the temperature and chemical make-up of the geothermal fluids can be estimated without drilling through the use of geothermometers (as discussed later in this chapter) and chemical sampling of hot springs, mudpots, or fumaroles. Even where there is no discharge of fluids at the surface, an estimate of reservoir temperatures can be determined from the mineralogy of hydrothermally altered rocks if exposed at the surface, as noted in Chapter 6. The chemical composition of the fluids, including total dissolved solids (TDS) and pH, also provides information on whether the geothermal reservoir is liquid or vapor dominated or is a two-phase mixture. Finally, this information will be crucial for designing the size and type of power plant (single or multiple flash, steam, or binary), if warranted. If no hydrothermally altered rocks or fluid discharges are present at the surface then drilling is the only way to sample potential reservoir fluids. Where to drill is based on results of geologic studies to locate favorable zones, such as zones of fault complexity (as discussed later in this chapter) and geophysical studies such as heat flow and electrical resistivity measurements.

After the exploration phase has been completed, including results of the targeted temperature-gradient holes, and data are fully analyzed, a decision can be made to either proceed or not continue to the next phase, which would be to drill a potential production well. Flow tests and temperature results obtained from such a well would be used to warrant drilling (or not) of additional wells from which power output and lifespan can be estimated. If the results fit a company's criteria for commercial viability, then the results are used to raise money from potential investors to drill additional production and injection wells and for the construction of a power plant of suitable size and type for the geothermal system (flash, binary, or steam). Typically, at least two to three full-size successful exploratory wells (for flow testing and reservoir power assessment), costing in the range of $5M to $15M, are needed prior to obtaining the financing for developing a geothermal production field and power plant.

LITERATURE REVIEW

A literature review is probably the most cost- and time-effective means of collecting information, especially in light of the development of online databases that house information on geology, drilling results, hot spring temperature and chemical data, and well fluid temperatures and chemical composition. Many of these online databases are housed at state geological surveys. Another important geothermal data repository is available at the Geo-Heat Center located at the Oregon Institute of Technology in Klamath Falls (Oregon Tech, 2016). The U.S. Geological Survey also hosts online databases applicable to geothermal development, including interactive maps and geographic information systems (GIS), as well as tabular data (USGS, 2016). A new effort is now underway in which all geothermal information is being collected and stored by the U.S. Department of Energy at a central site called the National Geothermal Data System (NGDS, 2016). Here, maps, publications, well data, etc., can be downloaded (or uploaded) for geothermal systems across the United States. As an example, Figure 8.1 shows a screen shot of an Excel® spreadsheet accessed at the NGDS website showing temperature with depth of a borehole in the Casa Diablo geothermal system located in the Long Valley caldera in eastern

C32

	A	B	C	D	E	F	G	H
1	CW-1_16JAN92							
2	CW-1, Logged January 16, 1992							
3	Source: USGS, Logged by Farrar using portable temperature logger with sensor 1560							
4	API: N/D							
5								
6	Depth_ft	Depth_m	Temp_C					
7	25	7.62	95.1					
8	50	15.24	101.8					
9	75	22.86	108.6					
10	100	30.48	114.3					
11	125	38.1	119.64					
12	150	45.72	123.95					
13	175	53.34	127.98					
14	200	60.96	131.45					
15	225	68.58	132.52					
16	250	76.2	131.8					
17	275	83.82	131.1					
18	300	91.44	130.63					
19	350	106.68	129.84					
20	400	121.92	129.61					
21	450	137.16	129.23					
22	500	152.4	127.92					
23	550	167.64	123.13					
24	600	182.88	122.06					
25	650	198.12	121.76					
26	675	205.74	121.87					
27								
28								

FIGURE 8.1 Screen shot of an Excel® spreadsheet for wellbore hole temperature data from a well in the Casa Diablo geothermal system in eastern California as accessed through the National Geothermal Data System website.

California. Production wells for the Casa Diablo geothermal power plant access a shallow outflow plume (Suemnicht et al., 2006) as reflected by the well data shown in Figure 8.1.

The bottom line is that for many prospective geothermal regions, someone has likely been there and collected data that are available for study. Review of such data can result in the recognition of features that were previously overlooked, which can have important implications for potential geothermal development. Literature review is really due diligence with respect to determining whether a region merits further attention and the spending of additional money (usually at considerably greater cost than that of conducting a thorough literature review) or saving that money for another prospect.

SPACEBORNE AND AIRBORNE STUDIES

In regions that have sustained some previous work or in greenfield plays (where little or no previous work has been conducted), the goal is to take a large area, covering hundreds of square kilometers or more, and narrow it down to more specific areas holding the greatest geothermal promise. To cover a large area, sensors on satellites or aircraft are used to map anomalies that can reflect localized geothermal phenomena. These studies mainly consist of remote sensing, aeromagnetic surveys, and aerial photography.

Remote Sensing Studies

Remote sensing studies can be both spaceborne, using satellite instruments, and airborne, using conventional aircraft. These methods include investigations using optical spectroscopy, InSAR, and LiDAR.

Optical Spectroscopic Investigations

Sensors on satellites or aircraft are used to map the ground in near-infrared and thermal infrared wavelengths. Good reviews of the application of remote sensing studies in geothermal exploration are provided by Calvin et al. (2015) and van der Meer et al. (2014). The data collected can be both multispectral, using several light wavelength channels, or hyperspectral, using hundreds of wavelength channels, depending on the instruments used. These datasets are sensitive to surface minerals, rock types, and surface temperature. Thus, this information can be used to detect areas of hydrothermally altered rock and surface thermal anomalies of warm ground overlying otherwise concealed geothermal systems. The technique is particularly useful to map mineralogy and possible thermal anomalies over broad and largely inaccessible regions. However, these methods work best in sparsely vegetated areas and are not very useful in tropical or densely forested regions. Optical and infrared spectroscopy has been used since the 1970s to identify rocks and minerals and map their occurrences on the ground. This is because different rocks and minerals have identifiable and commonly unique absorption spectra. For example, siliceous sinter (opal), calcite, and other hydrothermal minerals, such as clays, sulfates, and borates, can be distinguished by their distinctive spectral signatures (Figure 8.2).

Results of a hyperspectral remote sensing study identified a previously unrecognized and blind geothermal system in Columbus Salt Marsh in Esmeralda County of western Nevada (Kratt et al., 2009). The marsh lies within a right-directed step-over of northwest-striking, right-lateral faults within the Walker Lane* structural zone. A right step-over in right-lateral faults produces a localized region of extension (pull-apart basins, such as discussed for the Salton Trough and Cerro Prieto geothermal system in Chapter 7), which are known as prospective structural settings for hosting geothermal systems elsewhere in the Great Basin (Faulds et al., 2010) and discussed more later in this chapter. The occurrence of hydrothermal minerals in a favorable structural setting, however, does not necessarily indicate an active heat anomaly, only that at one time hydrothermal fluids were present. In this case, a follow-up program of drilling shallow (2-m-deep) temperature holes (a technique reviewed more later in this chapter) did disclose a thermal anomaly in the same

* The Walker Lane consists of a zone of diverse topography and right-lateral northwest-striking fault zones in western Nevada. The zone is considered to accommodate about 20% of the motion between the North American and Pacific tectonic plates, with the rest of the motion occurring along the San Andreas fault zone in California. See Chapter 7 for further details.

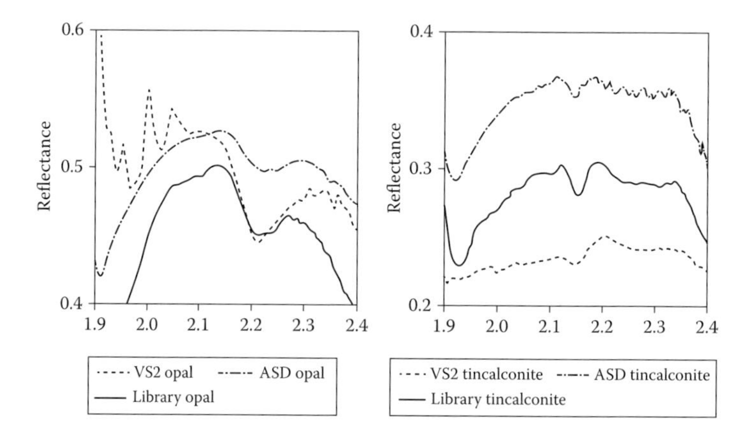

FIGURE 8.2 Mineral spectra for opal and tincalconite (a borate mineral commonly precipitated from geothermal fluids found in Columbus Marsh in west-central Nevada). The y-axis measures the strength of reflectance, and the x-axis is the wavelength measured in microns (μm). The solid line in both graphs is the reflectance spectra for the library reference of the indicated mineral. The dashed lines reflect the spectra signatures as recorded from two different sensing instruments. (Adapted from Kratt, C. et al., *Geothermal Resources Council Transactions*, 32, 153–158, 2009.)

region as the remote-sensing-mapped borate-containing evaporative crusts. Remote sensing mapping has also successfully detected hydrothermal alteration minerals, in some cases aligned along linear zones reflecting previously unrecognized fractures or faults. Within the Long Valley caldera in eastern California and in Dixie Valley in west-central Nevada are remote-sensing-detected, previously unrecognized faults that channeled hydrothermal fluids (Martini et al., 2003).

InSAR Studies

Interferometric synthetic aperture radar (InSAR) uses radar satellites to map ground deformation over large areas. Such satellites continuously emit microwave frequencies and record the reflected signal from the Earth. When features on the ground shift, the distance between the sensor and the ground changes, resulting in a change of the measured signal phase. Repeat passes of the satellite allow the acquisition of multiple signal phases that are used to measure changes in ground elevation. Integration of changes in signal phases over time produces an interferogram, which is used to detect ground movement between successive passes of the satellite. Using advanced data reduction algorithms, ground deformation, such as uplift, subsidence (vertical deformation), or movement along active faults (vertical and/or horizontal deformation), can be measured at the millimeter level of precision.

Because of its ability to detect small changes in ground deformation over time, InSAR studies can be particularly helpful for identifying and monitoring areas of subsidence and uplift in a producing geothermal fields, such as was done in the Salton Sea geothermal field in southern California (Eneva et al., 2009). To minimize surface disturbance, InSAR data can be used to help manage rates of production

FIGURE 8.3 Image showing a portion of the Salton Sea geothermal field in southern California. Compiled InSAR data indicate that the ground in the northeast portion of the image has an eastward component of movement, whereas that in the southwest part of the image has a westward component of movement. The northwest-striking red dashed line marks a hypothesized hidden fault that may explain the InSAR data and could represent an exploration target for additional drilling. (Adapted from Falorni, G. et al., *Geothermal Resources Council Transactions*, 35, 1661–1666, 2011.)

and injection and to help locate possible new injection and production wells that can improve reservoir performance and longevity. Also, because of the ability to detect horizontal ground movement, InSAR results can be potentially helpful for locating active (but possibly hidden) faults. For example, in Figure 8.3, the area on the east side of the map has an eastward horizontal component, whereas the area on the west side has a westward horizontal component. These InSAR results suggest that a possible hidden northwest-striking fault may exist near the middle of the map experiencing right-lateral strain, such that ground on the northeast side of the possible fault is moving southeast and ground on the southwest side of the fault is moving northwest. Such a fault could play a role in subsurface permeability; it could be sealed and act as a barrier to fluid flow, or, on the other hand, it could promote fluid flow if oriented favorably with respect to regional or localized extension as perhaps determined from geologic mapping of faults elsewhere.

At the San Emidio geothermal field in northwest Nevada, InSAR data indicate a sharp boundary between areas of uplift and subsidence (Figure 8.4). This abrupt change occurs in the middle of the valley and probably reflects structural control (by a hidden fault) on fluid movement, marking a potentially attractive target that may extend to the north and south of the existing well field.

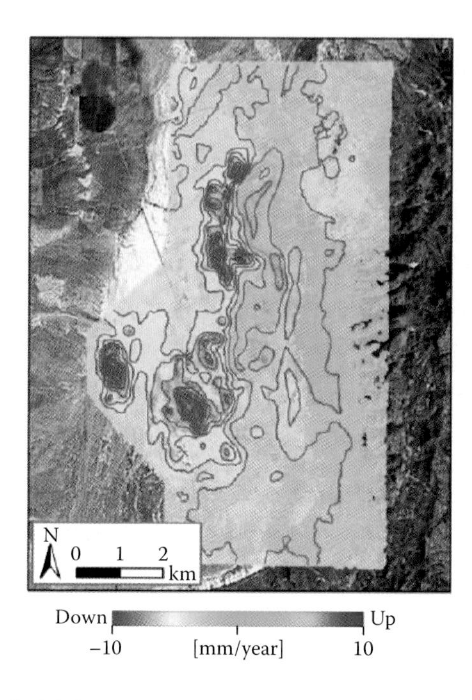

FIGURE 8.4 Compiled InSAR data over the San Emidio geothermal field in northwest Nevada. Note the abrupt transition between areas of subsidence and uplift suggesting structural control on subsurface geothermal fluids. (From Falorni, G. et al., *Geothermal Resources Council Transactions*, 35, 1661–1666, 2011.)

LiDAR Studies

Light detection and ranging (LiDAR) is a technique in which airborne sensors emit and detect reflected electromagnetic radiation (in the form of a pulsed laser) to produce very detailed maps of the Earth's surface. All collected data are in a digital format, which allows combining LiDAR results with other digitally collected data such as from gravity, aeromagnetic, and seismic surveys. Such data can then be superimposed to help reveal key geologic relationships and targets for additional follow-up study. LiDAR data are typically used to form digital elevation models (DEMs), which are extraordinarily detailed physiographic images of the Earth's surface. Because DEMs "see" through vegetation and are unaffected by the complicating interference of human-made structures, they can illuminate subtle but potentially important structural features that may not be evident from traditional aerial photo interpretation. Furthermore, the landscaped portrayed in a DEM, because of their digital format, can be illuminated from different sun angles. Illumination from low sun angles, in particular, can help highlight subdued geomorphic features, such as scarps in areas of low topographic relief that can mark active faults. Neotectonic studies in the Great Basin have shown that geothermal systems and fluid flow are commonly associated with young/active faults (e.g., Bell and Ramilli, 2009). Inspection of DEMs, because of their high resolution, can disclose subtle fault scarps that in conjunction with other data, such as geophysical and temperature surveys, can help target

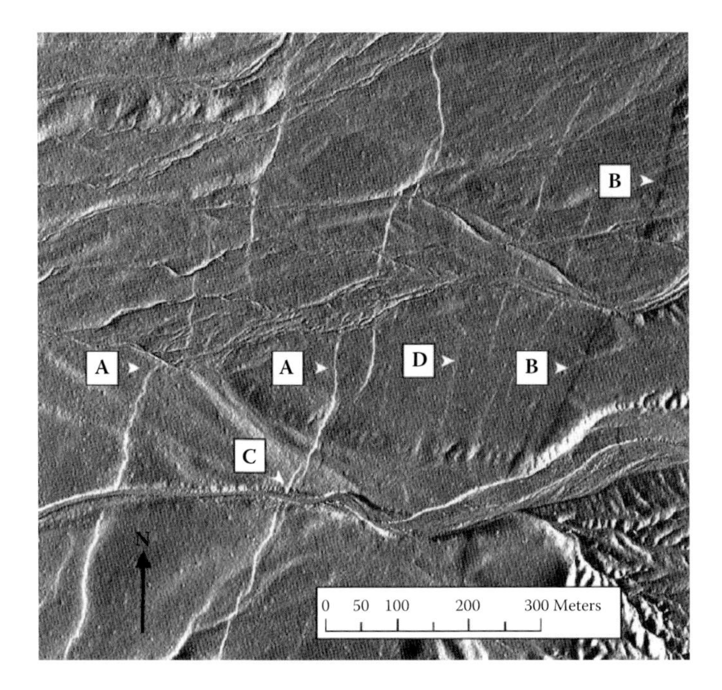

FIGURE 8.5 A digital elevation model produced from LiDAR data of the western side of Gabbs Valley in west-central Nevada. The DEM is illuminated from the east, clearly showing (A) the east-facing fault scarps produced by the 1954 Fairview Peak earthquake; (B) the west-facing fault scarps; (C) minor left lateral motion; and (D) faint north-northeast-trending east-facing fault scarps. (Adapted from Payne, J. et al., *Geothermal Resource Council Transactions*, 35, 961–966, 2011.)

promising geothermal prospects. As an example, a LiDAR study was conducted along the west side of Gabbs Valley in west-central Nevada, about 20 km northeast of the recently completed and operating 20-MWe (gross) Don A. Campbell geothermal power plant. The LiDAR study covered a geothermal lease in a tectonically active region spanning the transition from right-lateral shear of the Walker Lane to the southwest and Basin and Range extension to the northeast. A known hot spring, Rawhide Hot Springs, occurs about 5 km west of the study area. A low-sun angled DEM revealed both fault scarps produced from the 1954 Fairview Peak earthquake and also more faint fault scarps not evident from inspection of aerial photographs (Figure 8.5) (Payne et al., 2011).

Aerial Photography

Color stereographic aerial photographs, perhaps covering several hundred square kilometers, can be very useful for disclosing faults and zones of hydrothermally altered rock and for identifying specific regions requiring more detailed ground-based geologic field studies (discussed below). Indeed, stereo aerial photography is fundamental for geologic mapping, as it provides an aerial perspective of geologic features that may be difficult to discern from the ground. Aerial photograph

FIGURE 8.6 Google Earth image of the Casa Diablo geothermal system in the Long Valley caldera in California. Although this is a digital satellite image, it simulates a color aerial photograph that could be taken by a low-flying aircraft. Red ellipses outline areas of hydrothermally altered rock. The linear zone of altered rock located to the northeast of the Casa Diablo geothermal plant marks a northerly striking fault, shown by the orange dashed line, which channeled hydrothermal fluids.

interpretation is typically done prior to conducting on-ground geologic mapping as it identifies areas of primary geologic interest, including potential faults and areas of hydrothermally altered rock (Figure 8.6). Furthermore, aerial photograph interpretation helps the user formulate multiple working hypotheses that can be tested when mapping rock units and associated rock structures while on the ground.

Aeromagnetic Studies

Although this is a geophysical technique, it is discussed here because of its airborne application. Other geophysical techniques are discussed below. In an aeromagnetic study, a magnetometer is towed behind an airplane or helicopter that flies along parallel flight lines covering usually several hundred square kilometers. The magnetometer records the magnetic variations of rocks over the area surveyed. Regions found to be characterized by magnetic lows may reflect zones of hydrothermally altered rocks that otherwise may not be visible due to vegetative cover or burial by shallow soils, slope wash, or valley fill deposits (Figure 8.7). In the case of hydrothermally altered rocks, magnetic lows are produced because hydrothermal fluids alter and destroy original magnetic minerals in the rock, such as converting primary magnetite to nonmagnetic pyrite. This would be the situation mainly in areas of volcanic or

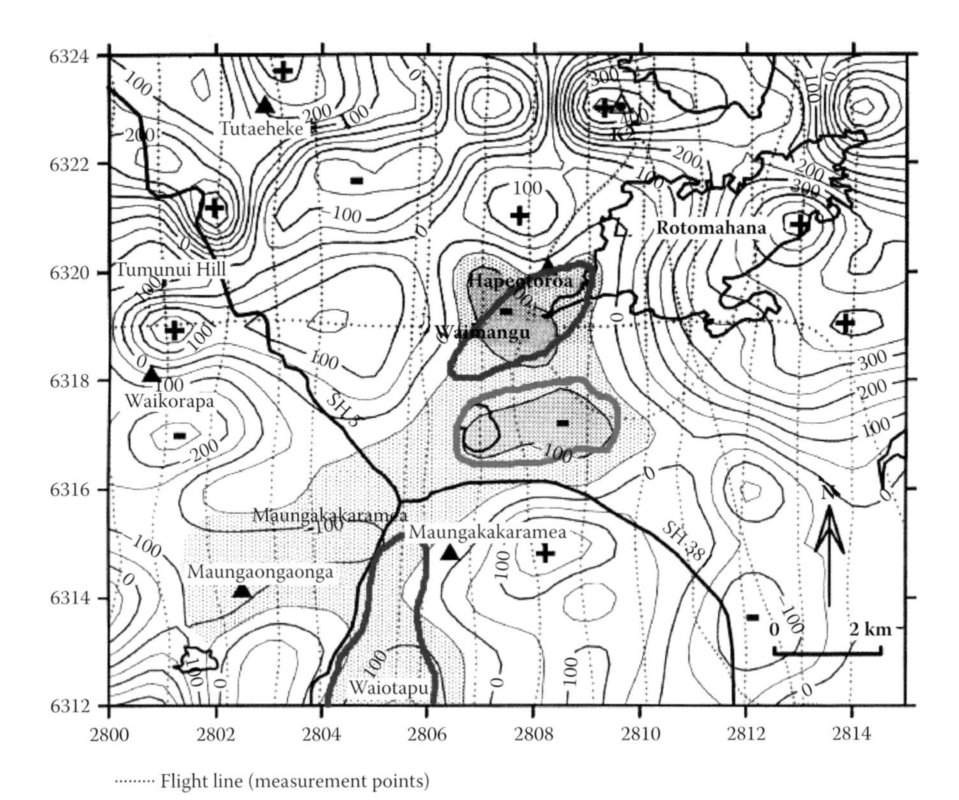

FIGURE 8.7 Aeromagnetic map of the Waimangu and Waiotapu geothermal regions in the Taupo volcanic zone of New Zealand. The shaded region reflects magnetic lows due to the demagnetization of volcanic rocks from hydrothermal alteration. The areas outlined in red contain surface thermal manifestations, while the area outlined in purple reflects hydrothermally altered rocks without much current surface geothermal activity. (Adapted from Soengkono, S., *Geothermics*, 30(4), 443–459, 2001.)

plutonic igneous rocks that contain original magnetic minerals. However, in regions dominated by sedimentary rocks, particularly limestone, magnetic low anomalies from hydrothermal alteration will not be as apparent because limestone contains essentially no original magnetic minerals. In fact, sometimes hydrothermal fluids affecting carbonate rocks may produce magnetic highs if the hydrothermal fluids are rich in iron, leading to the possible precipitation of magnetite. Moreover, magnetic lows can also reflect reversely magnetized rocks—that is, rocks containing magnetic minerals that formed during periods when the Earth's magnetic field was reversed compared to today's polarity. Therefore, an understanding of a region's geologic make-up and rock type, in part stemming from the literature review stage, should be obtained to determine if an aeromagnetic study is warranted and, if so, what kind of anomalies might be expected that could reflect zones of hydrothermal alteration.

GEOLOGIC STUDIES

Geologic studies actually begin with the literature review stage to identify the tectonic setting of the region of interest along with a review of the results of any previous more detailed geologic and geothermal studies that may exist. Geologic studies conclude with on-the-ground mapping to fill in any gaps of knowledge and to develop a conceptual model for the local geologic setting that helps measure the level of promise for hosting a potentially viable geothermal system.

Tectonic Setting

As discussed in Chapter 7, the tectonic setting exerts strong control on the characteristics of geothermal systems, such as magmatic and amagmatic (extension-related) systems. This in turn influences how temperature changes with depth and whether systems can be liquid or vapor dominated or a two-phase mixture. For example, in a continental or volcanic arc setting, geothermal systems are likely magmatic; therefore, relatively high temperatures at modest depths would be expected, making such settings potentially suitable for efficient flash-style geothermal power plants. However, liquid-dominated, magmatically heated systems can have high total dissolved solids (TDS) and/or low pH, resulting in scaling or potential corrosion of equipment. In an extensional or amagmatic geologic setting, on the other hand, fluids typically have low to modest TDS levels and near-neutral pH. This is good for equipment, but temperatures are commonly less than about 180°C, suitable for binary power facilities having lower power outputs (commonly in the range of 15 to 30 MWe).*

Geologic Mapping

The heart of geologic studies is on-the-ground mapping to determine the spatial and temporal distribution of rock types, geologic structures such as faults, and areas of hydrothermally altered rocks or actual geothermal deposits such as siliceous sinter or travertine (hot spring calcite). Geologic mapping is cost effective but somewhat time consuming; therefore, only the most promising areas, based on the previously discussed studies, are selected for mapping. Results of geologic mapping are crucial for assessing the level of geothermal promise and, in conjunction with concurrent geochemical and geophysical investigations, for targeting possible drill sites and identifying the areal extent of a geothermal system.

* The new binary McGinness Hills geothermal power facility is a notable exception as it has an installed capacity of 72 MWe, making it the second largest geothermal power facility in Nevada after the Steamboat complex of six power plants. Although fluid temperatures are modest (150° to 160°C), the high power output results from very high mass flow rates (50 to >100 gpm/psi) (Nordquist and Delwiche, 2013). Another example of a large binary plant is Ngatamariki in New Zealand, which has an installed capacity of 100 MWe, making it the largest all-binary geothermal power plant in the world (Legmann, 2015). At Ngatamariki, fluid temperature at the wellhead is 192°C, which could support a flash plant but a binary plant was built for environmental reasons and to promote longevity of the resource.

Geologic Environments

A primary purpose of mapping is to characterize the geologic environment. Important questions to be addressed while mapping include whether or not young volcanic rocks (<1 million years and preferably <0.1 million years) are present and, if so, their abundance and distribution. If volcanic rocks occur, their types and distribution tell us about the volcanic setting. For example, if volcanic rocks are mainly intermediate in composition, such as andesite, and are a few thousand feet thick, then a stratovolcano setting* having considerable topographic relief is probably indicated. On the other hand, if most of the volcanic rocks are more silicic in composition and consist of a series of rhyolite lavas, intrusive domes, and thick consolidated ash deposits (called *ash-flow tuffs*), then a caldera setting† or flow dome complex having low to modest topographic relief may be indicated. These two contrasting geologic environments have important implications for the type of any hosted geothermal systems. Although both can support high-temperature geothermal systems (>200°C), the fluid chemistry and associated hydrothermal alteration will likely be different, as discussed in Chapter 6 (Figure 8.8).

Hydrothermal Alteration Mapping

Mapping of hydrothermal alteration minerals becomes particularly important in areas where surface thermal features, hot springs, mudpots, or fumaroles are no longer active. Their effects, however, are still reflected by surrounding zones of hydrothermally altered rock. The distribution and size of the alteration zones further reflect the scale of the geothermal system. The types of alteration minerals are also mapped because different assemblages of alteration minerals, as discussed in Chapter 6, form at different temperatures and different conditions of acidity. For example, the mineral assemblage of alunite and pyrophyllite forms at relatively high temperatures (>200°C) and acidic fluid conditions (typically, pH <2). The mapping of alunite and pyrophyllite helped identify high-temperature zones in the Hachimantai geothermal region in Japan (Wohletz and Heiken, 1992) that hosts two geothermal power plants (Figure 8.9). The alteration minerals epidote and adularia also form under relatively high temperatures (>200°C and >150°C, respectively) (Figure 8.9), but under near-neutral conditions of pH.‡ Figure 8.10 illustrates the temperature stability range of a variety of minerals formed during hydrothermal alteration. Also, mapping the distribution of different minerals in hydrothermally altered rocks allows vectoring toward areas with the highest temperatures and hence the most promising targets for drilling.

* Stratovolcanoes, also called *composite volcanoes*, are the classic cone-shaped mountains like Mt. Fuji. They derive their name from an overlapping succession of interlayered lava flows and explosively derived ash deposits, which make up the volcano's edifice.

† A caldera is a volcanic depression measuring from 10 km to as much as 100 km across that forms from colossal eruption of rhyolitic magma from a shallow (4 to 6 km deep) magma reservoir. The rocks overlying the magma chamber founder and move downward into the partially evacuated magma chamber, producing a quasi-circular depression.

‡ It is worth noting that alteration minerals indicating temperatures >220°C commonly also indicate uplift and erosion, which may be good or bad for geothermal development potential. If the capping clay seal was breached by erosion, the system may have boiled down and lost permeability (J. Stimac, pers. comm., 2016).

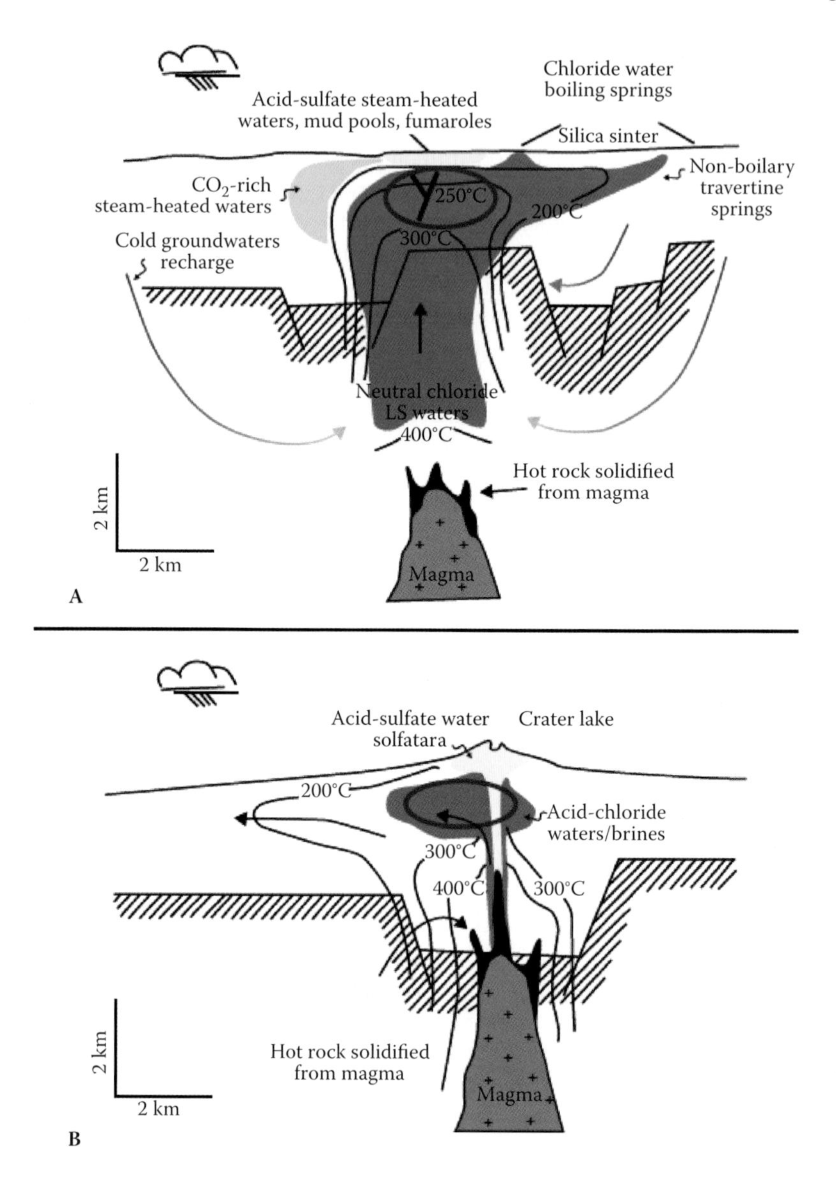

FIGURE 8.8 Schematic conceptual models of two different geologic settings hosting geothermal systems (enclosed by red ellipses). (A) This setting reflects a low-relief caldera or rhyolite flow dome complex in which geothermal fluids have near-neutral pH and have low to modest Cl contents. Thermal surface expressions include boiling springs with siliceous sinter and possible distal carbonate (travertine) springs. Steam-heated acid-sulfate alteration may occur over the main reservoir if fluids are allowed to boil at depth. (B) This setting depicts a stratovolcano with generally high topographic relief. Due to the close presence of magma, magmatic gases such as SO_2 and HCl result in acidic geothermal fluids that are rich in sulfate waters near the surface and acid-chloride brines in the main reservoir that typically contain significantly more TDS and Cl than the situation depicted in (A). (Adapted from Cooke, D.R. and Simmons, S.F., *Reviews in Economic Geology*, 13, 221–244, 2000.)

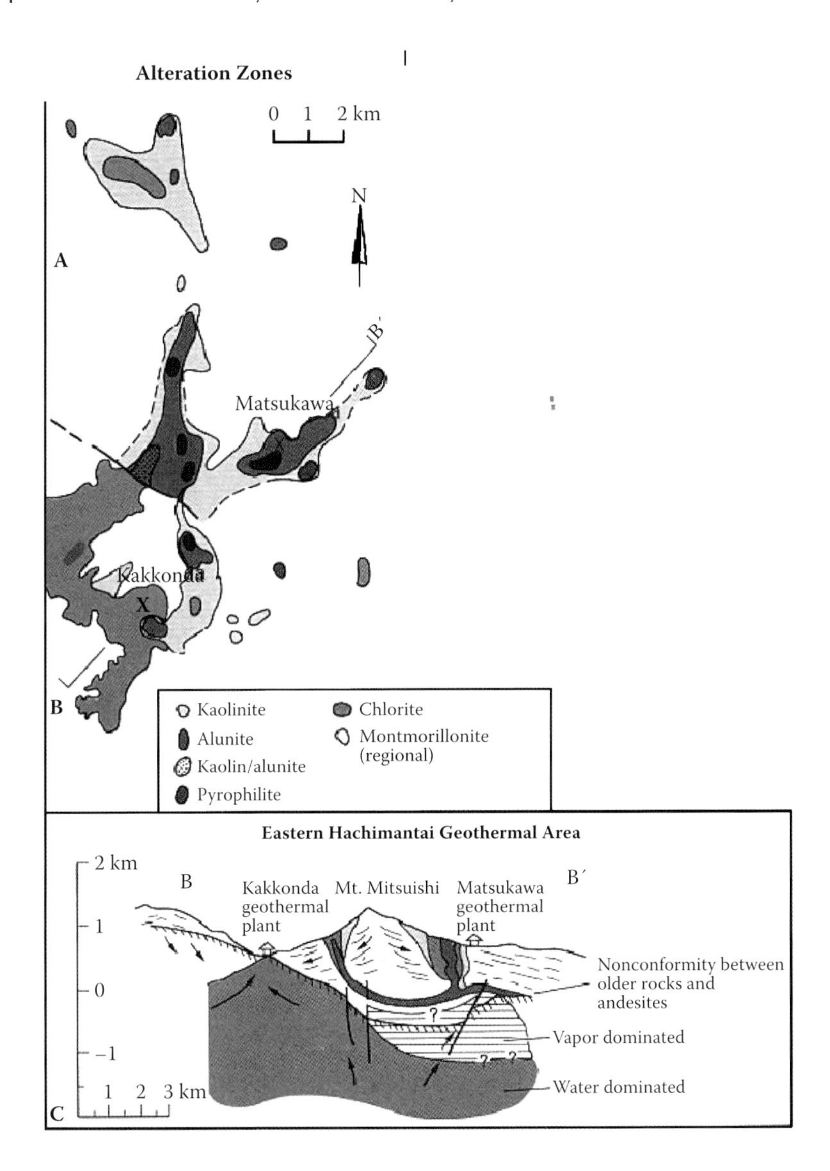

FIGURE 8.9 Map of alteration zones and schematic cross-section along B–B″ of the Hachimantai geothermal area in Japan. The well field for the Kakkonda geothermal power plant is located mainly to the south of the power plant in the area of alunite alteration. (Adapted from Wohletz, K. and Heiken, G., in *Volcanology and Geothermal Energy*, Wohletz, K. and Heiken, G., Eds., University of California Press, Berkeley, 1992, pp. 225–259.)

Zones of hydrothermal alteration can also serve as proxies denoting areas of elevated permeability. If rocks are relatively impermeable, fluids will have little access, and the rocks will remain relatively unaltered even at high temperatures. Therefore, linear-like zones of alteration probably reflect fluids flowing along fractures or faults that provide good fracture or secondary permeability.

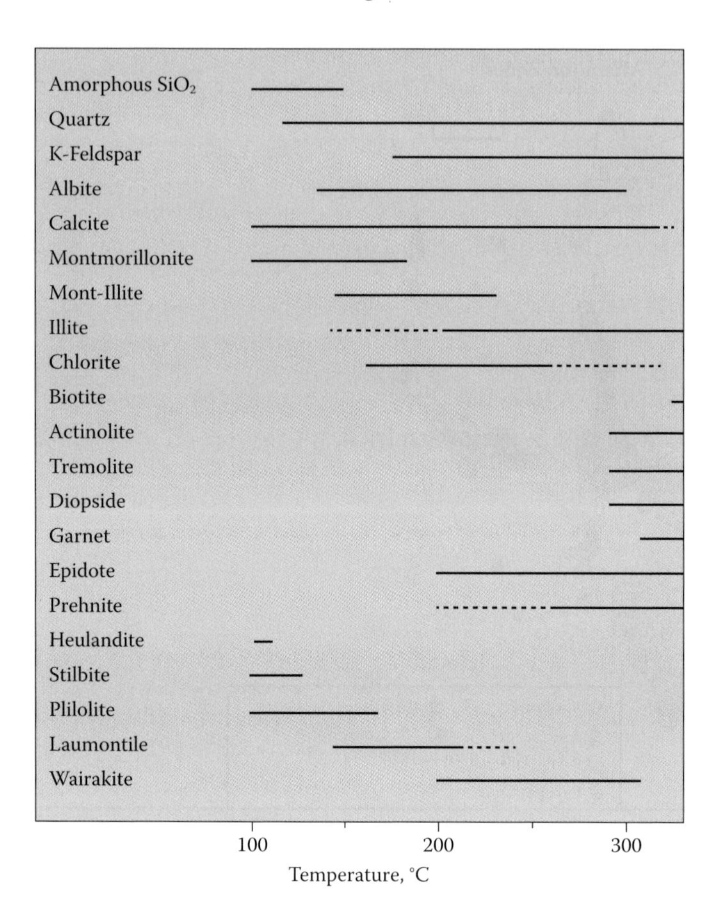

FIGURE 8.10 Temperature stability ranges of some important hydrothermal minerals. Note how some minerals are stable over a wide range of temperatures, such as quartz and calcite, whereas other minerals have a more restricted and definitive temperature range of stability, such as garnet. As a practical matter, epidote is one of the most useful of the hydrothermal minerals in helping determine the extent of geothermal reservoirs because it indicates temperatures greater than about 200°C and it is easy to identify. Dashed bars indicate extended zones of stability to higher or lower temperatures depending on fluid and wallrock compositions. (Adapted from Henley, R.W. and Ellis, A.J., *Earth-Science Reviews*, 19(1), 1–50, 1983.)

In the absence of active surface thermal features, the age of the hydrothermally altered rock must also be determined, because if the alteration is older than about 0.5 million years then the underlying geothermal system may have cooled to the point that it is no longer viable for development. The age of alteration should preferably be <0.25 million years to reflect a still hot and viable geothermal system (J. Stimac, pers. comm., 2016). The age of hydrothermal alteration can be determined in a few different ways. For example, if the age of the hydrothermally altered is known from an independent source, then the hydrothermal alteration must be younger than the age of the rock. In some cases, altered rock is overlain by unaltered lavas or tephras that can be dated, which would indicate a minimum age for the alteration. Another method is to date select alteration minerals directly. Hydrothermal alteration minerals containing potassium (K), such as alunite, adularia, and illite, can be dated radiometrically because one of the isotopes of potassium is radioactive, specifically K^{40}, which decays to argon-40 (Ar^{40}), the stable daughter isotope. By measuring the parent-to-daughter ratio of collected samples in the laboratory, the age can be determined from the decay rate of the parent isotope, which is well established.*

A good example of a blind geothermal discovery is the McGinness Hills geothermal power facility, which is one of Nevada's largest, with an installed capacity of 72 MWe. The developed geothermal reservoir has no active hydrothermal surface expressions; however, its presence was suggested by a conspicuous sinter terrace determined to be 2 to 3 million years old based on K/Ar dating of adularia. The sinter and adularia were found during a gold exploration program that preceded geothermal development (Casaceli et al., 1986; Nordquist and Delwiche, 2013). The earlier mineral exploration drilling also encountered hot water at depth, and, in conjunction with finding only low-grade gold values, the project was abandoned as a mineral play.

Mapping of Geothermal Deposits

These are deposits formed by minerals precipitating directly from the geothermal fluids due to changes in temperature and pressure. Sometimes underground pressure builds to the point that a hydrothermal eruption (phreatic explosion) occurs, creating a ring of ejecta or explosion debris outlining a generally shallow depression that may or may not be filled with water. Such eruption craters occur near the developed Dixie Valley geothermal system (Figure 8.11). One of the most important geothermal deposits to describe and map is siliceous sinter formed around some active or fossil hot springs (such as which helped lead to the discovery and development of the McGinness Hills geothermal field). When initially precipitated, the silica consists of opal, which is an amorphous (having no crystal or organized atomic structure), hydrated (contains included water molecules) form of quartz. With time (over a period of a few tens of thousands of years), the opal dehydrates and recrystallizes into a very fine-grained form of quartz called *chalcedony*. Siliceous sinter has two important implications:

1. It indicates reservoir temperatures of >175°C (Fournier and Rowe, 1966). As discussed further below, the solubility of silica increases with temperature (up to at least 350°C). To get enough to dissolve and then precipitate as amorphous silica (such as opal) to form hot spring sinter from cooling requires reservoir temperatures of at least 175°C—certainly a good sign for potential geothermal energy development.
2. It mainly forms from liquid-dominated systems having near-neutral pH and fluids of low to moderate TDS of the alkali-chloride type (as introduced in Chapter 6). Such systems would be of amagmatic type (e.g., McGinness Hills or Beowawe in Nevada), magmatic caldera or rhyolite flow dome

* This described process applies to potassium–argon (K/Ar) dating, which has now been superseded by Ar^{40}/Ar^{39} or argon–argon dating. In the latter case, superior accuracy and precision are achieved because it requires only a single measurement of Ar isotopes (produced via irradiation in a nuclear reactor), not two measurements of K^{40} and Ar^{40} as required for K/Ar dating.

FIGURE 8.11 View looking northeast across a hydrothermal eruption crater in Dixie Valley, Nevada. (Photograph courtesy of M. Coolbaugh.)

complexes of low relief (e.g., Casa Diablo in the Long Valley caldera in eastern California), or caldera-related systems such as in the Taupo Volcanic Zone in New Zealand. Hot spring sinter can also form in andesitic volcanic systems as long as neutral-pH fluid has rapid access to the surface, such as at Tiwi in the Philippines (Moore et al., 2000; J. Stimac, pers. comm., 2016).

Other important minerals deposited in and around hot springs are travertine and tufa. Although both consist of calcium carbonate or calcite ($CaCO_3$), they form in two different settings and have different implications for assessing geothermal potential. Travertine forms from hot springs on land and reflects low to moderate reservoir temperatures. This is because calcite has retrograde solubility; in other words, calcite becomes less soluble with increasing temperature (the opposite of silica or quartz). So, as temperature of a fluid increases, calcite will tend to precipitate rather than dissolve. This is because calcite solubility is strongly dependent on CO_2 solubility, and, as with most gases, CO_2 is less soluble in liquids with increasing temperature. This behavior can perhaps be best understood by looking at the following equilibrium chemical reaction:

$$Ca^{2+} + 2HCO_3^- \Leftrightarrow CaCO_3 + H_2O + CO_2 \tag{8.1}$$

The double arrow indicates an equilibrium chemical reaction so the reactants and products balance each other. When things get out of balance, such as when temperature is increasing and CO_2 becomes less soluble in solution and escapes, the reaction goes to the right to make more products, including calcite, to keep the reactants and products balanced or in equilibrium. On the other hand, when temperature falls, CO_2 becomes more soluble in solution, and the reaction moves toward the left to make more reactants to maintain equilibrium. Thus, in springs where travertine is being deposited, temperatures at depth must not be too hot; otherwise, calcite would deposit at depth and not at or near the surface.

Because the solubility of CO_2 is also a function of pressure, the loss of CO_2 due to reduced pressure at the surface leads to the precipitation of calcite, even though temperature may be decreasing as fluids rise. Travertine springs do not necessarily indicate, however, limited geothermal heat or power potential, as they can develop peripheral to the main geothermal system as exemplified by the beautiful travertine terraces of Mammoth Hot Springs in Yellowstone National Park. Other examples of travertine springs include Jemez Springs and Soda Dam, located distally to a potent geothermal system developed along and near the margin of the Valles Caldera in New Mexico (Goff and Gardner, 1994).

Tufa, unlike travertine, forms subaqueously where springs, hot or cool, discharge at the bottom of lakes whose waters are rich in bicarbonate ion (HCO_3^-). Excellent examples of tufa occur around Pyramid Lake, located about 30 to 40 miles northeast of Reno, Nevada (Figure 8.12). Pyramid Lake is a remnant of Lake Lahontan, which formed a large inland sea at the end of the last glacial period about 10,000 years ago. It is now the final destination of the Truckee River, which drains Lake Tahoe in the Sierra

FIGURE 8.12 Aerial view looking northwest across the north end of Pyramid Lake (The Needles) showing tufa towers aligned along northwest-striking faults (shown as red lines) having oblique normal right-lateral displacement. The tufa towers are as much as 300 feet high and therefore reflect much higher lake levels at the time of their formation, which has been in the last 10,000 years. The alignment of tufa deposits illustrates the structural or fault control of thermal fluids, yielding zones of increased permeability. Crescent-shaped, former shorelines can be discerned to the north of the tufa towers. (Adapted from Benson, L., *The Tufas of Pyramid Lake, Nevada*, Circular 1267, U.S. Geological Survey, Reston, VA, 2004.)

Nevada, and because there is no outlet evaporation has led to a rather high bicarbonate concentration as the lake level has lowered over time. Geologically, Pyramid Lake lies across the eastern margin of the northwest-trending Walker Lane, a structural zone of discontinuous strike-slip faults noted for hosting mineralized fossil geothermal systems (e.g., Comstock Lode) and several developed geothermal systems, including Steamboat Springs, San Emidio, Salt Wells, and Coso. At the north end of the lake, tufa deposits tower about 100 m high, reflecting much higher lake levels when they formed. Unlike in the formation of travertine, geothermal fluids forming tufa are relatively depleted in CO_2 but contain enough calcium to react with bicarbonate-rich lake water to deposit calcite as tufa. However, under these circumstances, the presence of tufa-formed calcite does not preclude high temperatures at depth because the precipitation of silica is suppressed due to dilution from lake water. Silica, instead, may be precipitated in loosely consolidated sediments underlying the lake and below where tufa is forming at the sediment–water interface. Consequently, the geologic setting of calcite-bearing thermal springs needs to be correctly identified as tufa vs. travertine, as this has important implications for characterizing an area's potential for geothermal development and where the best temperature potential may lie (Coolbaugh et al., 2009).

Structural Analysis

Many, if not most, exploited geothermal reservoirs require secondary or fracture-controlled permeability to make them commercially viable. As such, the structural setting of faults needs to be determined when assessing a region's geothermal potential. A considerable body of information has developed on this topic in the last decade or so from the work of Jim Faulds and coworkers at the Nevada Bureau of Mines and Geology and University of Nevada, Reno (Cashman et al., 2012; Faulds et al., 2011, 2013). Most of this work has been focused on characterizing the structural settings of geothermal systems in the Great Basin of the western United States, but findings from these studies are applicable to geothermal systems worldwide. A surprising result from this work is that only about 3% of the 400 or so geothermal systems studied in the Great Basin occur along the mid-segments of major normal faults that bound mountain ranges. Areas of greatest fault displacement are not necessarily conducive for hosting geothermal systems, probably because the presence of clay gouge on large offset faults can act like hydraulic barriers, restricting fluid flow. Furthermore, these faults move only every few hundred to a few thousand years to release stress in large earthquakes, and in the intervening time between earthquakes the fault becomes sealed with gouge and/or mineral precipitates. More frequent smaller earthquakes occurring on smaller, more numerous faults, on the other hand, appear to keep fluid pathways open by periodically reopening fractures that may have been clogged by minerals precipitated from circulating geothermal fluids. Six major types of structural settings have been recognized, four of which account for 85% of known geothermal systems in the Great Basin (Faulds et al., 2013). These four main structural settings are (1) step-over regions of normal faults, (2) horse-tailing ends of major normal faults, (3) fault intersections, and (4) accommodation zones (regions of overlapping faults having opposing directions of dip) (Figure 8.13).

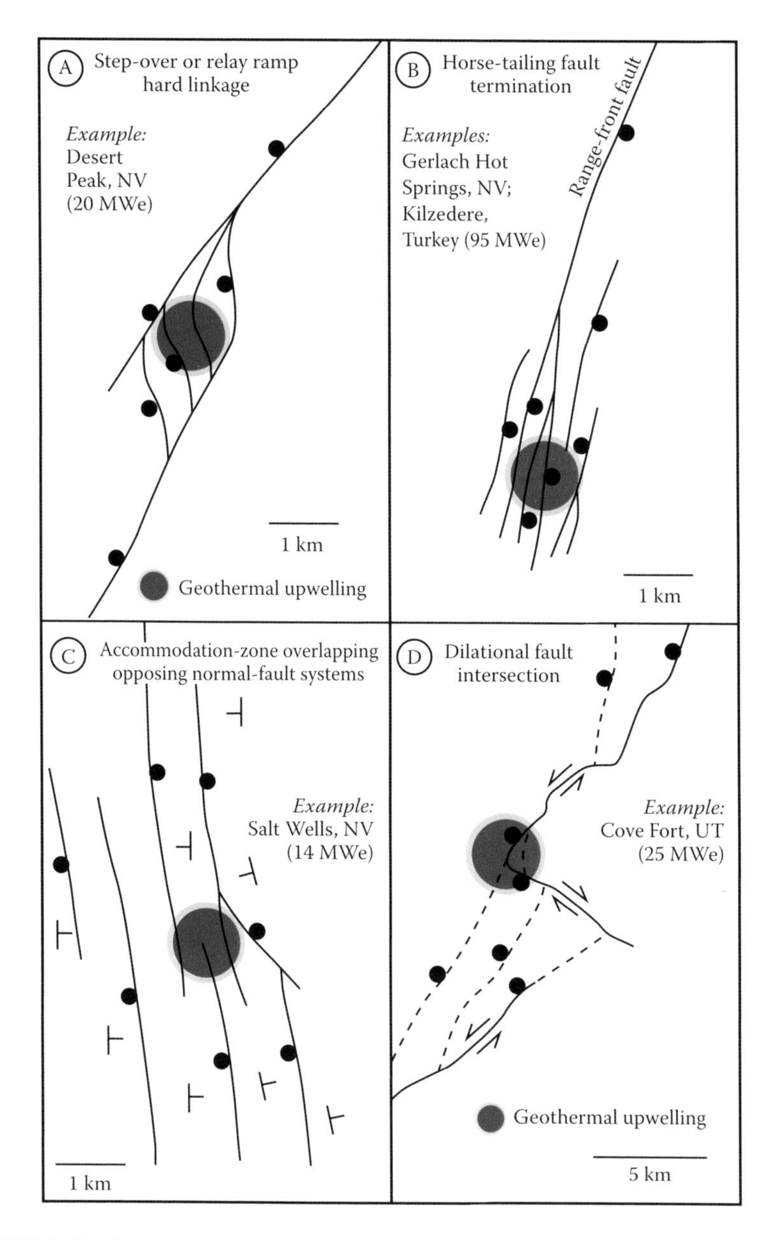

FIGURE 8.13 Four main structural settings of geothermal systems: (A) normal fault step-over; (B) fault termination zone with horse-tailing splays; (C) accommodation zone characterized by overlapping and intersecting oppositely dipping normal faults; and (D) dilational fault intersection. The red-filled circle represents main geothermal zones of upwelling in areas of high fracture/fault density and high permeability. Microearthquakes, too small to be felt, are common in such zones and help keep fluid pathways open by rupturing and clearing any precipitated minerals along fractures. (Adapted from Faulds, J.E. et al., *Geothermal Resources Council Transactions*, 37, 3–10, 2013.)

Of the 25 producing geothermal plants in the Great Basin, 11 occur in fault step-over zones, and 8 occur in fault intersections. Many of the larger geothermal power plants occur in areas where multiple fault type settings are evident, which are referred to as *hybrid settings*. For example, the Steamboat geothermal field (128-MWe installed capacity) has fault step-overs and intersecting fault zones, and it lies in a region in which west- and east-dipping normal faults overlap (accommodation zone). The Coso geothermal field, which hosts the largest geothermal power facility in the Great Basin (270-MWe installed capacity), lies in a zone of pull-apart dilation produced by right-stepping, right-lateral strike-slip faults and intersecting northeast-striking normal faults (see Figure 7.29 in Chapter 7). It is also important to note that the faults described here are active faults. That is, large faults have moved at least in the last 10,000 years or so, but smaller faults have moved more frequently. This is because less build-up of stress is required to cause rupture, which is critical for opening any sealing caused by minerals deposited from hydrothermal solutions, so permeability is restored or maintained and fluids can continue to circulate.

GEOCHEMICAL STUDIES

In geothermal systems exhibiting surface thermal manifestations, geochemical sampling of waters or gas vents helps characterize the chemistry of the system and resource type (vapor or liquid dominated), and it can also provide an estimate of fluid temperatures at possible reservoir depths through the use of select geothermometers.

Fluid Composition and Reservoir Type

Geochemical sampling of surface-discharged fluids allows characterization of fluid chemistry and pH. This in turn can yield insights into the type of geothermal reservoir present. For example, if sampling of thermal springs yields very low TDS and relatively acidic values, then a vapor-dominated system may be evident. Steam carries very small amounts of dissolved solids, the exception being volatile mercury, so when the steam condenses in near-surface cap rocks the water is basically "distilled." Moreover, because noncondensable gases such as hydrogen sulfide (H_2S) and carbon dioxide (CO_2) are partitioned into the vapor, acids are produced when gases are absorbed by shallow aquifers. Hydrogen sulfide combines with oxygen in the atmosphere or oxygenated shallow groundwaters to produce sulfuric acid (H_2SO_4), a strong acid, and CO_2 combines with water to produce carbonic acid (H_2CO_3), a weak acid. It is important to sample more than just a few springs, as findings can be misleading because similar results can be obtained in the steam-heated zone above an episodically boiling liquid-dominated reservoir. A more comprehensive sampling program in this case, however, might also disclose springs, at probably lower elevations, having near-neutral pH and low to high TDS values typical of a liquid-dominated reservoir. Results of the geochemical analysis in conjunction with mapping and characterizing hydrothermal alteration and thermal features can help lead to a reasonable interpretation of the resource type without involving expensive drilling at this stage in an exploration program.

Geochemical Thermometers (Geothermometers)

The measured chemical composition of select elements or ratios of select elements from sampled thermal springs can provide estimates of fluid temperatures at possible reservoir depths. This is because an empirical correlation exists between borehole temperatures and the concentrations or ratios of select elements from analyzed thermal spring waters. The basis for this has to do with chemical thermodynamics, which is beyond the scope of this book, except to say that the chemical signatures of sampled fluids are a function of rock–water interactions and temperature, and the higher the temperature and permeability the more quickly the fluid and rock will reach chemical equilibrium. In order for geothermometers to give reasonably accurate temperature estimates at depth, fluids and rock must be near chemical equilibrium. Moreover, on their way to the surface, hydrothermal fluids must flow quickly enough to prevent extensive re-equilibration with near-surface rocks and must not have mixed significantly with near-surface groundwater. Furthermore, geothermometers are not applicable to steam-heated waters, as discussed, above, because of partitioning of elements between the steam and residual liquid. They are used specifically for chloride-rich thermal waters. There are several chemical-based geothermometers. Three of the more commonly used are silica, Na–K, and Na–K–Ca geothermometers based on the following equilibrium reactions, respectively:

$$SiO_2 + 2H_2O \Leftrightarrow H_4SiO_4 \tag{8.2}$$
$$\text{(Silica quartz)} \Leftrightarrow \text{(Silicic acid)}$$

$$NaAlSi_3O_8 + K^+ \Leftrightarrow KAlSi_3O_8 + Na^+ \tag{8.3}$$
$$\text{(Albite feldspar)} \Leftrightarrow \text{(K-feldspar)}$$

$$2Ca_2Al_2Si_2O_6 + 4H_2CO_3 + 3O_2 \Leftrightarrow 4CaCO_3 + 2Al_2Si_2O_5(OH)_4 \tag{8.4}$$
$$\text{(Plagioclase feldspar)} + \text{(Carbonic acid)} \Leftrightarrow \text{(Calcite)} + \text{(Clay)}$$

All reactions are applicable where the interpreted reservoir rock contains quartz and feldspar, which are the two most common minerals in the Earth's crust. For waters with high concentrations of calcium and depositing travertine, the third reaction becomes applicable.

A fourth type of geothermometer, discussed below, is based on mineral equilibria of coexisting mineral species in geothermal fluids. It can yield satisfactory temperature estimates at depth when the aforementioned geothermometers fail because no *a priori* assumption of mineral equilibria is necessary. Finally, gas and isotope geothermometers can be applied where gas vents but no liquid springs are present at the surface.

Multiple geothermometers are typically employed as a cross-check on reliability. If there is little correspondence of results among geothermometers, then a problem with equilibration at depth or re-equilibration as fluids rise to the surface might exist. Also, even though quartz and feldspar are common minerals in the Earth's crust, if the inferred geothermal reservoir is in carbonate rocks, such as limestone or dolostone, then none of the first three geothermometers noted above would be

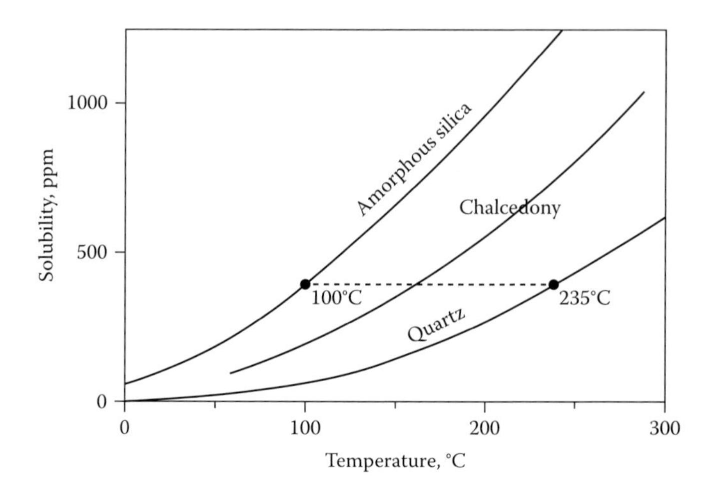

FIGURE 8.14 Solubility curves for three phases of silica: amorphous silica (opal), chalcedony, and quartz. In the example shown, fluid from an opal-depositing hot spring at 100°C contains about 370 ppm silica, indicating a reservoir temperature of 235°C. If the controlling form of silica is chalcedony, then a much lower reservoir temperature of about 150°C is indicated. (Adapted from Rimstidt, J.D. and Cole, D.R., *American Journal of Science*, 283(8), 861–875, 1983.)

applicable, as they assume that the dominant reservoir rocks are silicates, not carbonates. The mineral equilibria method, however, could be applied under these conditions using the calculated saturation indexes of appropriate alteration minerals that might occur in a carbonate reservoir, including calcite, gypsum, anhydrite, zeolites (e.g., laumontite), and epidote.

Silica Geothermometer

The solubility of silica increases with temperature, allowing the concentration of measured silica in solution to reflect the temperature at depth, provided the solution equilibrated with the reservoir rocks. Equilibration is reasonable considering that the rate of equilibrium reactions increases with temperature; equilibration takes only a matter of hours at a temperature of 250°C but several years at temperatures of around 100°C. Note that of the three solubility curves of silica shown in Figure 8.14, quartz has the lowest solubility at a given temperature compared to the other silica phases.* For reservoir temperatures greater than about 180°C, quartz is the main stable phase and controls solubility (Fournier, 1985). Below that temperature, however, chalcedony, rather than quartz, probably controls dissolved silica content, and its curve rather than that of quartz should be used for estimating temperatures at depth. Below a reservoir temperature of about 100°C, amorphous silica controls solubility and its curve should be used.

* Quartz is the most ordered or crystalline structure of silica. Chalcedony is a microcrystalline form of silica, and opal is an amorphous and hydrated form of silica having no internal crystalline or ordered atomic structure (i.e., the silicon and oxygen atoms are more or less randomly organized like in window glass).

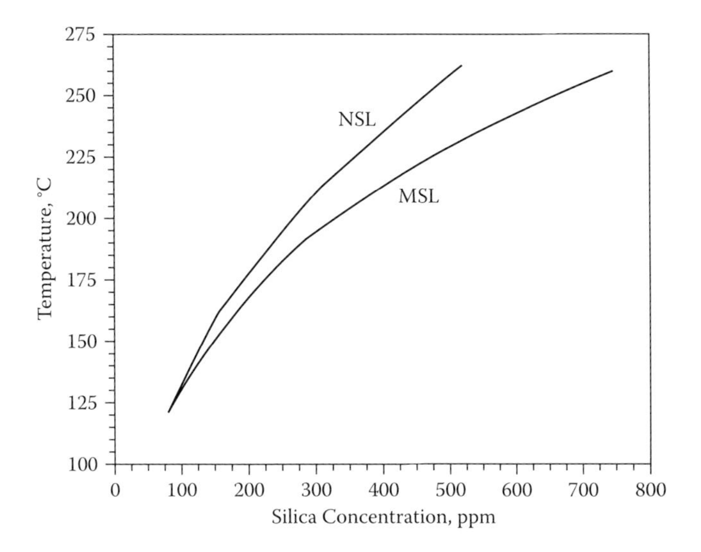

FIGURE 8.15 Solubility curves for quartz. NSL means no steam loss and MSL represents maximum steam loss. The lower temperatures indicated for the MSL curve reflects enhanced concentration of silica due to boiling and mass loss of steam as discussed in text. Comparing this figure with the preceding figure, the curves drawn for the preceding figure are for no steam loss, as 370 ppm silica for a boiling system would indicate a reservoir temperature of about 210°C. (Adapted from DiPippo, R., *Geothermal Power Plants: Principles, Applications, Case Studies, and Environmental Impacts*, 3rd ed., Butterworth-Heinemann, Waltham, MA, 2012.)

The experimentally determined effective temperature range for using the silica geothermometer is from ~50°C to ~300°C. Above 300°C, high concentrations of other dissolved species increasingly affect silica solubility, leading to potentially unreliable estimates of temperature. Another consideration that affects silica solubility is pH, and the silica geothermometer is calibrated for solutions having near-neutral pH (~5 to ~7). At lower and higher pH, silica solubility increases in addition to temperature effects, indicating that pH must be determined to see whether or not silica geothermometry analysis is applicable. Finally, if there is evidence of steam loss through boiling, the measured silica concentration will be higher than if there was no steam loss, because silica is largely retained in the liquid phase during boiling. Therefore, its concentration increases through mass loss of steam (Figure 8.15). Considering that the example noted in Figure 8.15 was from a hot spring at 100°C, boiling and steam loss were occurring, so the more realistic temperature would be about 210°C rather than 235°C.

A series of empirical equations has been developed to describe silica solubility as a function of temperature under varying conditions:

1. $$T\,(°C) = \left(\frac{1522}{5.75 - \log(\text{silica concentration in ppm})}\right) - 273.15$$

   This equation is for maximum steam loss for actively boiling springs with high discharge rates (>~2 kg/s). The equation takes into account cooling along the boiling point–depth curve as solutions rise to the surface.

2. $T(^{\circ}\mathrm{C}) = \left(\dfrac{1309}{5.19 - \log(\text{silica concentration in ppm})} \right) - 273.15$

   This equation is used for fluids that have cooled conductively on their way to the surface and have not boiled (no steam loss); it is best used for springs at sub-boiling temperatures.

3. $T(^{\circ}\mathrm{C}) = \left(\dfrac{1112}{4.91 - \log(\text{silica concentration in ppm})} \right) - 273.15$

   This equation is used if the above quartz geothermometer or another geothermometer (see below) indicates temperatures of 120° to 160°C, indicating that chalcedony rather than quartz controls silica solubility in the reservoir.

Comparisons of the measured water temperatures from bore holes and the calculated reservoir temperatures and various formulations of the silica geothermometer indicate that the silica geothermometer temperature estimates are generally higher than the measured temperature values (Figure 8.16). A better correlation

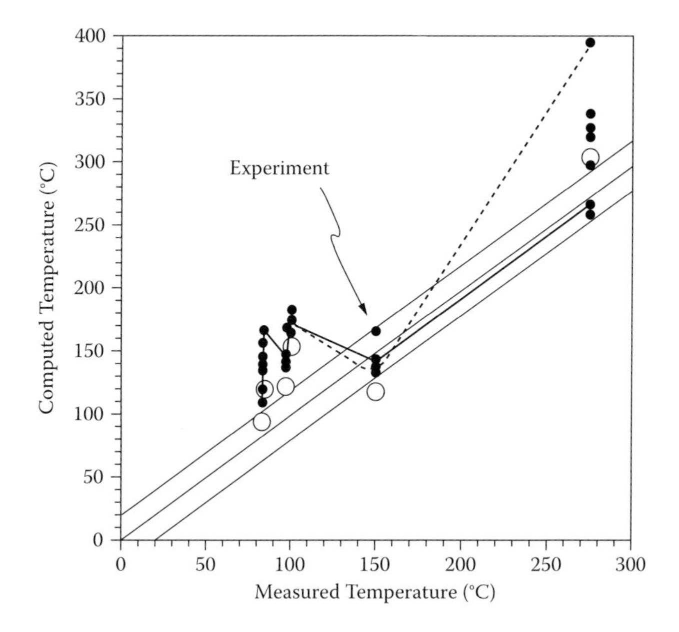

FIGURE 8.16 Comparison of measured water temperatures from well and calculated temperatures using the silica geothermometer. Solid circles are temperatures calculated assuming that quartz is the controlling silica phase, whereas open circles assume that chalcedony is the controlling silica phase in the reservoir. The one-to-one correlation is noted by the thin line within the shaded envelope. The lines on either side of the 1:1 correlation line mark the ±20°C limit of the correlation line. The bold and dashed lines connect temperature estimates using different formulations of the silica geothermometer. (Adapted from Glassley, W.E., *Geothermal Energy: Renewable Energy and the Environment*, 2nd ed., CRC Press, Boca Raton, FL, 2015.)

exists between measured temperatures and the chalcedonic geothermometer in lower temperature reservoirs, more or less consistent with the results of Fournier (1985). At the higher temperature end, estimates from silica geothermometry could reflect boiling and concentration of residual silica in the sampled solutions as shown in Figure 8.15.

Na–K Geothermometer

The Na–K geothermometer is based on the exchange reaction of Na and K between sodium and potassium feldspar (Reaction 8.2). In this case, the ratio of Na/K is used for the log term in the empirically derived equation shown below:

$$T\,(^\circ\mathrm{C}) = \left(\frac{1217}{1.438 + \log(\mathrm{Na/K})}\right) - 273.15 \qquad (8.5)$$

The ratio decreases with increasing temperature. The advantage of this geothermometer is that it is less affected by possible dilution (provided the diluting waters are low in Na and K) or steam loss because it is based on a ratio (both the numerator and denominator are equally impacted by either dilution or steam loss so the ratio changes little).

Empirical results comparing calculated values to well temperature data indicate that this geothermometer works best with high reservoir temperatures (>100°C) and is best applied where temperatures are interpreted, from other data, to be >160°C. Temperature range is applicable to as high as 350°C because possible re-equilibration upon ascent from such high temperatures is slower than that of the silica-quartz geothermometer. Also, geothermal fluids need to have near-neutral pH and be of the alkali-chloride type, not acid-sulfate rich waters, for the Na–K geothermometer to be used. This geothermometer is also unsuitable if waters have a high Ca content, such as for springs depositing travertine.

Comparisons of measured water temperatures to temperature estimates using the Na–K geothermometer also indicate that the computed temperatures are generally higher than the measured temperatures (Figure 8.17); however, the spread in calculated temperatures decreases and better matches the measured temperatures at higher temperatures, indicating that the Na–K geothermometer is better suited for higher temperatures (generally greater than about 160°C). Comparing temperatures determined by the silica and Na–K geothermometers shows that temperature estimates using the silica geothermometer are generally slightly higher than those determined using the Na–K geothermometer (Figure 8.18). Nonetheless, the correlation is still quite good, supporting the use of both for estimating temperatures of underlying liquid-dominated geothermal reservoirs.

Na–K–Ca Geothermometer

The Na–K–Ca geothermometer is applicable to thermal springs rich in Ca and possibly depositing travertine. Fluids at depth would be in equilibrium with alkali feldspar; however, plagioclase feldspar would be actively converting to calcite and clay, as indicated by Reaction 8.4. At higher temperatures (generally >200°C), plagioclase

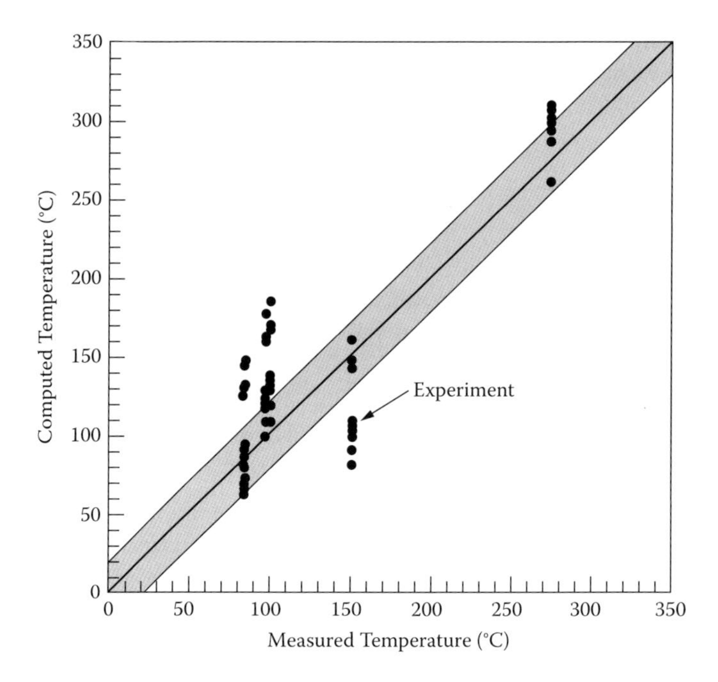

FIGURE 8.17 Comparison of measured reservoir temperatures with calculated temperatures using the Na–K geothermometer according to nine different formulations of five different samples of geothermal waters. The line running from the lower left to upper right marks the 1:1 correlation line, and the shaded zone is an envelope of ±20°C of the 1:1 correlation line. Experiment refers to a laboratory-produced water sample. (From Glassley, W.E., *Geothermal Energy: Renewable Energy and the Environment*, 2nd ed., CRC Press, Boca Raton, FL, 2015.)

would be replaced by epidote (a hydrated calcium-rich silicate) instead of calcite, unless the CO_2 concentration in the fluid is very high to keep the calcite stable. As with the silica and Na–K geothermometers, fluids must be of near-neutral pH and of the alkali-chloride type. Reduction of dissolved Ca, due to possible dilution or boiling causing rapid deposition of calcite from the loss of CO_2, can result in temperature estimates that are too high—the degree of overestimation being proportional to the amount of dissolved CO_2 in the fluid. So, in the case of travertine-depositing springs, reservoir temperature estimates determined using the Na–K–Ca geothermometer should be considered a maximum.

Mineral Equilibria Method of Geothermometry

As noted in the introduction of this section on geothermometry, the silica and alkali geothermometers assume that mineral equilibria are achieved in the geothermal reservoir and minimal re-equilibration occurs during ascent of the solutions to the surface. As a way to evaluate this assumption and obviate the limitation, Reed and Spycher (1984) developed a technique where the saturation characteristics of multiple mineral species can be examined simultaneously. This geothermometer involves

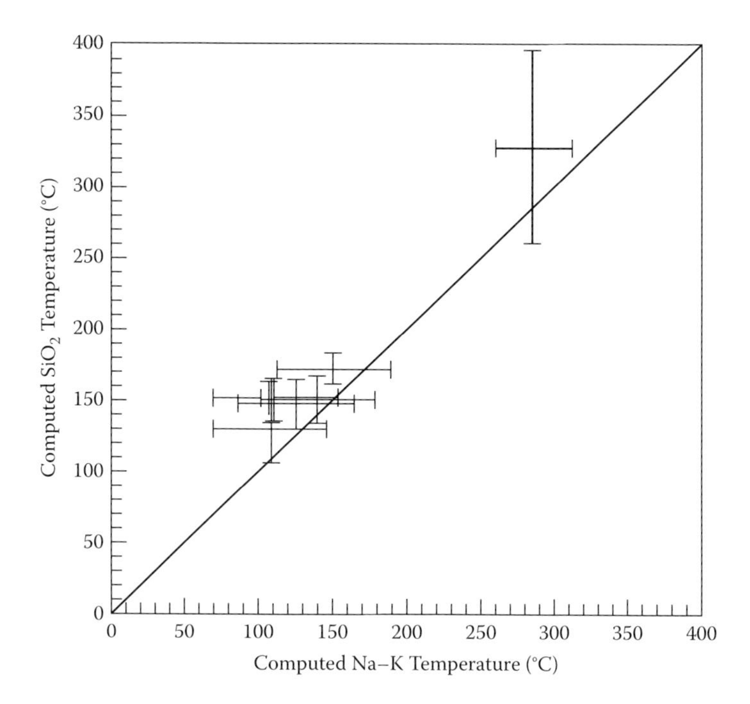

FIGURE 8.18 Comparison of silica and Na–K geothermometer temperature estimates with ranges for each geothermometer, horizontal bar for Na–K and vertical bar for silica. The crossing point of the two bars represents the median value for all computed temperatures of each sample. The line running from lower left to upper right is the 1:1 correlation line. (From Glassley, W.E., *Geothermal Energy: Renewable Energy and the Environment*, 2nd ed., CRC Press, Boca Raton, FL, 2015.)

the calculation of the saturation indexes as a function of temperatures of common alteration minerals using published thermodynamic data involving the activity ion products (Q) and equilibrium constants (K) of those minerals. The value of $\log(Q/K)$ is the saturation index. If $\log(Q/K)$ is >0, the solution is supersaturated with that mineral; if $\log(Q/K)$ is <0, the solution is undersaturated with that mineral. By plotting the saturation indexes of different minerals as a function of temperature, it is possible to determine (1) whether the geothermal fluid is in equilibrium with a given mineral assemblage in the reservoir, (2) which minerals are in equilibrium and which are not, and (3) the temperature of equilibrium. Furthermore, in cases where fluid and minerals are not in equilibrium, the most likely causes, whether boiling or dilution upon ascent, can be elucidated by inspection of the $\log(Q/K)$ vs. temperature plots of the different minerals.

Figure 8.19 shows plots of saturation indexes, $\log(Q/K)$, of various minerals as a function of temperature for the Sulphur Bank geothermal system and mercury mine located on the shore of Clear Lake, California. In Figure 8.19A, the estimated temperature is given where the saturation indexes of the various minerals converge at $\log(Q/K) = 0$. In this case, the indicated temperature is in the range of 140 to 165°C,

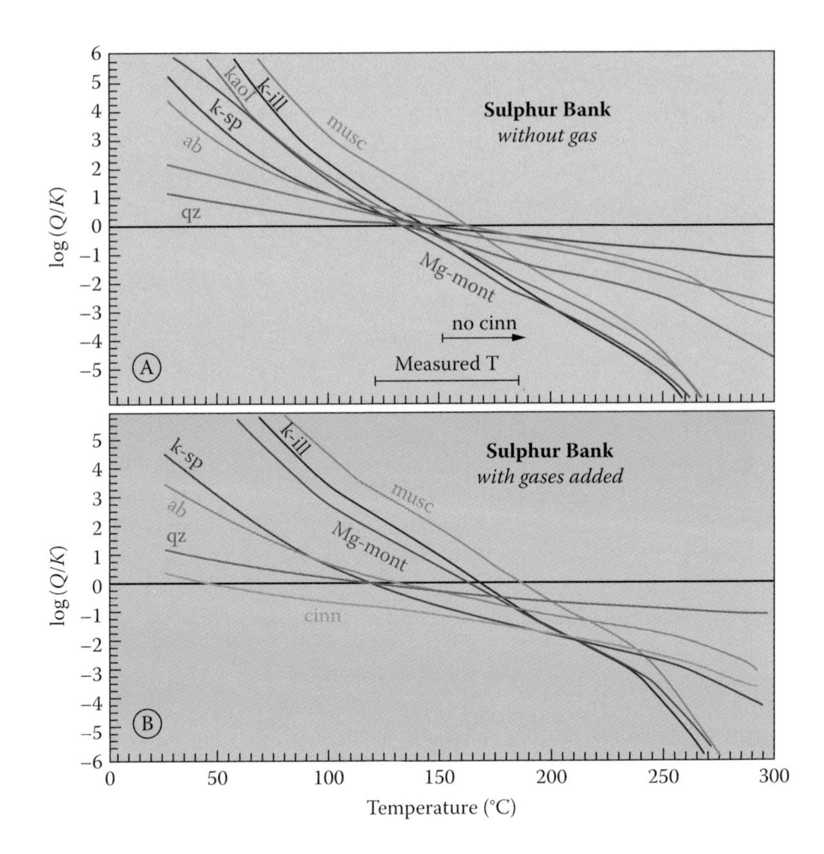

FIGURE 8.19 Saturation index, log(Q/K), as a function of temperature for waters from the Sulphur Bank geothermal system in California. (A) Apparent equilibration of subsurface minerals with boiled or degassed water that lies within the measured temperature range. (B) Saturation index curves of minerals from unboiled waters showing no convergence and hence a lack of equilibration. Mineral abbreviations: ab, albite; cinn, cinnabar; k-ill, K-rich illite; kaol, kaolinite; k-sp, potassium feldspar; Mg-mont, manganese-rich montmorillonite; musc, muscovite; qz, quartz. (Adapted from Reed, M. and Spycher, N., *Geochimica et Cosmochimica Acta*, 48(7), 1479–1492, 1984.)

and the minerals that converge in this temperature range consist of cinnabar (mercury sulfide), quartz, manganese-rich montmorillonite, kaolinite, albite, and potassium feldspar. This temperature range compares favorably with measured temperatures from various depths. In Figure 8.19B, the same saturation indexes with temperature are plotted but gases have been added, resulting in little convergence of the saturation indexes along the zero line. The superior convergence of saturation indexes in Figure 8.19A means that the fluids at Sulphur Bank have boiled, likely removing the gases from the liquid. Moreover, without boiling, cinnabar remains largely undersaturated (Figure 8.19B) and would not have deposited, and no mercury would have been mined.

To summarize, because the saturation indexes of several minerals are examined simultaneously, no prior assumption of individual mineral equilibria is required as is the case of the other aqueous geothermometers discussed above. Also, using the mineral equilibria method allows determination of which minerals species are in equilibrium with the fluid. Furthermore, if mineral equilibria are not attained due to such processes as boiling and dilution, each of those processes can be elucidated by modeling to see which of those processes best fits the data; by this means, the amount of boiling or dilution can be estimated as well.

Gas Geothermometers

In regions where thermal springs are lacking due to deep groundwater levels, but steaming ground or fumaroles are present, estimates of subsurface temperatures can be achieved using gas geothermometers based on temperature-dependent gas–gas or gas–mineral equilibria. This is particularly useful in acid-sulfate systems that commonly exhibit fumaroles or gas vents and any thermal springs that are too acidic and sulfate rich for water-based geothermometric analyses. For fumarolic steam, concentrations of CO_2, H_2S, and H_2 or their ratios (CO_2/H_2 or H_2/H_2S) can be measured and inserted into empirical-based equations. Temperature estimates from the various gas-type equations are then compared to check for correspondence and to discard any anomalous calculated results.

Isotope Geothermometers

Isotope exchange reactions, such as between noncondensable gases and steam, a mineral and a gas phase, water and a solute, or between different solutes, are temperature dependent. Isotopes are chemical elements that have the same atomic number (number of protons) but different atomic mass (different number of neutrons). For example, oxygen has three stable isotopes, O^{16} (the most common), O^{17}, and O^{18}, and hydrogen has two stable isotopes, H^1 (protium) and H^2 (deuterium). When steam separates from water, an isotopic fractionation occurs such that the water is enriched in O^{18} and deuterium, but the steam is depleted in O^{18} and deuterium, as the heavier isotopes favor the denser phase. One type of isotope geothermometer involves isotope equilibrium exchange of heavy and light hydrogen between steam and hydrogen gas:

$$^2H_2\,(\text{Deuterium}) + {}^1H_2O \Leftrightarrow {}^1H_2 + {}^2H_2O \tag{8.6}$$

where the valid temperature range is 100 to 400°C. The use of isotope geothermometers, although viable, is complicated because of the care necessary in sample collection and preparation, expense, variable rates of achieving isotopic equilibrium, and imperfect knowledge of equilibrium constants.

GEOPHYSICAL EXPLORATION TECHNIQUES

We have already discussed the use of aeromagnetic studies to help identify magnetic anomalies over a large area that might represent local areas of geothermal promise. For example, magnetic lows can form where igneous rocks have interacted with hot fluids that destroyed any original or primary magnetic minerals, such as might occur where igneous magnetite is converted to hydrothermal pyrite. However, magnetic lows can also represent reversely magnetized rocks, so results are not unique and must be corroborated through other geophysical and geological investigations to clarify anomalies.

Additional studies include resistivity and magnetotelluric surveys, gravity surveys, and seismic surveys. Before proceeding, these geophysical techniques do not directly indicate heat at depth; any found anomalies indicate only the possibility of heat. For example, a magnetic low anomaly in volcanic terrane may reflect a cold paleogeothermal system. Therefore, the primary or direct method for detecting heat is drilling holes and measuring temperature with depth as discussed earlier. Geophysical techniques allow isolating specific areas in a large swath of ground for more detailed on-the-ground geological and geochemical investigations. If results of those geological and geochemical studies are encouraging, then temperature probe holes are drilled to assess heat content at depth.

Resistivity and Magnetotelluric Studies

Rocks in general are poor conductors of electricity; in other words, they generally have high resistivity, which is measured in ohm-meters. In most geothermal reservoirs, however, hydrothermally produced clays are good conductors of electricity, and rock pores filled with hot water contain electrically charged dissolved ions, which increase the electrical conductivity (thus lowering the resistivity). Because of this behavior, electrical resistivity studies were utilized to help explore for geothermal resources in New Zealand. The results were very successful in identifying over 16 geothermal systems in the Taupo Volcanic Zone, many of which now support geothermal power plants (Figure 8.20).

For ground-based resistivity surveys, probes are inserted into the ground tens to hundreds of meters apart to record injections of electrical impulses. This technique is good for detecting the distribution of electrical resistors or non-resistors to several hundred meters below the surface. For deeper penetration to several kilometers or more, another technique called *magnetotelluric* is used. Magnetotelluric examines the variations in intensity and direction of Earth's magnetic field on a daily basis induced in part by the solar wind. Data are collected continuously or over short time intervals, typically carried out over several days. Sensors are placed over a region of interest, and after computer reduction of the collected data a reconstruction or model of the underlying geology can be made. This is because the behavior of electrical currents is governed by different properties of rocks in response to the varying magnetic field. Such a study was conducted across the geothermal system in Dixie Valley, Nevada, and the results are interpreted to reflect a zone of partially molten rock at a depth of about 15 to 20 km (Figure 8.21) (Wannamaker et al., 2013). In general, however, confidence for interpreting magnetotelluric data is greatest for depths less than 5 km.

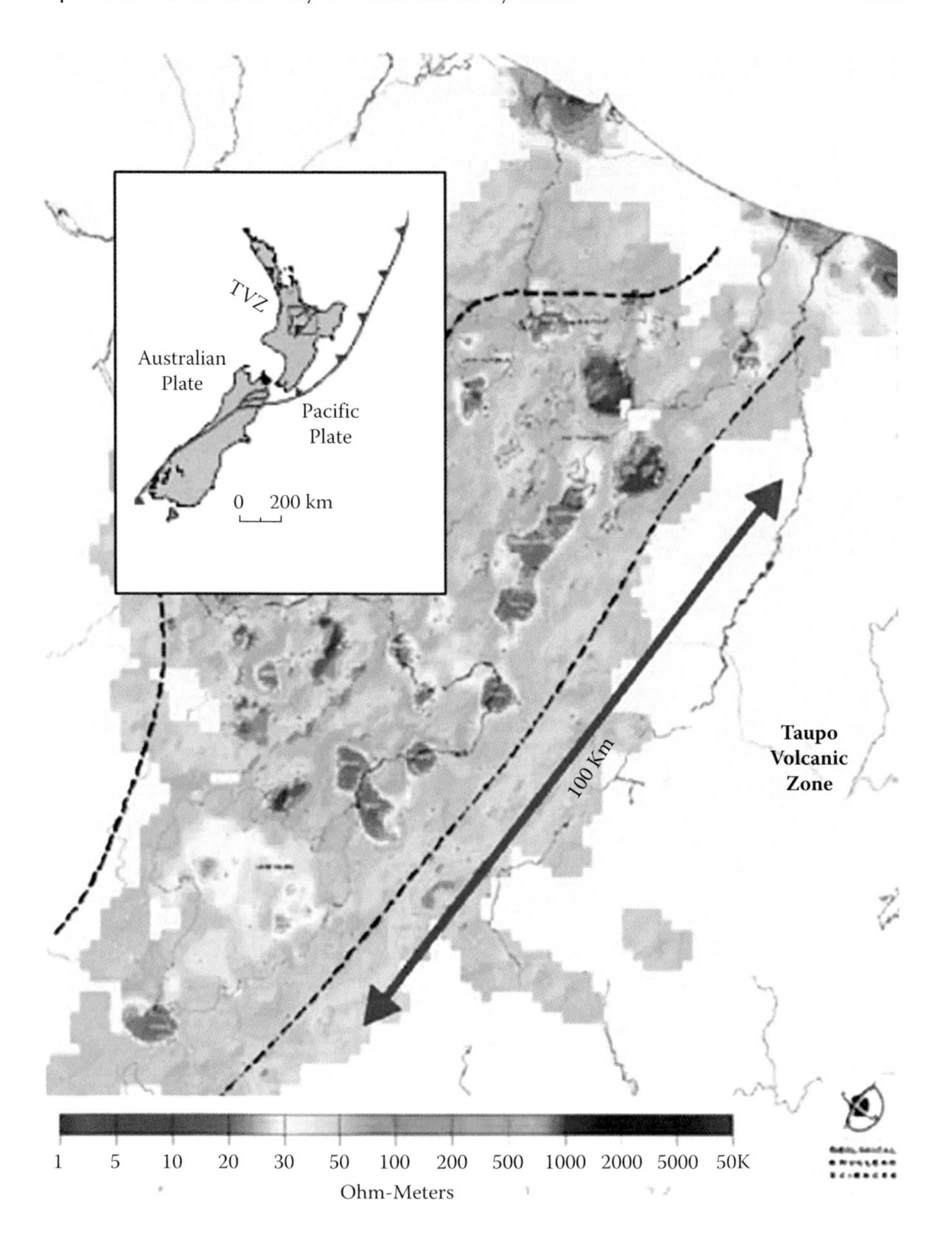

FIGURE 8.20 Composite resistivity map of the Taupo Volcanic Zone (TVZ) in New Zealand. Warm colors correspond to low-resistivity (electrically conductive) zones of geothermal systems. Cool colors denote zones of higher resistivity rocks outside of geothermal systems. Black dashed line indicates the approximate boundary of the TVZ. The location of the TVZ is shown in the inset lying to the west of a subduction zone indicated by red sawtoothed line. (Adapted from Bibby, H.M. et al., in *Proceedings of World Geothermal Congress 2005*, Antalya, Turkey, April 24–29, 2005.)

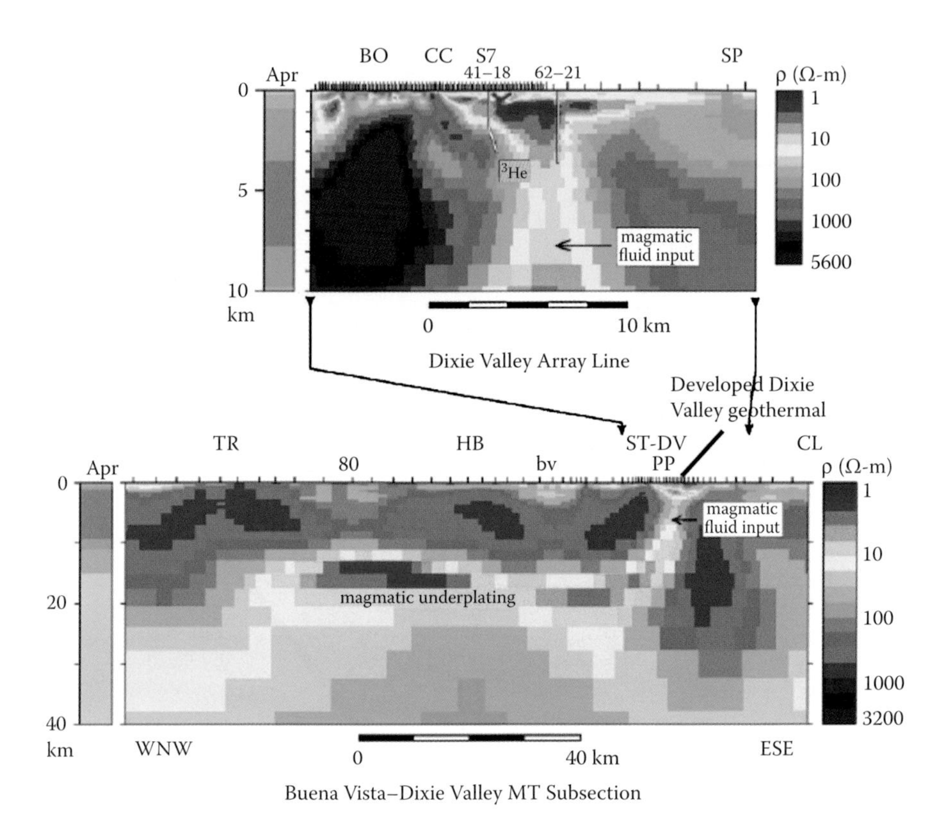

FIGURE 8.21 Resistivity cross-sectional model across Dixie Valley, Nevada, from collected magnetotelluric data. The low resistivity of the robust Dixie Valley geothermal system is imaged in the top cross-sectional view. The lower cross-sectional image is a smaller scale view across the entire valley with imaged data extending to more than 30 km (note that the pixel size increases or data resolution decreases with depth). The low resistivity zone at about 15 to 20 km is interpreted to reflect ponding of magma at depth. (Adapted from Wannamaker, P. et al., in *Proceedings of the 38th Workshop on Geothermal Reservoir Engineering*, Stanford, CA, February 11–13, 2013.)

Possible interpretations of resistivity signals as function of temperature gradients are shown in Figure 8.22. Because gas is a poor conductor, steam reservoirs, in contrast to liquid-dominated reservoirs, have relatively high resistivity. So a high resistivity anomaly in conjunction with high temperature gradients (as discussed further below) could be indicative of a steam reservoir at depth.

Gravity Studies

The gravitational field across the planet is not constant but varies slightly due to changes in the density of rocks at depth. A gravity study is one of the simplest and most cost-effective studies as it is passively registers changes in Earth's gravitational field without having to induce signals, such as in traditional resistivity surveys where

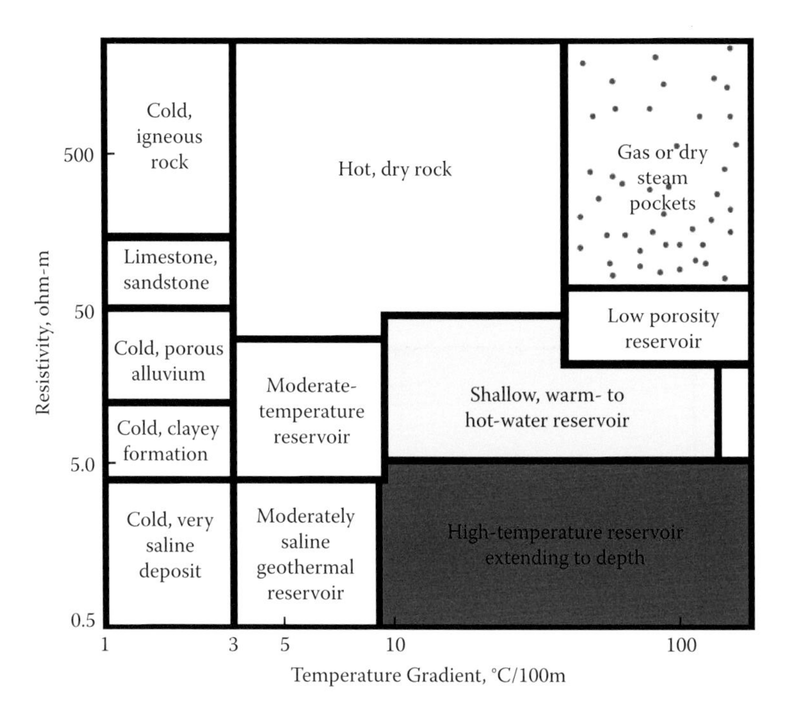

FIGURE 8.22 Matrix of resistivity vs. temperature gradient and possible reservoir type interpretations. (Adapted from DiPippo, R., *Geothermal Power Plants: Principles, Applications, Case Studies, and Environmental Impacts*, 3rd ed., Butterworth-Heinemann, Waltham, MA, 2012.)

an electric current must be introduced. For gravity surveys, a special instrument called a *gravimeter* can measure changes in Earth's gravitational acceleration in milligals (mGal), where 1 Gal equals 1 cm/s^2 or 0.01 m/s^2. Because the average acceleration of Earth's gravity is 9.8 m/s^2, the gravimeter is capable of tracking changes in gravity to 1 part in 98 million. Because gravity is also affected by topography, terrain corrections must be taken into account. The corrected reading is then compared to what the normal gravity should be for a given location, and the difference is referred to as the *Bouguer gravity anomaly*.

Areas characterized by strong gradients in gravity values could reflect exposed or buried faults that have juxtaposed rocks of different densities (Figure 8.23). Because faults occur in areas of broken ground they can serve as geothermal fluid pathways, possibly marking zones of upwelling, hot geothermal fluids or downwelling cool fluids. Buried faults channel hydrothermal fluids and also mark fluid boundaries along the edges of geothermal reservoirs as described in the Olkaria geothermal complex in the East African Rift Valley (Lagat et al., 2005).

A hydrothermal reservoir can be characterized by either a positive or negative Bouguer anomaly depending on conditions in the reservoir. If the reservoir consists of highly porous and permeable rocks, resulting in overall lower density, then the reservoir would be indicated as a negative Bouguer anomaly. This might

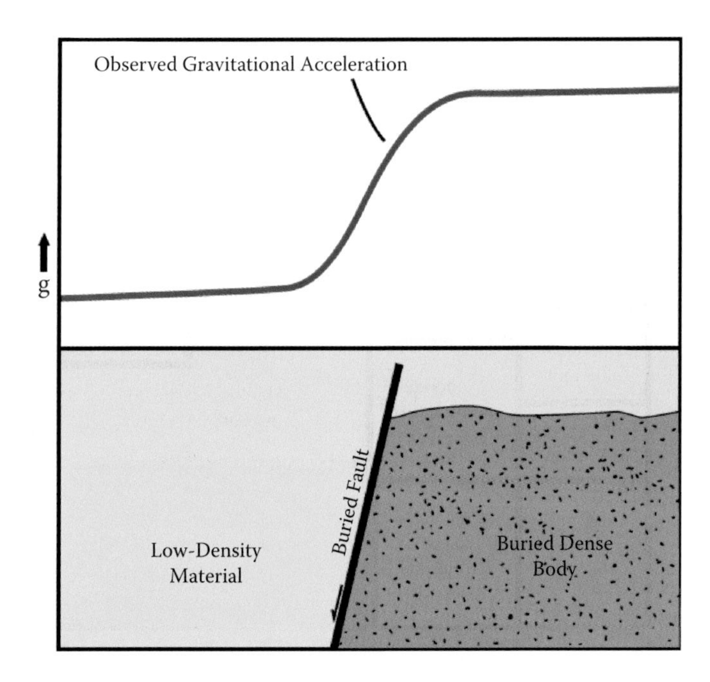

FIGURE 8.23 Schematic illustration of a gravity profile above an inferred buried "basement" fault. Because the "basement" rock is more dense than the adjacent rock, the gravity increases over the "basement" rock. The steep gravitational gradient suggests the two rock types are in fault contact.

occur where the original more dense minerals, such as feldspar, are replaced by low-density minerals such as clay or zeolites as part of the hydrothermal alteration products. On the other hand, if reservoir rocks have been affected strongly by hydrothermal mineral deposition, such as introduced silica or precipitation of epidote in fractures, then porosity is reduced, density is increased, and a positive Bouguer anomaly forms above the reservoir. Gravity studies are combined with other geophysical surveys, such as resistivity, along with results of geological and geochemical studies to lessen potential ambiguity of gravity survey results. In the Achuachapan geothermal field in El Salvador, the gravity survey was combined with a magnetotelluric resistivity survey, and the results showed that the field is located on the boundary of a magnetic high in a region with low resistivity (Figure 8.24) (Santos and Rivas, 2009).

To summarize, gravity studies, as with other geophysical techniques, can yield interpretations of results that are not unique. For example, during early development of The Geysers, a large negative Bouguer gravity anomaly east of the steam field was interpreted as a deep magma body and the source of heat for The Geysers geothermal system. Almost 40 years later, the gravity anomaly, in light of additional geophysical and geological studies, was determined to reflect relatively low-density shale of the Great Valley Sequence and not a buried magma body (Stanley and Blakely, 1995). Thus, the gravity anomaly bears no relationship to the high heat flow of The Geysers.

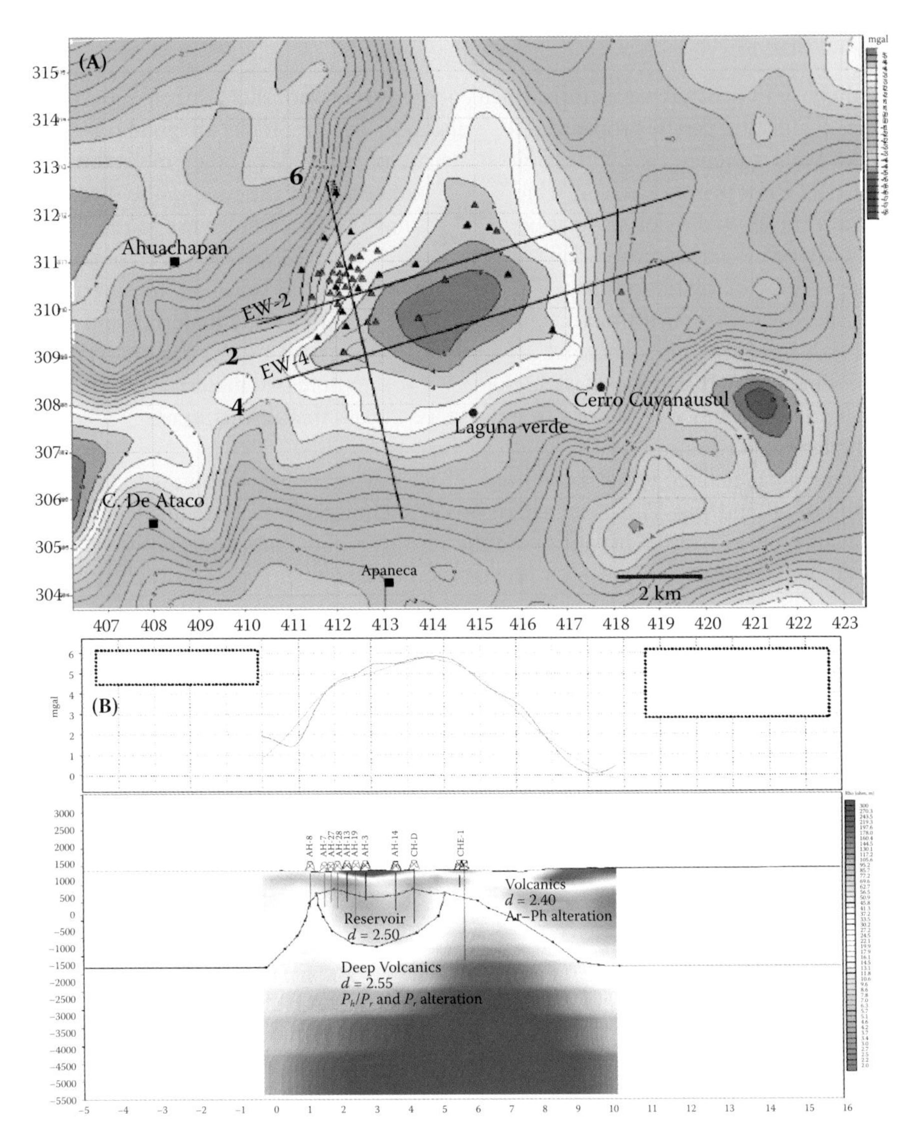

FIGURE 8.24 (A) Bouguer anomaly map of the Ahuachapan geothermal field in El Salvador. Triangles denote geothermal well sites. Note that the field is located in an area of a strong gravity gradient, possibly indicating fault-controlled fluid flow. Warm colors indicate areas of gravity highs and cool colors denote gravity lows. (B) Cross-section along line 2 in part A. Upper part of figure is the gravity profile of reduced measured values (red line) and modeled value (green line) for interpretation shown in lower cross-section where the geothermal reservoir has a modeled density value of 2.50 g/cm^3. The colors in the lower diagram reflect magnetotelluric resistivity values, with low resistivity values in red and high resistivity values in blue. (Adapted from Santos, P.A. and Rivas, J.A., Gravity Surveys Contribution to Geothermal Exploration in El Salvador: The Cases of Berlin, Ahuachapan, and San Vincente Areas, paper presented at Short Course on Surface Exploration for Geothermal Resources, Ahuachapan and Santa Tecla, El Salvador, October 17–30, 2009.)

Seismic Surveys

Earthquake or seismic waves form when energy is released, such as from movement along a fault, and can be propagated long distances because rocks are good transmitters of low-frequency energy. Seismic waves are of two types as they move through the earth: P- or primary waves, which dilate and compress the rock in the direction of wave travel, and S- or shear waves, which move the rock sideways to the direction of travel. P-waves move about 35% faster than S-waves, but the speed of both is strongly affected by the rigidity and density of the material in which they are moving. Speed increases with increasing density and rigidity of the material, and such materials have low impedance. Low-density materials, by contrast, transmit seismic energy more slowly and thus have high impedance. Where two materials of different density are in contact, a contrast in impedance occurs that results in reflection (and refraction) of seismic waves.

Reflection seismology has been used for tens of years to locate prospective reservoirs of oil and gas. This is an active survey technique in that seismic signals are induced, such as from explosions, and the resulting paths taken by the waves will reflect changes in subsurface rock properties. For example, a zone of high fracture density, which could channel geothermal fluids, would have overall lower density than surrounding rock and would have higher impedance. Such a contrast in impedance would be characterized as a preferred zone of seismic reflection. Other factors affecting the strength of reflection would be the angle of incidence of the incoming wave (more glancing angles are better reflected than steep incoming angles) and the degree to which pores are filled with liquid. For example, an impermeable cap rock above a geothermal reservoir consisting of clay might be expected to have relatively high impedance because clay has low density and rigidity.

A good illustration of using reflection seismology to help target drill holes was done at the Hot Pot geothermal prospect in north-central Nevada (Lane et al., 2012). Five seismic lines were conducted, each averaging 6 to 8 km in length. Two lines were oriented northwest-southeast, the rest were oriented northeast-southwest. The lines were oriented roughly perpendicular to known youthful faults that disrupt unconsolidated valley-fill and control the location of thermal springs (Figure 8.25). Known and concealed faults were disclosed on the seismic reflection profiles, whose construction resulted from innovative and sophisticated data-reduction programs (Figure 8.26). As a result of imaging the seismic data, a series of northeast-striking faults was disclosed that led to the identification of two priority drill targets in the hanging walls of the northeast faults (HP 101 and HP 102) (Figure 8.27). The proposed wells would intersect the faults at depths of about 800 m (Figure 8.28). The seismic imaging in this case assisted in development of a geologic conceptual model that could be tested with intermediate drill holes, such as 10-cm-diameter slim holes, reducing the risk of a drilling an expensive, much larger diameter geothermal production well. As Lane et al. (2012) concluded, making use of a viable and a continually refined geologic conceptual model, facilitated from seismic reflection studies, can significantly reduce exploration risk, increase chances of success, and minimize exploration costs.

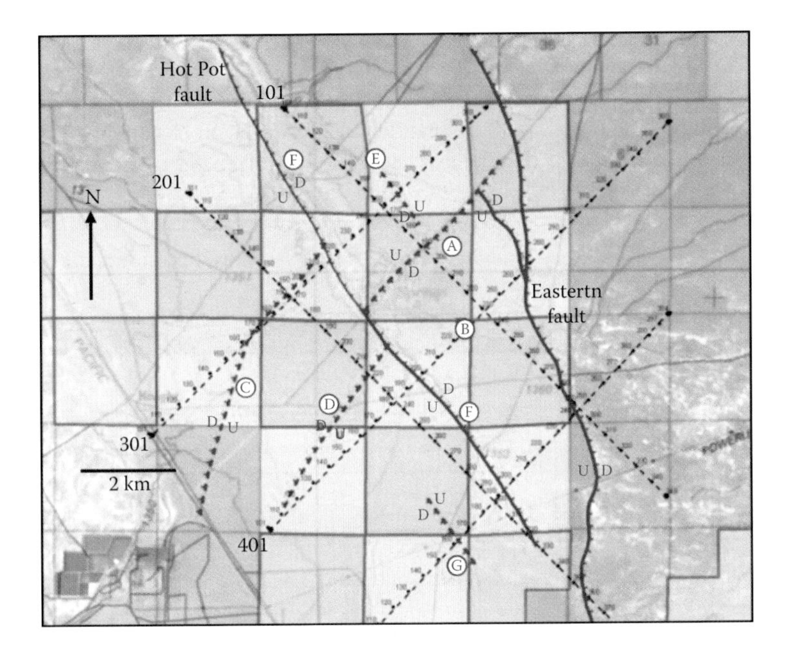

FIGURE 8.25 Map of Hot Pot geothermal prospect in north-central Nevada. Black lines denote seismic survey lines and red lines denote exposed (solid lines) and concealed (dotted lines) faults. Seismic lines are laid out orthogonal to fault strikes to best image their structure at depth. The blue and yellow squares represent lease holdings on private and public lands, respectively. (Adapted from Lane, M. et al., in *Proceedings of the 37th Workshop on Geothermal Reservoir Engineering*, Stanford, CA, January 30–February 1, 2012.)

TEMPERATURE SURVEYS

The purpose of exploration via geological, geochemical, and geophysical investigations is to determine whether a geothermal *resource* is present. If so, the next step is to see if the resource can be turned into a *reserve* and thus be viable for commercial development. To advance from resource to reserve requires expensive exploratory and development drilling, which entails the drilling of full-size potential production and injection wells (at the cost of several million dollars each) along with the necessary flow tests to see what size of power plant might be supported. Prior to drilling full-size exploratory and development holes, temperature-gradient holes are drilled to depths of about 150 m. Temperature is then measured over a time interval to allow for equilibration after disturbance from drilling muds. As a possible prior step to drilling expensive temperature-gradient holes,* drilling of inexpensive shallow holes (~2 m deep) can help document thermal anomalies (Coolbaugh et al., 2014; Kratt et al., 2008) and provide additional information for locating conventional temperature-gradient holes if warranted.

* It is worth noting that cost is relative to the information gained. Sometimes a 150- to 200-m-deep thermal gradient hole will have more value than 100 or so 2-m-deep temperature probe holes.

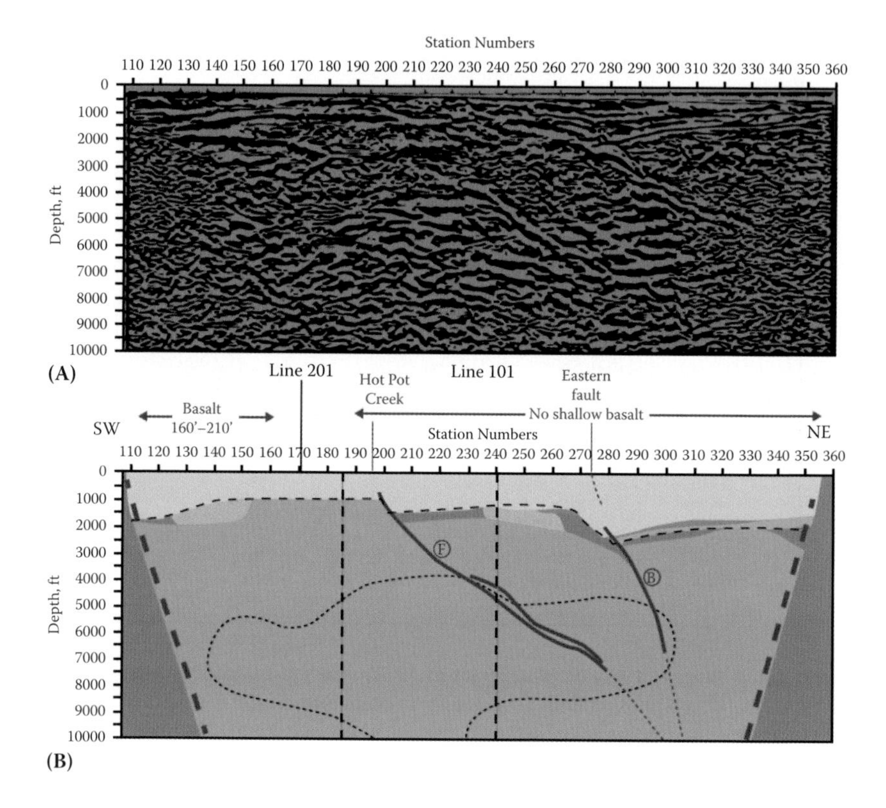

FIGURE 8.26 (A) Computer-processed seismic reflection profile along line 401 showing seismic reflectors as high-contrast zones. (B) Geologic interpretation of seismic reflection profile along line 401 as shown in Figure 8.25. The blue color represents old basement rocks (Ordovician Valmy Formation) overlain by unconsolidated to consolidated (green and orange colored zones) valley-fill deposits, including local basalt. The dotted blue line outlines an area of discontinuous reflective zones that are interpreted to be low-angle thrust slices of distinctive rock layers. Red lines denote interpreted faults that bound and truncate seismic reflectors. (From Lane, M. et al., in *Proceedings of the 37th Workshop on Geothermal Reservoir Engineering*, Stanford, CA, January 30–February 1, 2012.)

Shallow Temperature Surveys

Not only is drilling shallow temperature holes less expensive than drilling conventional temperature-gradient holes, but also a large area can be surveyed in a relatively short period of time (generally a few days but can vary depending on the size and topographic relief of the area surveyed). It should be noted that this technique is best suited for dry climates such as the Basin and Range in Nevada or Altiplano in the Andes. In wet climates, the saturated ground and high water table will likely mask detecting possible thermal anomalies from shallow probe holes. Where suitable, drilling shallow probe holes is also largely exempt from the potentially complex permitting required for drilling standard temperature-gradient holes. Shallow

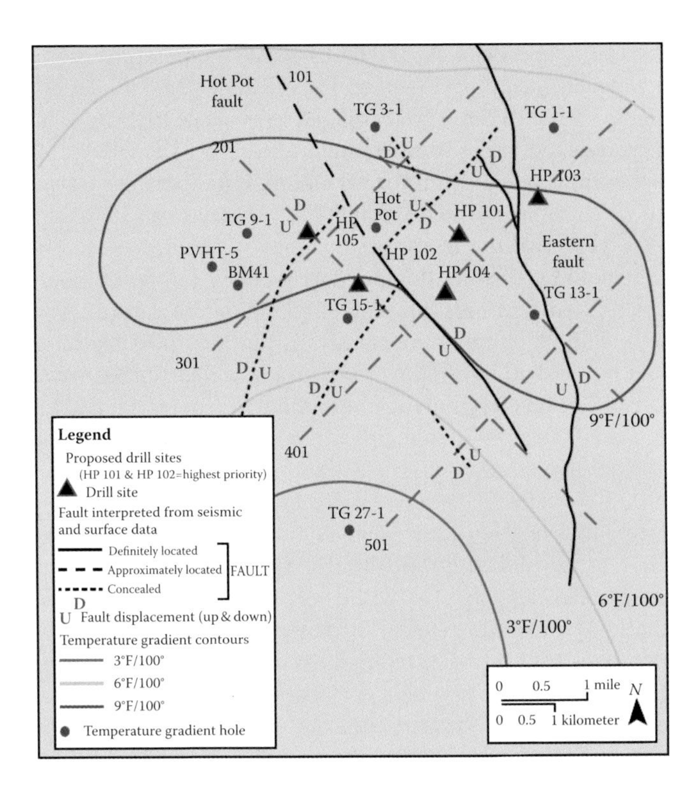

FIGURE 8.27 Proposed well locations at Hot Pot geothermal area. HP 101 and HP 102 are given highest priority as they lie in the hanging wall of seismically imaged but concealed normal faults and have easy access, with only limited road building required. (Adapted from Lane, M. et al., in *Proceedings of the 37th Workshop on Geothermal Reservoir Engineering*, Stanford, CA, January 30–February 1, 2012.)

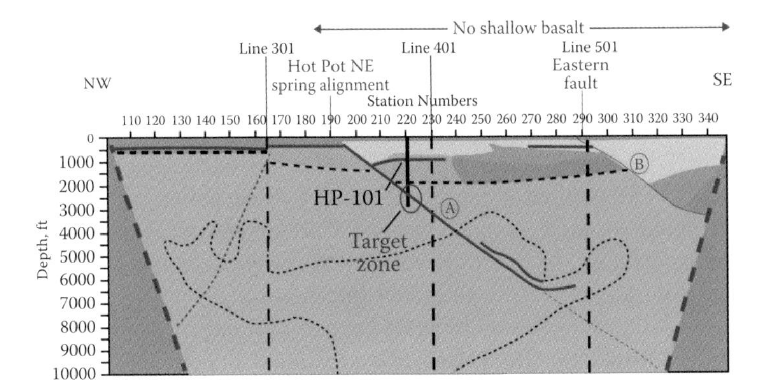

FIGURE 8.28 Geologically interpreted seismic profile along line 101. Red lines mark interpreted faults. Proposed drill hole HP 101 targets fault zone A at a depth between 2500 and 3000 feet. Blue dotted line encloses seismic reflectors interpreted to represent possible thrust slices of seismically contrasting rock units. (Adapted from Lane, M. et al., in *Proceedings of the 37th Workshop on Geothermal Reservoir Engineering*, Stanford, CA, January 30–February 1, 2012.)

temperature surveys have been used to better locate temperature anomalies at several sites in Nevada, including Emerson Pass, Rhodes Marsh, Teels Marsh, Salt Wells, and Gabbs Alkali Flat. The technique consists of drilling 2-m-deep holes using an auger mounted to the back of an all-terrain vehicle (ATV). At a depth of 2 m, daily solar influence is minimized, although temperature corrections must be conducted when returning to a site that has already been surveyed at a different time of the year. Results from these studies indicate that identified thermal anomalies are spatially distinct (temperatures values can be contoured) and are above background temperatures but remain qualitative in nature.* Another potential obstacle to shallow temperature surveys is that any possible underlying heat anomaly could be masked by shallow, cool groundwater. In the arid Great Basin, however, low rates of precipitation help minimize this potential masking effect.

Results of a shallow 2-m-deep temperature survey were used to help discover the "blind" (no obvious geothermal surface expressions, such as hot springs, fumaroles, etc.) Wild Rose geothermal system, located about 30 km west of the town of Gabbs in west-central Nevada at the south end of Gabbs Alkali Flat. This geothermal system is now developed by the Don A. Campbell geothermal power plant, which has an installed capacity of 20 MWe (16 MWe net) (Orenstein and Delwiche, 2014). The shallow temperature survey disclosed an area measuring about 1 km by 2 km in which temperatures were as high as 38.5°C (Figure 8.29). This shallow temperature anomaly coincides with a northeast-southwest-trending aeromagnetic low inferred to reflect hydrothermal clay and magnetite-destructive alteration at depth. Previously drilled mineral exploration holes encountered fluids as hot as 88°C in the vicinity of the shallow temperature anomaly and intercepted propylitically altered basin fill sands and gravels beginning 30 m from the surface (F. Koutz, pers. comm., 2015). In addition to the shallow temperature survey, other pre-drilling exploration studies consisted of geologic mapping and geologic study of previous drilling products, detailed gravity surveying, and resistivity studies including magnetotelluric and ground magnetic surveying. All of these data were synthesized into a geologic model that was used to select locations of five temperature-gradient holes, each drilled to depths of about 150 m. The temperature-gradient holes disclosed ground temperatures as high as 120°C within 60 m of the surface. Results of the temperature-gradient holes were integrated with the other pre-drilling data to select the first of three full-size deep exploratory geothermal wells. The depths of these wells range from about 200 to 380 m, and they all encountered good permeability and fluid temperatures of about 130°C. A month-long production–injection flow test then ensued, and the results were used to generate a numerical reservoir model that predicted that the reservoir could support sustained production of 16 MWe (net). Development drilling consisted of four additional production wells and one additional injection well. The Don A. Campbell geothermal power plant went into production in 2014, and the shallow 2-m temperature survey played an important role in the early exploration phase of this successfully discovered and developed geothermal system.

---

* This is because if indicated temperatures are at the high end of background then it can be difficult to resolve whether the measured temperature value is due to solar or geothermal influence.

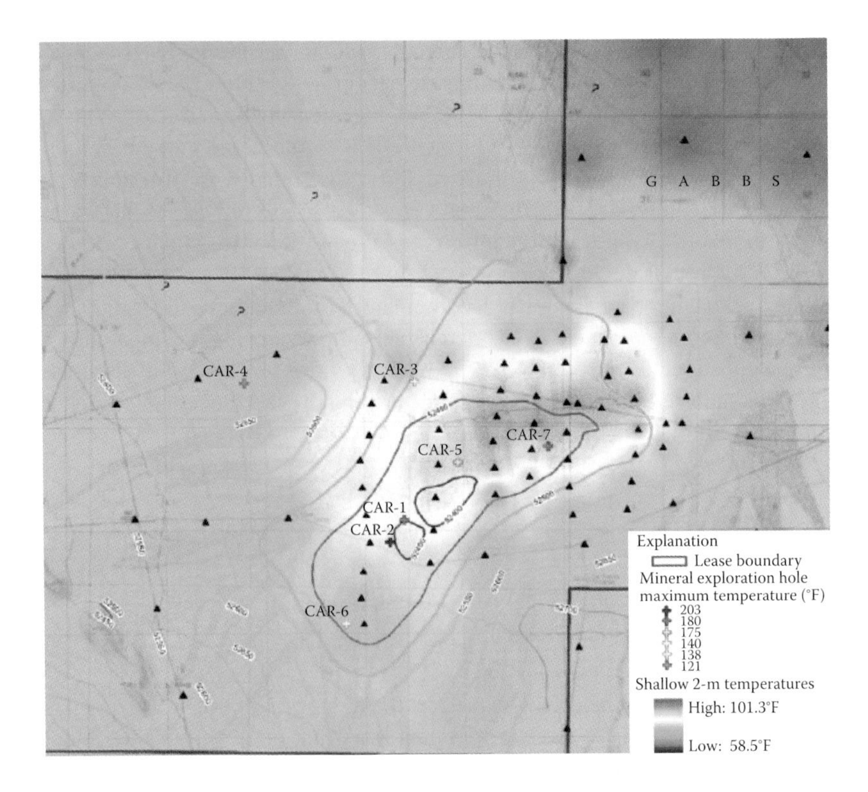

FIGURE 8.29 Map showing locations of mineral exploration holes (+ symbols), aeromagnetic contours, and locations of shallow 2-m temperature holes (solid triangles). The results of the shallow temperature survey are color shaded according to temperature, with the warmer colors indicating warmer ground. The magnetic contours are color coded; the blue contours outline the magnetic low area that overlaps with warmer temperatures found with the 2-m shallow temperature survey. The older mineral exploration wells are also color coded according to temperature. (Adapted from Orenstein, R. and Delwiche, B., *Geothermal Resources Council Transactions*, 38, 91–98, 2014.)

Temperature-Gradient Drilling

When a geothermal target area has been identified, from other exploration techniques already discussed, a program of drilling temperature-gradient holes is conducted to confirm the presence of a thermal anomaly at depth. This is the most important method of geothermal exploration, but it is expensive, typically consuming two-thirds or more of exploration budgets. As a result, most exploration projects are never drilled because the results of the aforementioned methods of exploration are not sufficiently compelling to warrant the expense of drilling. In the event that drilling is justified, any further financing for the project will depend on the results of drilling temperature-gradient holes. Generally, temperature-gradient holes are drilled using slim holes (hole diameters <15 cm) to depths averaging about 150 m. Such holes are much less expensive to drill than the large-diameter rotary drilling methods employed when drilling production or injection geothermal wells. Even then, if the results of a temperature-gradient drilling are encouraging, full-size exploratory production and injection wells will be drilled to document fluid flow and evaluate the power potential of the explored geothermal system, as was done in the development of the Wild Rose geothermal system discussed above. Finally, in slim-hole temperature-gradient drilling, drill cuttings and cores of rock are retrieved which allow a geologist to record changes with depth in rock types, intersected faults, and degree and type of hydrothermal alteration.

When the hole has been completed, equipment is lowered down the well on a wire to measure changes in temperature at depth. This allows downhole temperature profiles to be constructed. Furthermore, select samples from recovered drill cores and cuttings are analyzed in the laboratory to determine the thermal conductivity, from which the heat flow (Q) can be calculated as follows:

$$Q = k_{th} \times (\nabla T / \Delta X) \tag{8.7}$$

where

k_{th} = Measured thermal conductivity.
∇T = Temperature gradient.
ΔX = Specified depth interval.

Generally, calculated heat flow values on the order of 100 milliwatts per square meter (mW/m^2) and greater are positive indicators that power-producing temperatures can be reached within drillable depths (usually less than 4 km). As was evident from the temperature profiles of the Long Valley geothermal system, the subsurface thermal regime can be quite complex, such as temperature reversals indicating an influx of cold groundwater underlying buoyant, hot-water outflow plumes or multiple hot-water aquifers separated by impermeable rock layers or recurring cold-water recharge zones. Moreover, the area of highest surface heat flow may be displaced laterally from the zone of hottest, upwelling geothermal fluids (Figure 8.30). An ideal, simple temperature profile prospective for a potentially viable geothermal reservoir would be a rapid increase in temperature with depth in the upper rock layers. This is indicative of conductive heat flow of a thermal cap, followed below by a near isothermal profile, suggesting convective mixing and good permeability (see well M10 in Figure 8.31).

SUMMARY

In general, the successful discovery and development of geothermal resources require a systematic approach as outlined in a flow chart (Figure 8.32). This model allows for feedback loops to self-correct or revise conceptual models to maximize success and reduce risk. The first step in the exploration and development of a geothermal system is a review of the literature to learn what is known about a region. This would include identifying the geologic or tectonic setting and compiling and assessing all previous work to date (including any mineral, oil, or gas studies) to establish a baseline of knowledge. From that point, geological, geochemical, and geophysical

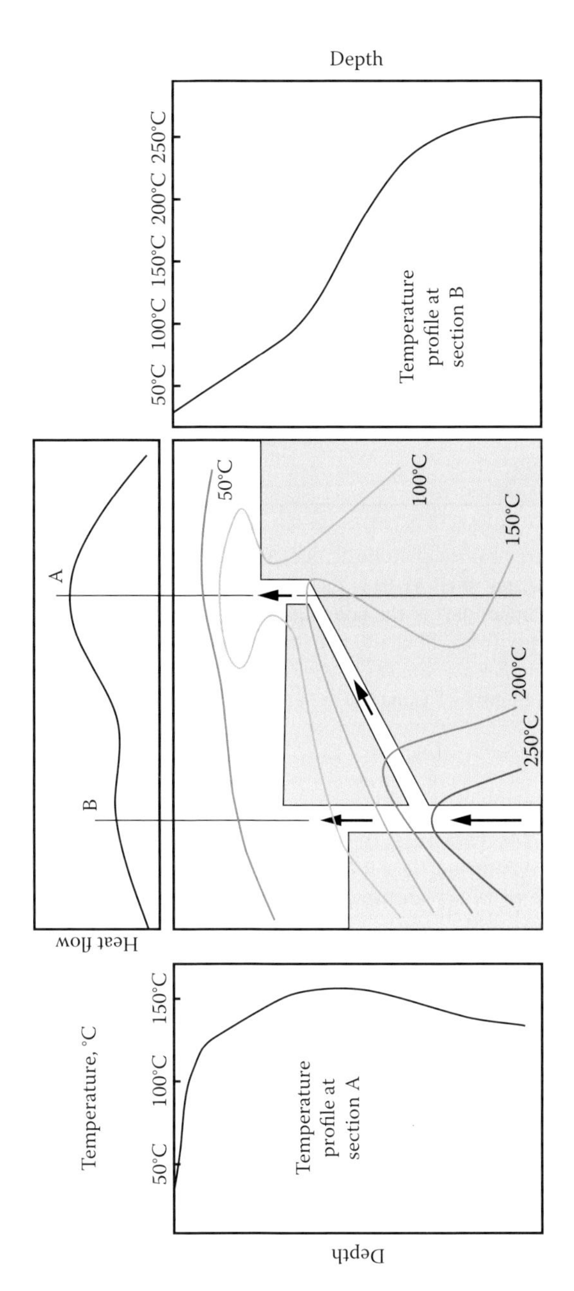

FIGURE 8.30 Hypothetical cross-section of a geothermal system illustrating lateral diversion of upwelling geothermal flow. Although location B is directly above the zone of hottest upwelling fluids, the surface heat flow anomaly is less than at A. Note the reversal in the temperature profile at A that would not occur at B. (Adapted from DiPippo, R., *Geothermal Power Plants: Principles, Applications, Case Studies, and Environmental Impacts*, 3rd ed., Butterworth-Heinemann, Waltham, MA, 2012.)

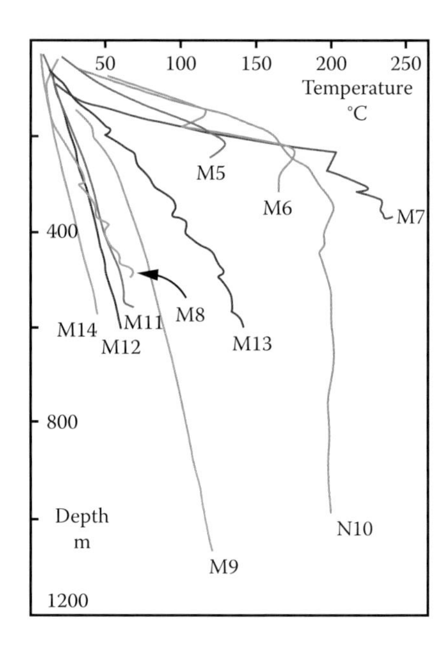

FIGURE 8.31 Temperature profiles of wells drilled at Meager Mountain geothermal prospect, British Columbia, Canada. Well M-10 is the most encouraging because it has a steep geothermal gradient in the upper part of the hole that then turns isothermal, indicating convection and good permeability from about 400 m to total depth (about 900 m). (From Jessop, A., *Review of National Geothermal Energy Program Phase 2—Geothermal Potential of the Cordillera*, Geological Survey of Canada Open File 5906, Natural Resources Canada, Ontario, 2008.)

programs would be undertaken. Geological studies would involve characterizing the regional and local geologic setting; mapping rocks, faults, and any surface geothermal features; and determining the character and distribution of hydrothermally altered rocks. In the case of blind geothermal systems, radiometric dating to determine the ages of hydrothermally altered rocks should be done to make sure the age of alteration is geologically young (less than a million years and preferably less than a few hundred thousand years). Also, evidence for recent faults and identifying the structural setting of those faults, such as fault step-overs or accommodation and transfer zones, can help identify prospective geothermal targets.

Geochemical studies are most helpful where extant geothermal surface expressions are exposed. The chemical analysis of tested fluids can help characterize the type of geothermal system, such as near-neutral pH, alkali-chloride type or more acid-sulfate type, such as might occur above either a boiling liquid-dominated reservoir or a vapor-dominated reservoir. The measured concentrations of silica and of alkali elements of sampled fluids can provide estimates of temperatures in the reservoir without having to drill.

Geophysical techniques, including aeromagnetism, gravity, seismic surveys, and electrical resistivity, can help locate areas of geothermal activity. Aeromagnetic lows can reflect demagnetized rocks due to possible hydrothermal alteration. Seismic and

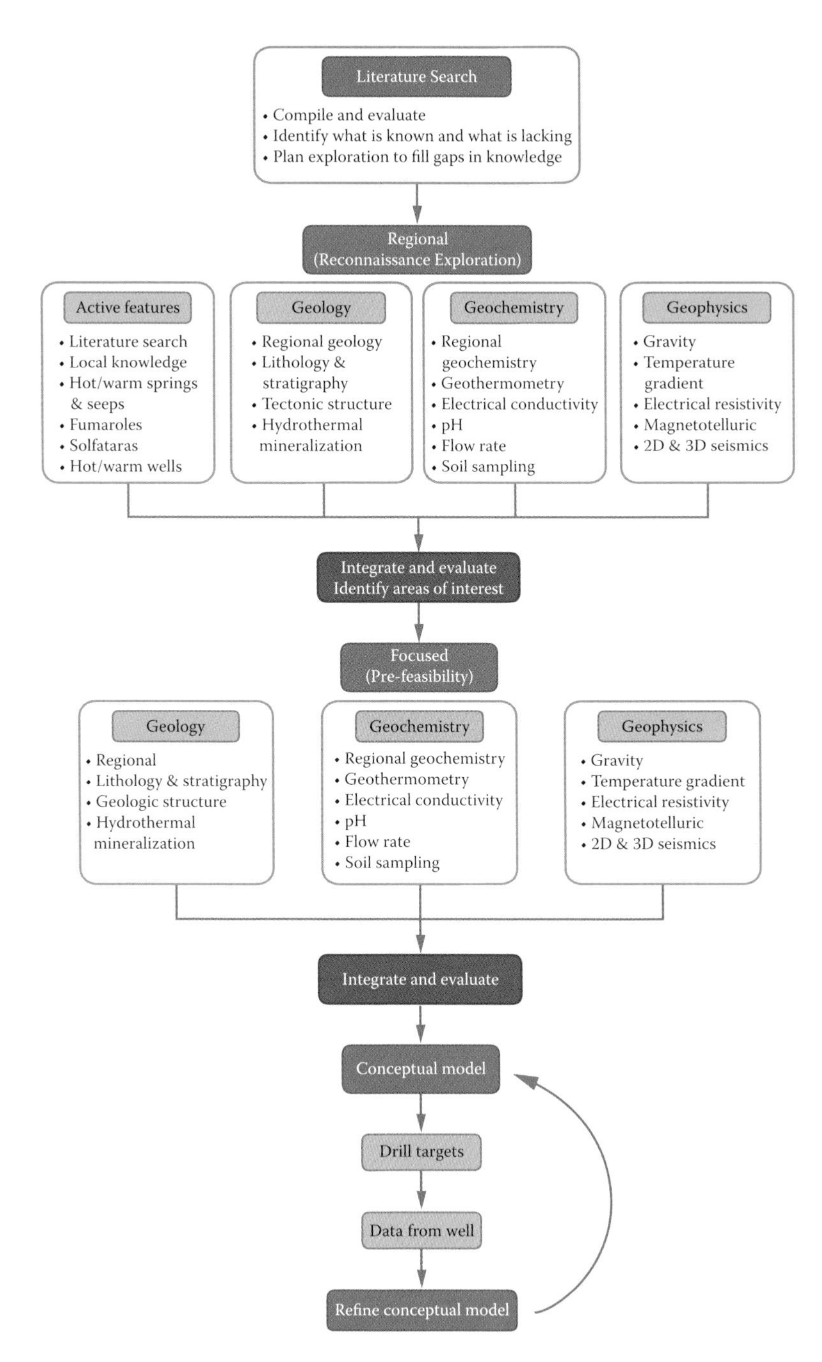

FIGURE 8.32 Flow chart showing exploration stages with types of data acquired and used to develop a conceptual model that will be revised as further data are gained and integrated. (From GeothermEx, Inc., and Harvey, C., *Geothermal Exploration Best Practices: A Guide to Resource Data Collection, Analysis, and Presentation for Geothermal Projects*, IGA Service GmbH, Bochum, Germany, 2013.)

gravity surveys can help locate concealed faults that can focus geothermal fluids, and low areas of electrical resistivity can reflect conducting geothermal fluids in a reservoir or low-resistivity, clay-rich cap rocks above a geothermal reservoir.

Shallow-temperature probe holes and follow-up temperature-gradient wells can help identify areas of greatest heat flow. When used in conjunction with the results of geological, geochemical, and geophysical studies, results of the temperature gradient program can lead to a decision as to whether or not a pilot production well should be drilled and where it should be drilled for the greatest chance of success.

SUGGESTED PROBLEMS

1. You sampled a vigorously boiling thermal spring and determined a silica content of 450 mg/kg or ppm back in the laboratory. (a) Using Figure 8.16, determine the estimated reservoir temperature and explain how you got the result you did. (b) Do you get the same result using Figure 8.15 (using the quartz solubility curve)? Why are the results the same or different?
2. Explain why a thermal spring at the surface may have no geothermal resource at depth directly below it. Please provide two explanations that could account for such behavior.

REFERENCES AND RECOMMENDED READING

Bell, J.W. and Ramelli, A.M. (2009). Active fault controls at high-temperature geothermal sites: prospecting for new faults. *Geothermal Resources Council Transactions*, 33: 425–430.

Benson, L. (2004). *The Tufas of Pyramid Lake, Nevada*, Circular 1267. Reston, VA: U.S. Geological Survey (http://pubs.usgs.gov/circ/2004/1267/).

Bibby, H.M., Risk, G.F., Caldwell, T.G., and Bennie. S.L. (2005). Misinterpretation of electrical resistivity data in geothermal prospecting; a case study from the Taupo volcanic zone. In: *Proceedings of World Geothermal Congress 2005*, Antalya, Turkey, April 24–29 (http://www.geothermal-energy.org/pdf/IGAstandard/WGC/2005/2630.pdf).

Calvin, W.M., Littlefield, E.F., and Kratt, C. (2015). Remote sensing of geothermal-related minerals for resource exploration in Nevada. *Geothermics*, 53: 517–526.

Casaceli, R.J., Wendell, D.E., and Hoisington, W.D. (1986). Geology and mineralization of the McGinness Hills, Lander County, Nevada. *Report—Nevada Bureau of Mines and Geology*, 41: 93–102.

Cashman, P.H., Faulds, J.E., and Hinz, N.H. (2012). Regional variations in structural controls on geothermal systems in the Great Basin. *Geothermal Resources Council Transactions*, 36: 25–30.

Cooke, D.R. and Simmons, S.F. (2000). Characteristics and genesis of epithermal gold deposits. *Reviews in Economic Geology*, 13: 221–244.

Coolbaugh, M., Lechler, P., Sladek, C., and Kratt, C. (2009). Carbonate tufa columns as exploration guides for geothermal systems in the Great Basin. *Geothermal Resources Council Transactions*, 33: 461–466.

Coolbaugh, M., Sladek, C., Zehner, R., and Kratt, C. (2014). Shallow temperature surveys for geothermal exploration in the Great Basin, USA, and estimation of shallow convective heat loss. *Geothermal Resources Council Transactions*, 38: 115–122.

DiPippo, R. (2012). *Geothermal Power Plants: Principles, Applications, Case Studies, and Environmental Impacts*, 3rd ed. Waltham, MA: Butterworth-Heinemann.

Eneva, M., Falorni, G., Adams, D., Allievi, J., and Novali, F. (2009). Application of satellite interferometry to the detection of surface deformation in the Salton Sea geothermal field, California. *Geothermal Resources Council Transactions*, 33: 284–288.

Falorni, G., Morgan, J., and Eneva, M. (2011). Advanced InSAR techniques for geothermal exploration and production. *Geothermal Resources Council Transactions*, 35: 1661–1666.

Faulds, J., Coolbaugh, M., Bouchot, V., Moeck, I., and Oguz, K. (2010). Characterizing structural controls of geothermal reservoirs in the Great Basin, USA, and western Turkey: developing successful exploration strategies in extended terranes. In: *Proceedings of World Geothermal Congress 2010*, Bali, Indonesia, April 25–30.

Faulds, J.E., Hinz, N.H., and Coolbaugh, M.F. (2011). Structural investigations of Great Basin geothermal fields: applications and implications. In: *Great Basin Evolution and Metallogeny* (Steininger, R. and Pennell, B., Eds.), pp. 361–372. Lancaster, PA: DEStech Publications. Copyright © Geological Society of Nevada.

Faulds, J.E., Hinz, N.H., Dering, G.M., and Silier, D.L. (2013). The hybrid model—the most accommodating structural setting for geothermal power generation in the Great Basin, Western USA. *Geothermal Resources Council Transactions*, 37: 3–10.

Fournier, R.O. (1985). The behavior of silica in hydrothermal solutions. *Reviews in Economic Geology*, 2: 45–61.

Fournier, R.O. and Rowe, J.J. (1966). Estimation of underground temperatures from the silica content of water from hot springs and wet-steam wells. *American Journal of Science*, 264(9): 685–697.

GeothermEx, Inc., and Harvey, C. (2013). *Geothermal Exploration Best Practices: A Guide to Resource Data Collection, Analysis, and Presentation for Geothermal Projects*. Bochum, Germany: IGA Service GmbH (http://www.geothermal-energy.org/ifc-iga_launch_event_best_practice_guide.html). This is an excellent reference for your library on the topic of geothermal exploration. This publication also discusses the financial aspects and risk, which are good to be aware of. Sections 3.4 to 3.11 and Appendixes A1.1 to A1.4 and A2.1 to A2.4 are most relevant to the information presented in this chapter.

Glassley, W.E. (2015). *Geothermal Energy: Renewable Energy and the Environment*, 2nd ed. Boca Raton, FL: CRC Press.

Goff, F. and Gardner, J.N. (1994). Evolution of a mineralized geothermal system, Valles Caldera, New Mexico. *Economic Geology*, 89(8): 1803–1832.

Henley, R.W. and Ellis, A.J. (1983). Geothermal systems ancient and modern: a geochemical review. *Earth-Science Reviews*, 19(1): 1–50.

Jessop, A. (2008). *Review of National Geothermal Energy Program Phase 2—Geothermal Potential of the Cordillera*, Geological Survey of Canada Open File 5906. Ontario: Natural Resources Canada.

Kratt, C., Coolbaugh, M., Sladek, C., Zehner, R., Penfield, R., and Delwiche, B. (2008). A new gold pan for the west: discovering blind geothermal systems with shallow temperature surveys. *Geothermal Resources Council Transactions*, 32: 153–158.

Kratt, C., Coolbaugh, M., Peppin, B., and Sladek, C. (2009). Identification of a new blind geothermal system with hyperspectral remote sensing and shallow temperature measurements at Columbus Salt Marsh, Esmeralda County, Nevada. *Geothermal Resources Council Transactions*, 33: 428–432.

Lagat, J., Arnorsson, S., and Franzson, H. (2005). Geology, Hydrothermal Alteration, and Fluid Inclusion Studies of Olkaria Domes Geothermal Field, Kenya. In: *Proceedings World Geothermal Congress, 2005*, Antalya, Turkey, April 24–29 (http://www.geothermal-energy.org/pdf/IGAstandard/WGC/2005/0649.pdf).

Lane, M., Schweikert, R., and DeRoacher, T. (2012). Use of seismic imaging to identify geothermal reservoirs at the Hot Pot area, Nevada. In: *Proceedings of the 37th Workshop on Geothermal Reservoir Engineering*, Stanford, CA, January 30–February 1 (http://www.geothermal-energy.org/pdf/IGAstandard/SGW/2012/Lane.pdf).

Legmann, H. (2015). The 100-MW Ngatamariki geothermal power station: a purpose-built plant for high temperature, high enthalpy resources. In: *Proceedings of World Geothermal Congress 2015*, Melbourne, Australia, April 19–24 (http://www.geothermal-energy.org/pdf/IGAstandard/WGC/2015/06023.pdf).

Martini, B.A., Silver, E.A., Pickles, W.L., and Cocks, P.A. (2003). Hyperspectral mineral mapping in support of geothermal exploration: examples from Long Valley Caldera, CA, and Dixie Valley, NV, USA. *Geothermal Resources Council Transactions*, 27: 657–662.

Monastero, F.C. (2002). Model for success. *Geothermal Resources Council Bulletin*, 31(5): 188–193.

Moore, J.N., Powell, T.S., Heizler, M.T., and Norman, D.I. (2000). Mineralization and hydrothermal history of the Tiwi geothermal system, Philippines. *Economic Geology*, 95(5): 1001–1023.

NGDS. (2016). National Geothermal Data System website, www.geothermaldata.org.

Nordquist, J. and Delwiche, B. (2013). The McGinness Hills Geothermal Project. *Geothermal Resources Council Transactions*, 37: 57–63.

Oregon Tech. (2016). *Geo-Heat Center.* Klamath Falls: Oregon Institute of Technology (http://geoheat.oit.edu/database.htm).

Orenstein, R. and Delwiche, B. (2014). The Don A. Campbell geothermal project. *Geothermal Resources Council Transactions*, 38: 91–97.

Payne, J., Bell, J., Calvin, W., and Spinks, K. (2011). Active fault structure and potential high temperature geothermal systems: LiDAR analysis of the Gabbs Valley, Nevada, fault system. *Geothermal Resource Council Transactions*, 35: 961–966.

Reed, M. and Spycher, N. (1984). Calculation of pH and mineral equilibria in hydrothermal waters with application to geothermometry and studies of boiling and dilution. *Geochimica et Cosmochimica Acta*, 48(7): 1479–1492.

Rimstidt, J.D. and Cole, D.R. (1983). Geothermal mineralization. I. The mechanism of formation of the Beowawe, Nevada, siliceous sinter deposit. *American Journal of Science*, 283(8): 861–875.

Santos, P.A. and Rivas, J.A. (2009). Gravity Surveys Contribution to Geothermal Exploration in El Salvador: The Cases of Berlin, Ahuachapan, and San Vincente Areas, paper presented at Short Course on Surface Exploration for Geothermal Resources, Ahuachapan and Santa Tecla, El Salvador, October 17–30.

Soengkono, S. (2001). Interpretation of magnetic anomalies over the Waimangu geothermal area, Taupo volcanic zone, New Zealand. *Geothermics*, 30(4): 443–459.

Stanley, W.D. and Blakely, R.J. (1995). The Geysers–Clear Lake geothermal area, California: an updated geophysical perspective of heat sources. *Geothermics*, 24(2): 187–221.

Suemnicht, G.A., Sorey, M.L., Moore, J.N., and Sullivan, R. (2006). The shallow hydrothermal system of the Long Valley Caldera, California. *Geothermal Resources Council Transactions*, 30: 465–469.

USGS. (2016). *Energy Resources Program: Geothermal.* Reston, VA: U.S. Geological Survey (http://geoheat.oit.edu/database.htm).

van der Meer, F., Hecker, C., van Ruitenbeek, F., van der Werff, H., de Wijkerslooth, C., and Wechsler, C. (2014). Geologic remote sensing for geothermal exploration: a review. *International Journal of Applied Earth Observation and Geoinformation*, 33: 255–269.

Wannamaker, P., Maris, V., Sainsbury, J., and Iovenitti, J. (2013). Intersecting fault trends and crustal-scale fluid pathways below the Dixie Valley geothermal area, Nevada, inferred from 3D magnetotelluric surveying. In: *Proceedings of the 38th Workshop on Geothermal Reservoir Engineering*, Stanford, CA, February 11–13 (http://www.geothermal-energy.org/pdf/IGAstandard/SGW/2013/Wannamaker.pdf).

Wohletz, K. and Heiken, G. (1992). Geothermal systems associated with basaltic volcanoes. In: *Volcanology and Geothermal Energy* (Wohletz, K. and Heiken, G., Eds.), pp. 225–259. Berkeley: University of California Press.

9 Environmental Aspects of Using Geothermal Energy

KEY CHAPTER OBJECTIVES

- Identify the main environmental advantages and challenges in the development of geothermal resources.
- Explain why geothermal power plants have such low air and particulate emissions.
- Compare and contrast the environmental impacts of geothermal operations relative to fossil-fuel and nuclear power plants and renewable sources of energy, such as solar, wind, and biomass.
- Describe how the development of geothermal resources can lead to land subsidence and how such effects might be mitigated.
- Describe the causes of induced seismicity from geothermal operations and how its effects can be mitigated.
- Relate how geothermal operations can impact existing geothermal surface phenomena, such as hot springs and geysers.

INTRODUCTION

As with any construction project, development of geothermal resources has environmental impacts; however, these impacts, compared to those produced from fossil-fueled power sources and even other renewable energy sources, are small and rather benign. The potential impacts explored include the following:

- Gaseous emissions to the atmosphere
- Land usage
- Solids emissions to the surface and atmosphere
- Water usage
- Noise pollution
- Alteration of natural views
- Land subsidence (sinking)
- Induced seismicity
- Disturbance of existing surface hydrothermal manifestations (geysers, hot springs, etc.)

These environmental effects can be categorized, based on their degree of impact, as either environmental advantages or environmental challenges. The environmental challenges are those mainly unique to developing geothermal resources and consist principally of potential land subsidence, induced seismicity, and potential disturbance of existing natural geothermal features.

A growing worldwide concern has emerged over the last decades about the increasing atmospheric emissions of carbon dioxide (CO_2)—the largest sources of which are coal-fired power plants. Concerns about rising atmospheric CO_2 are based on its heat-trapping properties, which amplify the Earth's greenhouse effect and thereby impact global climate. According to both the National Oceanic and Atmospheric Administration (NOAA) and the National Aeronautics and Space Administration (NASA), 2015 was the warmest year on record for the planet since records began in 1880. Indeed, the United Nations' Intergovernmental Panel on Climate Change (IPCC, 2014) considers warming of the global climate as unequivocal, and it is extremely likely* that most of the observed increase in global average temperatures since the mid-20th century is due to the observed increase in anthropogenic greenhouse gas concentrations, mainly as CO_2, but also including methane (CH_4) and nitrous oxides (NO_x).

ENVIRONMENTAL BENEFITS OF GEOTHERMAL RESOURCES

Environmental advantages of developing geothermal resources include limited gaseous emissions, small footprint, low amounts of emitted particulate matter, reduced water consumption, low noise impacts, and the ability of geothermal operations to blend in with their natural surroundings. These environmental assets are in addition to the general 24-hour availability of geothermal energy, be it for electrical power or direct use (such as space heating).

Gaseous Emissions

The main gases produced from dry- and flash-steam geothermal plants are CO_2, hydrogen sulfide (H_2S), and minor amounts of nitrous oxides (NO_x), ammonia, and possibly mercury (Hg). Air-cooled binary geothermal plants emit essentially no gases because only the heat of the geothermal fluid is transferred to the working fluid via a heat exchanger. Both the geothermal fluid and working fluid are confined to closed loops and neither vents to the atmosphere. Compared to fossil-fuel-fired power plants, the emissions of flash or dry-steam geothermal power plants represent a small amount (see Tables 9.1 and 9.2; Figures 9.1 to 9.5). According to a report by the Geothermal Energy Association (Kagel et al., 2007), a coal-fired power plant emits "24 times more carbon dioxide, 10,837 times more sulfur dioxide, and 3,865 times more nitrous oxides per megawatt hour than a geothermal steam plant." These observations are based on a comparison of emissions from the dry-steam power production facilities at The Geysers with a comparable coal-fired power plant (Table 9.1). Carbon dioxide

* The qualification of "extremely likely," according to the IPCC *Fifth Assessment Report*, indicates a 95 to 100% probability for a given outcome.

TABLE 9.1
Comparison of Emissions of Geothermal and a Coal-Fired Power Plant

				Emissions Rate (lb/MWh)		
Plant Name	**Year**	**Total MWh Produced[a]**	**Primary Fuel**	**NO_x**	**SO_2**	**CO_2**
Cherokee Station[b]	1997	4,362,809	Coal	6.64	7.23	2077
Cherokee Station	2003	5,041,966	Coal	4.02	2.33	2154
The Geysers[c]	2003	5,076,925	Geothermal (steam)	0.00104	0.000215	88.8
Mammoth Pacific (Casa Diablo)[d]	2004	210,000[e]	Geothermal (binary)	0	0	0

Source: Kagel, A. et al., *A Guide to Geothermal Energy and the Environment*, Geothermal Energy Association, Washington, DC, 2007.

[a] For year specified.

[b] Cherokee is a coal-fired, steam-electric generating station; data on Cherokee plant were provided by Xcel Energy.

[c] Values represent averages for 11 Sonoma County power plants at The Geysers; data were provided by Calpine Corporation as submitted to the Northern Sonoma County Air Pollution Control District for 2003 emissions inventory.

[d] Data were provided by Bob Sullivan, plant manager at Mammoth Pacific LP.

[e] Figure represents average yearly output rather than specific output for 2004.

emissions for flash geothermal power plants are generally higher, however, and average about 400 lb CO_2 per MWh, yielding a weighted average of steam and flash output of about 180 lb CO_2 per MWh (Figure 9.2). Using the weighted average of CO_2 emissions for steam and flash geothermal power plants, flash and steam plants produce less than a tenth of the CO_2 emissions of a comparable sized coal-fired power

TABLE 9.2
Air Emissions Summary of Different Power Plants

	Emissions Rate (lb/MWh)			
Source	**NO_x**	**SO_2**	**CO_2**	**Particulate Matter**
Coal	4.31	10.35	2191	2.23
Coal, life-cycle emissions	7.38	14.8	Not available	20.3
Oil	4	12	1672	Not available
Natural gas	2.96	6.04	1212	0.14
USEPA listed average of all U.S. power plants	2.96	6.04	1392.5	Not available
Geothermal (flash)	0	0.35	60	0
Geothermal (binary and flash/binary)	0	0	0	Negligible
Geothermal (The Geysers steam)	0.00104	0.000215	88.8	Negligible

Source: Kagel, A. et al., *A Guide to Geothermal Energy and the Environment*, Geothermal Energy Association, Washington, DC, 2007.

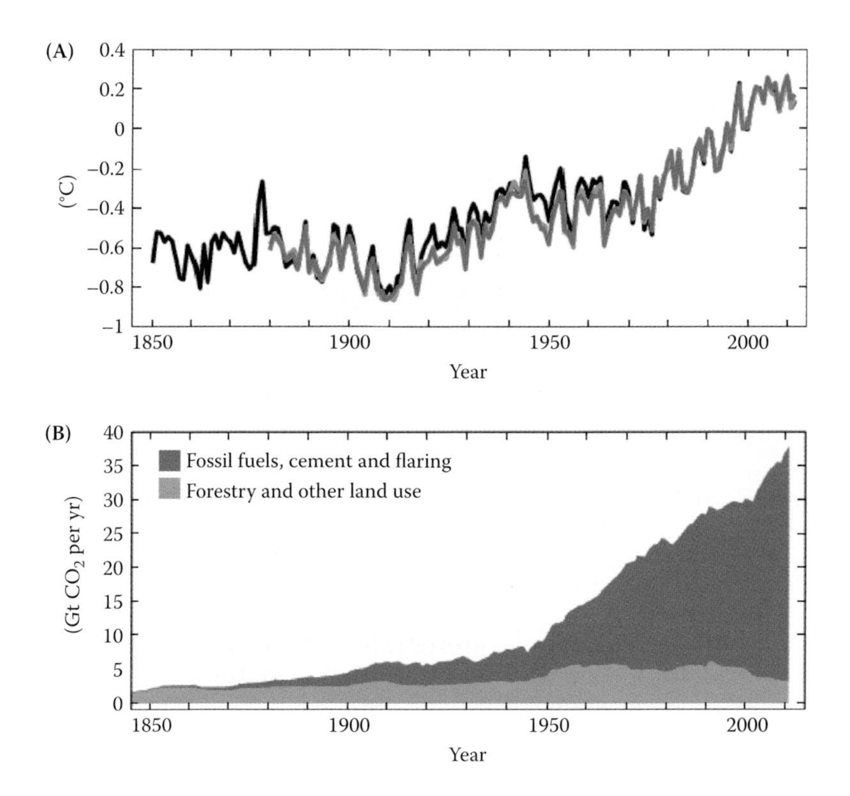

FIGURE 9.1 (A) Globally averaged combined land and ocean surface temperature anomaly referenced to the average for the period from 1986 to 2005. Different colors represent three different sets of data. (B) Global anthropogenic CO_2 emissions measured in gigatons per year (GT/yr). Note the abrupt increase of CO_2 emissions since about 1950. (Adapted from IPCC, *Climate Change 2014: Synthesis Report. Contribution of Working Groups I, II, and III to the Fifth Assessment Report of the Intergovernmental Panel on Climate Change*, Intergovernmental Panel on Climate Change, Geneva, 2014.)

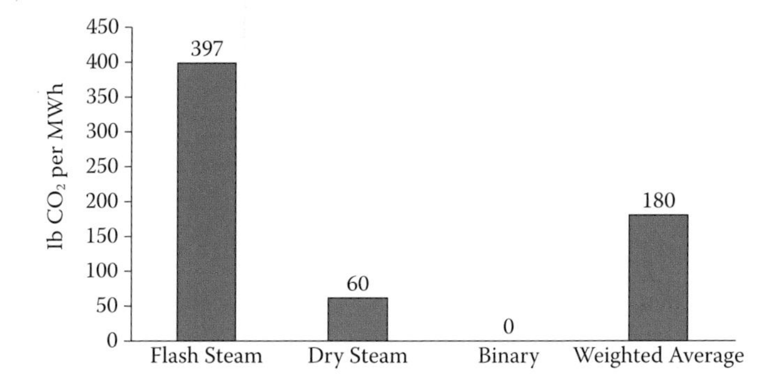

FIGURE 9.2 Average geothermal CO_2 emissions from different generating technologies as compiled from California facilities in 2010. (From Holm, A. et al., *Geothermal Energy and Greenhouse Gas Emissions*, Geothermal Energy Association, Washington, DC, 2012.)

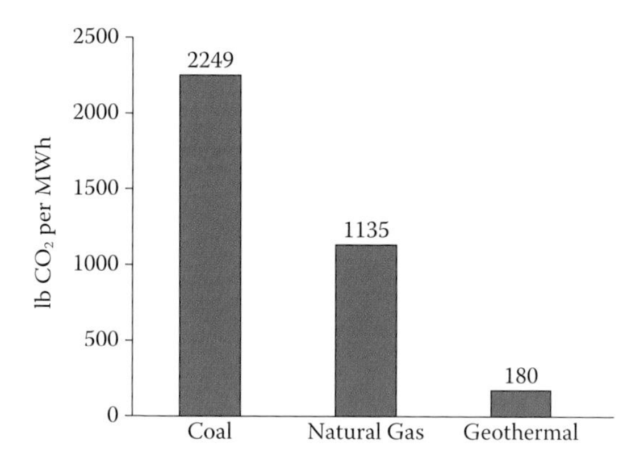

FIGURE 9.3 Comparison of CO_2 emissions from coal-fired, natural gas-fired, and geothermal power plants (flash and steam only) for California facilities. Data are from the California Air Resources Board, U.S. Environmental Protection Agency, California Energy Commission, and Geothermal Energy Association. (From Holm, A. et al., *Geothermal Energy and Greenhouse Gas Emissions*, Geothermal Energy Association, Washington, DC, 2012.)

plant (Figure 9.3). If the near-zero emissions of binary geothermal power plants are factored in, CO_2 emissions of geothermal power plants is less than 5% of that emitted from coal-fired sources having similar power output. Similar tabulations and graphs for other emitted gases show the low to nil output from geothermal operations compared to fossil fuel power installations (Table 9.2; Figures 9.4 and 9.5).

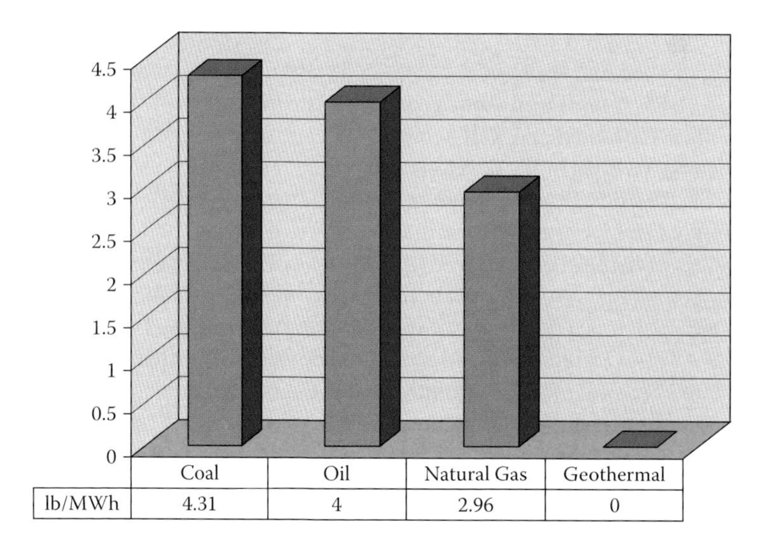

	Coal	Oil	Natural Gas	Geothermal
lb/MWh	4.31	4	2.96	0

FIGURE 9.4 Comparison of NO_x emissions from various power plants. Values reported are average existing power plant emissions; natural gas is reported as the average of existing steam-cycle, simple gas turbine, and combined-cycle power plant emissions. (From Kagel, A. et al., *A Guide to Geothermal Energy and the Environment*, Geothermal Energy Association, Washington, DC, 2007.)

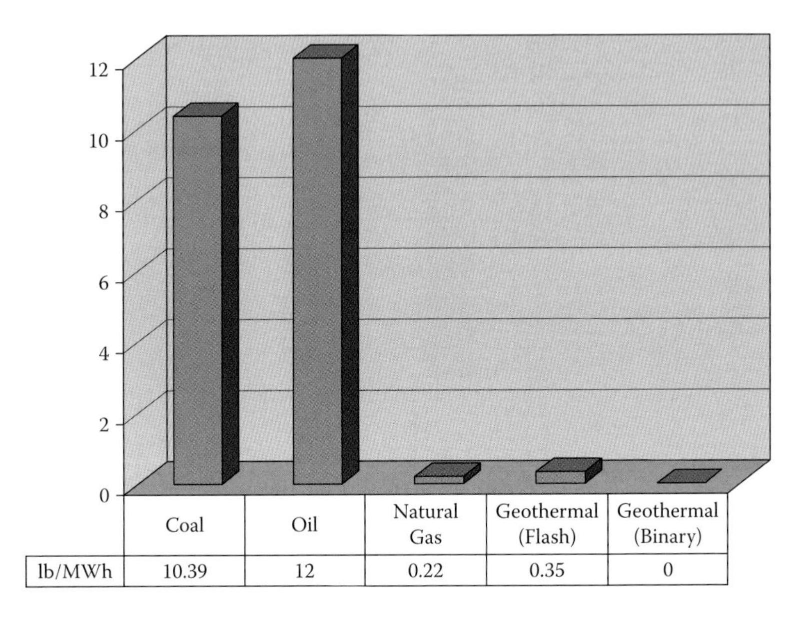

FIGURE 9.5 Comparison of SO_2 emissions of various power plants. For geothermal, the SO_2 amount reflects the conversion of H_2S as it enters the atmosphere, as little SO_2 is directly emitted from geothermal systems. Values reported are average existing power plant emissions; natural gas is reported as the average of existing steam-cycle, simple gas turbine, and combined-cycle power plant emissions. (From Kagel, A. et al., *A Guide to Geothermal Energy and the Environment*, Geothermal Energy Association, Washington, DC, 2007.)

Carbon dioxide and hydrogen sulfide (H_2S) are the most common noncondensable gases (NCGs) produced from geothermal steam, followed by minor amounts of ammonia (NH_3), methane (CH_4), nitrous oxides, and mercury depending on the geologic characteristics of the developed geothermal system. Typically, noncondensable gases make up only a few weight percent of the steam, and carbon dioxide usually represents in excess of 90% of the NCGs. At present, it is not required in the United States to capture or remove CO_2, but H_2S is rigorously regulated because of its unpleasant odor at low concentrations (30 parts per billion) and toxicity at higher levels (>500 ppb). It is a relatively straightforward to remove the H_2S through oxidation ($2H_2S + O^2 \rightarrow 2S^0 + 2H_2O$) and produce elemental sulfur that can be sold in the manufacture of fertilizers. This is routinely done at power stations at The Geysers.

Discussions are ongoing worldwide on how to stem global warming, such as placing carbon caps for power plants or a tax on carbon emissions. If restrictions are placed on carbon-producing power plants, then geothermal plants would be in a good situation to keep power costs low. Moreover, if a program of carbon emission credits were utilized, geothermal power plants could generate additional revenue by selling carbon credits in a trading market. The CO_2 concentration varies directly with reservoir temperature (Glassley, 2015). The concentration of CO_2 can be controlled by a chemical reaction involving the minerals prehnite, clinozoisite, calcite, and quartz plus water, as shown below (Glassley, 2015):

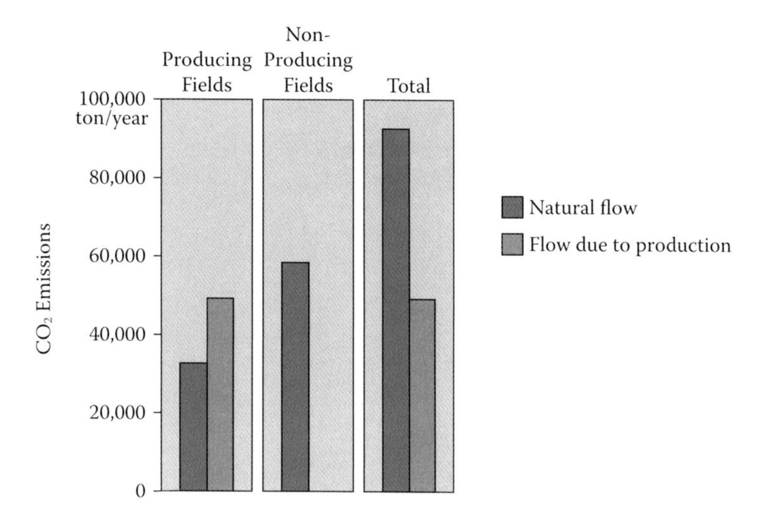

FIGURE 9.6 Carbon dioxide emissions from producing and nonproducing geothermal systems in Iceland. (From Ármannsson, H., in *Proceedings of International Geothermal Conference*, Reykjavik, Iceland, September 14–17, 2003.)

$$\text{Prehnite} + CO_2 \Leftrightarrow \text{Clinozoisite} + \text{Calcite} + \text{Quartz} + \text{Water} \qquad (9.1)$$

As the concentration of CO_2 increases with temperature, the reaction is driven toward the right (to maintain equilibrium, as noted by the opposing double arrows) favoring the formation of clinozoisite, calcite, and quartz as hydrothermal alteration products. However, in a convecting system, as the fluid circulates to cooler regions the reverse happens, and CO_2 escapes from the fluid, thus favoring the formation of prehnite at less deep, cooler levels in a geothermal system.

Finally, development of geothermal systems having surface expressions can actually impact the natural flow of carbon dioxide. From studies in Iceland, the CO_2 flux from producing fields exceeds the natural flow (Figure 9.6). This is probably due to accelerated flow rates provided by production wells and a corresponding reduction in natural flow rates. However, natural CO_2 emissions from non-producing geothermal fields having surface manifestation such as hot springs, fumaroles, and geysers are actually higher than producing fields (Figure 9.6). Combining the two sets of data indicates that the natural flow of CO_2 is about twice that resulting from flash geothermal plants (Figure 9.6).

Land Usage

Compared to other fossil-fuel-based power sources and even other renewable sources of energy (wind, solar, hydropower, and biomass), geothermal has the second lowest footprint behind nuclear (Figure 9.7). A nuclear power plant has the smallest footprint for power produced; however, if you consider the area of the uranium mines supplying fuel for the nuclear plants, then geothermal would actually have the smallest footprint

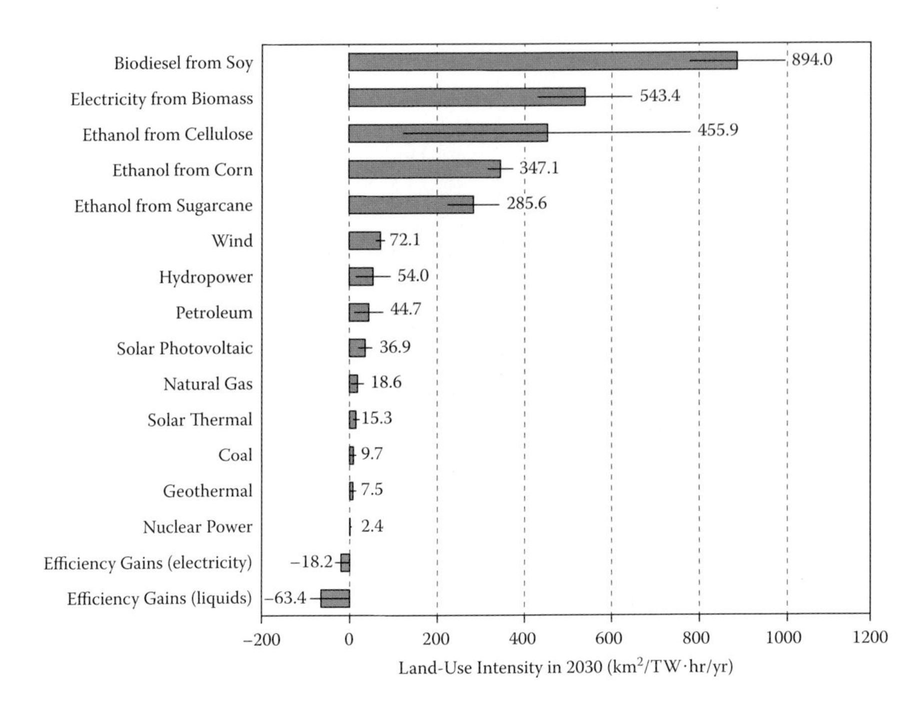

FIGURE 9.7 Land-use intensity for energy production for various conversion technologies calculated for the year 2030. Error bars show the most- and least-compact estimates of plausible future and current land-use intensity. Shown values are the midpoint of the most- and least-compact estimates. Note that the coal and nuclear values exclude the area of mines to supply fuel for the plants. (From McDonald, R.I. et al., *PLoS ONE*, 4(8), e6802, 2009.)

for all energy conversion technologies. Comparisons of land use for coal-fired power plants and renewable sources of wind and solar also demonstrate the small footprint of geothermal (Figure 9.8). The actual area required for a geothermal power plant depends on the size of the power plant, size of the supporting well field (both production and injection), access roads, pipelines, substation, and auxiliary buildings. The well field typically represents the largest amount of area, on the order of 5 to 10 km^2 for a 20- to 50-MWe plant. However, the well pads themselves typically consume only about 2% of that 5 to 10 km^2, an amount that can be reduced further if directional drilling is employed, allowing two or more wells to be drilled on a given pad.

In general, a geothermal flash or binary plant uses (per GWh) about 11% of the area required for a solar thermal plant and 12% of that for a solar photovoltaic (PV) farm (Figure 9.8). A coal plant, including 30 years of strip mining, requires 30 to 35 times the surface area necessary for a flash or binary plant on a per MWe or MWh basis. Of the geothermal plants, those developing hypersaline brines, such as those in the Salton Sea geothermal field, require about 75% more land than simple flash or binary plants, due to the additional chemical treatment facilities (flash crystallizer and reactor clarifier, or FCRC) to render the brines manageable for power production. Chemical treatment of hypersaline geothermal fluids is discussed more below in the section on discharge of solids to air and ground.

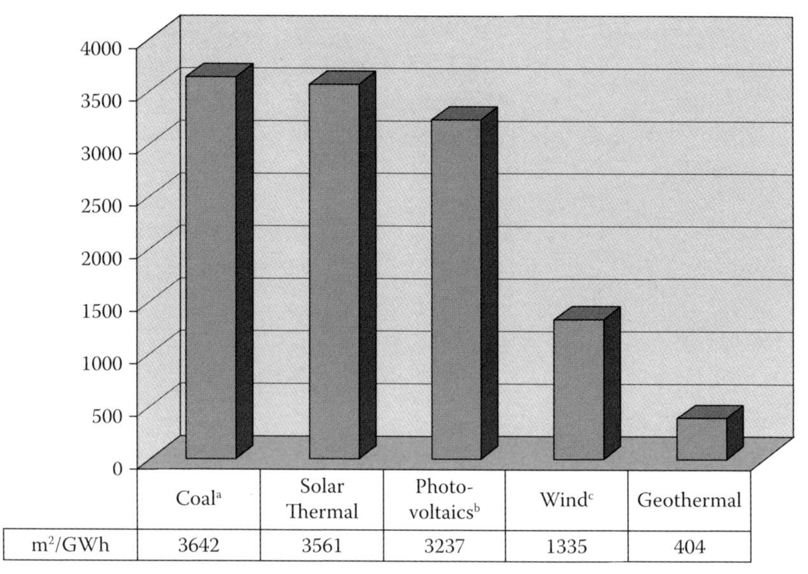

[a] Includes mining.
[b] Assumes central station photovoltaic project, not rooftop PV systems.
[c] Land actually occupied by turbines and service roads.

FIGURE 9.8 Area of land used per gigawatt-hour (GWh) for coal, solar thermal, solar photovoltaic, wind, and geothermal. (From Kagel, A. et al., *A Guide to Geothermal Energy and the Environment*, Geothermal Energy Association, Washington, DC, 2007.)

The pipelines carrying geothermal fluid from the wells to the power plant are compatible with many other land uses. They commonly rest on stanchions averaging 1 to 2 m above the ground and include vertical and horizontal expansion loops. Vertical loops in piping (to accommodate expansion) allow access to different parts of the field for vehicles or livestock (Figure 9.9). Geothermal well fields are also compatible with farming, as in the Salton Sea/Imperial Valley in southeastern California (Figure 9.10). The Wairakei geothermal field in New Zealand supports ranching and prawn farming, and locals and tourists enjoy soaking and relaxation at the famous Blue Lagoon at the Svartsengi geothermal plant in Iceland (Figure 9.11). At the producing Steamboat Springs geothermal field, near Reno, Nevada, a major freeway passes through the field, and power stations are located on either side (Figure 9.12). No other energy conversion technology, including other types of renewable power generation, affords such a variety of potential multiple uses of the land in the course of power generation.

Finally, it is worth noting that air cooling of many binary geothermal plants requires more area than the water-cooled evaporation used in most flash plants. This is because of water's better cooling and heat rejection characteristics compared to air. For example, the water cooling towers at the 92-MWe Heber 2 flash plant in the Imperial Valley of California cover about 5% of the land area for the plant and require about 61 m^2/MWe, whereas the air-cooled condensers at the 26-MWe Galena 3 plant occupy about 31.5% of the power station area and require about 209 m^2/MWe.

FIGURE 9.9 Cattle grazing at the Miravalles geothermal well field, Costa Rica. Steam pipelines are supported by stanchions that allow pipelines to slide in the event of an earthquake and allow access for cattle to graze within the well field. (From DiPippo, R., *Geothermal Power Plants: Principles, Applications, Case Studies, and Environmental Impacts*, 3rd ed., Butterworth-Heinemann. Waltham, MA, 2012.)

FIGURE 9.10 Agricultural land borders the Ormat 92-MWe Heber 2 geothermal plant in the Imperial Valley, California. Water cooling towers are shown in the middle distance bounding the cropland. (From NREL Image Gallery, http://images.nrel.gov/viewphoto.php?&albumId=207389&imageId=6312190&page=3&imagepos=12&sort=&sortorder=.)

FIGURE 9.11 Effluent from the Svartsengi geothermal power plant, shown in the background, supports the famous Blue Lagoon in Iceland where visitors can soak in the reputed therapeutic geothermal water. (Photograph by author.)

FIGURE 9.12 View looking northeast across the Steamboat geothermal field, near Reno, Nevada. The yellow arrow points to a newly constructed freeway that passes through the field. The red arrow points to the Galena 3 power plant, one of six power plants that develop the geothermal field. (Adapted from Stanford University's Geothermal Field Trip, Steamboat Spring, NV, May 19 2011, https://pangea.stanford.edu/ERE/life/photos/FieldTrips/SteamboatMay2011/index.htm.)

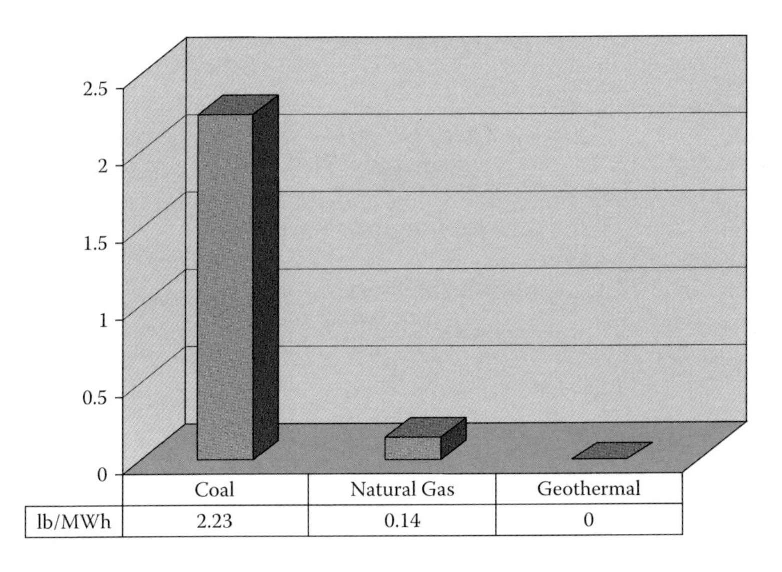

FIGURE 9.13 Graph showing comparison of particulate matter for coal, natural gas, and geothermal. Graphed values are for pulverized coal boiler, combined-cycle natural gas, and average existing geothermal plants. (From Kagel, A. et al., *A Guide to Geothermal Energy and the Environment*, Geothermal Energy Association, Washington, DC, 2007.)

Solids Discharge to Air and Ground

Particulate discharge to the air from geothermal power operations is minimal, especially compared to coal-fired power plants (see Table 9.2 and Figure 9.13). Solar, wind, and nuclear also have no particulate emissions to the atmosphere. Discharge of solids at the surface is generally not a problem in most geothermal systems; the solids remain dissolved in solution and then are reinjected into the ground after use to recharge the geothermal reservoir and prevent possible surface contamination. The main exception to this is the hypersaline brines of the Salton Sea geothermal field of southern California, which contain over 200,000 ppm dissolved solids (for comparison, average sea water contains about 33,000 ppm dissolved salts). Because of the corrosive and potentially clogging nature of these fluids, it took many years of research to successfully manage the hypersaline brines for power production and avoid surface contamination. The main method currently used involves a flash-clarifier/reactor-clarifier process, in which seed particles are introduced on which the supersaturated solids precipitate. Precipitation of dissolved solutes on the seed particles and their settling to the bottom of the reactor vessel can significantly reduce the scaling of brine on the sides of piping. The precipitated brines are rich in manganese (Mn), zinc (Zn), and lithium (Li). The long-term plan is to have the precipitated brines sent to an adjacent mineral recovery facility, which offers a potential added stream of revenue for the geothermal plant. Because of these innovative techniques for treating hypersaline brines, the Imperial Valley region of southern California may surpass The Geysers of northern California as the largest provider of geothermal power in the United States. In the next 5 to 7 years it and could also become a significant producer of Li for the growing battery market.

Water Usage

Water is used in the development and operation of geothermal facilities. Typically, the water use demands of geothermal operations are easy to satisfy and comparatively minor compared to, say, a nuclear power plant. The two main activities that use water are drilling wells during development and rejection of waste heat if water cooling is used to condense turbine steam exhaust. Water in drilling is used to cool the drill bit, drive cuttings back up the hole, and maintain the structural integrity of the well until the casing is installed. The water is actually mixed with minerals (e.g., barite, a high-density mineral) and chemicals to make a drilling mud that is recirculated during the course of drilling. Because the mud is recycled, only a small amount of make-up water is needed to compensate for evaporation during cooling of the mud prior to its reuse.

The main use of water is for rejection of heat to promote the condensation of turbine exhaust steam through evaporative water cooling towers. During evaporative cooling, typically more than 50% of the steam condensate and added cooling water is lost to the atmosphere in the cooling towers. Thus, fresh water is required to make up the difference, which can stress local surface and shallow groundwater supplies in arid regions. For example, the Coso geothermal facility in eastern California, having an installed capacity of 270 MWe, was experiencing a decline in reservoir productivity because the mass of fluid injected was about half of that produced due to losses in the evaporative cooling towers. To minimize the decline and help stabilize reservoir pressure, a 9-mile-long water pipeline was constructed that provides up to 4500 gallons per minute of water for reinjection. Before doing so, however, environmental impact reports were prepared for both federal and county agencies to demonstrate that the project would not adversely affect water availability for local ranchers in this arid region (BLM EA report, 2008).

No make-up water is needed for air-cooled condensers, as is the case with most binary geothermal power plants. In this scenario, water usage for plant operation is the lowest for all baseload power producers, consisting of fossil-fuel, nuclear, and hydroelectric power plants. However, air-cooled condensers, as noted above, occupy more land as a result of the lower heat-transfer properties of air compared to water. Moreover, air-cooled condensers have a higher parasitic load stemming from the use of electric motor-driven fans to blow air across heat exchangers to condense turbine steam exhaust. The efficiency of cooling and condensing of the steam varies inversely with temperature—the colder the ambient air, the greater the cooling and condensing efficiency and the higher the power output. DiPippo (2012) reported that in the case of the 15.5-MWe bottoming binary unit of the Miravalles geothermal power facility in Costa Rica, an air-cooled condenser would cost more than three times as much, cover more than three times the surface area, and consume more than three times the fan power than a water cooling tower. In a region where surface or groundwater supplies are ample, such as the moist climate at Miravalles, a water cooling tower makes the most practical sense. In the arid environment of Nevada, however, the extra expense and greater use of land and parasitic power may be justified considering the paucity of available water. The higher upfront costs of air-cooled condensers are more than compensated over time due to the improved efficiencies of binary turbines and sustaining reservoir pressure (and power output), as the mass of fluids injected equals that produced.

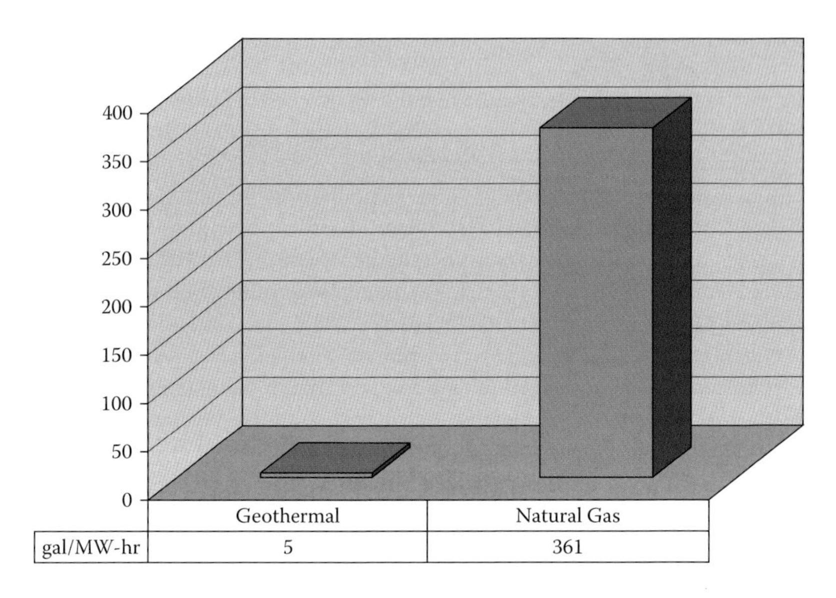

FIGURE 9.14 Water-use comparison of a water-cooled geothermal flash plant and a combined cycle natural gas power plant. (From Kagel, A. et al., *A Guide to Geothermal Energy and the Environment*, Geothermal Energy Association, Washington, DC, 2007.)

Water use at a water-cooled geothermal flash facility is still much lower than that of fossil-fuel-fired power sources. For example, a 500-MWe combined-cycle gas power plant would require 4 million gallons of water per day, whereas a 48-MWe water-cooled geothermal flash plant requires about 6000 gallons per day (Kagel et al., 2007). So, MWe for MWe, the gas plant would consume about 70 times more water than the geothermal plant (Figure 9.14).

In traditional flash geothermal plants, a novel method for reducing water loss through evaporative cooling might involve using a hybrid cooling system. In such a case, during warm weather evaporative cooling would still be used to condense steam, but during the colder months air cooling could be used, thereby reducing water loss and the need for make-up water for possible reinjection. An engineering analysis could determine the threshold air temperature at which air, instead of evaporative, cooling would be viable from energy and economic standpoints.

Noise

Numerous federal and local regulations apply to geothermal operations. At the federal level, the Bureau of Land Management, on which many of the geothermal facilities are developed, require that the noise level at 0.5 mile from a geothermal plant or lease boundary, whichever is closer, cannot exceed 65 units of A-weighted decibels (dBA). A-weighting is an electronic technique to mimic the human auditory response to sound at all frequencies. A comparison of various common noise

TABLE 9.3
Common Sound Levels

Noise Source	Sound Level (dBA)
Geothermal normal operation	15–28
Nearby leaves rustling in a breeze	25
Whisper at 6 feet	35
Inside at average suburban residence	40
Near a refrigerator	40
Geothermal plant construction	51–54
Geothermal well drilling	54
Inside average office, without telephone ringing	55
Talking at normal voice level at 3 feet	60
Car traveling at 60 mph at 100 feet	65
Vacuum cleaner at 10 feet	70
Garbage disposal at 3 feet	80
Electric lawn mower at 3 feet	85
Food blender at 3 feet	90
Automobile horn at 10 feet	100

Source: Kagel, A. et al., *A Guide to Geothermal Energy and the Environment*, Geothermal Energy Association, Washington, DC, 2007.

levels and the sound of normal geothermal operation is provided in Table 9.3. Note that the sound of normal geothermal operation is about half of that which occurs during drilling and construction and is comparable to the nearby rustle of leaves in a breeze. The only time the sound would exceed normal operation would be if the turbine should trip, possibly due to a problem in the transmission system, and the steam flow would have to be directed away from the turbine. Doing so allows the wells to remain open without sudden closure, which could damage well casings or wellhead valves. The steam is instead directed to rock mufflers or silencers where the velocity of the steam is drastically reduced. Because the noise associated with a moving gas stream is proportional to the velocity raised to the eighth power (DiPippo, 2012), if the steam velocity is reduced by half then the sound emitted is lowered by 256 times. Also, the intensity of sound drops rapidly with distance so that a geofluid issuing vertically from a wide-open well would register about 65 dBA at a distance of 1 km. Thus, the noise issuing from an operating geothermal plant will probably not disturb anyone living nearby, as is the case with people living in Pleasant Valley, Nevada, located within about a kilometer of the Upper Steamboat Hills geothermal plant (Figure 9.15). Indeed, typical noise from traffic on a newly constructed highway positioned below the Steamboat Hills geothermal power plant is estimated to be about 70 to 80 dBA, according to the Federal Highway Administration, or about two to five times more than that produced from normal geothermal operation.

FIGURE 9.15 View looking northeast toward Steamboat Hills geothermal power plant noted by the red arrow. The plant is located above the bedroom community of Pleasant Valley south of Reno, Nevada. Homes within the orange ellipse are within about 1 km of the plant and 0.5 km of a newly constructed freeway. (Adapted from Robertson, A., *Nevada Magazine*, June 25, 2012, http://nevadamag.blogspot.com/2012/06/northern-nevada-freeway-nears.html.)

Visual Elements

Any construction where none previously existed will alter the viewshed. Visual impacts of geothermal power facilities are generally minimal. This is because most geothermal power plants are not high standing. They are commonly built in topographically low-lying areas and are routinely painted to blend in with the landscape (Figure 9.16). This stands in marked contrast to high-standing wind turbines, conspicuous solar thermal power towers and heliostats, the tall smoke stacks of coal- and natural gas-fired power plants, and the formidable natural draft cooling towers of many nuclear power plants. The most common sight that belies the presence of a geothermal power facility is occasional rising steam plumes above evaporative cooling facilities on cold days. In general, most people do not find occasional white steam clouds objectionable, in that any nearby surface hot springs or fumaroles, if present, would be doing the same. With air-cooled binary geothermal plants no steam plumes are evident.

ENVIRONMENTAL CHALLENGES OF GEOTHERMAL OPERATIONS

Although geothermal development offers many environmental benefits, some issues can pose problems if not considered. The three main areas of potential concern are land subsidence, induced seismicity, and disruption of surface hydrothermal manifestations. Water usage could also be of concern with water-cooled flash geothermal power plants in arid environments, such as discussed for the Coso geothermal field.

FIGURE 9.16 Ormat's Mammoth Pacific binary geothermal plant at Casa Diablo hot springs near Mammoth in eastern California. The low-lying facility is built in a subdued valley and is virtually not visible from U.S. Highway 395 (shown at the tip of the orange arrow). The plant is painted green to blend in as much as possible with the natural surroundings. Because it is an air-cooled binary facility, steam plumes are not present. The small steam plume shown at the tip of the yellow arrow is the Casa Diablo fumarole. The plant causes little disruption to the spectacular viewshed. (Adapted from Bureau of Land Management, *News.bytes Extra*, Issue 209, December 6, 2005, http://www.blm.gov/ca/ca/news/newsbytes/xtra-05/209-xtra_mam_geothermal.html.)

Land Subsidence

Because fluid is withdrawn from a reservoir at depth, a potential exists for the overlying ground to subside. One of the best documented cases of subsidence in a geothermal field is at Wairakei, New Zealand—the site of the world's first geothermal power plant to develop a liquid-dominated reservoir (Bromley et al., 2015). Land surveys of established benchmarks determined that an area of about 50 km^2 had undergone subsidence and experienced as much as 15 m of settling over the past 50 years of production. From 1965 to 1985, the rate of subsidence averaged about 450 mm/year, and, from 1987 to 1997, the subsidence rate slowed to about 350 mm/year (Bromley et al., 2015). Throughout this period, none of the produced geothermal fluid was reinjected but instead was disposed of in the nearby Waikato River. Reinjection of about 25 to 30% of the produced fluids began in 1997, and since then the subsidence rate has slowed to about 55 mm/year.

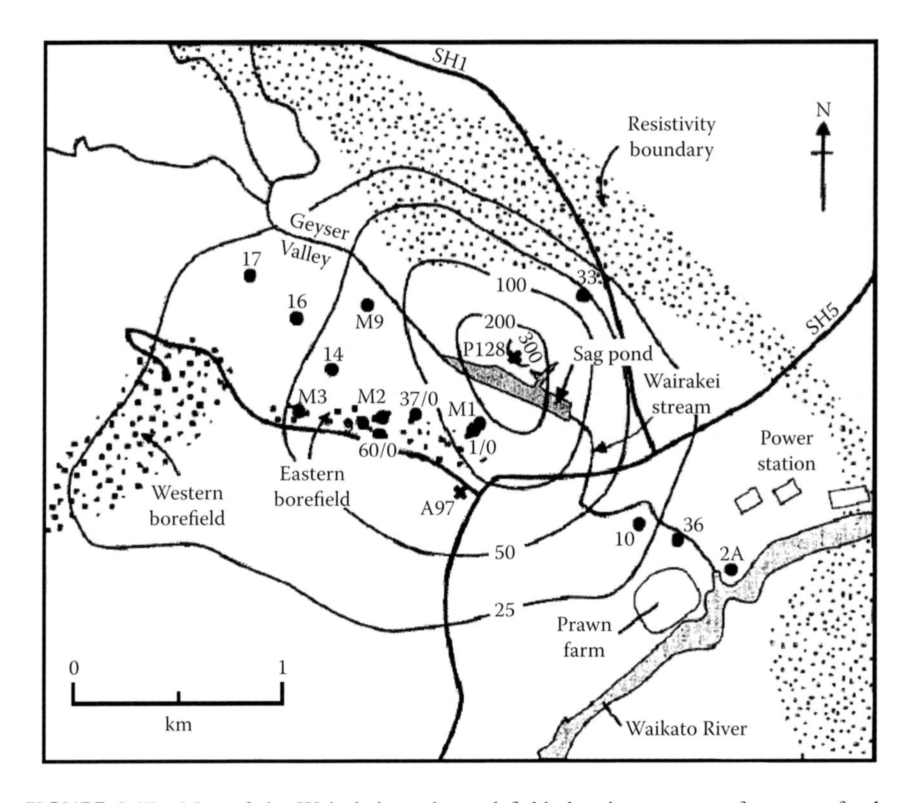

FIGURE 9.17 Map of the Wairakei geothermal field showing contours for rates of subsidence (mm/year) from 1986 to 1994 (prior to beginning reinjection). Filled circles denote selected geothermal wells. (From DiPippo, R., *Geothermal Power Plants: Principles, Applications, Case Studies, and Environmental Impacts*, 3rd ed., Butterworth-Heinemann. Waltham, MA, 2012.)

Rates of subsidence can be contoured over a time interval, and in this case doing so disclosed that the greatest rates of subsidence, interestingly, occurred about 0.5 km north of the eastern well field at Wairakei (Figure 9.17). The degree of subsidence was sufficient to create a small lake or sag pond on the southeastward-flowing Wairakei stream that drains into the Waikato River. In the late 1970s, the subsidence ruptured a flume carrying waste brine from the wells (separators there are located at the wellheads), resulting in a power shutdown lasting for 3 days until the flume could be repaired.

Although land leveling or airborne or satellite surveys (such as InSAR) can locate areas and rates of subsidence, they cannot determine the actual reason for changes in land elevation. The overall reason is that reservoir production rates greatly exceeded any natural recharge, and, as noted above, reinjection of waste brine did not begin until 1997. Geologic field studies and logging of select drill cores indicated that the greatest subsidence occurred where the cap rock overlying the reservoir was thickest. The cap rock consists of interlayered pumice breccias and mudstones having high compressibilities (i.e., weak rocks that tend to compress easily). Although displacement of greatest subsidence to the north of the well field is still imperfectly understood, a possible reason may involve the loss of pore fluid in the cap rock. As pressure declined in the underlying reservoir from fluid withdrawal, pore fluid in the cap rock may have drained downward, effectively reducing support and causing the cap rock to compress and overlying ground to subside. If so, compression and subsidence would be greatest where the cap rock is thickest.

Four main conditions favor subsidence:

- Withdrawal of geofluids that exceeds recharge either naturally or through piped reinjection
- Mechanically weak and compressible rocks within the geothermal reservoir and capping the reservoir
- Thermal contraction of rocks in the vicinity of injection wells, as may be the case in the Mokai geothermal field in New Zealand (Bromley, 2006)
- Geothermal reservoirs where fluid pressure is under lithostatic pressure (fluid bears the weight of the overlying rock column) rather than hydrostatic pressure (weight of water only)

In the last case, the fluid pressure contributes to the support of the weight of the overlying rock and its removal then leaves the overlying rock column partially unsupported, inducing it to collapse and promote surface subsidence. Subsidence is less likely in reservoir formations consisting of hard but fractured rock (e.g., metamorphic rocks), such as at The Geysers. For reservoirs developed in "soft" porous and permeable rock, on the other hand, such as the volcanic tuffs at Wairakei, subsidence becomes more of a concern. Thus, subsidence can be more of a problem in stratigraphically controlled geothermal reservoirs characterized by primary permeability compared to structurally controlled reservoirs in hard rock having secondary fracture-controlled permeability.

Probably the best technique for mitigation of subsidence is reinjection of waste brine back into the reservoir. Although reinjection does not guarantee the avoidance of subsidence, it does reduce the risk and also prolongs the life of the reservoir, if done properly to minimize thermal breakthrough (unwanted accelerated cooling of the reservoir). This is why subsidence is less a problem for air-cooled binary geothermal plants, where the geothermal fluid is fully recycled and no loss of fluid mass to the atmosphere from evaporative cooling towers occurs.

In the Imperial Valley region of southern California, which has over 500 MWe of installed geothermal capacity, subsidence has been detected from interferometric synthetic aperture radar (InSAR) studies (Eneva et al., 2013). Some of the subsidence probably reflects fluid loss from the producing geothermal reservoirs via evaporative cooling towers. Land surface deformation is important to identify, as uneven subsidence can adversely impact the flow of water in irrigation canals in this agriculturally important region. Areas of subsidence can be targeted for geothermal reinjection to help maintain reservoir pressure and minimize subsidence. Renewable power generation and agriculture are both major economic engines for the Imperial Valley, thus these two industries must remain compatible.

Induced Seismicity

Seismicity related to geothermal operations can develop from several factors:

- Injection of cool waters into hot rock causing thermal contraction and fracturing
- Fluid extraction from reservoirs causing a change in fluid pressure resulting in movement of fractured rocks (this happens in oil and natural gas fields at times during extraction)
- Injection of fluids under high pressure causing rocks to fracture and increase reservoir permeability (analogous to filling hydroelectric reservoir for the first time)

When rocks fracture, energy is released that can be transmitted to the surface and felt by people in the area. In large natural earthquakes with magnitudes >7, fractures (faults) rupture for hundreds of kilometers in length. For geothermal fields, any newly formed fractures are on the order of a few centimeters or less in length, although thousands or tens of thousands can form over time.

Nearly every geothermal field under exploitation has experienced induced seismicity to some extent. In most cases, such induced seismicity cannot be felt by people, and the resulting tremors, called *microearthquakes* (magnitudes generally <2), are detected only by sensitive seismic-measuring instruments. The largest induced seismic event associated with geothermal power production occurred at The Geysers. It was likely prompted by cold injection of municipal wastewater into hot rocks which is done to help maintain fluid pressure in the reservoir and power output. That event registered magnitude 4.6 and rattled buildings and the nerves of residents in the rural region. Most of the induced earthquakes in The Geysers region are less than 3 in magnitude and are either not felt or barely perceived by people. When cold water hits hot rocks, the rocks contract, causing fractures to form which generate earthquakes whose size or magnitude varies directly with the size and number of fractures formed. The extent of contraction and fractures developed is related to the coefficients of thermal expansion of the minerals making up the rocks. The greater the difference in temperature between injected fluids and reservoir rocks, the greater the coefficient of expansion, but because the rock is cooling when coming into contact with the injected fluid ΔT ($T_f - T_i$) is negative. So, in the geothermal case, the coefficient of expansion is negative or a positive coefficient of contraction (Figure 9.18). Furthermore, the addition of water increases pore water pressure and reduces the frictional strength of the rocks, making them easier to break. Added water can also chemically weaken rocks, which makes them more prone to slipping along existing fractures or forming new fractures, either of which results in seismicity (Figure 9.19).

Injection of waste brine at producing geothermal fields is normally done under low pressure, either hydrostatic head or slightly higher pressures, depending on the porosity and permeability of rock at the injection site. With engineered geothermal systems (EGSs), fluids are injected under higher pressure to stimulate fracture permeability in rocks at a depth where temperatures are high but existing or natural permeability is low (EGSs are discussed in more detail in Chapter 11). These newly created fractures create earthquakes (usually small), but in an experimental

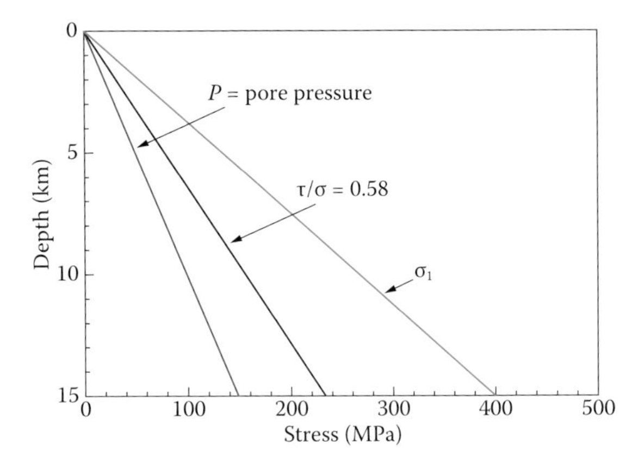

FIGURE 9.18 Graph showing change in rock strength for granite with depth for three different conditions: σ_1 is for unfractured granite, $\tau/\sigma = 0.58$ is the rock strength limit for fractured granite, and P is for water-saturated granite under hydrostatic pressure conditions. The strength in all cases increases with depth, reflecting increasing confining pressure which increases rock strength. The lower strength of P reflects water's ability to reduce effective stress and weaken chemical bonds. (From Glassley, W.E., *Geothermal Energy: Renewable Energy and the Environment*, 2nd ed., CRC Press, Boca Raton, FL, 2015.)

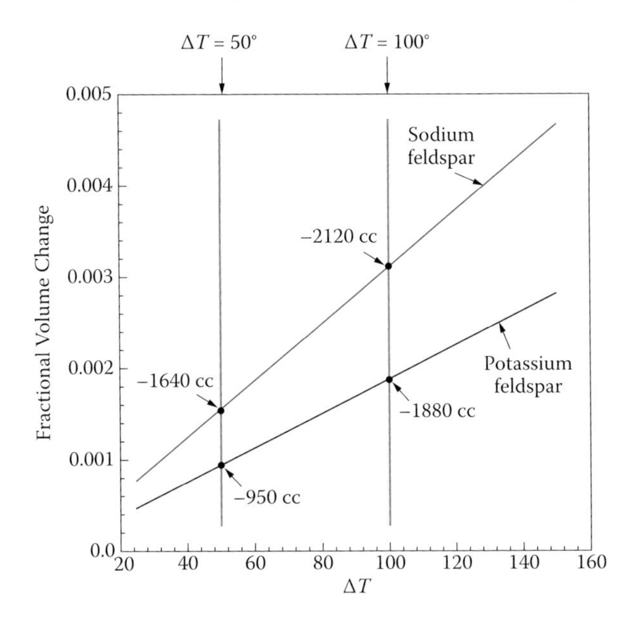

FIGURE 9.19 Graph showing fractional volume change of sodium and potassium feldspar as a function of temperature difference. Note that the degree of contraction increases with temperature difference between reservoir rock and injected fluid, increasing the tendency for fracture formation and seismic events. A rock consisting mainly of sodium feldspar would contract more for a given temperature change than a rock dominated by potassium feldspar and thus could generate larger induced seismicity. (From Glassley, W.E., *Geothermal Energy: Renewable Energy and the Environment*, 2nd ed., CRC Press, Boca Raton, FL, 2015.)

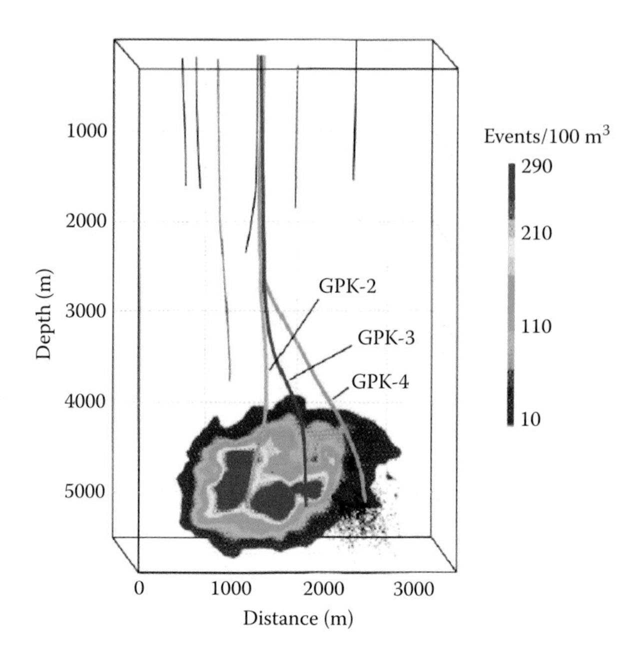

FIGURE 9.20 Density contour cross-section of seismic events associated with hydraulic stimulation of deep crystalline rocks at Soultz-sous-Forêts, France. The contour map reflects two periods of hydrofracturing of wells GPK-2 and GPK-3. The black dots are for individual seismic events stimulated from well GPK-4. (From Glassley, W.E., *Geothermal Energy: Renewable Energy and the Environment*, 2nd ed., CRC Press, Boca Raton, FL, 2015.)

geothermal well in Basel, Switzerland, the induced fracturing produced earthquakes with magnitudes as much as 3.4 that alarmed residents to the point that the operation was shut down (Deichmann and Giardini, 2009).

In general, induced seismicity is mostly at the microseismic level and can be beneficial for developing a geothermal reservoir. For example, the acoustic noise generated by microseismic events allows mapping of the fractures formed to determine the size of a newly formed reservoir. This was done at Soultz-sous-Forêts, France, which is an operating EGS geothermal plant producing from a reservoir at a depth of 4500 to 5000 m (Figure 9.20). It is also currently being done at Newberry Volcano in central Oregon, where a geothermal exploration company has successfully stimulated a small (~1 km^3) fractured controlled reservoir in otherwise hot dry (impermeable) rock. In conventional geothermal systems, the acoustic noise of seismic events can be mapped by sensitive, high-precision instruments to provide real-time information about fluid circulation through the reservoir and can delineate the more permeable portions of the reservoir as reflected by fractures and faults.

In that distinguishing between natural and induced seismic events can be difficult, a baseline of seismic data should be collected prior to and throughout operation of a geothermal power facility. Furthermore, if residents live in the vicinity, an educational program should be carried out to inform people of possible, albeit unlikely, disruptive seismic events and perhaps a hotline set up to report possible incidents. Maintaining a baseline of seismic events will also help distinguish activity associated with geothermal operations from potentially larger natural events. Doing so could prove helpful to show that the geothermal operation was not the cause of a possible naturally caused damaging earthquake.

Disturbance of Hydrothermal Surface Manifestations

Disruption of existing natural geothermal features (e.g., hot springs, geysers) at the surface is certainly an impact to be considered. Numerous examples exist where development has altered hydrothermal surface manifestations. A good example of such was the harnessing of the Wairakei geothermal system in New Zealand. Prior to development, the Wairakei stream valley hosted 22 active geysers along with numerous hot springs and was known as Geyser Valley. By the mid-1970s, the geysers and hot springs were extinguished and replaced by fumaroles and steaming ground, and the area has been renamed the Wairakei Thermal Valley. Another example of alteration of surface thermal features stemming from commercial development of geothermal power occurred at Beowawe in northeast Nevada. Beowawe was the site of the second largest geyser field in the United States, after Yellowstone, and now exists only as a large sinter terrace with some ephemeral thermal pools evident in wet years. Both Beowawe and Wairakei are liquid-dominated geothermal systems, and drawdown from production clearly impacted surface thermal features. At Wairakei, as we already discussed, reinjection did not begin until almost 40 years after the start of production. At Beowawe, although reinjection did occur with production, a net fluid loss from the reservoir occurred due to evaporative loss from the cooling towers and a more or less direct connection between shallow groundwater and deeper geothermal aquifers (Benoit and Stock, 1993).

Another commonly cited example of compromised or extinguished surface thermal features is at the producing Steamboat Springs geothermal field, just south of Reno, Nevada. Commercial geothermal production began there in 1986, and prior to that time the nearby silica terrace supported several fountaining hot springs (Figure 9.21). By the mid- to late 1990s, the fountaining hot springs no longer existed, and the fissures in the silica terrace only issued wispy, small plumes of steam condensate (Figure 9.22). However, the cause for the demise of the once vigorous surface thermal activity at Steamboat is more complicated because the decline also coincided with rapid residential development in the area and the drilling of shallow water wells to support the growth. Furthermore, except for the Steamboat Hills water-cooled flash plant, which is the farthest removed from the silica terrace, all other five power plants at Steamboat are binary air-cooled facilities in which geothermal fluids are fully recycled and evaporative loss of fluids is absent. Moreover, chemical tracer studies conducted by the current operator, Ormat, indicate that the injected tracer stays in the geothermal wells. None has been detected in shallow monitor or drinking water wells, arguing that the geothermal and shallow groundwater aquifers are separate and do not communicate with each other. Therefore, the degradation of surface thermal features at Steamboat appears less a result of commercial development of the geothermal resource than the drawdown of heated shallow groundwater by water wells to support surrounding residential growth.

FIGURE 9.21 View looking north across the main silica terrace at Steamboat Springs in 1986. The fountaining spring in the foreground is spraying water to a height of about 2 m. Since development of the geothermal power facilities in the late 1980s and continued suburban development of the nearby region, groundwater levels have fallen and the silica terrace no longer hosts active hot springs. Only fumarolic condensing vapors issue from fractures and fissures (see Figure 9.22). (Photograph courtesy of D.M. Hudson.)

Finally, it must be recognized that all surface geothermal features can also change from natural causes, as is well illustrated at Yellowstone National Park. Earthquakes, for example, can affect hydrothermal plumbing systems and hence the flow of fluids to hot springs or geysers—in some cases enhancing and in other instances decreasing flow rates and activity. To minimize potential adverse effects, baseline studies must be conducted prior to commercial development of geothermal resources. Such studies would include mapping the locations of existing surface thermal features; measuring temperatures, flow rates, and frequency of any geysering activity; and determining the fluid chemistry. Temperature-gradient wells can also be used to monitor the direction and flow rates of shallow groundwater and deeper geothermal fluids both prior to and during production. Reinjection of waste brine is probably the best technique not only to maintain long-term sustainability of the resource but also to mitigate disturbance of surface geothermal phenomena. Nonetheless, even reinjection may be insufficient to deter disturbance of surface thermal features, as noted above for Beowawe, Nevada. But, in other cases, such as at Wairakei, New Zealand, changes to surface thermal features may have been lessened by recharging the reservoir through reinjection of waste brines.

FIGURE 9.22 View of current conditions at Steamboat silica terrace. North-south-striking fissures now issue only wispy, clouds of steam condensate; prior to the mid-1980s, they issued flowing boiling water and fountaining springs or small geysers. (Photograph by author.)

SUMMARY

The environmental benefits of developing geothermal resources are considerable and include greatly reduced gaseous emissions compared to power facilities fueled by coal and natural gas. A small footprint, low potential water usage (especially for air-cooled binary power plants), low noise levels, and minimal visual impacts characterize geothermal power plants. When these environmental advantages are combined with the general 24-hour availability of geothermal energy and high capacity factor (typically >90%)of geothermal power plants, the attractiveness of developing geothermal energy, where available, is undeniable and compelling.

Nonetheless, a geothermal resource must be developed responsibly to protect the environment and to sustain the resource itself for long-term use (renewability and sustainability are discussed in Chapter 12). Potential problems center on land subsidence, induced seismicity, and reduction or loss of scenic hydrothermal surface manifestation, such as fountaining hot springs or even geysers. Land subsidence can be managed through a well-designed program of reinjecting spent geothermal fluid. Results of InSAR studies, which measure subtle changes in ground elevation, can be used to identify areas of subsidence (or tumescence) and thereby help locate placement of injection wells. Ground subsidence, for example, can have undesirable effects on irrigation canals in regions where agriculture and geothermal development overlap, such as in the Salton Sea region in southern California. Reinjection of spent geothermal fluids can help retard the loss of visually attractive hydrothermal surface features that may be present by maintaining fluid pressure in the tapped reservoir.

Induced seismicity can occur from reinjection of spent geothermal fluids but is more likely a potential problem where cold effluent is imported and injected into hot rocks, as the thermal contrast causes rocks to contract and break. Induced seismicity has been an issue at The Geysers, where water (as treated sewage effluent) is imported and injected. The importing of water is necessary because much of the geothermal fluid is lost to the atmosphere from evaporative cooling, leaving too little to be reinjected and causing a decline in reservoir pressure. From experience, workers at The Geysers found that, by controlling the rates of injected effluent, spreading the injection over a larger volume of rock, and injecting near the top of the steam reservoir, induced seismicity can be managed (M. Walters, pers. comm., 2015). On the flip side, induced seismicity can be advantageous in the formation of engineered geothermal systems. Newly formed fractures created by injecting cold water into hot rock leads to improved permeability and the possible generation of an artificially produced geothermal reservoir that could be viable for development (see Chapter 11 for further discussion on this topic).

SUGGESTED PROBLEMS

1. Consider at least two reasons why you think it took so long at Wairakei, nearly 40 years after production began, to begin injection of waste brine.
2. Why are binary geothermal plants the most environmentally benign? Please consider emissions, land use, water usage, subsidence, induced seismicity, and visual aspects in your response.
3. What causes possible induced seismicity during development and operation of a geothermal power plant? How might seismic activity be mitigated?
4. How is water used in the operation of a geothermal power plant? How might water consumption be reduced?

REFERENCES AND RECOMMENDED READING

Ármannsson, H. (2003). CO_2 emissions from geothermal plants. In: *Proceedings of International Geothermal Conference*, Reykjavik, Iceland, September 14–17 (http://www.jardhitafelag.is/media/PDF/S12Paper103.pdf).

Benoit, D. and Stock, D. (1993). A case history of injection at the Beowawe, Nevada, geothermal reservoir. *Geothermal Resource Council Transactions*, 17: 473–480.

Bromley, C.J. (2006). Predicting subsidence in New Zealand geothermal fields—a novel approach. *Geothermal Resource Council Transactions*, 30: 611–616.

Bromley, C.J., Currie, S., Jolly, S., and Mannington, W. (2015). Subsidence: an update on New Zealand geothermal deformation observations and mechanisms. In: *Proceedings of World Geothermal Congress 2015*, Melbourne, Australia, April 19–24 (http://www.geothermal-energy.org/pdf/IGAstandard/WGC/2015/02021.pdf).

BLM. (2005). Mammoth Pacific geothermal plant. *News.bytes Extra*, Issue 209, December 6, http://www.blm.gov/ca/ca/news/newsbytes/xtra-05/209-xtra_mam_geothermal.html.

BLM. (2008). *Environmental Assessment: Hay Ranch Water Extraction and Delivery System*, CA-650-2005-100. Ridgecrest, CA: Bureau of Land Management (http://www.blm.gov/style/medialib/blm/ca/pdf/ridgecrest/ea.Par.4165.File.dat/HayRanchEA.pdf).

Deichmann, N. and Giardini, D. (2009). Earthquakes induced by the stimulation of an enhanced geothermal system below Basel (Switzerland). *Seismological Research Letters*, 80(5): 784–798.

DiPippo, R. (2012). *Geothermal Power Plants: Principles, Applications, Case Studies, and Environmental Impacts*, 3rd ed. Waltham, MA: Butterworth-Heinemann.

Eneva, M., Adams, D., Falorni, G., and Morgan, J. (2013). Applications of radar interferometry to detect surface deformation in geothermal areas of Imperial Valley in Southern California. In: *Proceedings of the 38th Workshop on Geothermal Reservoir Engineering*, Stanford, CA, February 11–13 (http://www.geothermal-energy.org/pdf/IGAstandard/SGW/2013/Eneva.pdf).

Gemmell, J.B., Sharpe, R., Jonasson, I.R., and Herzig, P.M. (2004). Sulfur isotope evidence for magmatic contributions to submarine and subaerial gold mineralization: Conical Seamount and the Ladolam gold deposit, Papua New Guinea. *Economic Geology*, 99(8): 1711–1725.

Glassley, W.E. (2015). *Geothermal Energy: Renewable Energy and the Environment*, 2nd ed. Boca Raton, FL: CRC Press, Chapter 15.

Holm, A., Jennejohn, D., and Blodgett, L. (2012). *Geothermal Energy and Greenhouse Gas Emissions*. Washington, DC: Geothermal Energy Association (http://geo-energy.org/reports/GeothermalGreenhouseEmissionsNov2012GEA_web.pdf).

IPCC. (2014). *Climate Change 2014: Synthesis Report. Contribution of Working Groups I, II, and III to the Fifth Assessment Report of the Intergovernmental Panel on Climate Change*. Geneva: Intergovernmental Panel on Climate Change (http://ar5-syr.ipcc.ch/).

Kagel, A., Bates, D., and Gawall, K. (2007). *A Guide to Geothermal Energy and the Environment*. Washington, DC: Geothermal Energy Association (http://geo-energy.org/pdf/reports/AGuidetoGeothermalEnergyandtheEnvironment10.6.10.pdf).

McDonald, R.I., Fargione, J., Kiesecker, J., Miller, W.M., and Powell, J. (2009). Energy sprawl or energy efficiency: climate policy impacts on natural habitat for the United States of America. *PLoS ONE*, 4(8): e6802 (http://journals.plos.org/plosone/article?id=10.1371/journal.pone.0006802).

Robertson, A. (2012). Northern Nevada freeway nears completion. *Nevada Magazine*, June 25, http://nevadamag.blogspot.com/2012/06/northern-nevada-freeway-nears.html.

10 Geothermal Systems and Mineral Deposits

KEY CHAPTER OBJECTIVES

- Describe the similarities and differences between active geothermal systems and epithermal precious metal mineral deposits.
- Distinguish between magmatic and amagmatic mineralized geothermal systems.
- Explain how studying mineralized fossil geothermal systems helps finding and developing active geothermal systems and *vice versa*.

OVERVIEW

Today's geothermal systems have long been recognized as the modern analogs of high-crustal level (<2 to 3 km paleodepth) mineral deposits (Lindgren, 1933; White, 1955, 1981), including epithermal deposits (typically <1 to 2 km depth) of gold and silver (Henley and Ellis, 1983; Rowland and Simmons, 2012; Simmons and Browne, 2000) and slightly deeper porphyry-type deposits of copper, gold, and molybdenum (Gustafson et al., 2004; Heinrich et al., 2004; Sillitoe, 2010). Although most active geothermal systems contain some anomalous concentrations of metals, such as gold (Au), silver (Ag), arsenic (As), antimony (Sb), mercury (Hg), and copper (Cu), their concentrations and volumes within the rock reservoir are typically too low to be mined. Chemical analyses of fluids of many active geothermal systems disclose very low concentrations at or below limits of detection of economically important metals, such as Au, Ag, and Cu.* Considering that most paleogeothermal systems investigated in the search of economically minable mineral deposits (i.e., ore deposits) also have sub-economic concentrations of select elements, it is not surprising that most active geothermal systems are also not significantly mineralized. Nonetheless, modern geothermal systems afford opportunities to better understand the processes that form ore deposits, including transport, concentration, and deposition of sought-after elements. Also, mineralized paleogeothermal systems afford the opportunity to better understand fluid–rock interaction through the study of a large volume of exposed hydrothermally altered rocks made possible by erosion and/or mining.

The links between geothermal processes and ore formation have been the focus of numerous studies since the early 20th century (Lindgren, 1915, 1933). More recently, considerable attention has been focused on studying the geothermal systems hosted within the still active Taupo Volcanic Zone (TVZ) on the North Island of New Zealand

* Notable exceptions, as mentioned in Chapters 7 and 9, are the brines produced from the Salton Sea geothermal field that contain high concentrations of Mn, Li, Zn, and Ag.

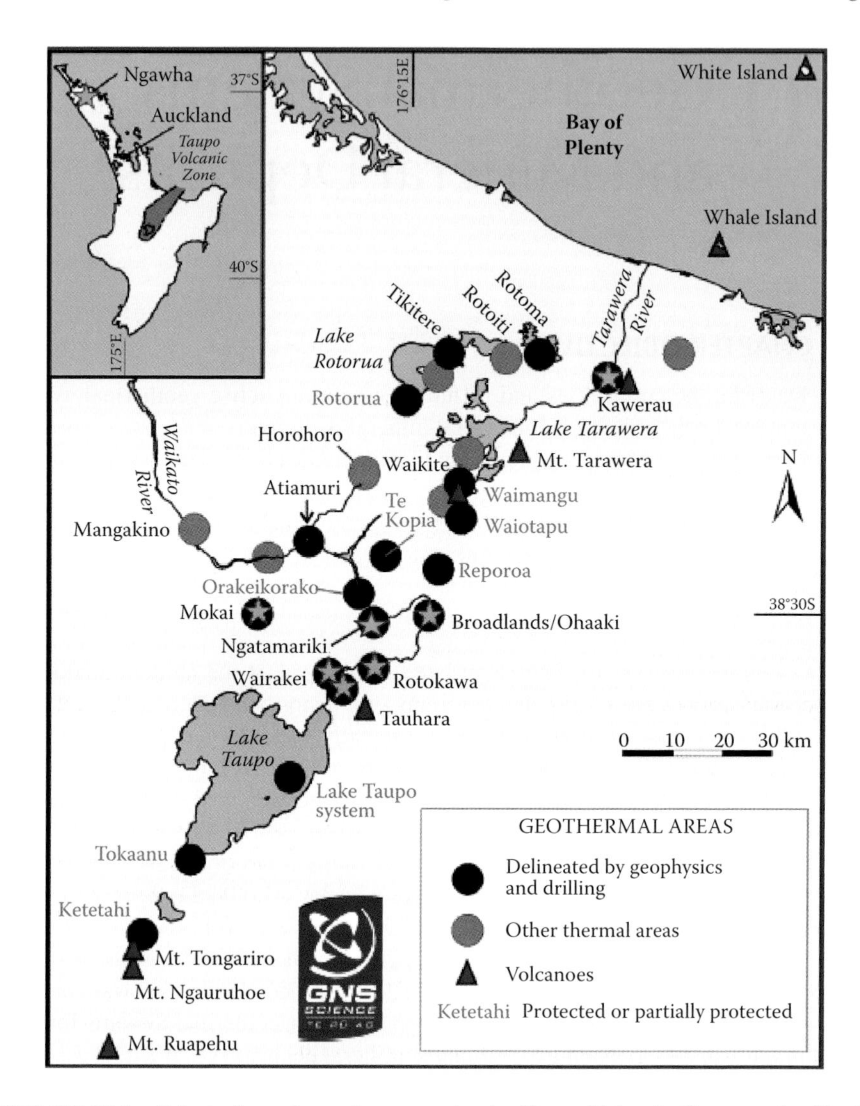

FIGURE 10.1 Principal geothermal systems in the Taupo Volcanic Zone on the North Island, New Zealand. Those with the green stars denote geothermal systems with operating geothermal power plants. Protected refers to those geothermal systems that occur as parks and are exempt from commercial development. (Adapted from Bignall, G. and Carey, B.S., *Proceedings of the New Zealand Geothermal Workshop*, 33, 5, 2011.)

(Figure 10.1) (Browne, 1969; Hedenquist, 1986; Rowland and Sibson, 2004; Rowland and Simmons, 2012; Simmons and Brown, 2007). Results of these studies indicate that, although precious and base metal values are low in fluids for most geothermal systems, ore-grade concentrations occur locally in geothermal fluid precipitates but are typically of small volume, such as at the Champagne Pool in the Waioitapu geothermal system (Hedenquist and Henley, 1985; Pope et al., 2005; Weissberg, 1969). At the Champagne Pool, amorphous, bright orange-colored precipitates of As, Sb, and

FIGURE 10.2 View of Champagne Pool at the Waiotapu geothermal field, New Zealand. The submerged vivid orange material rimming the edge of the pool consists of gold-bearing, As- and Sb-rich precipitate. The light gray rock bordering the pool consists of siliceous sinter formed during periods of high flow. The thermal pool gets its name from the CO_2 that bubbles out of the water (just like in champagne), not from boiling, as the temperature of the water is about 74°C. (Photograph by author.)

sulfur (S) contain >500 ppm Au and >700 ppm Ag, although the hot spring water itself contains only about 0.1 ppb Au (Figure 10.2). This disparity between low gold concentration in the spring water and its high concentration in the amorphous precipitates suggests that the gold in the spring water is being physically adsorbed by the As-, Sb-, and S-bearing precipitates, rather than chemically precipitating from the water. Another active geothermal system in the TVZ with notable gold and silver concentrations is at Rotokawa (Krupp and Seward, 1987). Au and Ag fluxes of fluids at Rotokawa are sufficient to form a large gold deposit (>30 tonnes or 1 million ounces) in as little as a few tens of thousands of years (Simmons and Brown, 2007). Just 10 km away at the large electrical power-producing Wairakei geothermal field, however, gold and silver values in geothermal fluids and deposits are low (as in many TVZ geothermal systems), even though temperatures and basic fluid chemistry (total dissolved solids and pH) are otherwise comparable. At the Broadlands geothermal system, located about 10 km northeast of Rotokawa, deep fluids also contain detectable gold and, unlike Wairakei, have a H-isotopic signature indicative of a low-salinity magmatic vapor component mixed with meteoric fluids.* Furthermore, the fluids at the Broadlands geothermal

* Meteoric fluids consist of water that originated in the atmosphere and entered the groundwater from precipitation or from melting of snow and ice. In contrast, fluids trapped in rocks at the time of formation, such as sediments accumulating on the seafloor, are termed *connate* and are commonly saline. Fluids originating during the late stages of crystallization of molten rock (magma) are termed *magmatic*, or sometimes juvenile, and can have high concentrations of dissolved metals, including Au, Ag, Cu, Pb, and Zn.

system are more gas rich (~2 wt%, mainly as CO_2 and H_2S) than those at Wairakei (<0.1 wt%) (Giggenbach, 1992). Although still not fully understood, the abrupt variation in Au and Ag content in fluids from adjacent geothermal fields may reflect one or more of the following:

- The presence of young rhyolite domes at Rotokawa suggest that magma exerted some control over the source and supply of metals (from either magmatically derived fluids or leaching of the rhyolite intrusions).
- The geothermal fluids at Rotokawa have an unusually high reduced sulfur content, which can transport gold as a bisulfide complex ($Au(HS)_2$) (Krupp and Seward, 1987).
- Numerous hydrothermal explosion craters indicate repeated episodes of catastrophic depressurization and flashing (boiling) of solutions, which can destabilize gold–bisulfide complexes and lead to gold deposition (Reed and Spycher, 1985).

As a result, 1.7 million to 3.3 million ounces of gold may have been deposited, based on metal concentrations and flux of geothermal fluids, in the upper 300 to 400 m below Lake Rotokawa that resides in a hydrothermal eruption crater formed about 6000 years ago (Krupp and Seward, 1987).

That more active geothermal systems are not more strongly mineralized argues that a special orchestration of physicochemical processes must come into play only at certain times (or perhaps not ever) during the lifespan of a geothermal system. These processes include focused fluid flow, repeated episodes of boiling, and mixing of rising and descending fluids to effectively concentrate metals in a limited volume of rock to make an ore deposit (Rowland and Simmons, 2012; Simmons and Brown, 2007). Such conditions must coincide in both space and time, which is generally not the case. Without the intersection of these processes, most geothermal systems are in a quasi-steady state in which economic metals remain largely dissolved in circulating fluids or precipitate in low concentrations (or repeatedly precipitate and redissolve) throughout a large volume of rock (the geothermal reservoir).

YOUNG MINERAL DEPOSITS AND ACTIVE GEOTHERMAL SYSTEMS

The source of the metals in both active geothermal systems and their fossil epithermal analogs has long been debated. In some cases, leaching of metals from deeply circulating geothermal fluids from wallrocks may be the main source. This would be where coeval igneous rocks are lacking, and stable isotope analyses of hydrogen and oxygen of fluids from active geothermal systems and of gangue minerals (such as calcite and quartz) associated with mineralization denote a meteoric origin. Good examples include the Dixie Comstock Mine (Vikre, 1994), located about 10 km south of the productive Dixie Valley geothermal system and power facility, in west-central Nevada, and the Hycroft Mine in northwest Nevada (Ebert and Rye, 1997). Leaching of metals from wallrocks would be enhanced if acid magmatic vapors can be entrained within deeply circulating meteoric groundwater (Hedenquist and Lowenstern, 1994). In other cases, a strong connection between magmatism and geothermal/epithermal mineralization is evident based on the flux of metals measured from active volcanoes, such as at White Island in New Zealand, and associations among the type of mineral deposit, tectonic setting, and composition of coeval igneous rocks, such as porphyry Cu–Mo deposits or epithermal high-sulfidation Au–Cu deposits and oxidized calc-alkaline magmas (Hedenquist and Lowenstern, 1994; John, 2001). From a geologic standpoint, young epithermal mineral deposits (≤~5Ma) can be classified, similar to active geothermal systems, as either magmatic or amagmatic depending on the presence or absence of coeval igneous rocks (Coolbaugh et al., 2005, 2011).

Young Magmatic Mineralized Geothermal/Epithermal Systems

For this discussion, young magmatic mineralized geothermal/epithermal systems are systems that have been mined, are currently being mined, or are mineralized but remain subeconomic due to low grade or volume. Examples discussed below consist of the Hishikari Gold Mine and Noya gold deposit, Japan; the McLaughlin gold–silver mine in northern California; the giant Ladolam gold deposit on Lihir Island in Papua New Guinea; the Long Valley gold–silver resource in east-central California; and Steamboat Springs (precious metal prospect and current geothermal power producer) near Reno, Nevada. Elsewhere, based on measured Cu/S ratios of the aerosols in the ash plume, gold and copper mineralization are inferred to be present under highly oxidizing and sulfidizing conditions beneath the active andesitic White Island volcano at the northern extension of the TVZ, New Zealand (Hedenquist et al., 1993).

Hishikari Gold Mine and Noya Gold Deposit, Kyushu, Japan

The Hishikari Gold Mine is a premier, bonanza-grade, gold–silver deposit containing 7.5 million tonnes* at 40 grams/tonne gold, yielding about 300 tonnes (about 10 million ounces) of gold (Tohma et al., 2010). Hishikari is one of several gold–silver deposits in the region, all of which lie 15 to 25 km west of an active volcanic arc containing Kirishima, Sakurajima, and Kaimondake volcanoes (Figure 10.3). Gold mineralization is distributed over three main vein systems covering only about 5 km^2, reflecting the high gold concentrations of the veins (Figure 10.4).

Veins are hosted by Pleistocene andesite and Late Cretaceous to early Tertiary shale and sandstone, with much of the high-grade gold mineralization focused in the vicinity of the unconformity separating the overlying andesite and underlying sedimentary rocks (Izawa et al., 1990). Radiometric ages on gold-associated adularia indicate mineralization extended over a range of about 700,000 years from 1.3 to 0.6 Ma (Sanematsu et al., 2006; Tohma et al., 2010). The northern veins generally yield older ages (1.3 to 1.2 Ma) compared to the southern veins (0.7 to 0.6 Ma), indicating that the focus of the mineralizing hydrothermal system migrated southward with time. Detailed dating of one of the southern mined veins indicates that the vein formed in less than 44,000 years (Tohma et al., 2010). Detailed study on another vein in the northern part of the deposit indicates that the vein grew in four to six distinct

* Tonne is the metric unit for 1000 kg, which is about 2200 pounds. Ton is the imperial unit and is equivalent to 2000 pounds.

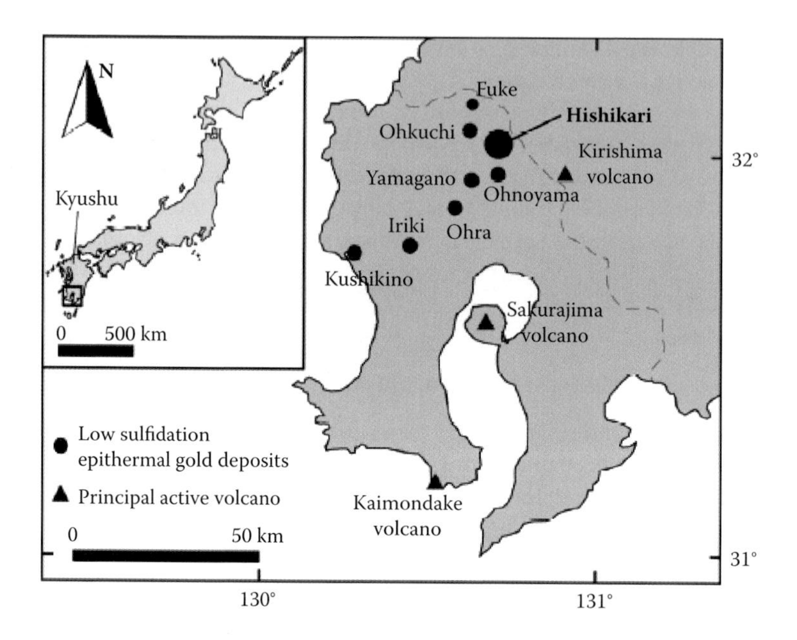

FIGURE 10.3 Map showing location of Hishikari and other epithermal gold deposits and active volcanoes. (From Tohma, Y. et al., *Resource Geology*, 60(4), 348–358, 2010.)

increments separated by intervals of 30,000 to 110,000 years over a total duration of 260,000 years. Inspection of error ranges on the reported ages yields ranges in age for vein formation from nil (Fukusen vein) to about 500,000 years for the Yamada zone. $^{40}Ar/^{39}Ar$-determined ages decrease more or less systematically from 1.044 ± 0.006 Ma, at the edge of that vein (oldest), to 0.781 ± 0.028 Ma, at the center of the vein (youngest) (Sanematsu et al., 2006).

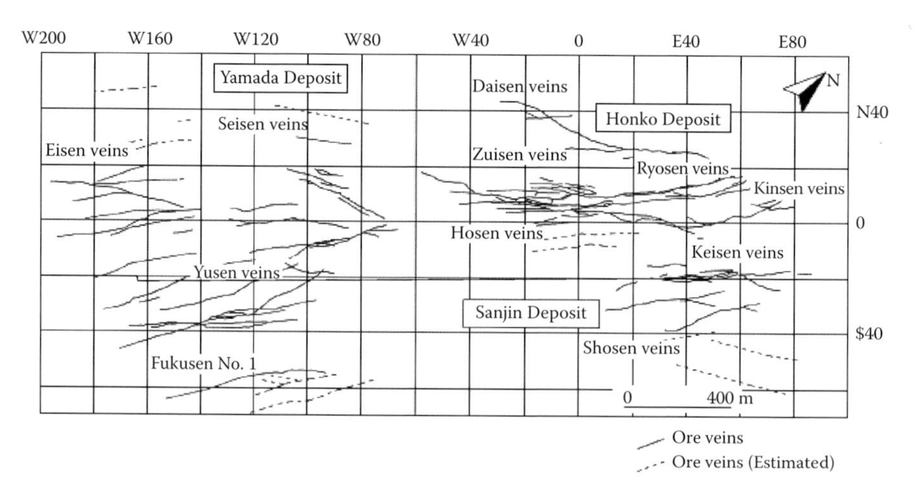

FIGURE 10.4 Map showing veins of the three main mined deposits of the Hishikari epithermal gold system. (From Tohma, Y. et al., *Resource Geology*, 60(4), 348–358, 2010.)

Corresponding to this decrease in age from vein margin to vein center, homogenization temperatures of trapped fluid inclusions* in vein quartz decrease from 190° to 210°C to less than 170°C, indicating that the mineralizing hydrothermal fluid cooled with time. Significantly, many of the veins are still bathed in hot geothermal fluids (65° to 92°C, as reported by Fuare et al., 2002), meaning that the Hishikari hydrothermal system has existed, at least episodically, for at least 1.3 Ma. Current geothermal fluids, however, contain only 0.6 parts per trillion (ppt) Au, suggesting that the present thermal water is not transporting gold nor remobilizing gold from the mineralized zones (Hayashi et al., 1997). These data suggest that the Hishikari geothermal system is now waning; however, as noted above, the Hishikari geothermal system has gone through several episodes of waxing and waning during its history, including incremental bonanza-grade gold deposition.

The undeveloped Noya gold deposit is located in the Hohi volcanic and geothermal zone in northern Kyushu about 120 km north-northeast of Hishikari (Figure 10.4). Three other gold deposits, Bajo, Hoshino, and Taio, occur within 40 km of the Noya gold deposit from which 50 tonnes of gold have been mined collectively (Figure 10.5) (Morishita and Takeno, 2010). Interestingly, the Noya gold deposit was discovered as a byproduct of geothermal exploration and drilling of geophysical anomalies that led to the development of the Takigami geothermal power plant. During the course of geothermal exploration well drilling, Pleistocene gold-bearing calcite–quartz–adularia veins were encountered with gold concentrations ranging from 0.1 to 400 ppm. These veins do not crop out, and surface expressions of the developed geothermal surface system are minimal, consisting of only minor and scattered occurrences of hydrothermally altered rock. The dearth of exposed altered rock and geothermal surface activity in the area is due in large part to burial by a young (89,000 years old) ash-flow tuff erupted from the Aso caldera that lies about 40 km to the southwest (see Figure 10.4) (Miyoshi et al., 2012). K–Ar dating of an adularia-bearing vein, containing 5.5 ppm Au, yielded an age of 0.37 ± 0.01 Ma.

The maximum temperature recorded at the bottom of a 700-m-deep geothermal well, in which gold-bearing calcite–quartz–adularia veins were discovered, was 177°C. The gold-bearing veins were found at depths of 164 to 214 m. In another well, silicified and adularized rock is more widespread than gold veins, which are limited to a narrower and shallower depth interval of 25 to 61 m where the gold content was found to vary from 0.1 to 4.2 ppm (Morishita and Takeno, 2010). Oxygen isotope equilibrium temperatures of calcite and quartz in veins and homogenization temperatures of fluid inclusions in calcite average about 20° to 25°C higher (170° to 185°C) than the measured well temperature (140° to 160°C) over the mineralized depth interval (Morishita and Takeno, 2010).

---

* As a mineral grows in the presence of a hydrothermal fluid, some of the fluid can be trapped within the mineral, resulting in a fluid inclusion. The trapped fluid contracts with declining temperature and as it does so a bubble forms. A mineral containing fluid inclusions can be heated in the laboratory, and the temperature at which the bubble disappears within an inclusion represents the homogenization or temperature of the fluid at the time of trapping. Similarly, a measure of a hydrothermal fluid's salinity can be determined by noting the freezing point depression of inclusions, such that the greater the depression in temperature to freeze an inclusion, the greater the salinity of the trapped fluid.

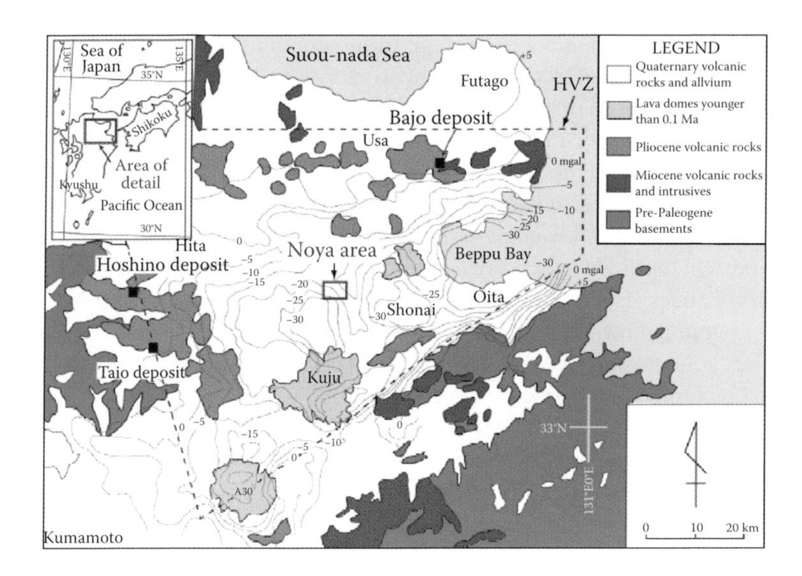

FIGURE 10.5 Map showing the location of the Noya area in the northeastern part of Kyushu (see inset). HVZ stands for the Hohi volcanic zone. The Noya area includes the Takigami geothermal power plant. The faint gray lines are Bouguer anomalies contoured in milligals. Two other geothermal plants, Otake and Hatchorbaru, lie on the west flank of Mount Kuju (pink-colored lava dome) about 25 km southwest of the Noya area. (From Morishita, Y. and Takeno, N., *Resource Geology*, 60(4), 359–376, 2010.)

The Takigami geothermal power plant went into production in 1996 and has a recently augmented installed capacity of 27.5 MWe. It is a single-flash facility served by five production wells located in the southwestern, hotter part of the field and by seven to ten injection wells located in the northern, cooler part of the field. Produced fluids are as hot as 250°C, with most in the range of 200° to 210°C. Subsurface temperatures in the injection field area range from 160° to 170°C (Furuya et al., 2000). The geothermal fluids are of the sodium chloride (NaCl) type and have near-neutral pH and low total dissolved solids (TDS) (400 to 600 ppm). These modern fluids have chemical characteristics similar to those indicated from the gold mineralized calcite–quartz–adularia veins (Morishita and Takeno, 2010). Current fluid temperatures, however, are similar or hotter than those indicated from homogenization temperatures of fluid inclusions and oxygen isotope fractionation temperatures from the gold mineralized veins. Calculated delta (δ) O^{18} values* of ore fluids are

* Delta ^{18}O (also designated as $\delta^{18}O$) reflects a ratio of the heavy oxygen isotope (^{18}O) to the light (and far more common) oxygen isotope (^{16}O) relative to that of a designated set of standard values for ocean water, referred to as standard mean ocean water (SMOW). Thus, a $\delta^{18}O$ of +5, for example, indicates that the sample is 5‰ (5 parts per thousand) or 0.5% richer in heavy oxygen than SMOW. Negative values, as in this case, indicate depletion of heavy oxygen relative to SMOW. As hot fluids move through rock, the $\delta^{18}O$ of the rock is lowered while that of the fluid is increased initially. In the case of low water-to-rock ratios (or low permeability), the fluid will have a relatively high $\delta^{18}O$ (low negative to slightly positive values), reflecting longer residence time and isotope exchange with the rock. If the water-to-rock ratios are high (or high permeability), however, the fluid will have low (typically negative) $\delta^{18}O$ values because water dominates the isotope exchange and is circulating through rocks already depleted in ^{18}O.

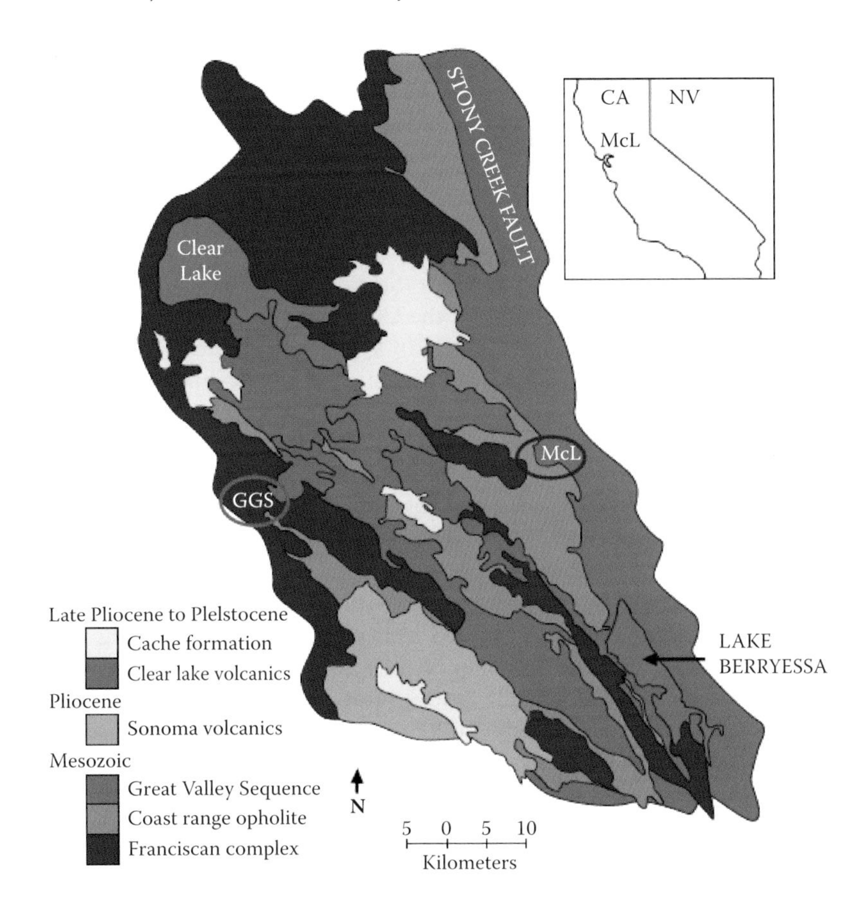

FIGURE 10.6 Location and simplified geologic map of the McLaughlin region. The McLaughlin mine (McL) and The Geysers geothermal system (GGS) are enclosed within the red and green ellipses, respectively. (Adapted from Sherlock, R.L. et al., *Economic Geology*, 90(8), 2156–2181, 1995.)

slightly heavier (~7.5‰), probably reflecting greater water–rock interaction (oxygen isotope shift) at the time of mineralization, than those of the modern geothermal fluids (~8.8‰) (Morishita and Takeno, 2010). The isotopically lighter value of the modern geothermal water probably indicates a high water-to-rock ratio (improved permeability and high flow rates) and that wallrocks have been previously depleted in δO^{18} from long-sustained hydrothermal fluid circulation.

McLaughlin Mine, California

Similar to Hishikari, the McLaughlin gold deposit is a world-class epithermal vein system containing local bonanza grades (100 to 1000 ppm Au) (Sherlock et al., 1995). The McLaughlin mine is located in the northern Coast Range of California and in the eastern part of the Pleistocene to Holocene Clear Lake Volcanic Field (CLVF). The Geysers geothermal field lies along the west margin of the CLVF, about 30 km west of the McLaughlin mine (Figure 10.6). The McLaughlin mine was operated by Homestake Mining from 1983 to 1996 and produced about 3.3 million ounces of gold. The McLaughlin deposit is considered an archetype example of a hot-spring gold deposit containing a preserved silica sinter terrace cut by auriferous banded veins of silica. The siliceous sinter is a fossil hot spring deposit formed at the paleosurface,* indicating that little erosion has taken place since hot spring activity. As a result, a mineralized, but no longer active, geothermal system is fully preserved. Gold mineralization is restricted to the upper 350 m, with Ag and base metal values increasing at depth (Sherlock et al., 1995). Mineralization locally cuts 2.2 Ma basaltic andesite of the Clear Lake Volcanic Field, providing a maximum age of mineralization. A vein of hypogene† alunite at the base of the sinter yielded a K–Ar age of 0.75 Ma (Lehrman, 1986), indicating a minimum age for gold mineralization.

The McLaughlin mine is located in the Knoxville mining district and is centered on the former Manhattan mercury (Hg) mine, which produced Hg intermittently for about 100 years prior to the discovery of gold. Most of the Hg mineralization there occurred in the silica sinter terrace, which is typically barren of gold unless cut by auriferous banded quartz veins. Several other Hg deposits occur in proximity to McLaughlin, but no gold has been discovered at the other deposits. At The Geysers geothermal field, steam contains Hg and hydrocarbons, and several Hg deposits occur along the southwestern boundary of the steam field (Peabody and Einaudi, 1992). Significantly, the McLaughlin gold deposit is the only original Hg deposit in the Knoxville district associated with local intrusions of basaltic andesite of the CLVF. Such associated magmatism could have helped sustain a robust geothermal system capable of producing multiple periods of boiling, as evidenced from fluid inclusion data (Sherlock et al., 1995), and development of gold mineralization.

As noted in Chapter 7, The Geysers steam field began as a fluid-dominated system about a million years ago (White et al., 1971) and evolved into a vapor-dominated system at about 0.28 to 0.25 Ma as a result of rapid depressurization and concomitant boiling (Hulen et al., 1997). Evidence for the liquid-dominated beginning of The Geysers steam field includes the presence of banded veins of quartz and adularia containing anomalous concentrations of Au, Ag, As, Sb, and Hg. In places, these veins have platy calcite suggestive of fluid boiling (Fournier, 1985; Simmons and Christenson, 1994). Platy calcite is commonly replaced by quartz, which can occur upon cooling of the fluids as calcite becomes more soluble and quartz less soluble with declining temperature. The quartz-replaced platy calcite is now overprinted by widespread steam-heated or solfateric (acidic) alteration, characteristic of surface rocks overlying the present vapor-dominated system.

---

* *Paleosurface* refers to the original Earth's surface at the time of hydrothermal activity and mineralization. In older systems, the paleosurface has typically been eroded away along with any surficial geothermal manifestations such as siliceous sinter, travertine, or hydrothermal eruption breccias.

† The term *hypogene* refers to processes forming underground from primary hydrothermal fluids. In contrast, the term *supergene* refers to secondary processes occurring at or near the surface under ambient surface temperature, such as what occurs during weathering. The mineral alunite, a hydrated potassium aluminum sulfate, can form under both circumstances. Hypogene alunite is typically coarser grained than later formed supergene alunite and is associated with other primary hydrothermal minerals. At the McLaughlin mine, the hypogene alunite probably reflects near-surface condensation of steam in the presence of acidic gases such as H_2S.

In addition to The Geysers steam field, there are numerous hot and mineral springs in many parts of the CLVF. Mineral springs have ambient temperatures, effervesce carbon dioxide (CO_2) only, precipitate travertine, are not gold bearing, and discharge from some known Hg deposits. Hot springs, on the other hand, are hotter and more saline. Some important mineral-bearing hot springs include the Sulphur Bank mercury deposit located on the southeast shore of Clear Lake and the Elgin mercury deposit in the Sulphur Creek district located about 20 km north of the McLaughlin mine. Hot springs can be associated with the young volcanic rocks of the CLVF, such as at Sulphur Bank, or are not directly associated with young volcanic rocks, as in the Sulphur Creek district. Some of the hot springs, such as the Elgin spring, precipitate a black sulfidic mud containing as much as 12 ppm gold, along with anomalous concentrations of Ag, Hg, Sb, and As (Peters, 1991). Hot springs typically effervesce CO_2, hydrogen sulfide (H_2S), and methane (CH_4). At the Cherry Hill hot spring gold deposit in the Sulphur Creek district, Pearcy and Petersen (1990) demonstrated, from fluid inclusion studies, that the present hot spring fluids are cooler and less saline than the auriferous vein-forming fluids.

Further comparisons of data between the McLaughlin gold deposit and active hot springs in the northern Coast Ranges of California indicate that the current active springs are losing mainly vapor (not liquid) because of a possibly lowered locally boiling water table or low permeability, preferentially allowing labile vapors to escape relative to liquid water. Although both the McLaughlin gold deposit and active hot springs are localized along fault zones, the faults controlling active hot springs are currently aseismic. Furthermore, reservoir rocks of the active springs involve tectonically juxtaposed graywacke and serpentinite (altered basalt and ultramafic rocks) that have low primary permeabilities. Active dilatant fault zones that can focus fluid flow and promote the rapid rise of geothermal fluids to the surface with resultant boiling, cooling, and deposition of gold, as was the case at the McLaughlin mine, are absent or limited in the active hot springs. Instead, these high-vapor-loss systems of active springs are somewhat similar to, but not as robust as, the conditions at The Geysers steam field. This helps explain the abundance of volatile Hg but relative dearth of gold and other largely nonvolatile metal species (Sherlock, 2005). Again, the formation of an ore deposit requires a special orchestration of geochemical processes, including focused high-volume fluid flow, multiple episodes of pressurization (from self-sealing due to mineral deposition) and depressurization (from possible seismic events), and resultant boiling or fluid mixing. Such mineralizing processes may occur episodically and be of relatively short duration (perhaps a few tens of thousands of years or even less) in an otherwise long-lived geothermal system (1 to 2 Ma or even longer).

Ladolam Gold Deposit and Geothermal System, Lihir Island, New Guinea

The Ladolam epithermal gold deposit is located on Lihir Island, part of a volcanic island arc off the east coast of New Ireland, Papua New Guinea (Figure 10.7). Ladolam is one of the world's largest gold mines, containing a gold resource of 37 million ounces (428.9 million tonnes at 2.69 g/tonne) (Carman, 2003). Ladolam is remarkable not only for its size and high gold grades but also for its occurrence in an active geothermal system (White et al., 2010). To help offset the energy needs of the mining operation, a 56-MWe (installed capacity) geothermal power plant was

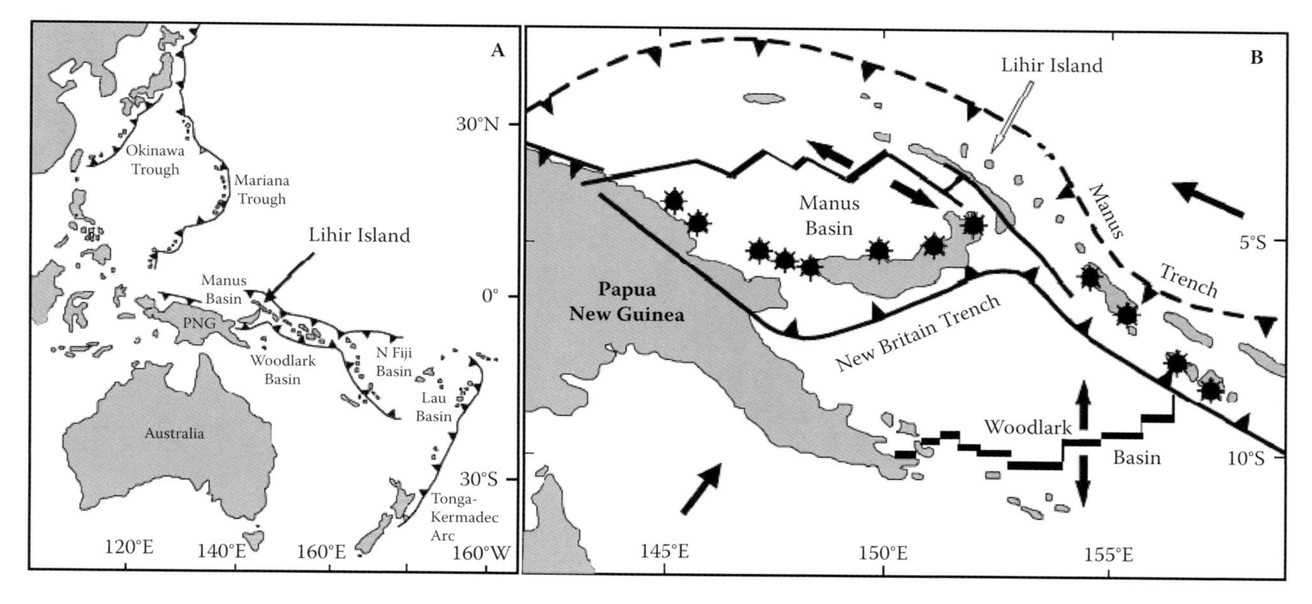

FIGURE 10.7 (A) Location map and (B) tectonic setting of Lihir Island, which hosts the giant Ladolam gold deposit and active power-producing geothermal system. Black starred circles denote active volcanoes. Thick bold lines denote divergent (spreading) tectonic boundaries, and thinner bold lines mark transform fault boundaries connecting divergent boundary segments. Dashed and solid lines with sawteeth indicate subduction zones associated with convergent tectonic boundaries, with the sawteeth on the overriding plate and pointing in the direction of downgoing subducted slab. (Adapted from Gemmell, J.B. et al., *Economic Geology*, 99(8), 1711–1725, 2004.)

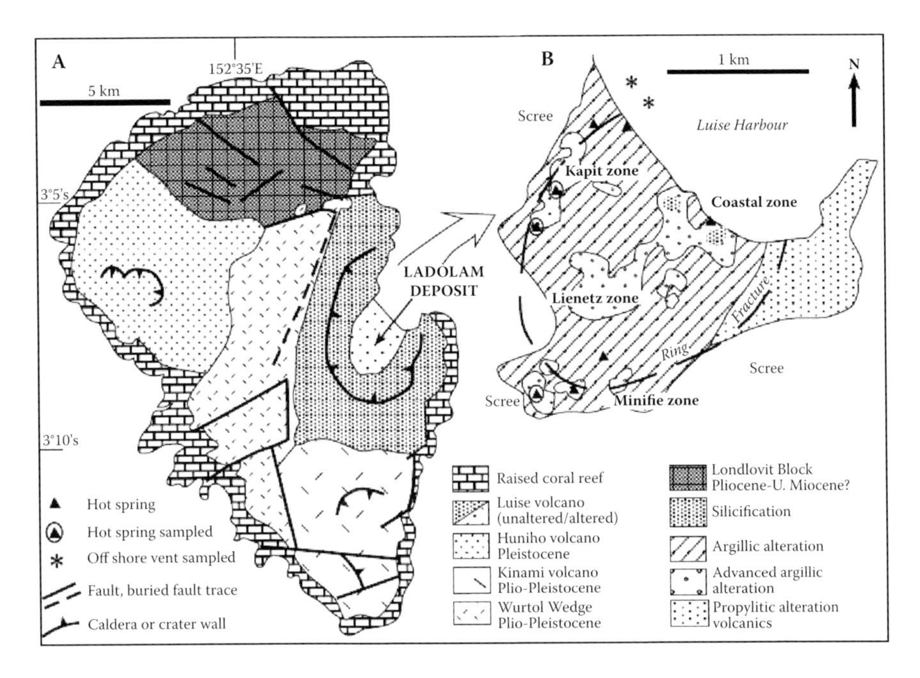

FIGURE 10.8 (A) Simplified geologic map of Lihir Island and Ladolam gold deposit and (B) mapped zones of alteration and locations of thermal features in the mine and offshore. (From Gemmell, J.B. et al., *Economic Geology*, 99(8), 1711–1725, 2004.)

constructed that allows the haul pack trucks to be battery operated. Indeed, mining cutoff is not only a function of grade but also of temperature because parts of the deposit are still simply too hot (or too dangerous from potential flashing of hot water to steam upon excavation) to mine. For example, geothermal zones along the western margin of the mine reach temperatures of 240°C at a depth of 300 m, and rock temperatures near the bottom of the mine and along the western walls are as high as 150°C. Ladolam is the only active major mine on the planet in which geothermal power is used to support the mining of gold. An attempt was made to help supply geothermal electrical power (75 kWe) at the Florida Canyon Mine in western Nevada, but the well to supply the power unit failed (J. Barta, pers. comm., 2016).

The Ladolam gold deposit is exposed in an amphitheater opening on the side of the extinct Luise volcano or caldera (Figure 10.8). The amphitheater formed by sector collapse of a volcanic edifice that removed an estimated 1100 m of overlying rock as inferred by the hydrothermal alteration mineral assemblage now exposed at the current surface (White et al., 2010). Three stages of mineralization are recognized at Ladolam (Carman, 2003). Stage I consists of minor porphyry-style copper–molybdenum–gold mineralization that developed prior to sector collapse and includes hydrothermal biotite and orthoclase. Stage II mineralization developed upon explosive decompression due to sector collapse of the volcanic edifice and resulted in refractory* gold-bearing pyrite mineralization and widespread adularia–pyrite–anhydrite

* *Refractory* means the gold is difficult to separate from ore minerals by traditional metallurgical practices.

alteration. Due to the abrupt pressure drop from the sector collapse, numerous mineralized hydrothermal eruption breccias produced from explosive boiling of the hydrothermal system are associated with this stage of mineralization. Stage II mineralization overprints earlier stage I mineralization and contains the bulk of shallow bulk-minable ore. Dating of secondary potassium feldspar indicates that the sector collapse of the Luise volcano occurred prior to 0.3 Ma (unpublished report by T.M. Leach, as reported by White et al., 2010). Stage III veins of calcite–quartz–adularia–pyrite–marcasite–electrum cross cut earlier anhydrite–adularia veins and are mostly subeconomic. A silicic breccia of stage III is capped by advanced argillic alteration (also referred to as acid-sulfate alteration), reflecting oxidation and condensation of H_2S-rich vapor above boiling hydrothermal fluids. Fluid inclusion and stable isotope data indicate that stage I fluids are mainly of magmatic origin (Carman, 2003). Stage II fluids appear to reflect mixing of magmatic ore fluid with cool meteoric groundwater along with episodic boiling. Another period of mixing of moderately saline (5 ± 0.5 wt% NaCl equivalent) magmatically derived fluids with dilute (near 0 wt% NaCl equivalent) meteoric groundwater produced the quartz and calcite veins associated with Stage III mineralization.

Radiometrically determined ages on biotite from hydrothermally altered intrusions at Ladolam range from 0.9 ± 0.1 to 0.34 ± 0.04 Ma (using the K–Ar method) (Moyle et al., 1990). Using the more precise $^{40}Ar/^{39}Ar$ dating method, adularia from sulfide-rich ores of stage II mineralization yielded ages of 0.61 ± 0.25 and 0.52 ± 0.11 Ma (Carman, 2003). A whole-rock K–Ar age on alunite (of probable steam-heated origin and part of the advanced argillic alteration assemblage) yielded an age of 0.15 ± 0.02 Ma (Moyle et al., 1990). These data indicate that sector collapse of the Luise volcano and onset of stage II mineralization occurred around 0.5 to 0.6 Ma. Moreover, the geothermal system and earlier porphyry-style mineralization formed by at least 1 Ma. An important question is whether the current geothermal system at Ladolam is the same that formed the Ladolam gold deposit or a younger superimposed system unrelated to mineralization. Two observations bear on this question. One is that calcite–dolomite scale on exploration drill pipes contains as much as 2.8 ppm gold (Moyle et al., 1990). The other is that submarine hot springs currently venting in Luise harbor are depositing auriferous iron sulfide minerals (Pichler et al., 1999).

The current Luise volcano geothermal system is liquid dominated in its deeper part with temperatures greater than 275°C in wells at a depth of 1 km. It also contains a shallow vapor- or steam-dominated zone within close proximity of the gold ore bodies (Melaku, 2005). Geothermal power production comes mainly from the geothermal wells tapping the deeper liquid-dominated portion of the system. The shallow steam zone appears to have developed mainly in response to dewatering and earth removal to access mining of the gold orebodies. Both dewatering and excavation lowered and depressurized the groundwater table, which induced boiling of underlying geothermal fluids. The walls of the open-pit mine are potmarked by plumes of condensed steam from steam-relief wells. These wells serve to depressurize the shallow steam zone and prevent steam blowouts that would likely occur as overlying rock is removed during mining (Figure 10.9). Indeed, some of the current

FIGURE 10.9 Mining of ore in the Ladolam mine. Note the steam plumes rising from wells to help depressurize and prevent explosive blowouts in the area being mined. (Illustration from http://www.ninefinestuff.com/wp-content/uploads/2015/07/94.jpg.)

power-producing geothermal wells began as deeper, intermediate-depth (200 to 700 m) depressurizing wells. In addition to providing access for mining in areas of higher temperature, these wells also provided steady and good volume flow rates of fluid, characterized by high steam-to-liquid ratios and enthalpy contents (2400 to 2700 kJ/kg) suitable for power production (Melaku, 2005).

Active thermal areas occur along the southern and western peripheries of the areas being mined with hot water ascending at a rate of 50 kg/s in geothermal wells (Simmons and Brown, 2006). Sampling of deep fluids below the ore deposits disclose a gold content of about 15 ppb (almost 3 orders of magnitude less than the average grade of mined ore of about 3 ppm) which, when applied to the current flow rate, yields a gold flux of 25 kg/year. Assuming 100% deposition of the noted aqueous gold concentration, 1200 tonnes of gold (the approximate size of the Ladolam deposit) could be deposited in as little as 55,000 years (Simmons and Brown, 2006). Of course, the time required could be longer or shorter depending on the efficiency of gold deposition and variable fluid concentrations of gold and mass flux rates of geothermal fluids.

According to a 2012 press release, Newcrest Mining reported that the 56-MWe geothermal power plant currently provides about 40% of the power needed for the mining operation, and an expansion of the geothermal power plant is planned. The current geothermally produced electricity saves about 250,000 tonnes of CO_2 per year, which represents about 4% of Papua New Guinea's annual CO_2 emissions (UNFCC, 2006). The produced geothermal energy allows selling of carbon credits on the world market that provide as much as $US5 million in annual revenue. Moreover, the 52.8 net MWe of geothermally produced power (411 GWh per year of

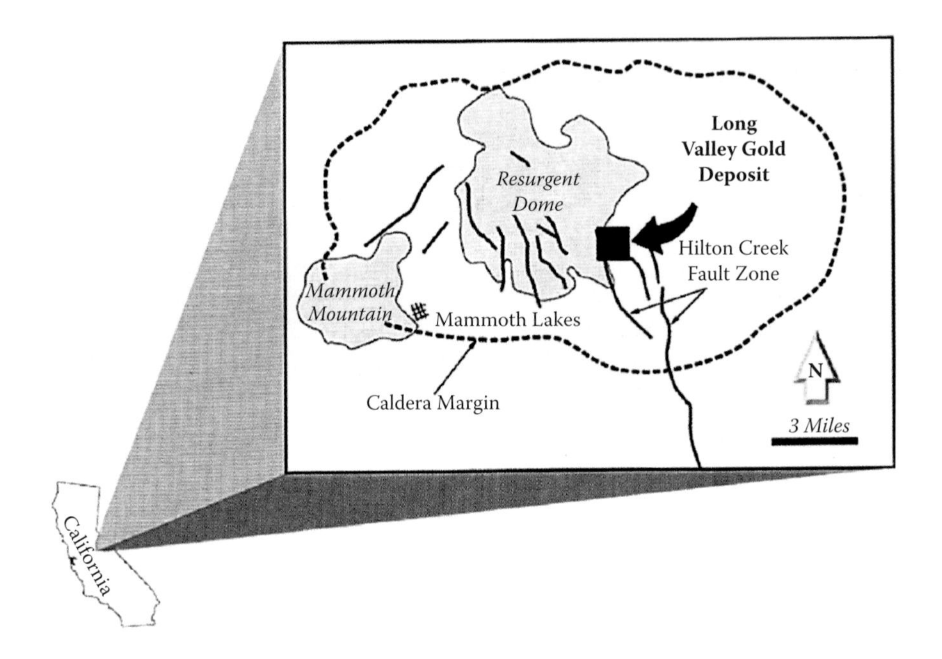

FIGURE 10.10 Location map of the Long Valley gold deposit. (From Steininger, R.C., in *Volcanic Geology, Volcanology, and Natural Resources of the Long Valley Caldera, California* (Leavitt, E.D. et al., Eds.), Geological Society of Nevada, Reno, 2005.)

electrical energy) displaces about 10 million gallons of heavy fuel oil per year* that would be otherwise consumed for diesel-powered generators. This alone saves the mine, depending on the price of fuel, $US20M to $US30M per year.

Long Valley Gold Deposit and Casa Diablo Geothermal System, California

The Long Valley gold deposit, located within the 760-ka Long Valley caldera in east-central California (Figure 10.10), is a resource of about 60 million tons (54.5 million tonnes) of 0.02-ounce-per-ton gold (Steininger, 2005). The gold deposit is located about 5 km northeast of the Casa Diablo geothermal power plant, which also lies within the Long Valley caldera. The 18-km by 30-km caldera formed upon massive and explosive eruption (about 600 km^3) of silicic magma that deposited the Bishop Tuff and whose distal equivalents are found as far away as Kansas and Nebraska. Much of the erupted Bishop Tuff ponded within the newly formed caldera. About 100,000 years after formation of the caldera, the central part of the caldera rose to form a resurgent dome as new magma moved into the underlying magma reservoir to replace what was erupted (Bailey, 1989). A central highland with a surrounding trough characterized the caldera at this time. A lake formed in this trough, forming a moat that trapped volcaniclastic conglomerates and sandstones along with interbedded tuffs and lava flows derived from silicic domes that erupted locally along the structural margin of the caldera. This

* Conversion factors are from the U.S. Energy Information Administration, using 3412 Btu per kWh and 5,800,000 Btu per barrel of heavy fuel oil.

moat igneous activity has continued to the present as evidenced by seismic activity indicating the intrusion of magma at depth (Hill et al., 2003) and the release of copious CO_2 (Sorey et al., 1998) that killed thousands of trees from 1989 to 1990 (Farrar et al., 1995). These post-caldera, moat-filling volcanic and sedimentary rocks are the principal host rocks for gold mineralization at the Long Valley gold deposit (Steininger, 2005) and also comprise the principal geothermal reservoir for fluids tapped by wells feeding the Casa Diablo geothermal power facility.

The Long Valley gold deposit lies along the southeastern flank of the resurgent dome and at the northern end of the north-striking Hilton Creek fault zone (Figure 10.11). The age of mineralization may predate 280-ka moat rhyolite domes that occur nearby as they appear largely unaltered (Steininger, 2005). However, volcaniclastic

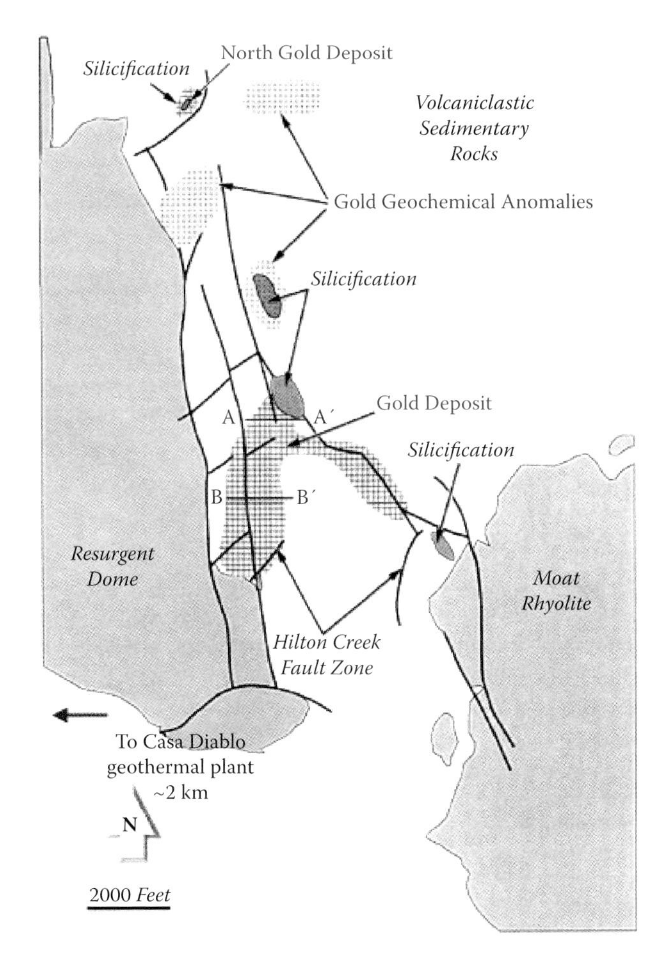

FIGURE 10.11 Generalized geology of the Long Valley gold deposit. Note the structural control of mineralization at the north end of the Hilton Creek fault zone as it splays into numerous branches. (Adapted from Steininger, R.C., in *Volcanic Geology, Volcanology, and Natural Resources of the Long Valley Caldera, California* (Leavitt, E.D. et al., Eds.), Geological Society of Nevada, Reno, 2005.)

debris that would appear shed from these domes are altered and weakly mineralized, suggesting that the mineralization may be younger or about the same age as the moat rhyolite (S. Weiss, pers. comm., 2013). Although adularia is present, its intergrowth with similar fine-grained quartz prevented efforts to separate and date the adularia directly when the deposit was explored (S. Weiss, pers. comm., 2013). In an earlier study that predates discovery of the gold deposit, two samples from a silicified outcrop collected near the north end of the gold deposit yielded uranium–thorium disequilibrium ages of 260,000 and 310,000 years (Sorey et al., 1991). These ages agree well with an earlier estimate of the age of silicified lacustrine sedimentary rocks, a main host of the identified gold deposit, that interfinger with the 280-ka-old Hot Creek moat rhyolite and its breccia equivalents (Bailey et al., 1976). Ages determined from other hot spring deposits within the caldera are consistent with two main periods of hydrothermal activity: one from 130,000 to 300,000 years ago and the current one that began about 40,000 years ago (Sorey et al., 1991). Although hot water was encountered during exploration drilling of the gold deposit, the gold mineralization appears to have formed during the earlier episode of hydrothermal activity, as indicated from geological observations and reported age determinations. Initiation of the latest geothermal system, now developed by the Casa Diablo geothermal power facility, coincides with formation of the Inyo–Mono Craters volcanic belt that extends from the caldera's west moat to 25 km north of the caldera. Thus, available data suggest that the gold mineralization is related to the waning stages of magmatism associated with the Long Valley caldera, whereas the current geothermal system developed possibly by heat from a new pulse of magmatism focused along and extending to the north of the caldera's western moat. No data are available as to whether fluids of the currently exploited geothermal system are gold bearing or whether or not any possible scale in the geothermal piping at Casa Diablo contains gold.

The Casa Diablo geothermal power facility is situated on the southern base of the resurgent dome (Figure 10.12) and has an installed capacity of 40 MWe provided by three binary plants. The first plant was installed in 1984 with a 10-MWe capacity. Two additional plants went online in 1990, each providing 15 MWe. The plants are served by six production and five injection wells. Until 2005, production came from a shallow outflow plume (~200 m deep), covering only about 165 acres (0.7 km^2), which flows to the east; total production flow rates are ~750 kg/s (Suemnicht, 2012). The temperature of the production fluids averages 160° to 170°C (Campbell, 2000). In 2005, deeper (~450 m) production wells were drilled west of the current production field and produced 185°C fluids. A 2.9-km pipeline was built to transport the 185°C fluids at 225 L/s to sustain production at the Casa Diablo power plants (Suemnicht, 2012). As a result, several of the former shallow production wells were closed. The shallow nature of the Casa Diablo geothermal reservoir results, in part, from a large buried and impermeable landslide block (~3 km^2). This block keeps cold recharge water from the caldera's southern margin from invading the shallow, exploited hydrothermal outflow plume (Figure 10.12) (Suemnicht et al., 2006). To avoid adversely cooling the shallow hydrothermal reservoir, injection wells are completed in the deeper (~700 m) and permeable Bishop Tuff that underlies the landslide block (such as well 38-32 in Figure 10.13). Natural recharge of the shallow hydrothermal plume must be sufficient so that reinjection into the reservoir

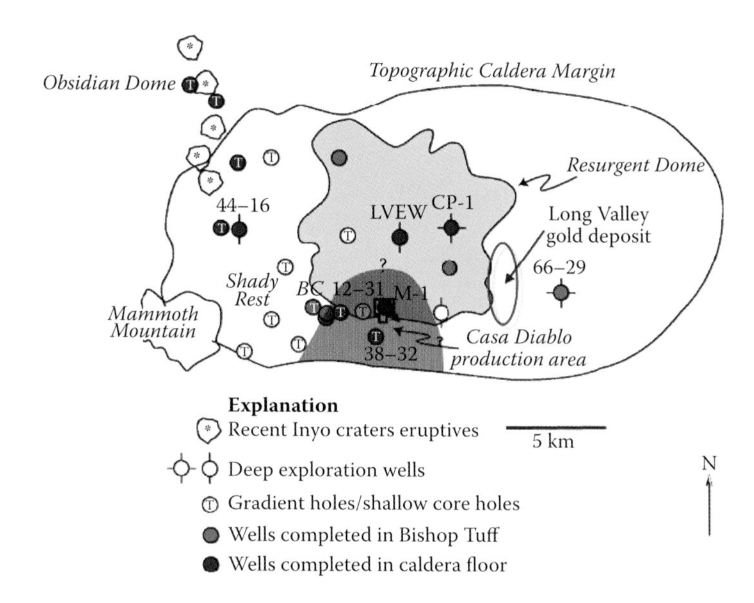

FIGURE 10.12 Map of Long Valley caldera showing principal geologic and geothermal features, including geothermal wells, locations of the Long Valley gold deposit and Casa Diablo production area, and subsurface distribution of landslide block (shown in blue). See Figure 10.13 for cross-section view (looking west) of landslide block. (From Suemnicht, G.A. et al., *Geothermal Resources Council Transactions*, 30, 465–469, 2006.)

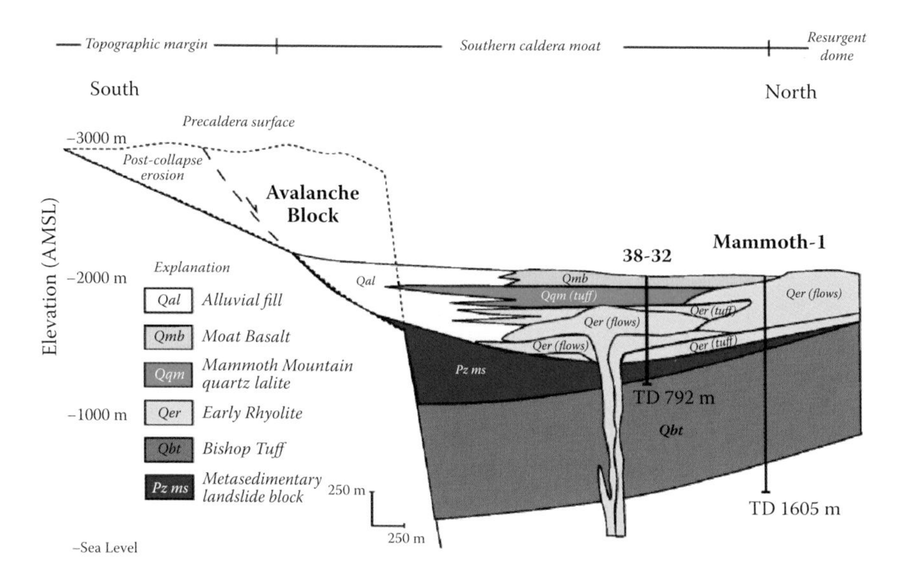

FIGURE 10.13 Structural cross-section looking west across the southern margin of the Long Valley caldera in the area of the Casa Diablo production field. The blue unit is the landslide block that slid off the caldera wall during formation of the caldera and now serves as a hydrologic barrier. See text for discussion. (From Suemnicht, G.A. et al., *Geothermal Resources Council Transactions*, 30, 465–469, 2006.)

of spent fluid, normally done to maintain fluid pressure and flow, is unnecessary. Reinjection below the producing reservoir (below the landslide block) serves to help minimize any possible subsidence and possible cooling of the reservoir.

Ormat, the current operator of the Casa Diablo geothermal facility, recently received federal approval to expand its geothermal complex at Casa Diablo and plans to build an additional 40-MWe binary geothermal plant. This plant will be served by as many as 16 new production and injection wells. This new well field will be located west or up-hydraulic gradient of the current production field near the inferred upflow zone that feeds the tapped, shallow thermal reservoir to the east. Whether drawing fluids from the inferred upflow zone will compromise the flow rates or temperature from the existing production holes in the shallow reservoir is uncertain at this time but will be resolved as the project proceeds.

Geologically, Casa Diablo is in an area of numerous hot springs and local fumaroles situated on the south flank of the resurgent dome of the Long Valley caldera. Hydrothermally altered rocks are exposed just to the north of the existing power plant. Gold mineralization at Long Valley, located about 2 to 3 km to the east of Casa Diablo, probably represents a discrete event within a geothermal system engendered by caldera-related moat rhyolite volcanism. The older geothermal system may have waned until a more recent flux of magmatism resulted in development of the Mono Craters and Inyo Domes exposed to the north and near the northwestern margin of the Long Valley caldera. This younger pulse of magmatism began about 50,000 years ago, with the most recent volcanism occurring only 650 years ago (Hildreth, 2004). As a result, the older geothermal system may have been reinvigorated or a separate geothermal system may have formed; either scenario could provide the heat to support the geothermal system tapped by the Casa Diablo power facility.

Steamboat Springs Geothermal System, Western Nevada

The Steamboat Springs geothermal field, located about 16 km south of downtown Reno, Nevada, hosts six operating power plants (five air-cooled binary plants and one evaporative water-cooled flash plant) having an installed electrical power capacity of about 140 MWe with an average year-round net output of about 90 MWe (electrical net power output ranges from about 75 MWe to about 115 MWe, reflecting seasonal temperature changes and associated cooling efficiencies and variations in parasitic load). The Steamboat Springs geothermal system has been of long-standing interest as a possible modern analog of epithermal precious metal deposits. Don White of the U.S. Geological Survey, who carefully studied this geothermal system and its geology (Thompson and White, 1964; White, 1968), stated that Steamboat Springs "is now viewed as the present-day equivalent of geothermal systems of Tertiary age that formed epithermal gold–silver deposits throughout the Great Basin of the Western United States and elsewhere" (White, 1985). The Steamboat Springs geothermal system is also one of the longest lived, documented geothermal systems in which the oldest silica sinter was deposited about 3 million years ago and another episode of sinter deposition occurred about 1.1 million years ago. Younger episodes of sinter formed during at least the last 0.1 million years and possibly longer.

A magmatic connection to the geothermal activity at Steamboat Springs is supported by (1) the occurrence of four nearby rhyolite domes whose ages range from 1.1 to 3.0 million years, (2) the prolonged duration of hydrothermal activity, and (3) elevated He^3/He^4 ratios of geothermal fluids (R/R_a values from 3.7 to 5.9*) (Torgersen and Jenkins, 1982). The small volume of erupted rhyolite domes is clearly insufficient to provide the necessary heat to sustain the long, albeit perhaps episodic, duration of the Steamboat Springs geothermal system. However, as noted in Chapter 8, the domes could represent apophyses from a large crystallizing silicic magma chamber undergoing multiple cycles of crystallization and magma replenishment and that underlies the area but at considerable depth—too deep to erupt wholesale (such as during caldera-forming eruptions) but shallow enough to sustain a long-lived geothermal system (White, 1985).

Prior to geothermal power development in the mid- to late 1980s, the main sinter terrace contained many flowing and even fountaining hot springs depositing vuggy or porous opaline sinter† (as shown in Figure 9.21 in Chapter 9). Siliceous sulfidic muds coating some of these springs contained as much as 15 ppm Au and 150 ppm Ag. Samples of metastibnite and opal from an erupting spring contained 60 ppm Au and 400 ppm Ag along with significant As, Sb, and Hg (Hudson, 1987; White, 1985). Chalcedonic veins intersected in drill holes contained as much as 1.5 ppm Au and 30 ppm Ag (White, 1985). Analyzed samples of recently deposited opaline sinter, however, contained less than detectable amounts of Au (<100 ppb or <50 ppb Au) (Hudson, 1987; White, 1985). Some analyzed samples of opaline sinter in bore holes, on the other hand, contained as much as 300 ppb Au and 2 ppm Ag, but other samples of opaline sinter had less than detectable concentrations of Au (<100 ppb) (White, 1985). The no to low gold–silver contents in recently deposited opaline sinter compared to the relatively metal-rich sulfidic muds lining some springs suggest that the Au and Ag are not being significantly precipitated at the surface. They were probably precipitated at depth and carried upward as a gelatinous suspension during periods of high flow or turbulence. These sulfidic muds also contain generally the highest concentrations of base metals (Cu, Pb, and Zn), further suggesting precipitation at depth followed by physical upward transport. However, locally weakly auriferous opaline sinter and chalcedonic veins, with few or no base metals, indicate some continued precipitation of gold to high levels in the system.

Solfateric alteration (acid leaching) of both silica sinter and wallrocks occurs where the water table drops below the surface and vapors rich in H_2S oxidize upon ascending to form sulfuric acid ($H_2S + 2O_2 \rightarrow H_2SO_4$). The descending acidic fluid can leach both silica sinter and wallrocks, consisting in many places of basalt, an

* R/R_a is the helium isotope ratio normalized to the same ratio for air, such that $R/R_a = He^3/He^4$ of sample divided by He^3/He^4 of air. Low R/R_a values (generally <1) indicate a strong crustal component reflecting the high U and Th content of crust which decays to produce He^4. A high R/R_a (generally >3) implies a strong mantle or primordial component that can be inherited by mantle-derived magmas and their derivative hydrothermal fluids.

† Opaline sinter consists mainly of opal, which is a solidified, hydrated form of noncrystalline silica. With time, opaline sinter dehydrates to form microscrystalline silica or chalcedonic sinter.

granodiorite; in the latter case, the rock is decomposed to a mixture of clays and residual quartz. Oxidation of H_2S also leads to local deposition of native sulfur ($2H_2S + O_2 \rightarrow 2S + 2H_2O$ or $2H_2S + 2O_2 \rightarrow S + H_2SO_4 + H_2$). Mercury is precipitated as cinnabar in solfaterically altered wallrocks and in sinter but is found only within 15 m of the current topographic surface (White, 1985). This distribution reflects the volatile nature of mercury and its preferential partitioning into the gas phase from either an underground boiling water table or boiling and silica-depositing hot springs at the surface. Studies to date indicate that gold, arsenic, mercury, thallium, and boron are strongly concentrated in near-surface deposits on the order of 10 to 100 times higher than deposits from deeper levels of the explored system. Silver tends to increase with depth but occurs in shallow level deposits as well.

In spite of a likely magmatic heat source for the Steamboat Springs geothermal system, stable isotope analyses of thermal waters indicate a dominant meteoric origin, although as much as 10% of magmatic water could have been present but not detectable isotopically (White et al., 1963). The δO^{18} shows a modest shift of 2 to 3 per mil to heavier O^{18} compared to cold meteoric water of the region, reflecting limited exchange between high O^{18} wallrocks and low O^{18} meteoric water. This limited shift is probably due to high water-to-rock ratios and high rock permeability.

Young Amagmatic Mineralized Geothermal Systems

These are geothermal systems in which the thermal energy comes from elevated heat flow due to mainly crustal thinning from extension bringing hot rocks close to the surface. As such these systems are also termed extensional-related geothermal systems. Young or coeval volcanic and intrusive rocks are lacking in the vicinity of the active geothermal systems and nearby mineralized zones. Whether the metals came from deeper or distal residing magma bodies (Breit et al., 2010; Hunt et al., 2010) or from mainly leaching of wallrocks by deeply circulating hydrothermal fluids remains uncertain in most cases. In the case of gold-depositing active geothermal systems at Kawerau, Broadlands–Ohaaki, and Rotokawa in New Zealand (see Figure 10.1), geological and geochemical studies argue for a magmatic source for the gold (Giggenbach, 1995; Simmons and Brown, 2008). Such a magmatic connection of geothermal fluids appears lacking, however, in the active, but gold-barren geothermal system of Wairakei, located only about 15 km west of Rotokawa.

Case studies of mineralized geothermal systems discussed below are all located in Nevada, which is a type locality for extension- or amagmatic-related geothermal systems. Examples of recently or currently active gold mines lacking coeval igneous rocks but containing associated hot water include Florida Canyon, Wind Mountain, Crowfoot–Lewis (Hycroft), Willard and Colado, and Dixie Comstock (Figure 10.14). Other geothermal power facilities in Nevada, in particular those at Blue Mountain, McGinnis Hills, and the recently commissioned Don A. Campbell (formerly known as Wild Rose) (Figure 10.14), started out as gold exploration plays which evolved into geothermal discoveries. These relationships underscore the fact

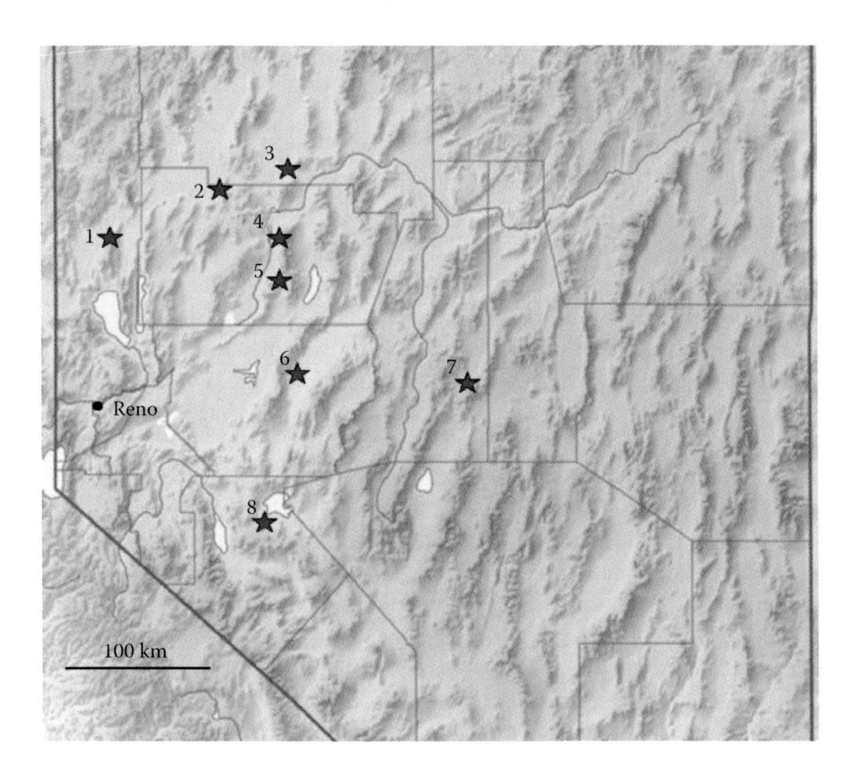

FIGURE 10.14 Map of north-central Nevada showing locations of geothermal power plants and young precious metal deposits, denoted by red stars (see text). (1) Wind Mountain precious metal deposit (San Emidio geothermal power plant located a few kilometers to the south), (2) Hycroft precious metal deposit, (3) Blue Mountain geothermal electric power plant, (4) Florida Canyon mine, (5) Willard and Colado precious metal deposits, (6) Dixie Comstock precious metal deposit, (7) McGinness Hills geothermal power plant, and (8) Don A. Campbell geothermal power plant.

that active geothermal systems are indeed the modern analogs of precious metal epithermal ore deposits (White, 1981), although most fluids from active geothermal systems contain sub-ore grade (or much sub-ore grade) concentrations of gold and silver. As pointed out by Coolbaugh et al. (2011), young ($\leq$7 million years, with many less than 2 million years) mineralized geothermal systems in the Great Basin are typified by low grades but high tonnages. The significant oxidized zones makes the contained gold amenable to extraction by economically favorable heap leaching techniques,* thus avoiding more costly milling and flotation procedures for recovering mined gold.

* *Heap leaching* refers to placing crushed ore onto a lined pad and spraying a cyanide solution that percolates through the crushed ore and extracts the gold. The gold-enriched cyanide solution collects atop the lining at the bottom of the pad and is sent to a processing plant that recovers the gold from the solution.

Florida Canyon Gold Deposit and Humboldt House Geothermal System

The Florida Canyon gold deposit, located in northwest Nevada, about 250 km northeast of Reno, Nevada, is hosted by metamorphosed sedimentary rocks of early Mesozoic age (250 to 200 million years) and contains in excess of 7.6 million ounces of gold. It has been mined more or less continuously since 1986. Between 1986 and 2008 about 180 million tons of ore grading 0.018 ounce per ton gold Au (or 0.62 grams/tonne or 620 parts per billion Au) were mined, which is about half of the contained resource. Detailed age determinations indicate that the geothermal system that produced the gold deposit has been active, at least episodically, for the last approximately 5 million years. The gold mine lies adjacent to the active and locally mineralized Humboldt House geothermal system. Radiometric dating of the mineral adularia, a common hydrothermal mineral associated with the gold mineralization, yielded ages ranging between 4.6 and 5.1 million years (Fifarek et al., 2011). This mineralization was overprinted by steam-heated, acid-sulfate alteration, which formed between 1.8 and 2.2 million years ago based on radiometric dating of alunite associated with the acid-sulfate alteration. These ages of mineralization and hydrothermal alteration appear to coincide with episodes of major movement along faults that opened fluid pathways. The development of the overprinting steam-heated alteration reflects uplift of the deposit, again along faults, and a probable change in the paleoclimate to more arid conditions, resulting in a corresponding lowering of the water table. The resulting decrease in hydrostatic pressure associated with the falling water table induced boiling, at least at times, leading to the overlying and overprinting acid-sulfate alteration. By about 0.9 to 0.4 million years ago, the Florida Canyon deposit was being weathered and further oxidized, and the locus of faulting and hydrothermal activity shifted westward to the current and adjacent Humboldt House geothermal system (Figure 10.15).

The Humboldt House geothermal system has reservoir temperatures of 219° to 252°C and gold-bearing quartz–adularia veins similar to those in the Florida Canyon mine—locally cut partially silicified basin-fill gravels at depth. Siliceous sinter and travertine are exposed at or near the surface (Breit et al., 2011; Coolbaugh et al., 2005). Igneous or volcanic rocks having comparable ages to the hydrothermal alteration and mineralization are not exposed nor have they been encountered in drilling at Florida Canyon or Humboldt House. This is consistent with both being classified as amagmatic systems. Although a magnetic anomaly adjacent to the Florida Canyon deposit could reflect a buried body of igneous rock, the age of the suspected igneous body is not known. It could be easily much older and thereby unrelated as a possible heat source for either the young gold mineralization or current geothermal activity at Humboldt House. Indeed, nearby local mafic and hornblende-rich dikes and small intrusions of probable Jurassic age may represent apophyses of a larger, more deeply buried intrusion that could be the source of the magnetic anomaly (D. John, pers. comm., 2015).

In 2013, a small geothermal power plant (75 kWe), using the organic Rankine cycle,* was installed at Florida Canyon. The plant was built by ElectraTherm, a small renewable energy company headquartered in Reno, Nevada. The power unit

* This is the same process as is used in binary geothermal plants in which the geothermal fluid boils a working fluid, usually a hydrocarbon, having a low boiling point. The produced steam spins a turbine that feeds an electrical generator.

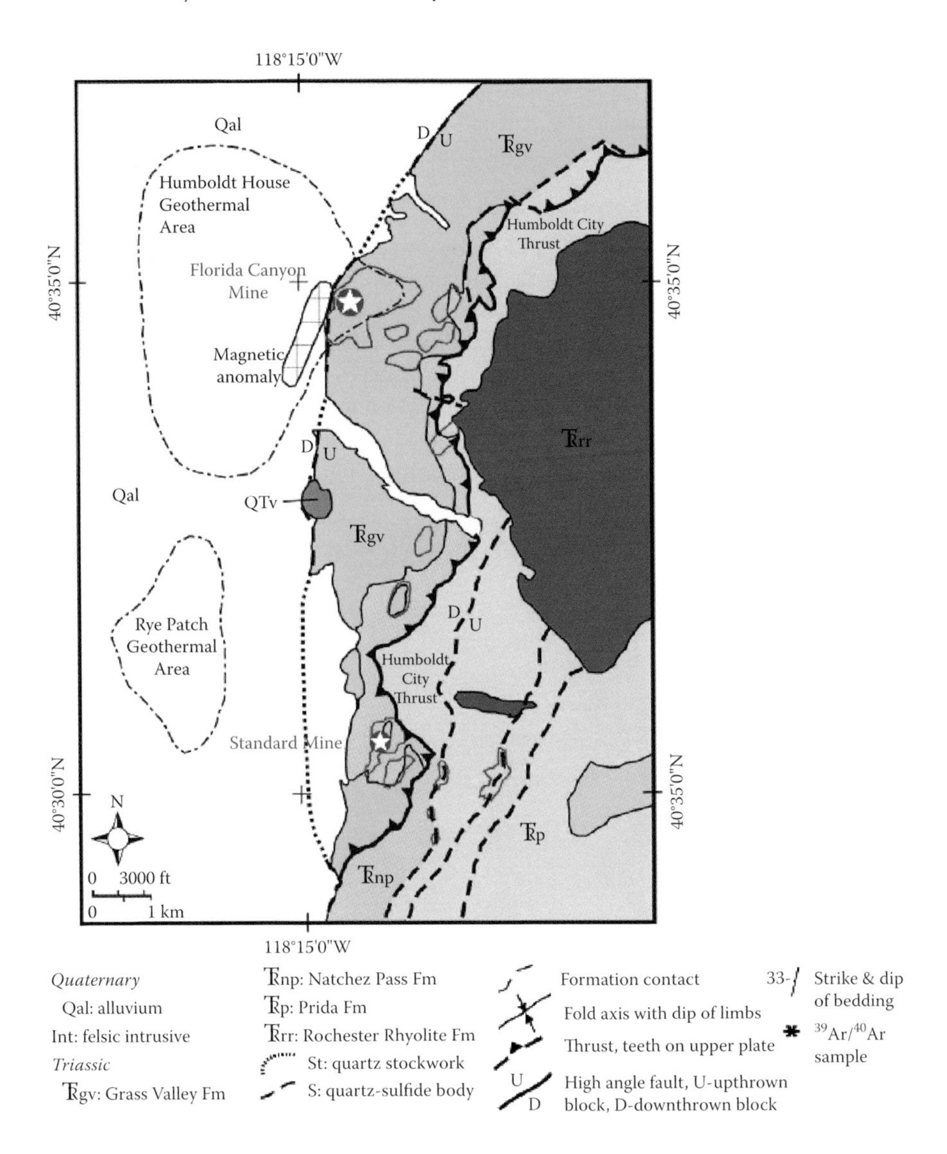

FIGURE 10.15 Geologic setting of the Florida Canyon mine and locations of adjacent or nearby active geothermal systems of Humboldt House and Rye Patch. (Adapted from Fifarek, R.H. et al., in *Great Basin Evolution and Metallogeny*, Steininger, R. and Pennell, B., Eds., DEStech Publications, Lancaster, PA, 2011, pp. 861–880. Copyright © Geological Society of Nevada.)

was to use about 150 gallons per minute of 105° to 110°C geothermal well fluid. Unfortunately, the well never produced and the facility is currently idle (J. Barta, pers. comm., 2016). Another side benefit of the power plant was to cool fluids so they could be used immediately in the heap leaching of mined ore. Currently, pumped water from the wells goes to cooling ponds before being used in heap leaching. If the adjacent Humboldt House geothermal resource, with its indicated reservoir temperatures of >200°C, proves to be viable, then geothermal power at Florida Canyon could still be a viable option in the future. If so, any possible excess or unused power could be sold to the grid and provide added revenue for the mine and possibly beyond after the deposit is mined out.

Hycroft Mine and Geothermal System

The Hycroft or Crowfoot–Lewis gold–silver mine is located in northwest Nevada about 45 km northwest of the Florida Canyon mine (Figure 10.16). The mine produced about 1.1 million ounces of gold from 1987 to 1998. The mine reopened in 2008 and through 2013 had produced an additional 500,000 ounces of Au and about 2.5 million ounces of Ag. Proven and probable reserves based on $800 per ounce Au as of the end of 2012 were 1.1 billion tons at 0.011 ounce per ton (opt) Au (or 370 ppb Au) and 0.46 opt Ag (13,800 ppb Ag) for a minable content of 11.8 million ounces of Au (Allied Nevada Gold Corp, 2013). Although these figures along with the low strip

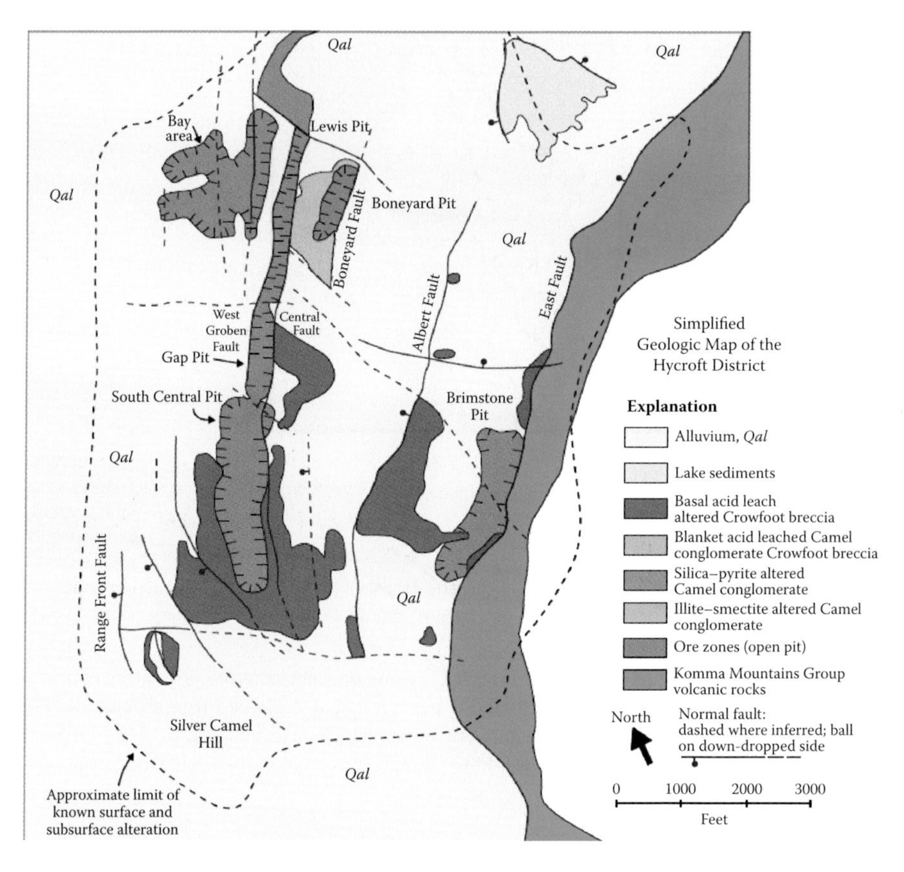

FIGURE 10.16 Simplified geologic map of the Hycroft mine. The Bay Area pit at the north end of the mine develops gold mineralization, which is in part hosted by siliceous hot spring sinter deposits marking the original paleosurface at the time of mineralization (3.8 to 3.9 million years ago). (Adapted from Ebert, S.W. and Rye, R.O., *Economic Geology*, 92(5), 578–600, 1997.)

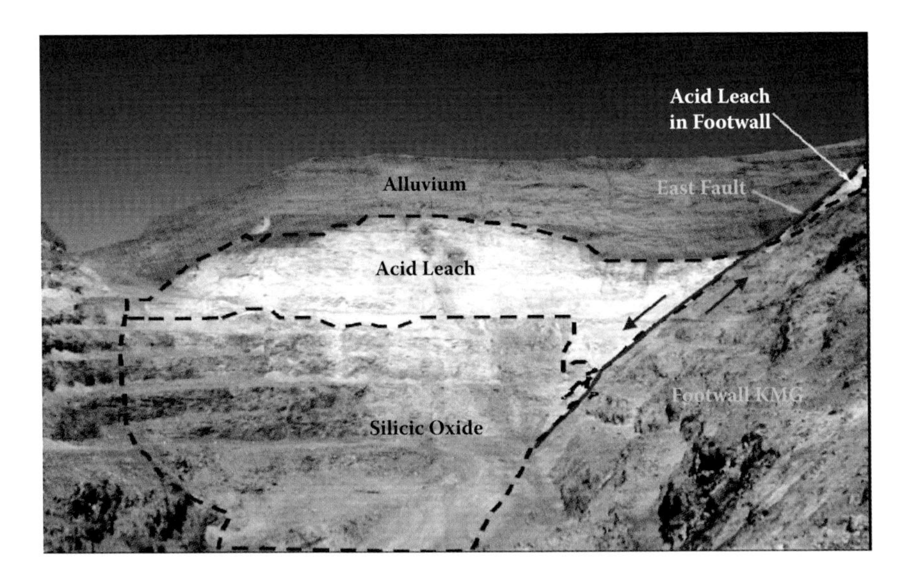

FIGURE 10.17 View looking north of the Brimstone open pit in the Hycroft mine. Note how the steam-heated, acid-leached zone overlies and overprints underlying Au-mineralized silicic oxide zone. The acid-leached and silicic oxide zones are truncated along the East fault, with a portion of the offset acid-leached zone visible near the top of the footwall block. Bench heights are about 20 feet. (From Allied Nevada Gold Corp., *Technical Report—Hycroft Mine, Winnemucca, Nevada, USA*, Allied Nevada Gold Corp., Reno, 2013.)

ratio of 1.15:1 are encouraging, the operation has suffered from poor metallurgical recoveries, which, combined with the current depressed price of gold and silver, have led to suspension of the mining operation.

The Hycroft gold–silver deposit is of the low sulfidation type characterized by adularia and sericite alteration, and, similar to Florida Canyon, the early Au–Ag-bearing low-sulfidation alteration is overprinted by younger steam-heated acid-sulfate mineralization that has locally remobilized the Au and Ag (Figure 10.17). The youthful age of mineralization is geologically indicated by a well-preserved paleosurface that includes abundant opaline (amorphous silica) sinter; high-level, steam-heated acid-sulfate altered rocks; and bedded hydrothermal eruption breccias (coarse angular debris formed during explosive venting of overpressured hydrothermal fluids). Other paleosurface indicators include the presence of native sulfur deposits that formed around fumaroles and the deposits of cinnabar—a mercury sulfide mineral whose elemental constituents of Hg and S partition preferentially into the steam or vapor phase above a boiling liquid-dominated reservoir and then precipitate at or near the surface as cinnabar. Radiometric ages on adularia in gold ore range from 3.8 to 3.9 million years old, whereas those on alunite, associated with the younger overprinting steam-heated acid-sulfate alteration, range from 1.2 to 2.1 million years old (Ebert and Rye, 1997). Additional weathering and oxidation occurred around 0.7 million years ago based on a radiometric age from jarosite (a hydrated potassium iron sulfate), a common weathering product of pyrite (an ancillary mineral of iron sulfide that is typically associated with Au–Ag mineralization). Although the bulk of the mineralization is hosted in relatively young volcanic and sedimentary rocks deposited between about 5 and 20 million years ago, unlike the much older metasedimentary rocks hosting gold mineralization at Florida Canyon, igneous rocks of comparable age to the mineralization and alteration are absent, consistent with a largely amagmatic heritage of the fluids.

Published information on the nature of geothermal fluids within the mine area is rather limited. Based on about 40 or so water wells drilled in the mine area, the water table ranges from 800 to 1300 feet below the surface, and water temperatures in many of the wells range between 100°F and >150°F. Some wells also yielded significant concentrations of H_2S (D. Hudson, pers. comm., 2015), possibly reflecting the reaction of hot water with iron sulfide to produce H_2S, native S, and iron hydroxide ($2H_2O + FeS_2 \rightarrow H_2S + S + Fe(OH)_2$). Chemical analyses of those fluids are not published, nor is it clear whether a directional trend for increasing temperature exists as it does at Florida Canyon, where fluid temperatures increase northwestward of the mine toward the Humboldt House geothermal system. Nonetheless, the modest geothermal temperatures encountered in wells at Hycroft indicate that the geothermal system is still active and has been so, at least intermittently, for perhaps as much as 4 million years.

San Emidio Geothermal System and the Wind Mountain Mine

This geothermal system and adjacent mine are located about 80 km southwest of the Hycroft Mine and about 100 km north of Reno, Nevada. Original mine reserves (1988) at Wind Mountain consisted of 13.7 million metric tons (tonnes) grading 0.72 g/tonne Au (0.022 opt Au or 730 ppb Au) and 11.4 g/tonne Ag. The deposit was mined from 1989 to 1999 and produced almost 300,000 ounces of gold and over 1.7 million ounces of silver. The mine is currently idle, but recent exploration work indicates a resource of about 42 million tons grading 0.011 opt Au and 0.26 opt Ag within the previously mined open pits (about 462,000 contained ounces of gold).

The mine lies in the footwall or upthrown block of the San Emidio and northern Lake Range faults, whereas the producing San Emidio geothermal system lies in the downthrown hanging wall block about 7 km south of the mine (Figure 10.18). The current San Emidio geothermal system lies in a right-lateral step region of a major northerly striking, west-dipping fault zone that bounds the San Emidio valley on the east (Rhodes et al., 2010). A hot spring style of mineralization at Wind Mountain is indicated by deposits of native sulfur and cinnabar, bedded hydrothermal breccia deposits, silicified plant debris in sinter, and travertine. Although no direct radiometric age determination of mineralization has been done at Wind Mountain, its age is believed to be <5 million years old based on the presence of silicified plant reeds in sinter (preservation of the original paleosurface).

The San Emidio geothermal facility, located about 7 km south of the Wind Mountain mine, was recently renovated and upgraded to an installed capacity of 11.8 MWe. Reservoir temperatures calculated from thermal fluid chemistry average about 187°C (Coolbaugh et al., 2005); the temperatures measured from two wells were 144°C and 152°C, and production pressures were 132 psi and 140 psi, respectively (Breit et al., 2011). The geothermal system also supports a vegetable drying facility.

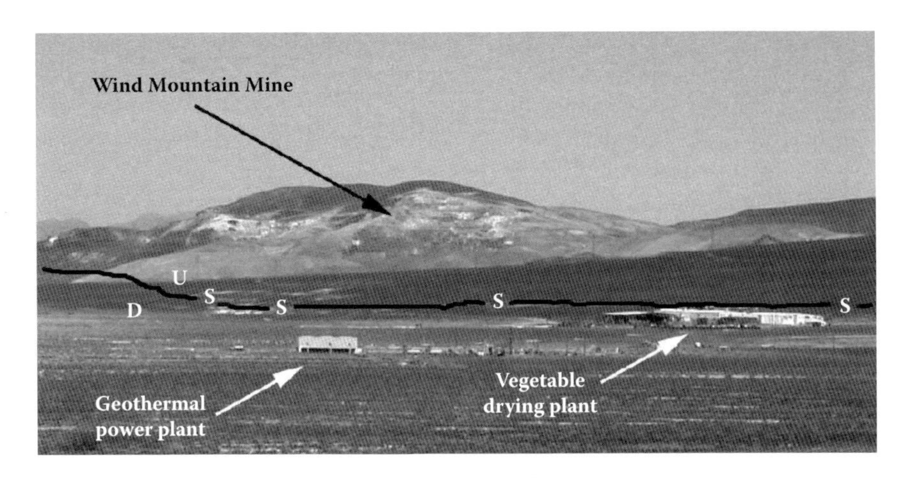

FIGURE 10.18 View looking northeast across San Emidio valley at the geothermal power plant and vegetable drying facility in the near distance and Wind Mountain mine in the far distance (about 7 km north of the geothermal plant). The geothermal power plant shown is the older 4.8-MWe unit prior to its recent upgrade to 11.8 MWe. The bold black line running across the image traces the San Emidio fault, with the "D" on the downdropped or hanging wall side and the "U" on the upthrown footwall side of the fault. The letter "S" marks areas of solfataric alteration, expressed by native sulfur, gypsum, warm to steaming ground, and local cinnabar, along the San Emidio fault. (Adapted from Coolbaugh, M.F. et al., in *Geological Society of Nevada Symposium 2005: Window to the World Proceedings*, Rhoden, H.N. et al., Eds., Geological Society of Nevada, Reno, 2005, pp. 1063–1082.)

The active San Emidio geothermal system in the hanging wall and the geologically young Wind Mountain epithermal precious metal deposit in the footwall of the San Emidio and northern Lake Range faults argue for a long-lived (at least episodically) geothermal system. Both owe their existence to regions of enhanced permeability due to complex faulting patterns and modest dilation associated with a right-step in a northerly trending fault zone (Rhodes et al., 2010).

SUMMARY

Other examples of magmatically related young mineral deposits include the high-temperature Hudson Ranch geothermal plant in the Salton Sea geothermal system whose brines are sufficiently rich in manganese, zinc, and lithium salts that a separate mineral processing facility is planned to recover and sell these metals—a form of liquid mining of metals. Also within the Great Basin of the western United States, Coolbaugh et al. (2005, 2011) described other magmatic and amagmatic young mineral deposits associated with young or still active geothermal systems, including Como in Lyon County, Nevada, and Quartz Mountain in Oregon, both of which are magmatically heated, and Colado (Willard Mine) and Blue Mountain, both seemingly amagmatic. Altogether, about 8.8 million ounces of produced gold and inventory are associated with active hydrothermal systems and nearby gold deposits in northwestern Nevada (Breit et al., 2011; Coolbaugh et al., 2011). In general, the young magmatically heated mineral deposits contain overall more gold at higher grades than the amagmatic young mineral deposits. Furthermore, geothermal systems can be relatively long lived (as much as 5 million years), but significant mineral deposition appears to occur only during discrete episodes of disequilibrium conditions, such as rapid boiling and cooling, leading to the deposition of metals in a localized volume of rock to make an ore deposit. During the more typical quiescent periods of a geothermal system, metals dissolved in geothermal fluids precipitate in a more limited and dispersed manner. In other words, hydrothermal perturbations appear to be an important factor in whether or not a mineral deposit forms in a geothermal system.

SUGGESTED PROBLEMS

1. Imagine you are an exploration geologist and are drilling a siliceous sinter deposit looking for gold and silver mineralization. During the course of drilling you encounter some low-grade mineralization but also hot water (~90°C). (a) What does this tell you about the possible age of the siliceous sinter, and (b) how might your thinking about the intended target change as a result?
2. Do you think The Geysers geothermal system could be actively forming a mineral deposit? Why or why not? If so, what kinds of deposits might form there?
3. If active geothermal systems are the modern analogs of epithermal precious metal deposits, then why do you think most active geothermal systems that have been explored or developed, although they may contain some anomalous metal concentrations, are not forming potentially minable gold–silver deposits?

REFERENCES AND RECOMMENDED READING

Allied Nevada Gold Corp. (2013). *Technical Report—Hycroft Mine, Winnemucca, Nevada, USA*. Reno: Allied Nevada Gold Corp. (http://www.alliednevada.com/wp-content/uploads/130306-tech-report.pdf).

Bailey, R.A. (1989). *Geologic Map of the Long Valley Caldera, Mono-Inyo Craters Volcanic Chain, and Vicinity, Eastern California*, Map I-1933, scale 1:62,500. Reston, VA: U.S. Geological Survey (http://ngmdb.usgs.gov/Prodesc/proddesc_15.htm).

Bailey, R.A., Dalrymple, G.B. and Lanphere, M.A. (1976). Volcanism, structure, and geochronology of Long Valley caldera, Mono County, California. *Journal of Geophysical Research*, 81(5): 725–744.

Bignall, G. and Carey, B.S. (2011). A deep (5 km) geothermal science drilling project for the Taupo Volcanic Zone: who wants in? *Proceedings of the New Zealand Geothermal Workshop*, 33: 5.

Breit, G.N., Hunt, A.G., Wolf, R.E., Koenig, A.E., Fifarek, R.H., and Coolbaugh, M.F. (2010). Are modern geothermal waters in northwest Nevada forming epithermal gold deposits? In: *Great Basin Evolution and Metallogeny* (Steininger, R. and Pennell, B., Eds.), pp. 833–844. Lancaster, PA: DEStech Publications. Copyright © Geological Society of Nevada.

Browne, P.R.L. (1969). Sulfide mineralization in a Broadlands geothermal drill hole, Taupo Volcanic Zone, New Zealand. *Economic Geology*, 64(2): 156–159.

Campbell, R. (2000). Mammoth geothermal: a development history. *Geothermal Resources Council Bulletin*, 29(3): 91–95.

Carman, G.D. (2003). Geology, mineralization, and hydrothermal evolution of the Ladolam gold deposit, Lihir Island, Papua New Guinea. *Economic Geologists Special Publication*, 10: 247–284.

Coolbaugh, M.F., Arehart, G.B., Faulds, J.E., and Garside, L.J. (2005). Geothermal systems in the Great Basin, western United States: modern analogues to the roles of magmatism, structure, and regional tectonics in the formation of gold deposits. In: *Geological Society of Nevada Symposium 2005: Window to the World Proceedings* (Rhoden, H.N., Steininger, R.C., and Vikre, P.G., Eds.), pp. 1063–1082. Reno: Geological Society of Nevada.

Coolbaugh, M.F., Vikre, P.G., and Faulds, J.E. (2011). Young (<7 Ma) gold deposits and active geothermal systems of the Great Basin: enigmas, questions, and exploration potential. In: *Great Basin Evolution and Metallogeny* (Steininger, R. and Pennell, B., Eds.), pp. 845–860. Lancaster, PA: DEStech Publications. Copyright © Geological Society of Nevada.

Ebert, S.W. and Rye, R.O. (1997). Secondary precious metal enrichment by steam-heated fluids in the Crofoot–Lewis hot spring gold-silver deposit and relation to paleoclimate. *Economic Geology*, 92(5): 578–600.

Farrar, C.D., Sorey, M.L., Evans, W.C. et al. (1995). Forest-killing diffuse CO_2 emission at Mammoth Mountain as a sign of magnatic unrest. *Nature*, 376(6542): 675–678.

Faure, K., Matsuhisa, Y., Metsugi, H., Mizota, C., and Hayashi, S. (2002). The Hishikari Au–Ag epithermal deposit, Japan: oxygen and hydrogen isotope evidence in determining the source of paleohydrothermal fluids. *Bulletin of the Society of Economic Geologists*, 97(3): 481–498.

Fifarek, R.H., Samal, A.R., and Miggins, D.P. (2011). Genetic implications of mineralization and alteration ages at the Florida Canyon epithermal Au–Ag deposit, Nevada. In: *Great Basin Evolution and Metallogeny* (Steininger, R. and Pennell, B., Eds.), pp. 861–880. Lancaster, PA: DEStech Publications. Copyright © Geological Society of Nevada.

Fournier, R.O. (1985). The behavior of silica in hydrothermal solutions. *Reviews in Economic Geology*, 2: 45–61.

Furuya, S., Aoki, M., Gotoh, H., and Takenaka, T. (2000). Takigami geothermal system, northeastern Kyushu, Japan. *Geothermics*, 29: 191–211.

Gemmell, J.B., Sharpe, R., Jonasson, I.R., and Herzig, P.M. (2004). Sulfur isotope evidence for magmatic contributions to submarine and subaerial gold mineralization: Conical Seamount and the Ladolam gold deposit, Papua New Guinea. *Economic Geology*, 99(8): 1711–1725.

Giggenbach, W.F. (1992). Magma degassing and mineral deposition in hydrothermal systems along convergent plate boundaries. *Economic Geology*, 87(7): 1927–1944.

Giggenbach, W.F. (1995). Variations in the chemical and isotopic composition of fluids discharged from the Taupo Volcanic Zone, New Zealand. *Journal of Volcanology and Geothermal Research*, 68(1–3): 89–116.

Gustafson, L.B., Vidal, C.E., Pinto, R., and Noble, D.C. (2004). Porphyry–epithermal transition, Cajamarca region, northern Peru. *Economic Geologists Special Publication*, 11: 279–299.

Hayashi, S., Nakao, S., Yokoyama, T., and Izawa, E. (1997). Concentration of gold in the current thermal water from the Hishikari gold deposit in Kyushu, Japan. *Shigen Chishitsu*, 47(4): 231–233.

Hedenquist, J.W. (1986). Geothermal systems in the Taupo Volcanic Zone: their characteristics and relation to volcanism and mineralisation. *Bulletin of the Royal Society of New Zealand*, 23: 134–168.

Hedenquist, J.W. and Henley, R.W. (1985). Hydrothermal eruptions in the Waiotapu geothermal system, New Zealand: their origin, associated breccias, and relation to precious metal mineralization. *Economic Geology*, 80(6): 1640–1668.

Hedenquist, J.W. and Lowenstern, J.B. (1994). The role of magmas in the formation of hydrothermal ore deposits. *Nature (London)*, 370(6490): 519–527.

Hedenquist, J.W., Simmons, S.F., Giggenbach, W.F., and Eldridge, C.S. (1993). White Island, New Zealand, volcanic–hydrothermal system represents the geochemical environment of high-sulfidation Cu and Au ore deposition. *Geology*, 21(8): 731–734.

Heinrich, C.A., Driesner, T., Stefansson, A., and Seward, T.M. (2004). Magmatic vapor contraction and the transport of gold from the porphyry environment to epithermal ore deposits. *Geology*, 32(9): 761–764.

Henley, R.W. and Ellis, A.J. (1983). Geothermal systems ancient and modern: a geochemical review. *Earth-Science Reviews*, 19(1): 1–50.

Hildreth, W. (2004). Volcanological perspectives on Long Valley, Mammoth Mountain, and Mono Craters: several contiguous but discrete systems. *Journal of Volcanology and Geothermal Research*, 136(3): 169–198.

Hill, D.P, Langbein, J.O., and Prejean, S. (2003). Relations between seismicity and deformation during unrest in Long Valley caldera, California, from 1995 through 1999. *Journal of Volcanology and Geothermal Research*, 127(3): 175–193.

Hudson, D.M. (1987). Steamboat Springs geothermal area, Washoe County, Nevada. In: *Bulk Mineable Precious Metal Deposits of the Western United States* (Johnson, J.L., Ed.), pp. 408–412. Reno: Geological Society of Nevada.

Hulen, J.B., Heizler, M.T., Stimac, J.A., Moore, J.N., and Quick, J.C. (1997). New constraints on the timing of magmatism, volcanism, and the onset of vapor-dominated at The Geysers steam field, California. In: *Proceedings of the 22nd Workshop on Geothermal Reservoir Engineering*, Stanford, CA, January 27–29 (http://www.geothermal-energy.org/pdf/IGAstandard/SGW/1997/Hulen.pdf).

Hunt, A.G., Landis, G.P., Breit, G.N., Wolf, R., Bergfeld, D., and Rytuba, J.J. (2010). Identifying magmatic versus amagmatic sources for modern geothermal system associated with epithermal mineralization using noble gas geochemistry. In: *Great Basin Evolution and Metallogeny* (Steininger, R. and Pennell, B., Eds.), pp. 899–908. Lancaster, PA: DEStech Publications. Copyright © Geological Society of Nevada.

Izawa, E., Urashima, Y., Ibaraki, K., Suzuki, R., Yokoyama, T. et al. (1990). Hishikari gold deposit: high-grade epithermal veins in Quaternary volcanics of southern Kyushu, Japan. *Journal of Geochemical Exploration*, 36(1–3): 1–56.

John, D.A. (2001). Miocene and early Pliocene epithermal gold–silver deposits in the northern Great Basin, western USA. *Economic Geology*, 96: 1827–1853.

Krupp, R.E. and Seward, T.M. (1987). The Rotokawa geothermal system, New Zealand; an active epithermal gold-depositing environment. *Economic Geology*, 82(5): 1109–1129.

Lehrman, N.J. (1986). The McLaughlin Mine, Napa and Yolo counties, California. *Nevada Bureau of Mines and Geology Report*, 41: 85–89.

Lindgren, W. (1915). *Geology and Mineral Deposits of the National Mining District, Nevada*, USGS Bulletin 601. Reston, VA: U.S. Geological Survey.

Lindgren, W. (1933). *Mineral Deposits*. New York: McGraw-Hill.

Melaku, M. (2005). Geothermal development at Lihir—an overview. In: *Proceedings World Geothermal Congress, 2005*, Antalya, Turkey, April 24–29 (http://www.geothermal-energy.org/pdf/IGAstandard/WGC/2005/1343.pdf).

Miyoshi, M., Sumino, H., Miyabuchi, Y. et al. (2012). K–Ar ages determined for post-caldera volcanic products from Aso Volcano, central Kyushu, Japan. *Journal of Volcanology and Geothermal Research*, 229–230: 64–73.

Morishita, Y. and Takeno, N. (2010). Nature of the ore-forming fluid at the Quaternary Noya gold deposit in Kyushu, Japan. *Resource Geology*, 60(4): 359–376.

Moyle, A.L., Doyle, B.J., Hoogvliet, H., and Ware, A.R. (1990). Ladolam gold deposit, Lihir Island. In: *Geology of the Mineral Deposits of Australia and Papua New Guinea*, Vol. 2 (Hughes, F.E., Ed.), pp. 1793–1805. Victoria, Australia: Australasian Institute of Mining and Metallurgy.

Peabody, C.E. and Einaudi, M.T. (1992). Origin of petroleum and mercury in the Culver–Baer cinnabar deposit, Mayacmas District, California. *Economic Geology*, 87(4): 1078–1103.

Pearcy, E.C. and Petersen, U. (1990). Mineralogy, geochemistry and alteration of the Cherry Hill, California, hot-spring gold deposit. *Journal of Geochemical Exploration*, 36(1–3): 143–169.

Peters, E.K. (1991). Gold-bearing hot spring systems of the northern Coast Ranges, California. *Economic Geology*, 86(7): 1519–1528.

Pichler, T., Giggenbach, W.F., McInnes, B.I.A., Buhl, D., and Duck, B. (1999). Fe sulfide formation due to seawater–gas–sediment interaction in a shallow-water hydrothermal system at Lihir Island, Papua New Guinea. *Economic Geology*, 94(2): 281–288.

Pope, J.G., Brown, K.L., and McConchie, D.M. (2005). Gold concentrations in springs at Waiotapu, New Zealand: implications for precious metal deposition in geothermal systems. *Economic Geology*, 100(4): 677–687.

Reed, M.H. and Spycher, N.F. (1985). Boiling, cooling, and oxidation in epithermal systems; a numerical modeling approach. *Reviews in Economic Geology*, 2: 249–272.

Rhodes, G.T., Faulds, J.E., and Teplow, W. (2010). Structural controls of the San Emidio Desert Geothermal Field, Northwestern Nevada. *Geothermal Resources Council Transactions*, 34(2): 753–756.

Rowland, J.V. and Sibson, R.H. (2004). Structural controls on hydrothermal flow in a segmented rift system, Taupo Volcanic Zone, New Zealand. *Geofluids*, 4(4): 259–283.

Rowland, J.V. and Simmons, S.F. (2012). Hydrologic, magmatic, and tectonic controls on hydrothermal flow, Taupo Volcanic Zone, New Zealand: implications for the formation of epithermal vein deposits. *Economic Geology*, 107(3): 427–457.

Sanematsu, K., Watanabe, K., Duncan, R.A., and Izawa, E. (2006). The history of vein formation determined by $^{40}Ar/^{39}Ar$ dating of adularia in the Hosen-1 vein at the Hishikari epithermal gold deposit, Japan. *Economic Geology*, 101(3): 685–698.

Sherlock, R.L. (2005). The relationship between the McLaughlin gold–mercury deposit and active hydrothermal systems in the Geysers–Clear Lake area, northern Coast Ranges, California. *Ore Geology Reviews*, 26(3–4): 349–382.

Sherlock, R.L., Tosdal, R.M., Lehrman, N.J. et al. (1995). Origin of the McLaughlin Mine sheeted vein complex: metal zoning, fluid inclusion, and isotopic evidence. *Economic Geology*, 90(8): 2156–2181.

Sillitoe, R.H. (2010). Porphyry copper systems. *Economic Geology*, 105(1): 3–41.

Simmons, S.F. and Brown, K.L. (2006). Gold in magmatic hydrothermal solutions and the rapid formation of a giant ore deposit. *Science*, 314(5797): 288–291.

Simmons, S.F. and Brown, K.L. (2007). The flux of gold and related metals through a volcanic arc, Taupo volcanic zone, New Zealand. *Geology*, 35(12): 1099–1102.

Simmons, S.F. and Brown, K.L. (2008). Precious metals in modern hydrothermal solutions and implications for the formation of epithermal ore deposits. *SEG Newsletter*, 72(1): 9–12.

Simmons, S.F. and Browne, P.R.L. (2000). Hydrothermal minerals and precious metals in the Broadlands–Ohaaki geothermal system: implications for understanding low-sulfidation epithermal environments. *Economic Geology*, 95(5): 971–999.

Simmons, S.F. and Christenson, B.W. (1994). Origins of calcite in a boiling geothermal system. *American Journal of Science*, 294(3): 361–400.

Sorey, M.L., Suemnicht, G.A., Sturchio, N.C., and Nordquist, G.A. (1991). New evidence on the hydrothermal system in Long Valley caldera, California, from wells, fluid sampling, electrical geophysics, and age determinations of hot-spring deposits. *Journal of Volcanology and Geothermal Research*, 48(3–4): 229–263.

Sorey, M.L., Evans, W.C., Kennedy, B.M., Farrar, C.D., Hainsworth, L.J., and Hausback, B. (1998). Carbon dioxide and helium emissions from a reservoir of magmatic gas beneath Mammoth Mountain, California. *Journal of Geophysical Research*, 103(B7): 15,303–15,323.

Steininger, R.C. (2005). Geology of the Long Valley gold deposit, Mono County, California. In: *Volcanic Geology, Volcanology, and Natural Resources of the Long Valley Caldera, California* (Leavitt, E.D. et al., Eds.), pp. 189–196. Reno: Geological Society of Nevada.

Suemnicht, G.A. (2012). Long Valley caldera geothermal and magmatic systems. In: *Long Valley Caldera Field Trip Guide*, NGA Long Valley Field Trip, July 5–7. Reno, NV: National Geothermal Academy, Great Basin Center for Geothermal Energy.

Suemnicht, G.A., Sorey, M.L., Moore, J.N., and Sullivan, R. (2006). The shallow hydrothermal system of the Long Valley caldera, California. *Geothermal Resources Council Transactions*, 30: 465–469.

Thompson, G.A. and White, D.E. (1964). *Regional Geology of the Steamboat Springs Area, Washoe County, Nevada*, Professional Paper 458-A. Reston, VA: U.S. Geological Survey.

Tohma, Y., Imai, A., Sanematsu, K. et al. (2010). Characteristics and mineralization age of the Fukusen No. 1 Vein, Hishikari epithermal gold deposits, southern Kyushu, Japan. *Resource Geology*, 60(4): 348–358.

Torgersen, T. and Jenkins, W.J. (1982). Helium isotopes in geothermal systems: Iceland, The Geysers, Raft River and Steamboat Springs. *Geochimica et Cosmochimica Acta*, 46(5): 739–748.

UNFCC. (2006). *Project 0279: Lihir Geothermal Power Project*. Geneva: United Nations Framework Convention on Climate Change (http://cdm.unfccc.int/Projects/DB/DNV-CUK1143246000.13).

Vikre, P.G. (1994). Gold mineralization and fault evolution at the Dixie Comstock Mine, Churchill County, Nevada. *Economic Geology*, 89(4): 707–719.

Weissberg, B.G. (1969). Gold–silver ore-grade precipitates from New Zealand thermal waters. *Economic Geology*, 64(1): 95–108.

White, D.E. (1955). Thermal springs and epithermal ore deposits. *Economic Geology*, Fiftieth Anniversary Volume, pp. 99–154.

White, D.E. (1968). *Hydrology, Activity, and Heat Flow of the Steamboat Springs Thermal Area, Washoe County, Nevada*, Professional Paper 458-C. Reston, VA: U.S. Geological Survey.

White, D.E. (1981). Active geothermal systems and hydrothermal ore deposits. *Economic Geology*, Seventy-Fifth Anniversary Volume, pp. 392–423.

White, D.E. (1985). Summary of the Steamboat Springs geothermal area, Nevada, with attached road-log commentary (USA). *U.S. Geological Survey Bulletin*, 1646: 79–87.

White, D.E., Hem, J.D., and Waring, G.A., Eds. (1963). *Chemical Composition of Subsurface Waters*, USGS Professional Paper 440-F. Reston, VA: U.S. Geological Survey.

White, D.E., Muffler, L.J.P., and Truesdell, A.H. (1971). Vapor-dominated hydrothermal systems compared with hot-water systems. *Economic Geology*, 66(1): 75–97.

White, P., Ussher, G., and Hermoso, D. (2010). Evolution of the Ladolam geothermal system on Lihir Island, Papua New Guinea. In: *Proceedings World Geothermal Congress 2010*, Bali, Indonesia, April 25–30 (http://www.geothermal-energy.org/pdf/IGAstandard/WGC/2010/1226.pdf).

11 Next-Generation Geothermal

KEY CHAPTER OBJECTIVES

- Distinguish between enhanced and engineered geothermal systems and provide examples of each.
- Explain how hydroshearing and hydraulic fracturing used in reservoir stimulation are different.
- Contrast the potential benefits and obstacles of exploiting supercritical fluids, CO_2 and H_2O, in the development of geothermal systems.
- Describe deep, hot sedimentary aquifers and how they are different from currently developed geothermal reservoirs.
- Relate advantages and disadvantages of developing geothermal systems below the brittle–ductile transition zone.

OVERVIEW

This chapter explores new, promising means of developing geothermal systems. A main arena for this form of development centers on engineered or enhanced geothermal systems (EGSs). Unlike conventional hydrothermal systems, where the geothermal fluid (either liquid or vapor) is naturally circulating in the reservoir and can be easily produced and reinjected due to good inherent permeability, in an EGS fluid circulation must be stimulated artificially. This is because the rock reservoir has little or no water or the permeability is so low that the water cannot be efficiently removed to carry heat (energy) to the surface to support a power plant. As such, systems of low water and/or permeability are also referred to as *hot dry rock* (HDR) or petraheat. To develop them for possible use requires artificial stimulation (enhancement or engineering). In EGSs, rock permeability is commonly enhanced by pumping cold water under pressure into the target region. This process has the effect of expanding any existing fractures and producing additional fractures due to thermal contraction. The injected water flows through the enhanced existing fractures, becoming heated as it moves toward production wells. It then flows to the surface to support either a flash or a binary geothermal power plant, depending on the fluid's attained temperature and flow rate (Figure 11.1).

The terms *enhanced geothermal systems* and *engineered geothermal systems* are commonly used interchangeably. For our purposes here, making a distinction seems worthwhile. An enhanced geothermal system generally applies to an existing conventional geothermal system, either liquid- or vapor-dominated, where a portion of the field has subpar permeability and fluid flow. In these areas, water is injected to

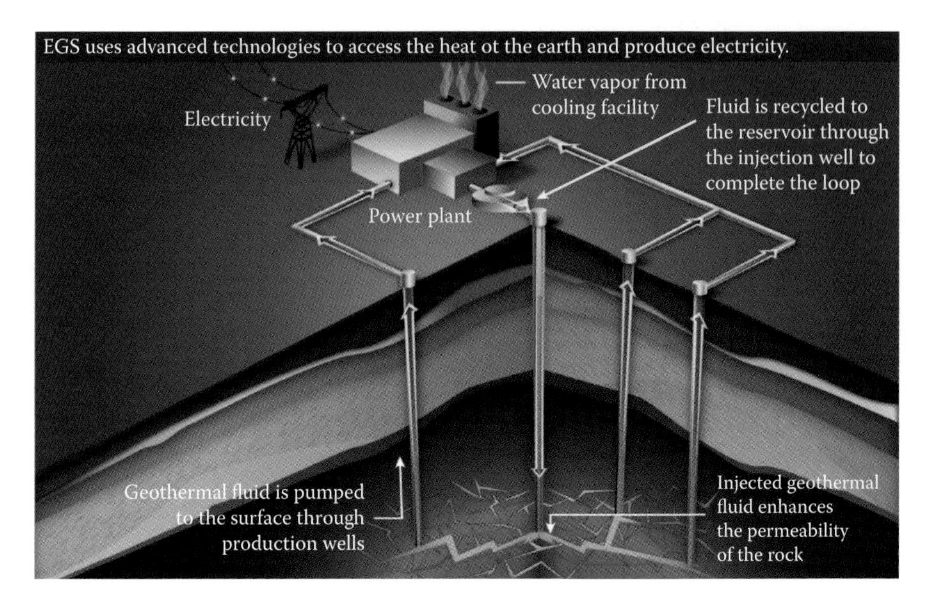

FIGURE 11.1 Diagram of an engineered geothermal system. Cool water is injected in the middle and is heated as it flows through stimulated fractures. The heated fluid is then pumped to the surface in production wells that feed the fluid to the power plant. The spent fluid is then reinjected and the process repeats. (From EERE, *What Is an Enhanced Geothermal System (EGS)?*, DOE/EE-0785, Office of Energy Efficiency and Renewable Energy, Washington, DC, 2012.)

stimulate existing fractures so subcommercial wells become productive, *enhancing* the existing geothermal system. Examples include Raft River, Idaho; Desert Peak, Nevada; and The Geysers, California. At The Geysers, a program was implemented (the reinjection project) where municipal waste from Lake County and Santa Rosa is delivered via two pipelines and reinjected to sustain reservoir steam pressure that had been rapidly diminishing. An engineered geothermal system, as adopted or defined here, is applied to regions where no natural convecting geothermal system exists, but rock is hot. In these circumstances, specific procedures (such as hydroshear, discussed below) attempt to create an artificial geothermal reservoir in which existing fractures are dilated and extended around the injection wellbore. Examples of engineered geothermal systems include the small (~5 MWe) geothermal power plants at Landau and Insheim in the Rhine Graben, Germany, and the exploratory Newberry EGS project in Oregon.

Another area of intriguing promise is drilling into reservoirs where the temperature and pressure are high enough that the fluid exists in a supercritical state (temperature, >374°C; pressure, >220 bars). Under these conditions, the fluid has not only very high enthalpy but also much higher transmissivity. Supercritical fluids can have a density similar to that of a liquid but a mobility similar to that of a gas, which can result in very high mass transport rates. A major effort to develop a supercritical system is being pursued at the Iceland Deep Drilling Project (IDDP); these efforts are summarized in a dedicated volume of *Geothermics* (49(1), 2014). Other ventures exploring the possibility of developing supercritical systems are the Japan

Beyond-Brittle Project (JBBP), which is designed to test the potential of exploiting fluids below the brittle–ductile transition zone in the crust, and the Hotter and Deeper Exploration Science (HADES) endeavor in New Zealand.

Finally, an area of EGSs that perhaps is closest to realizing significant growth potential (next 5 to 10 years) could be the development of deep (3 to 4 km), hot (160° to 180°C) sedimentary aquifers. Good examples include Steptoe Valley in eastern Nevada and the Black Rock Desert in western Utah, where previous drilling for oil and gas has disclosed significant volumes of hot water under hydrostatic conditions (Allis and Moore, 2014). Another region having hydrothermal aquifer potential is the Molasse Basin in southern Germany.

HYDROSHEARING VS. HYDRAULIC FRACTURING

Hydroshear is the process in which cold water is injected into hot rock under low to modest pressures (generally less than 500 to 600 psi) to stimulate (dilate) existing fractures. Due to the thermal contrast between cold injected water and hot rock, the rock thermally contracts, creating additional fractures (as also discussed under induced seismicity in Chapter 9). As the fractures dilate from increased fluid pressure, the normal force (σ_n) is reduced, which causes the sides of the fracture to slip past each other (Figure 11.2). Because the sides of the fractures are no longer aligned, the fractures remain relatively open due to the offset of fracture wall irregularities or asperities. This largely eliminates the need to use proppants, such as silica sands, to keep fractures open. Furthermore, introduced proppants can, over time, degrade and can potentially clog the fractures.

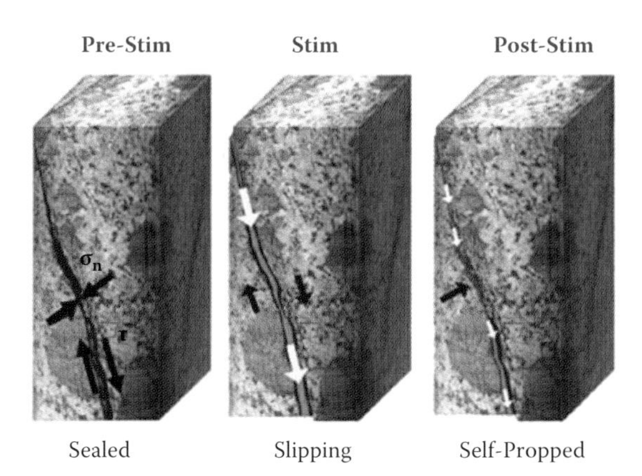

FIGURE 11.2 Illustration showing the process of hydroshear stimulation of existing fractures. The diagram on the left shows partially sealed existing fractures. Injection of water dilates fractures, as shown in the central diagram, causing them to slip from the weight of the overlying rock column. Due to the slip, the sides of the fracture no longer line up, which keeps the fracture open after stimulation as illustrated by the figure on the right (post-stimulation). See text for details. (From AltaRock, *Thermally Degradable Zonal Isolation Materials (TZIMs)*, AltaRock Energy, Seattle, WA, 2014.)

AltaRock Energy has developed a method in which the size of the geothermal reservoir can be increased over what could be achieved from simple injection alone (AltaRock, 2014). After the initial injection and hydroshearing, a thermally degradable zonal isolation material (TZIM) is injected which temporarily seals the newly opened fractures. A second pulse of injected fluid follows and dilates fractures below those dilated in the initial injection. This is because the injectate is prevented from entering the earlier dilated fractures due to the TZIM. The process is repeated until the bottom of the bore is reached. The TZIM eventually degrades as temperature rises, allowing injected fluid to access a volume of rock three to four times greater than could have been developed without using the TZIM, resulting in superior heat transfer and eventually improved production efficiency. Finally, AltaRock Energy reports that the TZIM is biodegradable and contains no deleterious chemical byproducts or residues; no acid washes or gel breakers are needed. AltaRock Energy has successfully used this technique to develop a potential geothermal reservoir of about 0.8 km^3 at its Newberry project in central Oregon.

By contrast, hydraulic fracturing or hydrofracking requires injecting fluids under very high pressures (≥5000 psi) to break rock deep underground; the technique is used mainly for the extraction of oil and gas contained in tight or impermeable rock layers. Through pumping of water under high pressure, hydraulic fracturing breaks the rock to liberate oil and gas in shale formations (the source rocks for oil and gas). In most circumstances, chemical and solid additives are pumped with the water to reduce fluid friction and to prop fractures open to maximize extraction of oil and gas. Furthermore, if the source rock is also calcareous, the injected solution is typically acidified to further improve permeability by dissolving some of the calcite cement. Although hydraulic fracturing has received considerable attention lately because of environmental concerns, contamination of shallow groundwater aquifers from hydraulic fracturing is not widespread and appears isolated to a few cases according to a study by the U.S. Environmental Protection Agency (USEPA, 2015). Potential contamination would likely not be due to the actual breaking of rock that occurs thousands of feet below shallow groundwater aquifers but instead would be due to possible spillage or leaking of the backwash as it returns to the surface. From there, the backwash is disposed of or recycled at a special waste treatment facility. Perhaps a more imortant environmental concern of hydraulic fracturing is the high volume of water required, compared to hydroshearing, for which the water can be largely recycled between injection events.

ENHANCED AND ENGINEERED GEOTHERMAL SYSTEMS

Enhanced and engineered geothermal systems (EGSs) pertain to augmenting or restoring production in conventional geothermal systems and developing potentially new producing geothermal systems from hot dry rock, respectively. Collectively, only about 2% of geothermal projects under development are EGSs. Yet, development of EGSs could increase the current U.S. geothermal power output of 3500 MWe by about one to two orders of magnitude (Figure 11.3) (Tester et al., 2006). A recent report on EGSs by Southern Methodist University (2016) indicated that EGSs could expand geothermal power production to more than 3 million MWe, or about three times that of current installed power generation in the United States. The reason for this great increase in

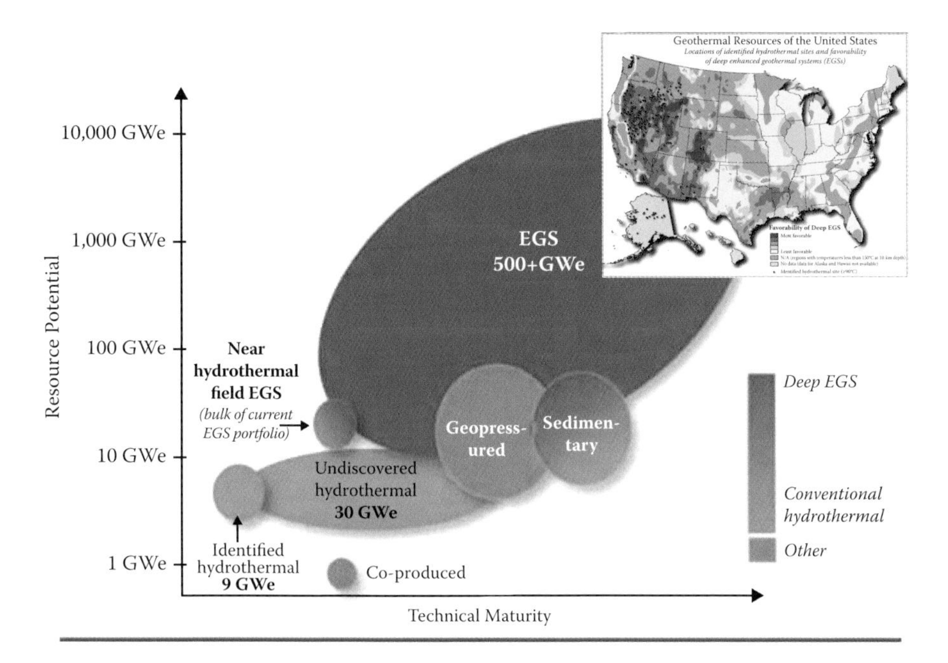

FIGURE 11.3 Graph plotting various geothermal resources as a function of resource potential vs. technical maturity. Identified and undiscovered conventional hydrothermal systems could increase current U.S. geothermal output of 3500 MW nearly ten times or to about 3.5% of current United States installed power capacity. If fully realized, EGSs could provide upwards of 3 times the current U.S. electrical generation of 1 million MWe (1000 GWe). Inset shows a heat flow map of the United States with warm colors indicating regions of high heat flow. (Adapted from MIT, *The Future of Geothermal Energy: Impact of Enhanced Geothermal Systems (EGS) on the United States in the 21st Century*, Massachusetts Institute of Technology, Cambridge, MA, 2006; NREL, *Dynamic Maps, GIS Data, & Analysis Tools*, National Renewable Energy Laboratory, Washington, DC, 2015.)

power output is that many more regions become available for geothermal power development using EGSs. Hot dry rocks (petra-heat) are much more widespread than the currently utilized conventional covnvecting geothermal systems, which are restricted mainly to active tectonic boundaries and geologic hot spots. An important limiting factor, however, is financial; it is certainly more expensive to develop an EGS resource in Kansas, due to the greater drilling depths involved and added costs of materials, than a conventional geothermal system in northern Nevada, for instance. Another issue concerns the availability of needed water to make an EGS viable.

Enhanced Geothermal Systems

Some examples of enhanced geothermal systems include Raft River, ID; Desert Peak, NV; and the Northwest Geysers project in northern California. All of these are producing conventional geothermal systems that show much promise for enhancing current power output by stimulating permeability in select previously subcommercial wells.

Desert Peak, Nevada

The producing Desert Peak geothermal system is located in west-central Nevada, about 50 miles east-northeast of Reno, Nevada. It represents the first discovered blind geothermal system (no geothermal surface manifestations) found in the Basin and Range Province of the western United States in the late 1970s (Benoit et al., 1982). A 9-MWe, dual-flash plant was commissioned in late 1985 supported by two producing and one injection well. Initial well temperatures averaged 210°C. In 2006, the dual-flash plant was replaced by a 23-MWe binary plant that required the drilling of several new production and injection wells. Desert Peak also represents one of the first commercially successful enhanced geothermal projects in the United States; an additional 1.7 MWe of power (a 38% increase) was obtained through stimulation of a previously unproductive well (Ormat, 2013).

Until 2013, the Desert Peak power plant was served by seven producing wells, averaging about 1000 to 1200 m deep, and two injection wells located about 700 to 1400 m north of the main producing wells (Figure 11.4). Well 27-15, which lies on the margin of the producing geothermal field, was selected for enhanced stimulation because of favorable bottom-hole temperatures of 180° to 196°C, but permeability was too low for use as either an injection or production well. The initial targeted zone for stimulation lies at a depth of 900 to 1150 m in what has been identified as the lower rhyolite unit of lower Tertiary age (Chabora et al., 2012; Zemach et al., 2010).

Initial stimulation of well 27-15 took place over a 7-month period and involved three phases: initial hydroshearing at low to modest fluid pressures, chemical stimulation using chelating and mud acid agents, and final hydraulic fracturing at high fluid pressures. The stimulated zone around well 27-15 was determined by recording the locations of microearthquake events as existing fractures moved and new fractures formed (Figure 11.5). During the hydroshearing phase, wellhead injection pressures were increased from 250 psi to 650 psi in steps of 100 psi to extend shear-stimulated dilation of fractures progressively outward from the wellbore. Each step lasted about a week. Results of the hydroshearing phase returned an increase in injectivity of about 15 times over the original value prior to stimulation (from ~0.01 to 0.15 gpm/psi). Follow-up chemical stimulation, which involved pumping 12,000 gal of 12% hydrochloric acid and 3% hydrofluoric acid, was conducted to dissolve any carbonate fracture fillings and remove any residual silica and clays from near-wellbore fractures. Wellbore stability, however, deteriorated, requiring extensive wellbore cleanout of the lower part of the well, and any resulting gains in permeability were difficult to quantify. The purpose of the final phase of hydraulic fracturing was to extend fracturing farther away from the wellbore and to further dilate and promote additional shear of existing fractures. Wellhead pressures during this phase ranged generally between 800 and 1000 psi with injection rates as high as 725 gpm. Several days after concluding the hydraulic fracturing phase, a step-rate test was conducted to monitor gains in permeability or injectivity. The test disclosed that injection of as much 321 gpm was achieved at a wellhead pressure of 450 psi, which is 300 psi below the pressure necessary for new fractures to form. Thus, the further increase in injectivity of about four times more than that achieved after the hydroshearing phase is mainly the result of self-propping shear failure of existing and newly created fractures.

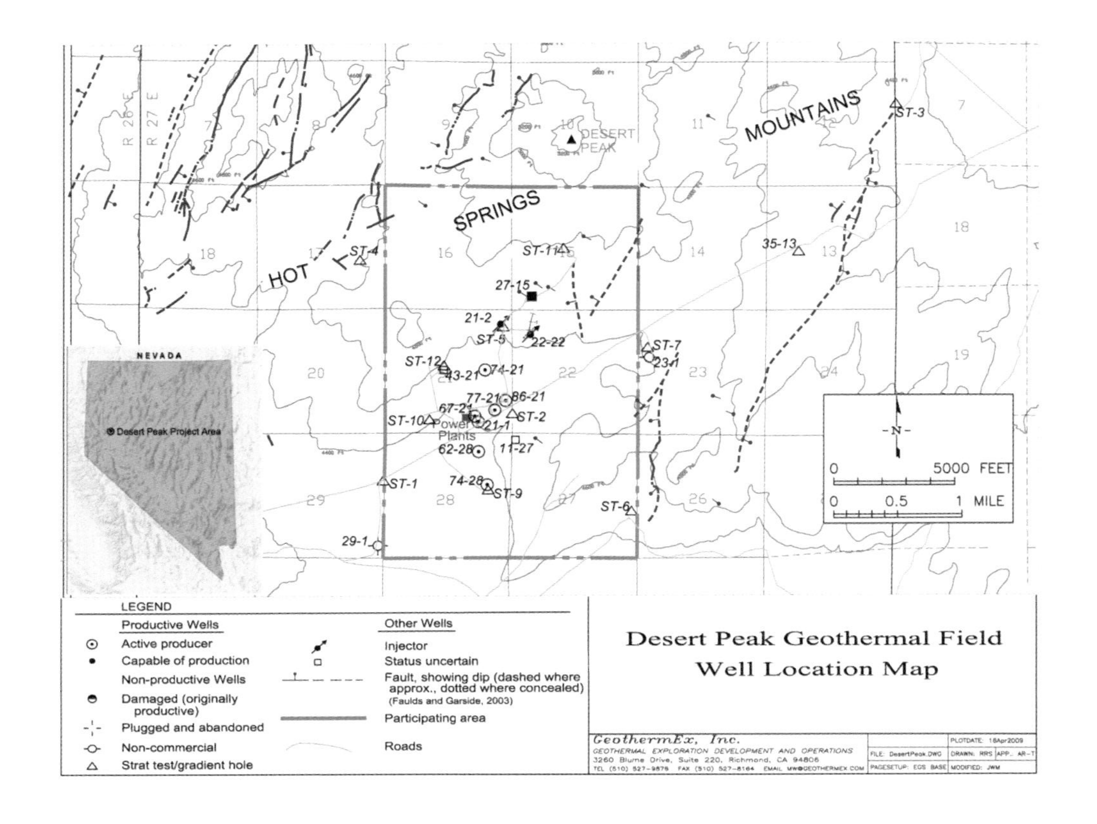

FIGURE 11.4 Location map of wells in Desert Peak geothermal field showing producing wells in amber, injection wells in green, and the targeted stimulation well 27-15. Short dashed blue lines denote faults. (From Chabora, E. et al., in *Proceedings of the 37th Workshop on Geothermal Reservoir Engineering*, Stanford, CA, January 30–February 1, 2012. Faults from Faulds et al., 2003.)

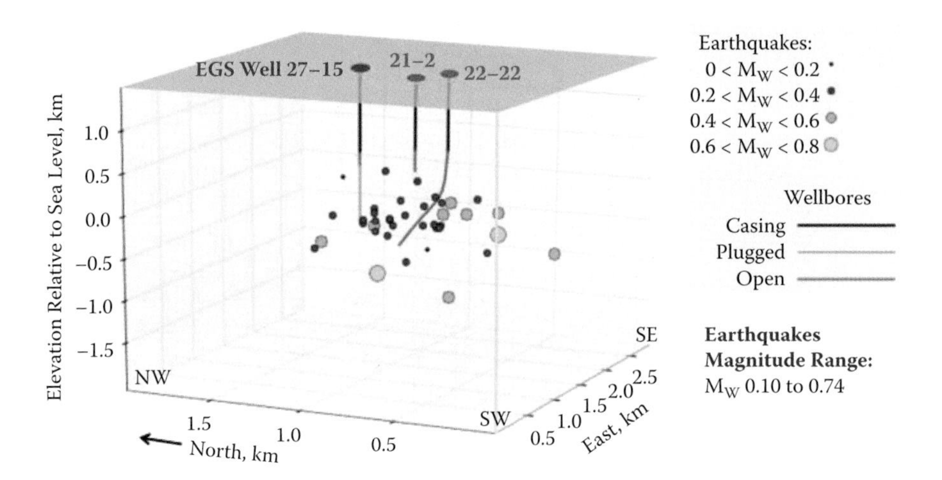

FIGURE 11.5 Three-dimensional illustration of 42 microseismic events formed during the hydraulic fracturing phase of EGS stimulation of well 27-15. Wells 21-2 and 22-2 are operating injection wells. The distribution of earthquake foci demonstrates fracture connectivity between well 27-15 and the current injection wells. (From Chabora, E. et al., in *Proceedings of the 37th Workshop on Geothermal Reservoir Engineering*, Stanford, CA, January 30–February 1, 2012.)

After the 7-month stimulation program, injectivity of well 27-15 had increased over 60-fold, from ~0.01 gpm/psi to an estimated stabilized injectivity of 0.63 gpm/psi. Although this was still less than the 1.0-gpm/psi injectivity rate considered necessary for commercial viability, it was a marked improvement. Moreover, increased connectivity to other wells, including injection wells 21-2 and 22-22, and production well 74-21, was demonstrated through tracer tests conducted at times during the stimulation and by the foci locations of microseismic events produced during stimulation (Figures 11.5 and 11.6).

Because of these encouraging results, well 27-15 was reconditioned in late 2012, a process that involved drilling out the cement plug between 1150 and 1210 m and removing mud and debris to a total depth of 1770 m. A slotted liner was installed from 925 to 1770 m, and high-flow and long-term stimulations were conducted from January to March 2013. During these last phases of stimulation, more than 300 microseismic events, reflecting shearing of existing and newly formed fractures, were recorded. Because of this added fracture stimulation and the increased depth interval of stimulation in the wellbore, from 150 m to 845 m, the injectivity rate increased to 2.1 gpm/psi, about a 175-fold increase compared to the well's initial injectivity rate. Currently, the well is accepting about 1500 gpm at a wellhead pressure of 700 psi resulting in a 38% increase in power output (1.7 MWe). Total expenditure to accomplish this improvement was about $8 million, of which $5.4 million was funded by the Department of Energy and $2.6 million was funded privately. This price (about $4.7 million per megawatt) is comparable to the cost of building a new 25- to 30-MWe binary geothermal plant, which currently averages about $4M to $4.5M per megawatt.

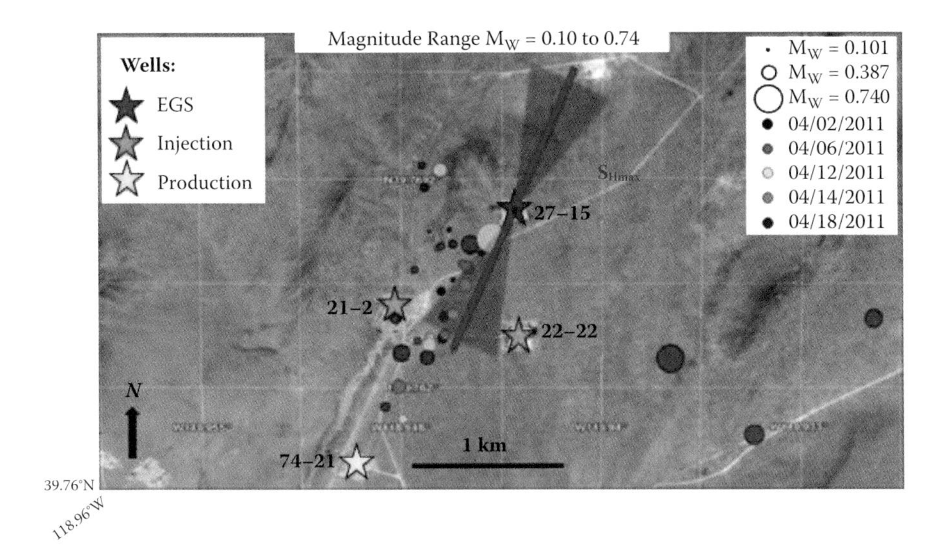

FIGURE 11.6 Map view of microseismic events shown in Figure 11.5. The near linear distribution of microseismic epicenters likely reflects a northeast-striking fault as also indicated by the maximum horizontal stress direction (S_{Hmax}). The minimum horizontal stress direction is oriented orthogonal to S_{Hmax} and indicates the direction of extension that is west-northwest/east-southeast. Mapped normal faults strike north-northeast and form perpendicular to direction of extension or parallel to S_{Hmax}. (From Chabora, E. et al., in *Proceedings of the 37th Workshop on Geothermal Reservoir Engineering*, Stanford, CA, January 30–February 1, 2012.)

Raft River, Idaho

The Raft River binary, water-cooled geothermal plant is located in southern Idaho, approximately 6 miles north of the Utah/Idaho border; it has a gross power output of 13 MWe. The plant is serviced by four production wells supplying about 5000 gpm total flow of 135° to 150°C geothermal fluid and three injection wells. The fluids have near-neutral pH and low total dissolved solids (TDS) (1200 to 6800 ppm). Production zones lie at depths of 1500 to 2000 m in the Elba Quartzite—a metamorphosed quartzose sandstone. The facility is currently operated by United States Geothermal, Inc., which is in partnership with the Department of Energy as part of an enhanced geothermal project. Their goal is to assess the suitability of using a previously noncommercial well, RRG-9, as a potential new injection well and thereby bring more of the geothermal resource into production. RGG-9 was drilled to test the intersection of the Narrows and Bridge fault zones (Figure 11.7). RGG -9 has a total depth of 1644 m and bottoms in quartz monzonite, which directly underlies the Elba Quartzite. The well is currently undergoing thermal and hydraulic stimulation to improve permeability and injectivity.

Results of initial tests indicated that the well accepted 43 gpm of plant injectate (temperatures ranging from 39° to 46°C) at a wellhead pressure of 280 psig for an injectivity rate or index of only 0.15 gpm/psig, well below what is typically considered commercially viable (~1 gpm/psig). A series of seven test periods were

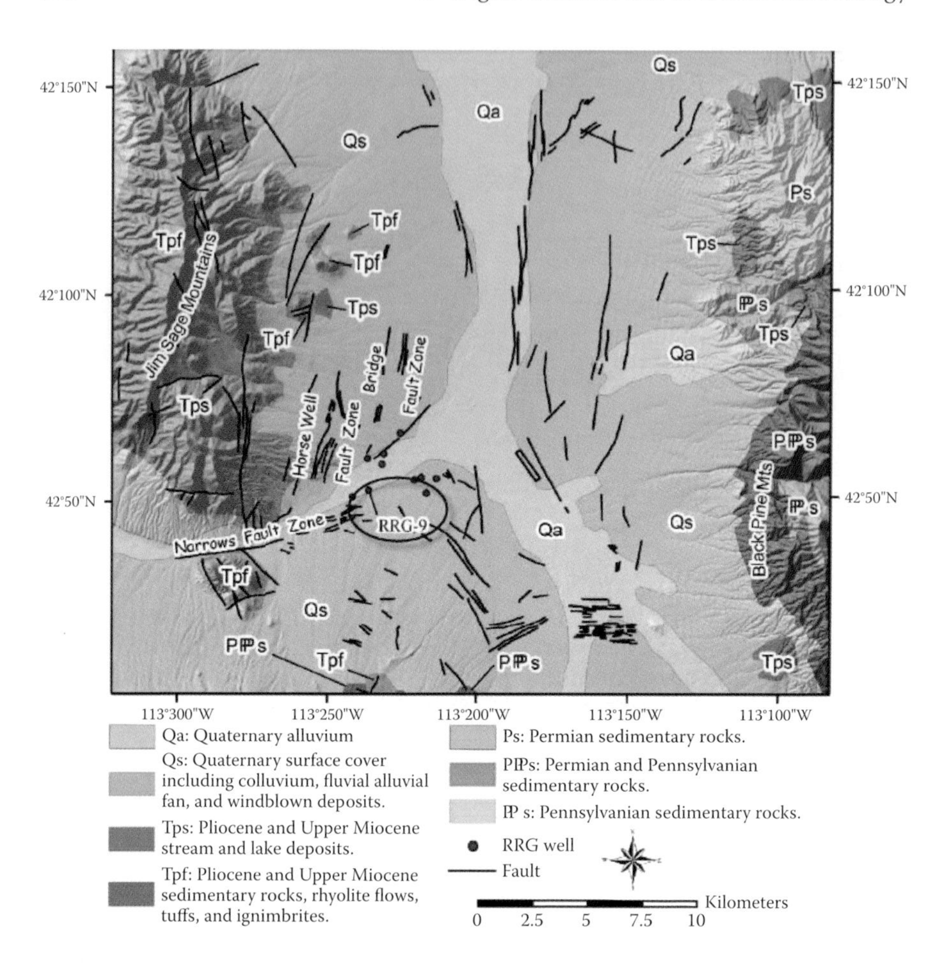

FIGURE 11.7 Geologic map of the Raft River region. The well selected for stimulation (RRG-9) is outlined by the blue ellipse. (Adapted from Nash, G.D. and Moore, J.N., *Geothermal Resources Council Transactions*, 36, 951–958, 2012.)

conducted beginning in June 2013 and ending in February 2014, ranging in length from 3 days to just over 2 months. During this period, a maximum flow of 283 gpm at a wellhead pressure of 862 psig (injectivity index of 0.33 gpm/psig) was achieved using plant injectate averaging 40°C. However, after a 12-day period of injecting cold water (12° to 13°C), the injectivity index of the plant injectate increased to ~0.45 to 0.48 gpm/psig, about a threefold increase from initial values (Figure 11.8). This result indicates that the injection of cold water provided additional thermal stresses to dilate or shear existing fractures and generate new fractures due to thermal shocking and contraction of hot rock. High-pressure hydraulic stimulation was scheduled to begin in spring 2014 with the goal of achieving continued gains in permeability and injectivity to make well RGG-9 viable. If so, an increase in power output comparable to that achieved after stimulation of a previously nonserviceable well at Desert Peak (1.7 added MWe) is possible, if not likely.

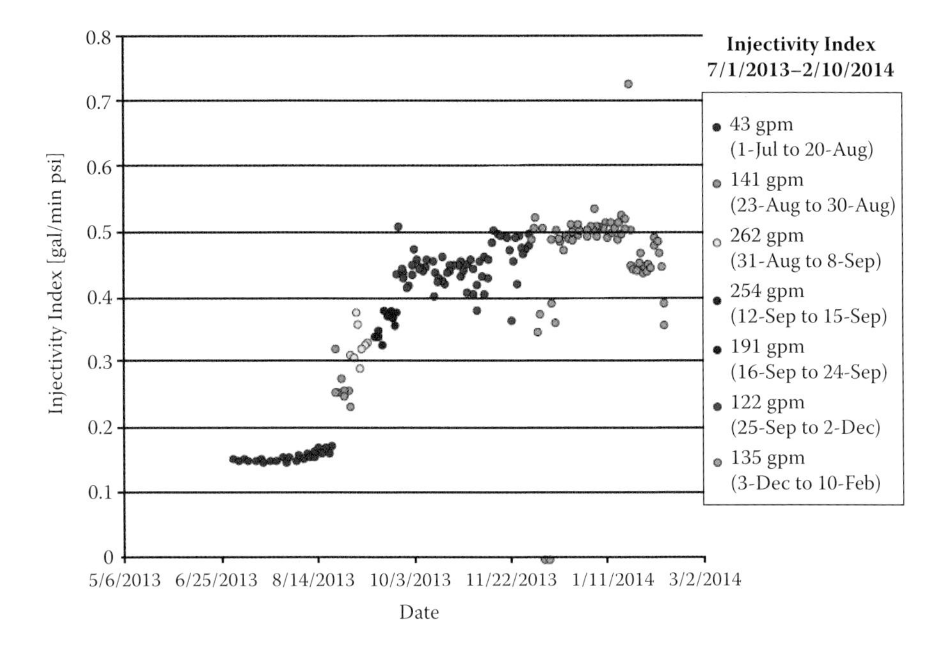

FIGURE 11.8 Injectivity index for well RRG-9. Warmer colors indicate plant injectate injection; cooler colors, cold well water injection. Thermal stimulation at low to modest pressures yielded a threefold increase in injectivity. (From Bradford, J. et al., in *Proceedings of the 39th Workshop on Geothermal Reservoir Engineering*, Stanford, CA, February 24–26, 2014.)

Northwest Geysers Project, California

The Geysers geothermal field is the world's largest geothermal power producing region. It is also the planet's largest vapor-dominated reservoir. Some of the highest temperatures encountered in drilling occur in the northwest portion of the geothermal field. However, in spite of the high temperatures (as much as 400°C), production steam wells in the area have been abandoned due to the high concentrations of noncondensable gases and corrosive hydrogen chloride gas in the steam. As part of the exploration of the northwest region of The Geysers in the 1980s, a high-temperature zone (280° to 400°C) was found to underlie the 240°C steam zone in impermeable rocks. In a joint effort between Calpine, the principal operator in The Geysers geothermal field, and the Department of Energy's Geothermal Technologies Office, the Northwest Geysers EGS Demonstration Project is underway to test the viability of stimulating the hot impermeable rock below the steam reservoir. The main objectives of the EGS project are as follows:

- Develop an enhanced geothermal system in hot, conductively heated, and impermeable rock capable of producing 5 MWe of added power.
- Improve permeability by injection of cold effluent under low pressure from the Santa Rosa Reinjection Project Pipeline.
- Reduce the high noncondensable gas content of native steam to produce a high-quality, less corrosive steam for power production.

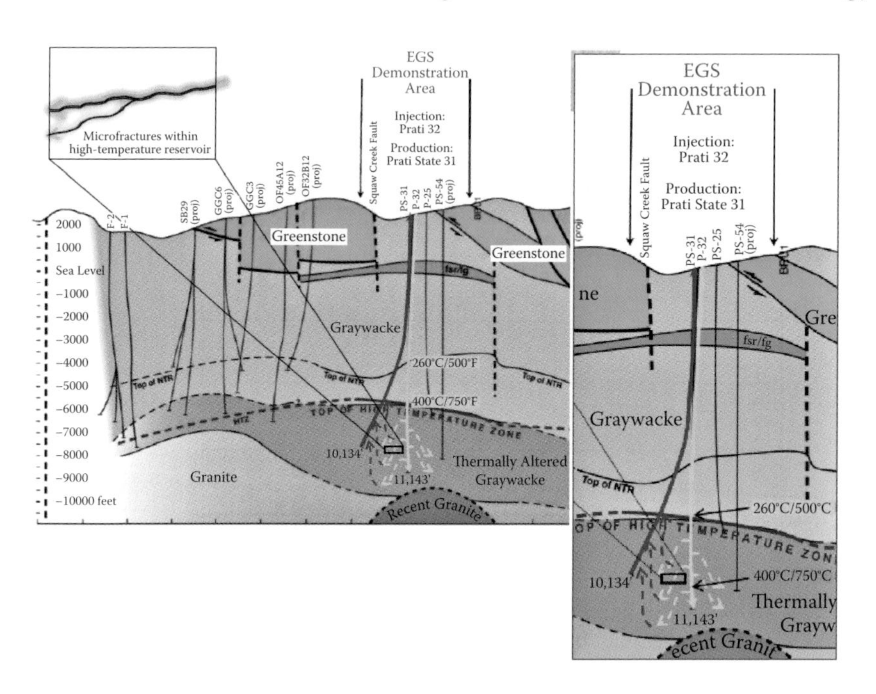

FIGURE 11.9 Southwest–northeast geologic cross-section showing rock units, upper normal steam-dominated reservoir and lower high-temperature reservoir, the target for the EGS demonstration project. Deepened and reconditioned injection well P-32 and production well PS-31 are shown. (From Garcia, J. et al., in *Proceedings of the 37th Workshop on Geothermal Reservoir Engineering*, Stanford, CA, January 30–February 1, 2012.)

The project consists of three phases: (1) a pre-stimulation phase, (2) a stimulation phase, and (3) a monitoring phase. As part of the pre-stimulation phase, two earlier drilled exploration wells were reopened and deepened (P-32 and PS-31) to form an injection/production doublet (Figure 11.9) (Garcia et al., 2012). The proposed injection well reached a vertical depth of 3326 m, with a bottom-hole temperature of 400°C. The stimulation phase began in October 2011 using treated effluent from the Santa Rosa Recharge Pipeline and ran for just under a year and a half (October 2011 through March 2013). During the first 3 months of stimulation, wellhead pressures in production wells PS-32 and P-25 increased appreciably (Figure 11.10). Moreover, the injection of effluent in P-32 significantly reduced the noncondensable gas content in P-25 from 3.7 wt% to 1.1 wt%. The rate of increase in wellhead pressure of PS-31 began to decline after the onset of production from well P-25.

The analysis of seismic events discloses a strong correspondence between injection wells and epicenters of earthquakes (Figure 11.11). Also, the frequency of seismic events increases with the rate of injection (Figure 11.12). Hypocenters of monitored earthquakes form a steeply dipping pattern indicating movement along steeply dipping fractures (Figure 11.13). The frequency of earthquakes also helps explain the abrupt increase in wellhead pressure of PS-31 on November 29, 2011, when injection was increased from 400 gpm to 1000 gpm (Figure 11.14). Results to date include the following:

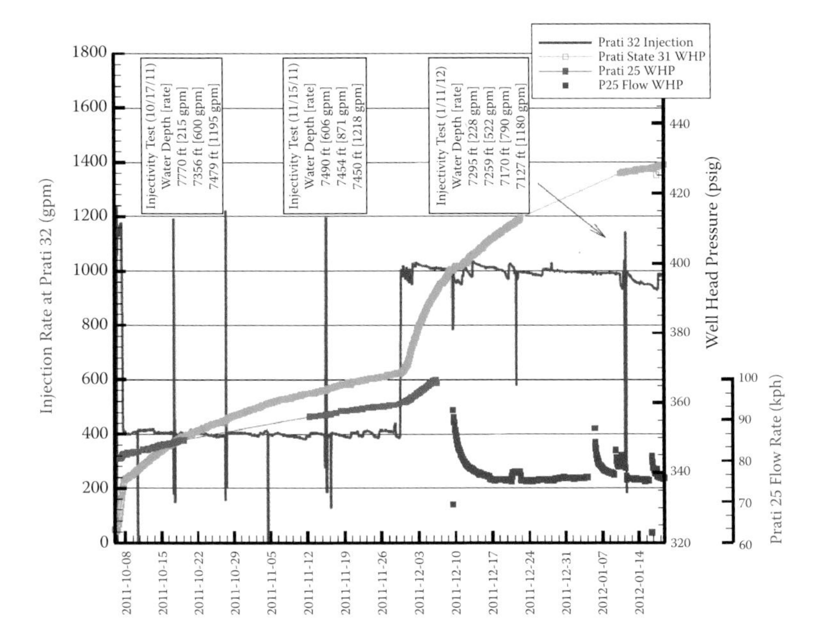

FIGURE 11.10 Figure showing injection rates in P-32 (blue lines) and corresponding increases in wellhead pressures in production wells PS-31 and P-25. Also shown in red is the flow rate of P-25 in thousand pounds per hour (kph). (From Garcia, J. et al., in *Proceedings of the 37th Workshop on Geothermal Reservoir Engineering*, Stanford, CA, January 30–February 1, 2012.)

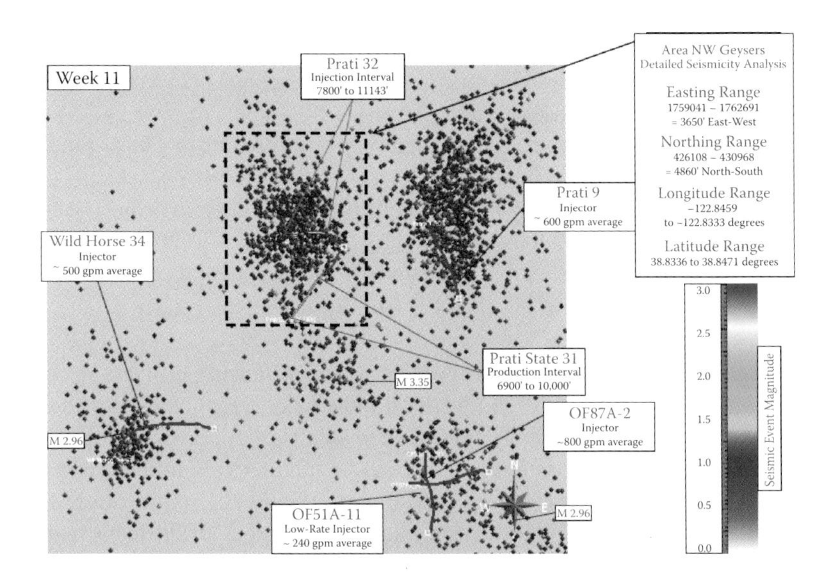

FIGURE 11.11 Map view of earthquake epicenters over the first 75 days of stimulation. Note the clustering of earthquakes around injection wells reflecting thermal contractive stresses inducing fracture dilation and shear failure. (From Garcia, J. et al., in *Proceedings of the 37th Workshop on Geothermal Reservoir Engineering*, Stanford, CA, January 30–February 1, 2012.)

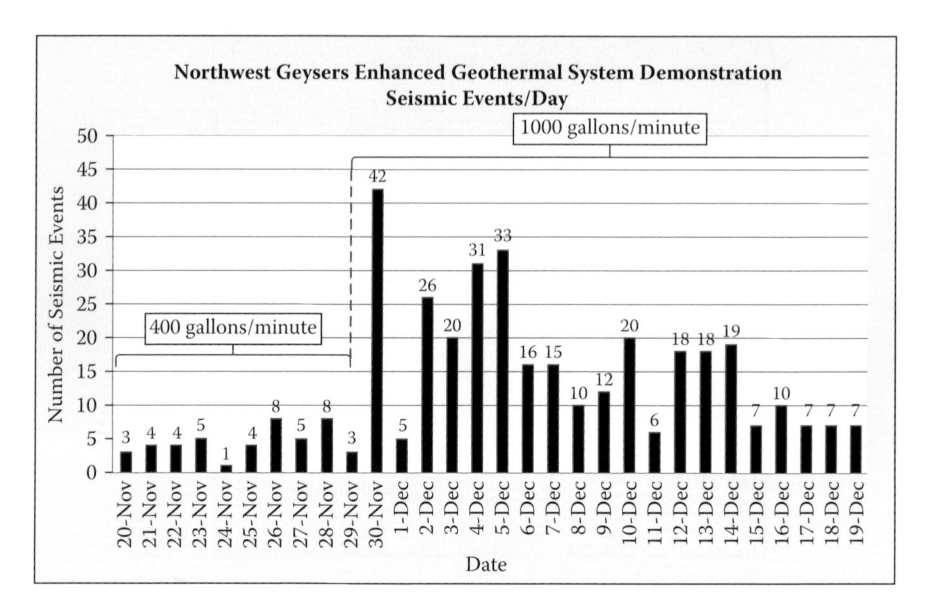

FIGURE 11.12 Frequency of seismic events as a function of flow injection rate. (From Garcia, J. et al., in *Proceedings of the 37th Workshop on Geothermal Reservoir Engineering*, Stanford, CA, January 30–February 1, 2012.)

- Wellhead pressure increased in producing well PS-31, from 323 psi to 465 psi, where injection rates were 1000 gpm; for producing well P-25, wellhead pressure increased from 345 psi to 365 psi during injection.
- An 85% to 90% reduction in noncondensable gas contents in wells P-25 and PS-31, respectively, was obtained.
- A negligible reduction in Cl content was observed during high rates of injection (1000 gpm) in well PS-31.
- Rates of steam production of 45% and 90% were derived from injectate in P-25 and PS-31, respectively, as indicated from stable isotopic analyses.
- Measured wellhead pressures indicated an added potential power capacity of 1.75 MWe for P-25 and 3.25 MWe for PS-31.

Although the noncondensable gas content was significantly reduced in the produced steam in both well PS-31 and well P-25, the Cl content (averaging about 130 ppm) did not decrease appreciably in produced steam from PS-31 during injection into P-32. The elevated Cl content led to corrosion and leaking in the upper 800 m of the liner of well PS-31, forcing it to be shut-in in early 2013. Injection into P-32 and production from P-25, however, continue as the Cl content in steam from well P-25 is only about 20% of that from PS-31. Further monitoring and testing of PS-31 will be delayed until a new high-grade alloy steel or titanium production liner is installed. Eventual plans to produce the EGS steam for power production remain uncertain, as no power purchase agreement from a major utility has yet been negotiated.

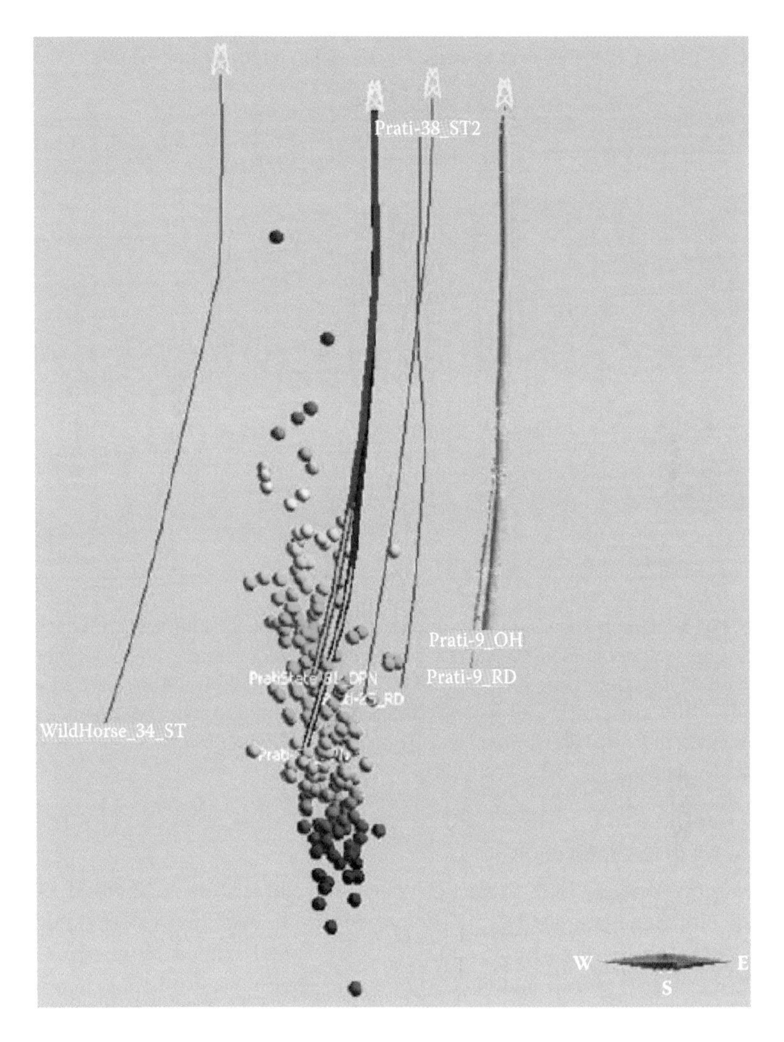

FIGURE 11.13 Plotted hypocenters of earthquakes during the first 75 days of stimulation. Note the steeply dipping nature of the hypocenter distribution, probably reflecting a similarly oriented fracture zone. Color coding of hypocenters reflects depth range. (From Garcia, J. et al., in *Proceedings of the 37th Workshop on Geothermal Reservoir Engineering*, Stanford, CA, January 30–February 1, 2012.)

Engineered Geothermal Systems

The formation of a potentially producing geothermal system where no previous conventional or convecting geothermal reservoir exists is the arena of engineered geothermal systems. Specifically these are regions of hot, conductively heated rocks having limited permeability. Fluids are injected under low to high pressure to induce dilation and shear failure along existing fractures. Some high-pressure hydraulic fracturing might be done to enhance formation of new fractures produced by thermal stresses caused by cold fluid interacting with hot rock.

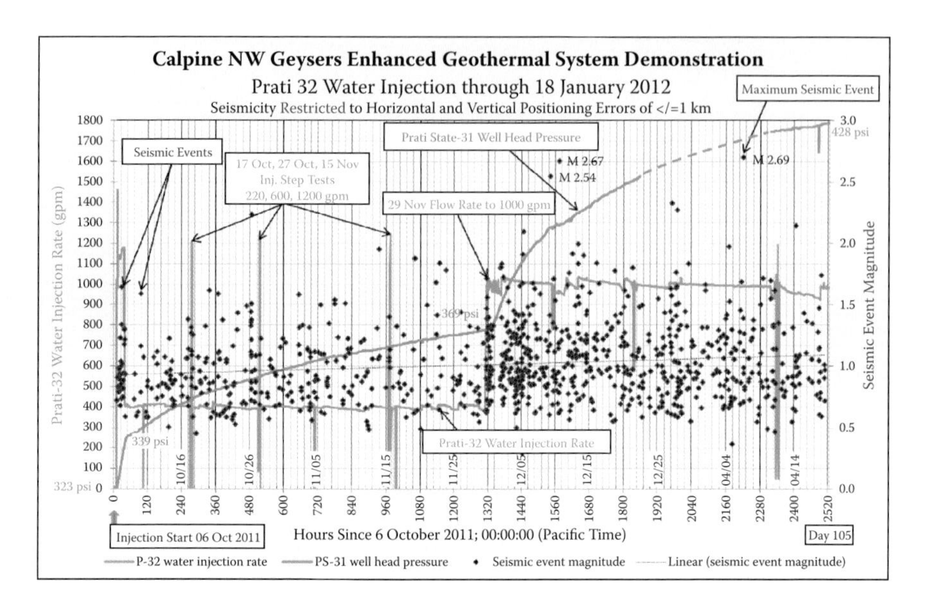

FIGURE 11.14 Graph showing injection rate (blue line), wellhead pressure of PS-31, and seismic events (red dots). Note the increase in frequency of seismic events with the onset of the increase of injection from 400 gpm to 1000 gpm. Also, the abrupt increase in wellhead pressure in PS-31 correlates with increased rate of injection and induced seismicity. (From Garcia, J. et al., in *Proceedings of the 37th Workshop on Geothermal Reservoir Engineering*, Stanford, CA, January 30–February 1, 2012.)

Newberry Volcano, Oregon

The Newberry Volcano EGS project is located in central Oregon, about 35 km south of the town of Bend (Figure 11.15). The project site lies on the western flanks of the large (115 km north-south by 45 km east-west), shield-shaped Newberry Volcano. The volcano is still active, and lava erupted as recently as 1300 years ago (the Big Obsidian flow). The summit of the volcano is marked by a caldera measuring 6.5 km by 8 km, which formed upon eruption of silicic pumice and ash about 75,000 years ago. The caldera now hosts two lakes that contain hot springs with temperatures as much as 60°C. In the late 1980s, the U.S. Geological Survey drilled a bore hole to 900 m near the middle of the caldera and encountered temperatures up to 260°C. Geothermal exploration activities in the region began in the 1970s, and most recently AltaRock Energy, in partnership with Davenport Newberry, was awarded a grant by the U.S. Department of Energy in 2010 to demonstrate the feasibility of EGS technology (Cladouhos et al., 2013).

A previous contractor drilled a well (55-29) in 2008 to a depth of 2.7 km and encountered a bottom-hole temperature of 300°C; however, flow tests indicated limited permeability. AltaRock Energy stimulated the well in three stages resulting in multi-stage fracturing through hydroshearing (injection of fluids at modest pressures, below the minimum principle stress to cause existing fractures to slip and

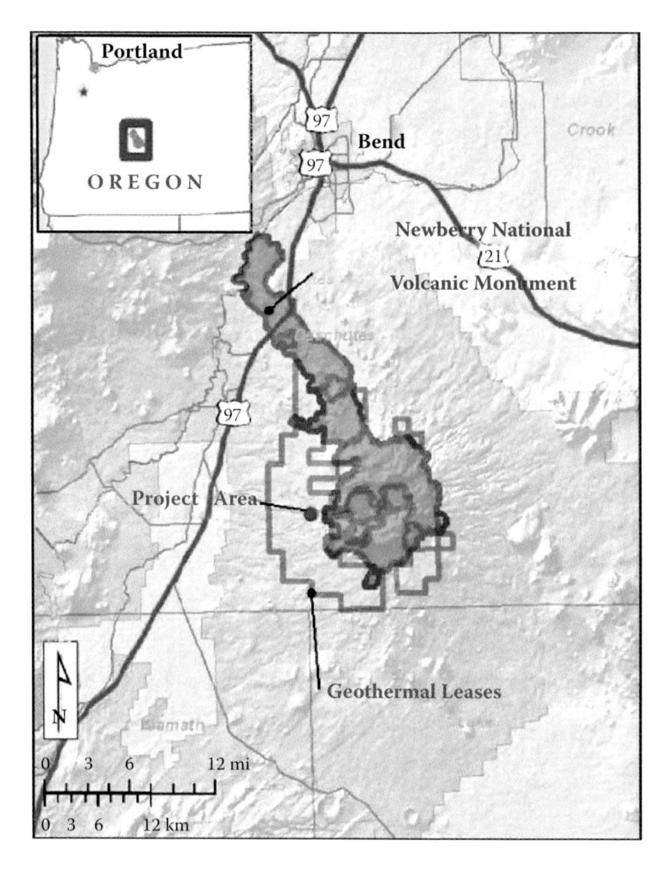

FIGURE 11.15 Location map of the Newberry EGS project. (Adapted from Cladouhos, T.T. et al., *Geothermal Resource Council Transactions*, 37, 133–140, 2013.)

dilate). Unlike hydrofracking, which requires pressure much above the minimum principle stress, no proppants are used in hydroshearing as fracture wall irregularities, after slip, help keep the fractures open to fluid flow. In the stimulation tests, cold water was injected at pressures up to 16 MPa (160 bars) (Figure 11.16). Injection of cold water under pressure causes the rock to fail both by hydroshearing and thermal contraction. As a result of slip along existing fractures from hydroshearing and opening of new fractures from thermal contraction, microseismicity occurred. The foci and epicenters of the microseismic events map out the location of the potential new geothermal reservoir.

Near the end of each stimulation phase, thermally degradable material was injected to plug the newly opened fractures. Further stimulation events then opened a new set of fractures. This was repeated three times, leading to three enhanced zones of permeability. The injected material used to plug newly opened fractures is nontoxic and results in a larger permeable reservoir than what could have been accomplished by multiple pressurized injection events alone. Results

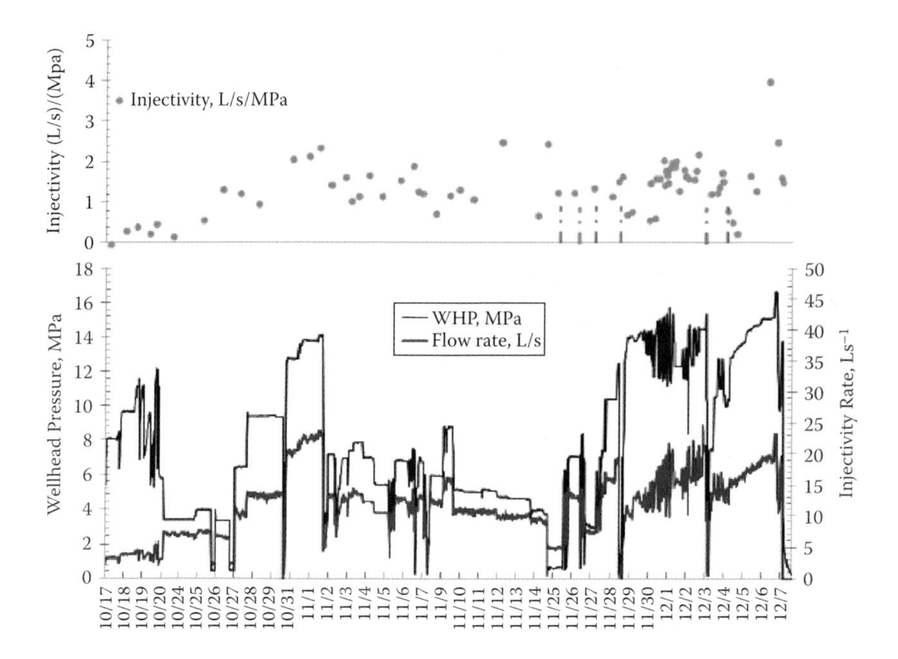

FIGURE 11.16 Stimulation results of well 55-29, tracking wellhead pressure, injection rate, and injectivity. (From Cladouhos, T.T. et al., *Geothermal Resource Council Transactions*, 37, 133–140, 2013.)

indicate injection rates as high as about 5 liters per second per MPa, which is about 0.6 gallon per minute (gpm) per psi (pound per square inch). An injectivity index of about 1 gpm per psi is generally considered the minimum necessary for commercially viable injection rates (as noted above for the Desert Peak enhanced geothermal system). Although permeability increased by two orders of magnitude after the performed hydroshearing stimulation of well 55-29, some further improvement in permeability (injectivity) appears necessary to attain potential commercial viability.

Next planned steps will include a flow test, involving an air lift of the cold water column to initiate flow of steam and hot water. Results of this test will be used to see if further stimulation is required to secure desired flow rates for potential commercial viability. If flow tests are positive and can be maintained over time, the next step would be to drill production wells, which would complete the EGS system. A demonstration power plant having a power output on the order of 5 to 10 MWe would probably be constructed initially.

Rhine Graben, Germany and France

The Rhine Graben is an extensional and seismically active corridor stretching from Basel, Switzerland, in the south and running along the border between eastern France and southwestern Germany and into west-central Germany near Frankfurt (Mannheim) (Figure 11.17). Several small geothermal power plants (~5 MWe) have been built to take advantage of the elevated heat flow resulting from thinned crust

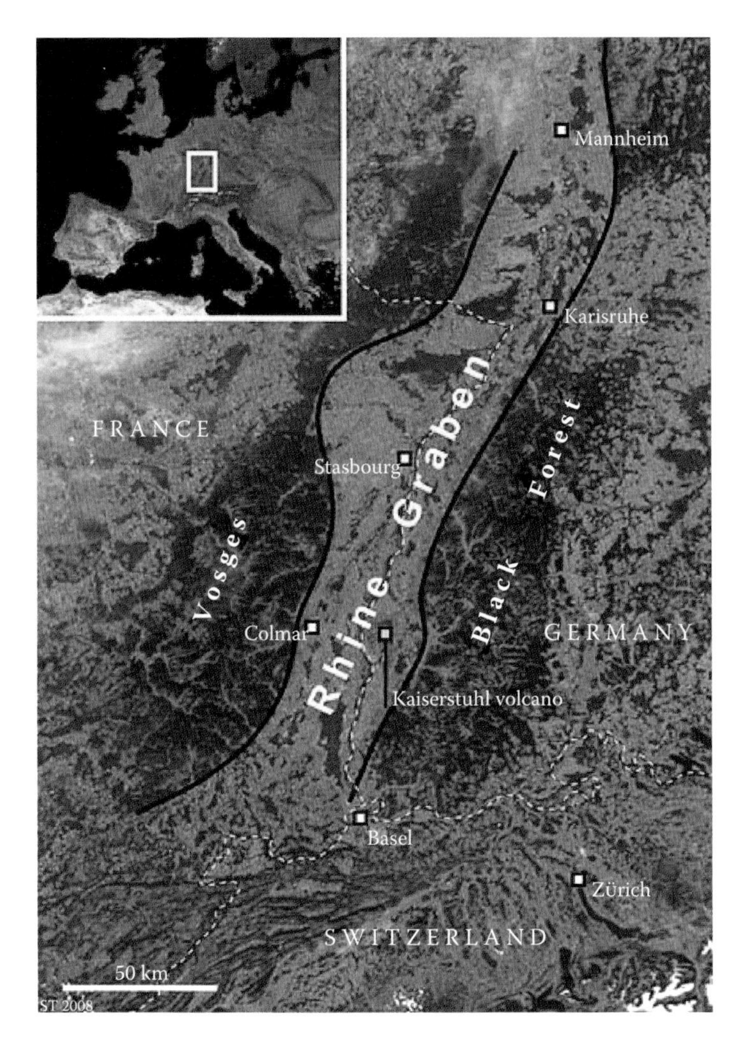

FIGURE 11.17 Location of the upper Rhine Graben. (From Wikipedia, *Upper Rhine Plain*, https://en.wikipedia.org/wiki/Upper_Rhine_Plain, 2015.)

within the graben. These plants include Soultz-sous-Forêts (pilot test plant) in France and Landau, Bruchsal, and Insheim in Germany. The Landau plant is under renovation and the Insheim and Bruchsal plants are currently operational. These plants operate using two wells—an injection and a production well couplet. Although at the surface the wellheads are only a few tens of meters apart or less (Figure 11.18), the wells diverge with depth so that the bottoms of the wells are at least a few hundred meters apart. Well depths are typically about 3.5 to 4.5 km.

Insheim is a small geothermal power plant that came online in October 2012. The plant has an installed capacity of 4.8 MWe, apparently sufficient to power about 8000 homes. (In the United States, the conversion factor of 1 MWe for about 1000 homes is typically applied, suggesting that the German homes are more power efficient

FIGURE 11.18 View of Insheim geothermal power plant showing wellhead pumps for production well at left and injection well at right. (Photograph by author.)

or smaller, or a combination of both.) The Insheim power plant uses a doublet well system of one production well and one injection well that extend to depths of 3600 m and 3800 m, respectively. The temperature of produced fluids is about 165°C (the highest for operating geothermal power plants in Germany) and the flow rate averages about 80 L/s (Figure 11.19).

Insheim is a binary geothermal power plant using a working fluid with a low boiling point (about 30°C) that is flashed to steam by the geothermal fluid in a heat exchanger or vaporizer. Because the working fluid is an organic hydrocarbon, it works on the organic Rankine cycle. The nearby Bruchsal geothermal plant uses the Kalina cycle, which uses an ammonia–water mixture as the working fluid. The Kalina cycle offers improved turbine efficiency for lower temperature geothermal fluids (about 123°C at Bruchsal) because the boiling point can be adjusted, unlike with a hydrocarbon. The Kalina cycle, however, requires careful monitoring to avoid pre-condensing as fluid expands through the turbine, which would reduce turbine efficiency and power output.

The German geothermal plants use hydraulic stimulation of existing fractures in reservoirs to promote permeability and fluid flow. Fracture stimulation occurs via hydroshearing at relatively low pressure, just above the critical pressure for existing fractures to shift or shear. Reservoir stimulation, prior to construction of the Insheim plant, resulted in two earthquakes having magnitudes of 2.2 and 2.4. Because induced earthquakes occur at such shallow depths (2 to 4 km), magnitudes as low as 1.3 to 1.5 can be felt by people in the immediate vicinity. To reduce induced seismicity in proximity to the injection well at Insheim, a side-leg concept was used in drilling the well. The side leg distributes the injected fluid over two separated ends of the injection well. Spreading the injected fluid over a larger volume of rock reduces the size and frequency of induced quakes.

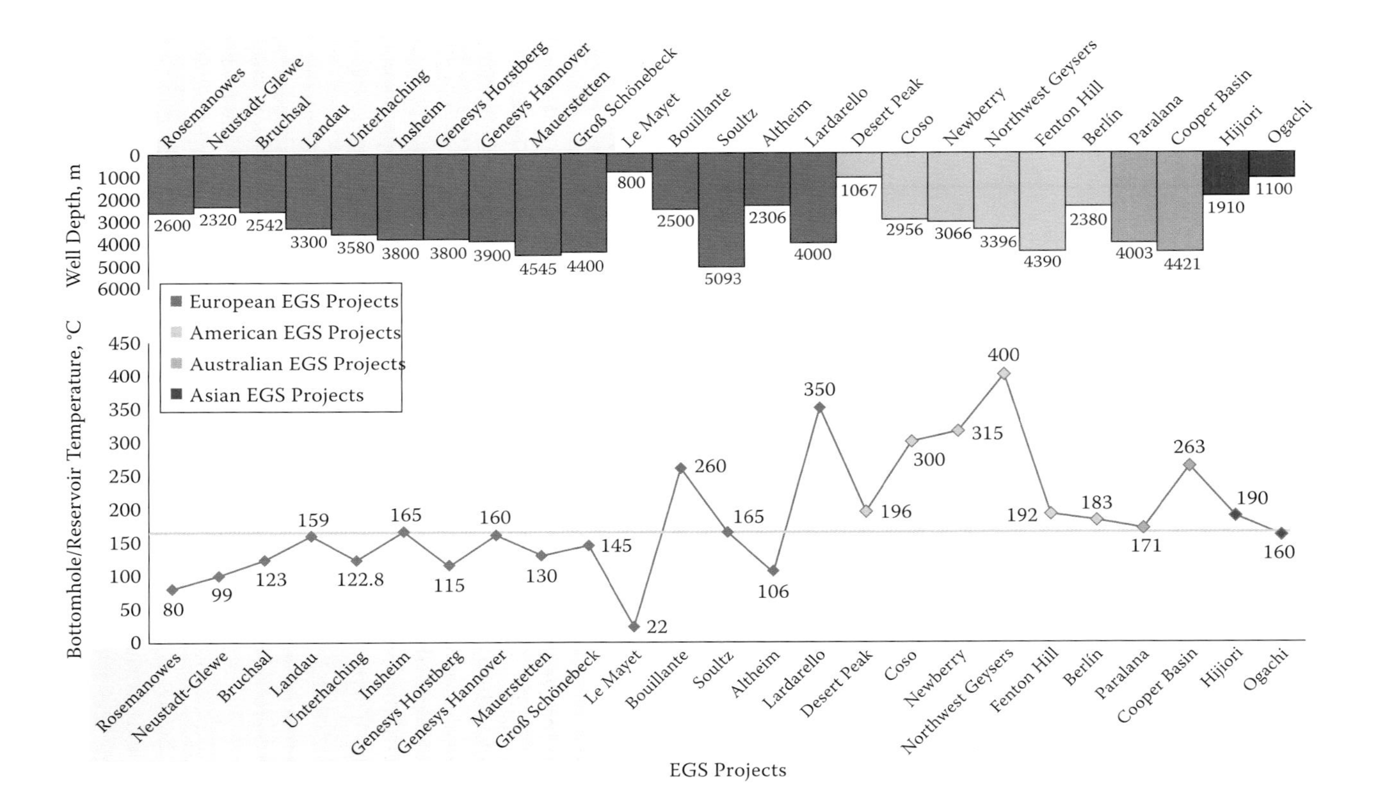

FIGURE 11.19 Reservoir temperatures and reservoir depths of electricity-producing EGS facilities worldwide. The highlighted names are for those in Germany. Note that EGS includes both engineered and enhanced existing geothermal systems for American projects. (Adapted from Breede, K. et al., *Geothermal Energy*, 1, 4, 2013.)

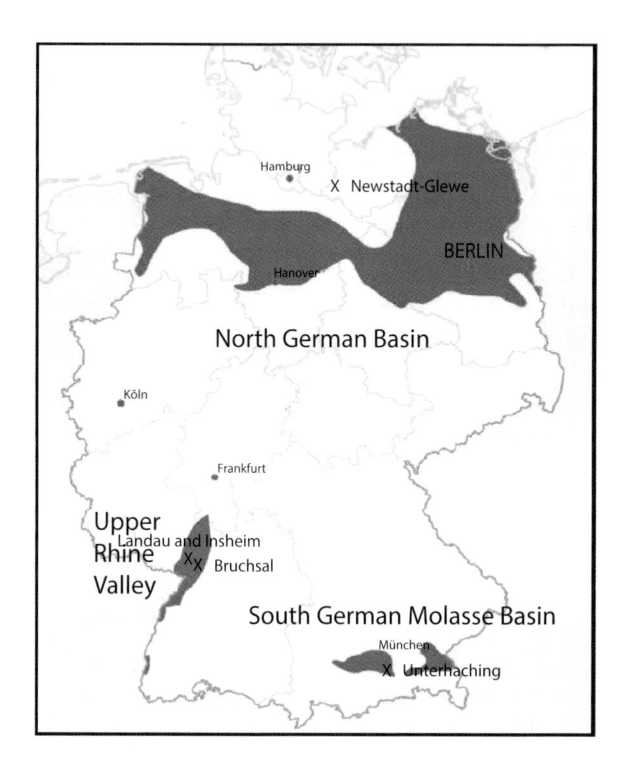

FIGURE 11.20 Map showing locations of the three regions of geothermal potential (red regions) and four operating geothermal power or combined heat plants as of late 2012. (Adapted from Herzberger, P. et al., *Geothermal Resources Council Transactions*, 33, 352–354, 2009.)

Other attractive geothermal regions in Germany consist of deep (3 to 4 km), hot (>100° to 150°C) sedimentary aquifers, such as the North German Basin and the Molasse Basin in southern Germany (Figure 11.20). Those in southern Germany include combined heat and geothermal power facilities at Unterhachting and the recently completed Durrnhaar and Kirchstockach geothermal facilities. At Kirchstockach (http://www.tiefegeothermie.de/projekte/kirchstockach), production and injection well depths are 3750 m. The production well produces fluid at 139°C at a flow rate of 145 L/s; the installed power capacity is 5.5 MWe.

Supercritical CO_2 in Engineered Geothermal Systems

Rather than using injected and pressurized water to stimulate fractures and transport heat energy, some researchers have suggested using supercritical carbon dioxide ($ScCO_2$). $ScCO_2$ is a fluid-like phase that exists above the critical point of temperature and pressure, where distinct liquid and gas phases no longer exist (Figure 11.21). The critical point for CO_2 is at a temperature of about 31°C and a pressure of 73 atmospheres (about 75 bars). A supercritical fluid has the ability to move through solids similar to a gas but can dissolve constituents similar to a liquid. Because of its low critical point temperature, $ScCO_2$ is used, for example,

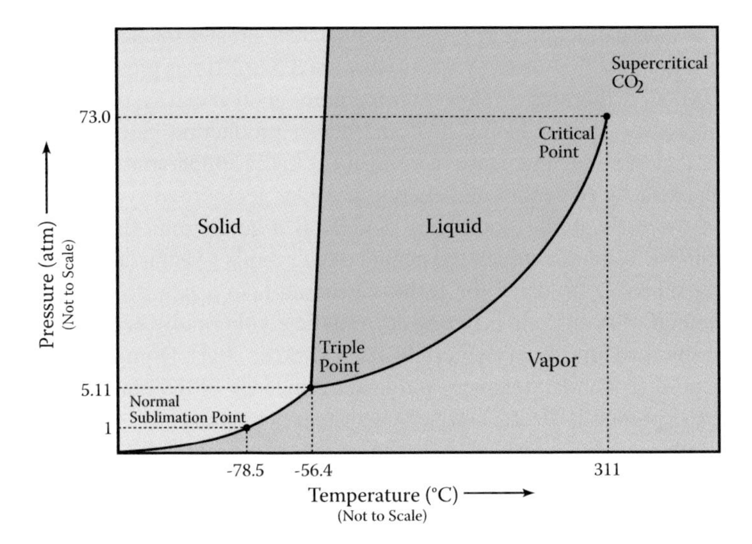

FIGURE 11.21 Phase diagram for CO_2 illustrating the temperature and pressure range for supercritical conditions where CO_2 exhibits both gas-like and liquid-like properties.

to decaffeinate coffee without destroying or removing other desirable qualities. As pointed out by Brown (2000) and Pruess (2006, 2007), using $ScCO_2$ has several potential advantages as a medium for stimulating and extracting heat energy from hot rocks, including the following:

1. A large difference in wellbore density between cold injected $ScCO_2$ (0.96 g/cc) and hot produced $ScCO_2$ (0.39 g/cc) results in a large buoyant force to circulate fluids through the potential geothermal reservoir and reduces the need for significant pumping and power consumption.
2. Because $ScCO_2$ is not an ionic solvent, its ability to dissolve and transport mineral species is greatly reduced which in turn significantly diminishes mineral scaling problems of surface equipment or reduction in reservoir permeability due to the precipitation of mineral species carried in a hot aqueous phase.
3. As part of geothermal energy development, sequestration of captured CO_2 from fossil-fuel-fired power plants would reduce greenhouse gas emissions and climate disruption.

Although the mass heat capacity of $ScCO_2$ is about 40% that of water, the ratio of fluid density to viscosity, which is a measure of reservoir flow potential, is approximately 1.5 times that of water (Brown, 2000). So, even though the rate of geothermal energy produced from $ScCO_2$ is approximately 60% that of water, when the reduced needs of pumping and power consumption are considered (reflecting the buoyant drive and superior flow capacity of $ScCO_2$ from its low viscosity), power production of $ScCO_2$ would be about equal to that of a water-based system. Numerical simulations indicate that heat extraction may be as much as 50% greater for $ScCO_2$ compared to water, based on mass flow rates of $ScCO_2$ that are 3.5 to 5 times greater than for water (Pruess, 2007). Furthermore, in lower temperature geothermal systems, the viscosity of water increases, thus reducing flow and power production. $ScCO_2$, on the other hand, experiences no such increase in viscosity at lower temperatures, making $ScCO_2$ an efficient extractor of geothermal energy in cooler geothermal systems.

Most of the early studies looking at $ScCO_2$ as a geothermal fluid medium considered a binary system in which the heated $ScCO_2$ boils a secondary working fluid, via a heat exchanger, to drive the turbine-generator. In a new study, however, the heat-exchanged working fluid is bypassed, and the geothermally heated $ScCO_2$ is fed directly to the turbomachinery (Freifeld and Hawkes, 2011). Doing so would eliminate the capital costs, maintenance, and lower efficiency of the binary cycle system. Testing of the proposed direct-feed $ScCO_2$ system is ongoing.

A main obstacle in using $ScCO_2$ is the expense in capturing a steady supply of CO_2. Capturing emitted CO_2 from fossil-fired power plants can be done but adds a cost that ranges between $20 and $80 per metric ton of produced CO_2 (IEA, 2006). Furthermore, carbon capture reduces overall power plant efficiency as the process is energy consuming. Considering that a typical coal-fired power plant emits about 2.78 million tonnes of CO_2 per year according to the U.S. Energy Information Administration (EIA, 2015, 2016), the annual cost of CO_2 capture for an average coal-fired power plant, using pulverized solid coal as fuel, is about $150M ($55/tonne of CO_2 avoided) (Finkenworth, 2011). Such a high cost is economically untenable at the present time and does not include the cost of transport and sequestration of the CO_2, which is estimated to be about an extra $10/tonne of CO_2 avoided (David and Herzog, 2000). Other potential problems are unexpected effects of chemical interaction of $ScCO_2$ and minerals in rocks that could adversely affect permeability. This could include precipitation of alkali salts and carbonates as CO_2 content increases in any residual aqueous phase, potentially reducing permeability.

Deep, Hot Sedimentary Aquifers

Potentially attractive targets for future development are deeply buried (3 to 4 km) sedimentary reservoirs in regions having elevated heat flow, such as the Great Basin in the western United States (Allis and Moore, 2014; Allis et al., 2011, 2012) and the Molasse Basin in southern Germany (Homuth and Sass, 2014). For the Great Basin, geothermal gradients indicate temperatures in the range of 150° to 250°C at depths of about 3 to 4 km for buried stratigraphic reservoirs (Figure 11.22). These reservoirs have surface areas generally at least a magnitude larger (>100 km^2) than currently developed structurally controlled hydrothermal reservoirs (generally <10 km^2) in the Great Basin. Although deeper than most currently producing fault-controlled geothermal systems, the large surface areas of buried sedimentary stratigraphic reservoirs make for much better drilling targets. If good permeability is encountered, the large volume of these reservoirs could provide a power potential of hundreds of megawatts. A key feature of these potential stratigraphic geothermal reservoirs is that they underlie 2 to 3 km of loosely consolidated basin fill that acts like a thermal blanket, allowing for temperatures of 170° to 230°C to occur at

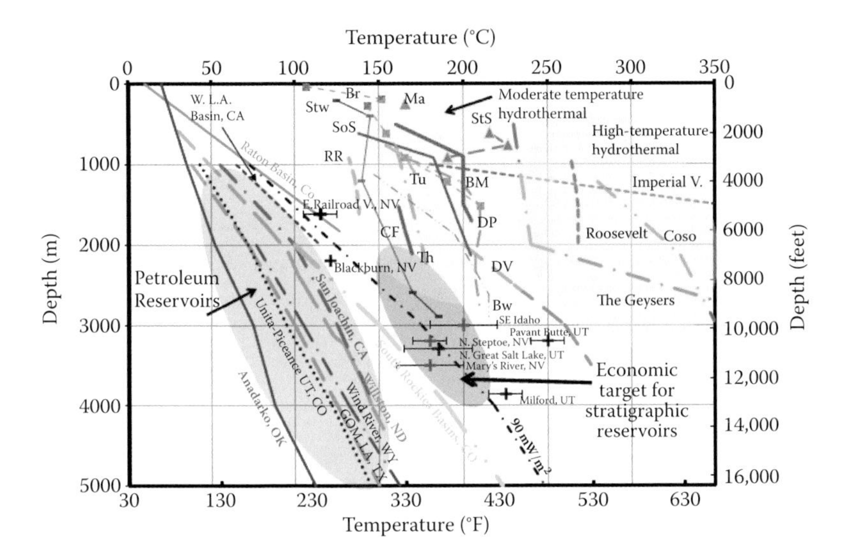

FIGURE 11.22 Compilation of temperature and depth data of geothermal systems in shades of yellow, deep stratigraphic reservoirs in mauve, and petroleum reservoirs in blue–gray. The economic target zone for deep stratigraphic reservoirs occurs at depths of 3 to 4 km and a temperature range from 160° to 220°C using a geothermal gradient of 55° to 60°C per km corresponding to a heat flow of 90 mW/m^2 typical for the Great Basin in which low thermally conductive, unconsolidated basin fill rocks overlie the potential stratigraphic reservoirs. Colored lines are measured geothermal gradients from well data for geothermal systems as noted by abbreviations: BM, Blue Mountain, NV; Br, Bradys, NV; Bw, Beowawe, NV; CF, Cove Fort, UT; DP, Desert Peak, NV; DV, Dixie Valley, NV; Ma, Mammoth, CA; RR, Raft River, ID; SoS, Soda Springs, NV; StS, Steamboat Springs, NV; Stw, Stillwater, NV; Th, Thermo, UT; Tu, Tuscarora, NV. (From Allis, R. and Moore, J., *Geothermal Resources Council Transactions*, 38, 1009–1016, 2014.)

depths of about 3 km. Because of their size and power potential, Allis et al. (2012) suggested that they could serve as a bridge toward development of fully engineered geothermal systems.

Potential Stratigraphic Reservoirs in the Great Basin, Western United States

Current geothermal production in the Great Basin mainly occurs along upwelling hydrothermal fluids localized along range-bounding normal faults. In the eastern Great Basin, tabular and sub-horizontal basins of carbonate rocks occur in a region of high crustal heat flow (~80 to 100 mW/m^2) (Figure 11.22). A key question is whether good permeability (~>50 millidarcys [mD]) can be found at depths of 3 to 4 km. Studies from oil and gas wells drilled in Wyoming, Colorado, New Mexico, and Utah showed that the permeabilities averaged about 100 mD, with some approaching 1000 mD even at depths of about 5 km (Figure 11.23). Also notice in Figure 11.23 that the permeability of carbonate rocks seems to increase slightly at depths from 3 to 5 km, suggesting that the higher temperatures of fluids at those depths might result in some dissolution (karstification) or volume contraction due to dolomitization or possible formation of calc-silicate minerals (such as prehnite, wairakite, and iron-rich epidote).

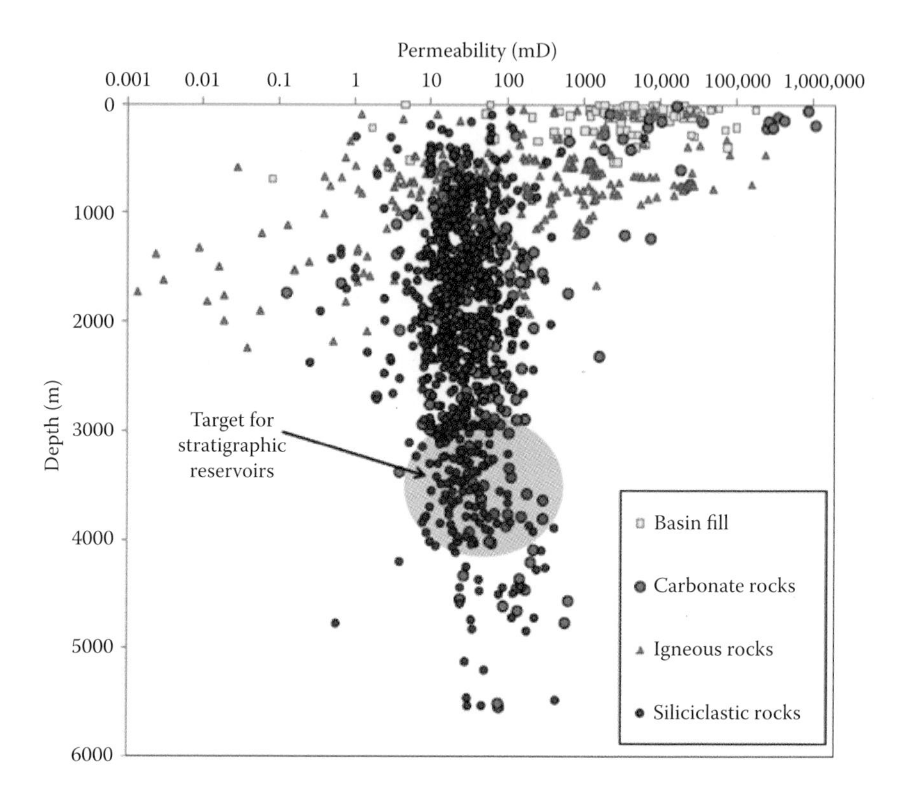

FIGURE 11.23 Permeability data from drill holes (oil, gas, and water) for Utah, New Mexico, and Colorado. (Adapted from Allis, R. and Moore, J., *Geothermal Resources Council Transactions*, 38, 1009–1016, 2014.)

Two potential basins for geothermal development that have been examined are the Black Rock Desert in west-central Utah and Steptoe Valley in eastern Nevada (Figure 11.22) (Allis et al., 2012). The Black Rock Desert is underlain by about 3 km of sedimentary basin fill lying on top of carbonate rocks of lower Paleozoic age. The basin fill acts like a thermal blanket (low thermal conductivity) due to its high porosity. Three oil and gas exploration wells drilled in the basin encountered bottom-hole temperatures ranging from about 160° to 230°C at depths from 3300 to 5300 m, resulting in calculated heat flows of 85 to 100 mW/m^2. Tests on the matrix permeability of basement carbonate rocks from one well were low (0.1 to 7 mD), but this represents a minimum as several zones in the carbonate rocks are broken to highly fractured. A drill-stem test, which would take into account fracture permeability, indicated permeability of about 42 mD (Allis et. al., 2012). However, results of drill-stem tests of deep exploration oil wells in the Great Basin typically indicate hydrostatic conditions for deep stratigraphic rock units, suggesting possible laterally extensive high permeability for these zones (Allis et al., 2013).

In northern Steptoe Basin, Nevada, the deepest well extended to 3600 m with a bottom-hole temperature of 200°C, yielding a calculated heat flow of 95 mW/m^2. The unconsolidated basin fill ranges from 1600 to about 2100 m thick and overlies basement rocks of limestone and dolostone, known regionally to have high permeability where they serve as major groundwater aquifers at shallower depths.

Deep Carbonate Reservoirs of the Molasse Basin, Germany

A 1600-m-deep research well and a 6020-m-long geothermal well (total vertical depth of 4850 m) were drilled to help characterize deep carbonate units as possible stratigraphic geothermal reservoirs. The bottom-hole temperature of the deeper geothermal well was about 170°C. Samples were collected during drilling to characterize rock permeability and thermal conductivity under laboratory conditions. Determined rock (matrix) permeability ranges from 0.001 to 10 mD (Homuth and Sass, 2014). Such permeability values would generally be too low to support flow rates necessary for geothermal power production; however, no *in situ* drill-stem tests appear to have been done. For this reason, the laboratory-measured permeability values could be a minimum, as analyzed samples probably represent regions of less fractured rock between fracture zones or cavities due to karstification. Therefore, a likely next step would to re-enter the wells and conduct drill-stem tests at different intervals within the perceived reservoir target zone to assess actual *in situ* permeability.

Paris Basin Direct Use

The deep sedimentary aquifer of the Paris Basin has been tapped for geothermal heating of buildings since the early 1970s. At present, approximately 150,000 buildings are heated from 40 operating geothermal facilities consisting mainly of doublet boreholes (one well for production and one well for injection). The geothermal reservoir is the Dogger aquifer that lies at a depth between 1500 and 2000 m and has temperatures ranging between 65° and 85°C. Flow rates average about 200 m^3/hr but can be as much as 600 m^3/hr (Boissier et al., 2009). A single doublet well system with a 250-m^3/hr flow rate, a production temperature of 70°C, and an injected fluid temperature of 45°C can serve about 4000 dwellings. Despite the long production of some of the doublet systems (35 to 40 years), production well temperatures have remained steady. Nonetheless, the net heat flux is considered insufficient to maintain the current production temperatures indefinitely. Most numerical models indicate cooling of doublet systems by 1.5° to 3.5°C after 40 years due to the continued reinjection of 40° to 45°C cool brine (Lopez et al., 2010). As a way to mitigate the cooling effect, a method of deep aquifer heat storage has been proposed in which waste heat from other industrial activities is injected into the reservoir during the summer months when heat withdrawal is minimal. Conceptually, such a process would help thermally recharge the producing aquifer and store the heat for exploitation during winter.

Supercritical Water Systems

Exploiting supercritical water systems is being explored by the Iceland Deep Drilling Project (IDDP), Japan Beyond-Brittle Project (JBBP), and Hotter and Deeper Exploration Science (HADES) program in New Zealand. Of those three projects, the Iceland effort has done the most work in trying to develop a supercritical system as described in more detail below. Reasons for developing supercritical water systems center on the much greater enthalpy and mass transfer abilities of supercritical water systems compared to subcritical or conventional hydrothermal systems. At supercritical conditions for water, similar to that discussed for CO_2, a distinct boundary between liquid and steam phases no longer exists. For pure water, this change occurs at a temperature of 374°C and 221 bars called the *critical point* (Figure 11.24). The temperature and pressure of the critical point increase with increased dissolved solutes. For example, the critical point for seawater containing 33‰ NaCl is about 298 bars and about 407°C. Because of the high temperatures of supercritical fluids, any acids, such as HCl or H_2SO_4, that might be present do not form reactive H^+ because there is no liquid water in which to dissolve and dissociate. This avoids the potential acid problem that can plague high-temperature, but subcritical, liquid- and vapor-dominated geothermal systems, such as in the northwestern part of The Geysers geothermal field in northern California.

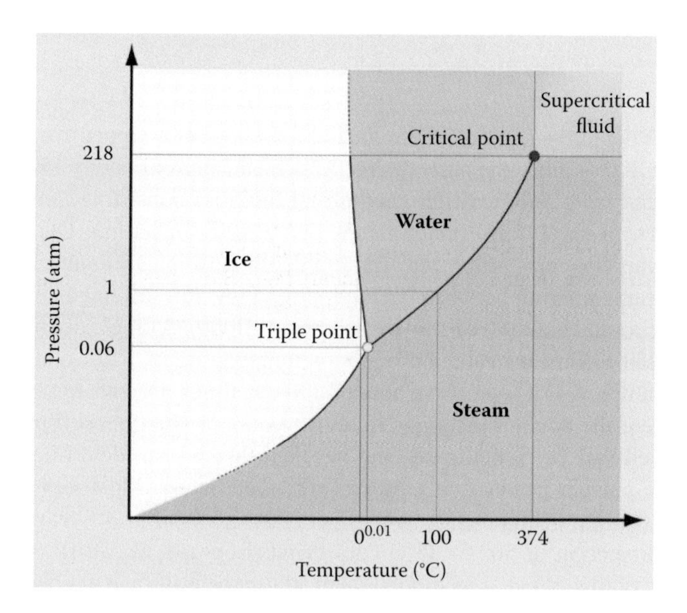

FIGURE 11.24 Phase diagram for pure water showing the region of supercritical fluid, whose properties vary according to temperature and pressure (218 atm is equal to about 220 bars of pressure). At higher temperature for a given pressure, a supercritical fluid is more gas like, and at higher pressure for a given temperature the supercritical fluid is more liquid like. Note that with increasing dissolved salts the critical point increases with temperature and pressure, so for seawater the critical point is 407°C and 298 bars.

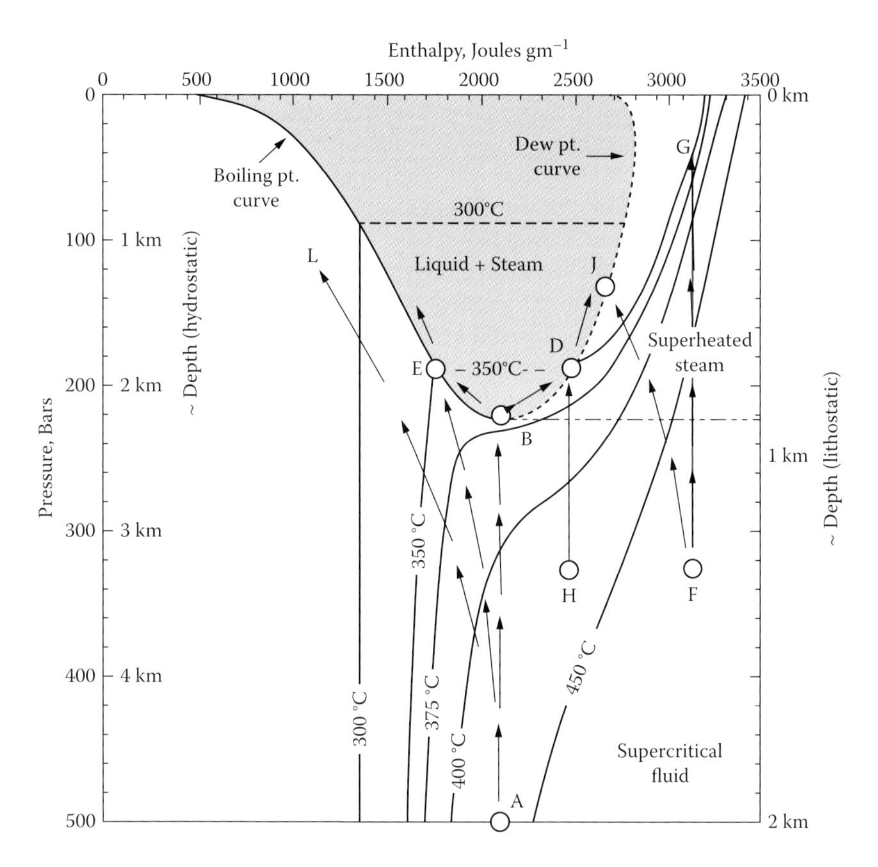

FIGURE 11.25 Pressure–enthalpy diagram showing conditions for super- and subcritical conditions for pure water. Shaded region is the two-phase field of liquid plus steam and is bounded on the left by the boiling point with depth curve and on the right the condensation or dew point curve (where steam condenses to liquid). Point B represents the critical point, and arrows denote various paths for ascending geothermal fluids under different starting conditions. See text for details. (From Elders, W.A. and Fridleifsson, G.O., in *Proceedings of World Geothermal Congress 2010*, Bali, Indonesia, April 25–30, 2010.)

Another benefit is that the diffusivity of supercritical water is 10 to 100 times that of the subcritical state because surface tension is negligible as water becomes nonpolar under supercritical conditions. As a result, the ratio of buoyancy to viscous forces is greatly improved, leading to enhanced rates of mass transport. Furthermore, the enthalpy of supercritical fluids is much higher than that of subcritical fluids (Figure 11.25). Figure 11.25 is an inverted variation of the pressure–enthalpy diagram that was examined Chapter 6 but is expanded here to emphasize the supercritical region. Flow paths to the left of vertical line A–B reflect paths of ascending supercritical geothermal fluids experiencing conductive cooling. These fluids transition into subcritical fluids typical of liquid-dominated reservoirs with boiling (at E) or without boiling at L. To the right of line A–B, path H–D represents a transition from supercritical conditions to subcritical conditions in which the fluid is partitioned into wet steam at D and liquid at E, a situation characteristic of vapor-dominated geothermal reservoirs. The goal of developing a supercritical system would be to follow path F–G in which the fluid delivered to the power plant arrives as superheated steam (no liquid present) and with higher enthalpy than wet steam, such as at D, which might be found in a subcritical vapor-dominated reservoir. The path F–G would be facilitated by high flow rates, dominated by adiabatic and limited conductive cooling.*

Combining the improved rates of mass transport and high enthalpy content means that supercritical fluids contain 5 to 10 times the power output than a typically developed subcritical liquid-dominated resource. For example, an aqueous fluid at 400°C and pressure of 250 bars (which would be at supercritical conditions) contains more than five times the power-producing potential of a subcritical geofluid at 225°C (Tester et al., 2006). By comparison, a typical geothermal well tapping a geofluid at temperatures ranging from 200° to 300°C at a depth of 2000 to 3000 m would have a power output ranging between about 5 and 10 MWe. Drilling such a well at current prices would cost about $5 to $7 million. Drilling 1 to 2 km deeper to access supercritical fluids might cost an additional $1 to $2 million. However, it could increase power output to 25 to 50 MWe, or about the aggregate equivalent of five wells or more tapping a subcritical or conventional geothermal system and costing $25 to $30 million to drill.

Nonetheless, significant challenges remain in developing such systems. For one, supercritical conditions are typically found below the brittle–ductile transition zone, where rock deforms plastically, greatly restricting the formation of any open-space fracture due to rock creep. Therefore, rock permeability is reduced significantly. For silicic igneous rocks or siliciclastic sedimentary rocks, the brittle–ductile transition occurs at temperatures of about 370 to 400°C, a range close to the critical temperature. For mafic rocks, such as basalt and gabbro, the transition to ductile deformation occurs at temperatures of 500° to 600°C, reflecting the higher temperature stabilities of minerals making up mafic igneous rocks. Thus, for mafic volcanic rocks, which make up much of Iceland, supercritical fluids may be encountered in still brittle rocks than can support fracture permeability.

Even in brittle rocks containing supercritical fluids, another potential difficulty concerns the solubility of quartz or silica. As discussed by Fournier (1999, 2007), at temperatures above about 350°C, silica or quartz solubility no longer increases but decreases with temperature, which can lead to precipitation of quartz that can clog any existing or newly formed fractures. Furthermore, at temperatures above about 350°C, silica solubility is sensitive to pressure, such that decreasing pressure also reduces silica solubility (Figure 11.26). This means that a supercritical fluid at, say, 450°C that undergoes adiabatic (i.e., minimal heat transfer) expansion while rising in a well could deposit significant silica and potentially clog the well. In silica-poor mafic reservoir rocks, fluid silica content may be lower, which could reduce potential silica deposition in fractures or scale in wells.

* Vertical paths such as A–B and H–D in Figure 11.25 reflect adiabatic cooling or no transfer of heat to the surroundings as occurs during conductive cooling, such as illustrated by paths A–L or F–J. As a parcel of fluid ascends adiabatically it decompresses or expands, so the temperature declines but the overall enthalpy remains constant because transfer of heat to the surroundings is nil.

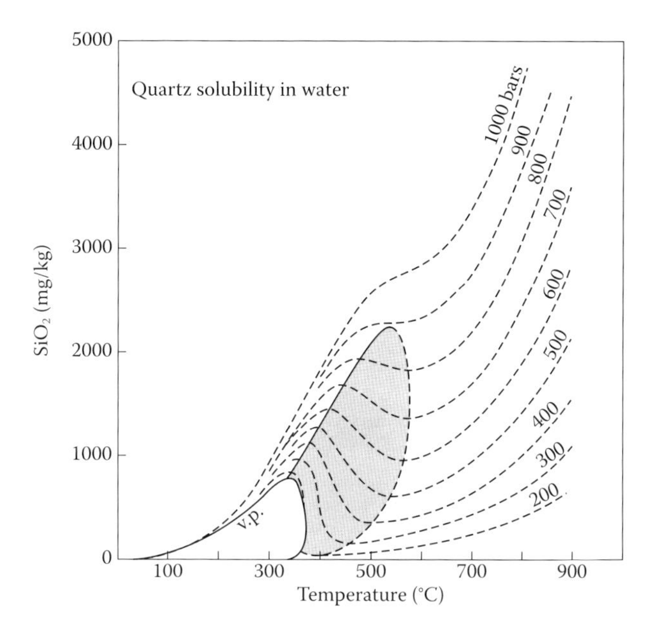

FIGURE 11.26 Calculated solubilities of quartz as a function of temperature and pressure with isobars ranging from 200 to 1000 bars. The stippled region denotes retrograde solubility of quartz in which solubility decreases with increasing temperature which can lead to precipitation and clogging of fractures. Similarly, at temperatures above 350°C, decreasing pressure, as would be experienced with a fluid rising in a well, leads to lower quartz solubility, which can also promote quartz precipitation or scaling in pipes. (From Fournier, R.O., *Economic Geology*, 94(8), 1193–1211, 1999.)

Iceland Deep Drilling Project

The Iceland Deep Drilling Project (IDDP) is a consortium of Icelandic power companies, the Icelandic government, and international partners to investigate the potential of developing geothermal systems under supercritical conditions as a source of energy. From 2008 to 2009, IDDP-1 was drilled in the Krafla geothermal system located in north-central Iceland, about 1 km north of the 60-MWe Krafla geothermal power plant. The plan was to drill to a depth of about 4500 m and to reach temperatures >450°C and pressures of about 300 bars (Figure 11.27) (Elders and Fridleifsson, 2010; see also special edition of *Geothermics*, 49(1), 2014). These conditions of temperature and pressure would coincide with the supercritical region in the vicinity of point F in Figure 11.25, and fluids would move up the well adiabatically along path F to G. As the fluid rises, it would transition from a supercritical aqueous phase to superheated (but subcritical) steam, without intersecting the dew point curve that bounds the high enthalpy side of the two-phase liquid–steam region shaded in green. Path F to J would indicate conductive cooling that might occur with slower flow rates. At J, the fluid would partition into a mixture of steam and liquid and any acids that are present could then dissociate, as discussed above, and result in potentially acidic and corrosive two-phase flow.

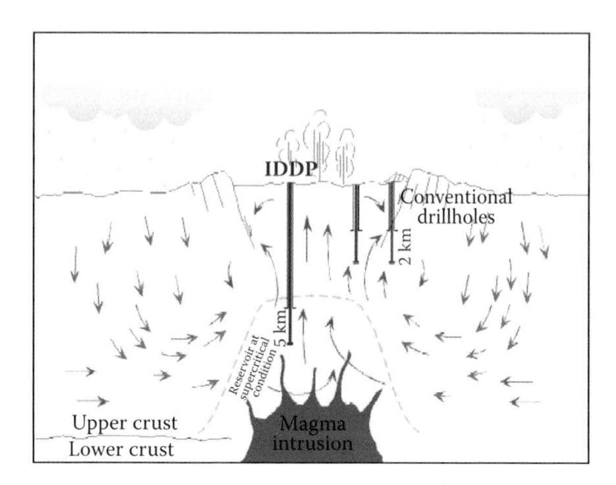

FIGURE 11.27 Conceptual model of target zone of IDDP-1 to intersect reservoir of supercritical fluids below a depth of 4 km. (From Elders, W.A. and Fridleifsson, G.O., in *Proceedings of World Geothermal Congress 2010*, Bali, Indonesia, April 25–30, 2010.)

Drilling was aborted at a depth of 2.1 km when rhyolitic magma (>900°C) flowed into the well (Zierenberg et al., 2013). The well, however, was successfully cased to that depth to examine the possibility of producing from the >500°C contact zone of the intrusion. A 2-year flow test ensued, involving a nearby well that served as an injection well, and the IDDP well produced >450°C superheated steam at high pressures of 40 to 140 bars (Elders et al., 2014). While it was flowing, IDDP-1 was the hottest geothermal well on the planet and was a successful demonstration of an EGS system developed in the chilled contact zone of an underlying silicic intrusion. Unfortunately, the master valves at the wellhead eventually failed due to the harsh conditions, and the well is currently being quenched with cold water injection until new master valves are obtained. If put back online, the well could add about 25 to 30 MWe (about the equivalent of five or six conventional wells combined) to the current 60-MWe Krafla geothermal plant—a potential 50% increase in power output from one well.

Japan Beyond-Brittle Project

Although in a tectonically favorable region for developing geothermal energy (subduction-zone magmatism related to oceanic–oceanic plate convergence), Japan has had limited geothermal development for the last 12 to 15 years for several reasons:

1. Geothermal energy is perceived as having high relative costs.
2. Geothermally produced power is less than that produced by fossil- and nuclear-fueled power plants.
3. Uncertainty and high risks are associated with developing underground resources (e.g., risk of dry wells).
4. Thermal springs, considered culturally important, could potentially become degraded.
5. Existing power plants are of relatively small capacity (10 to 20 MWe) and have experienced declines of temperature over time (sustainability concerns).

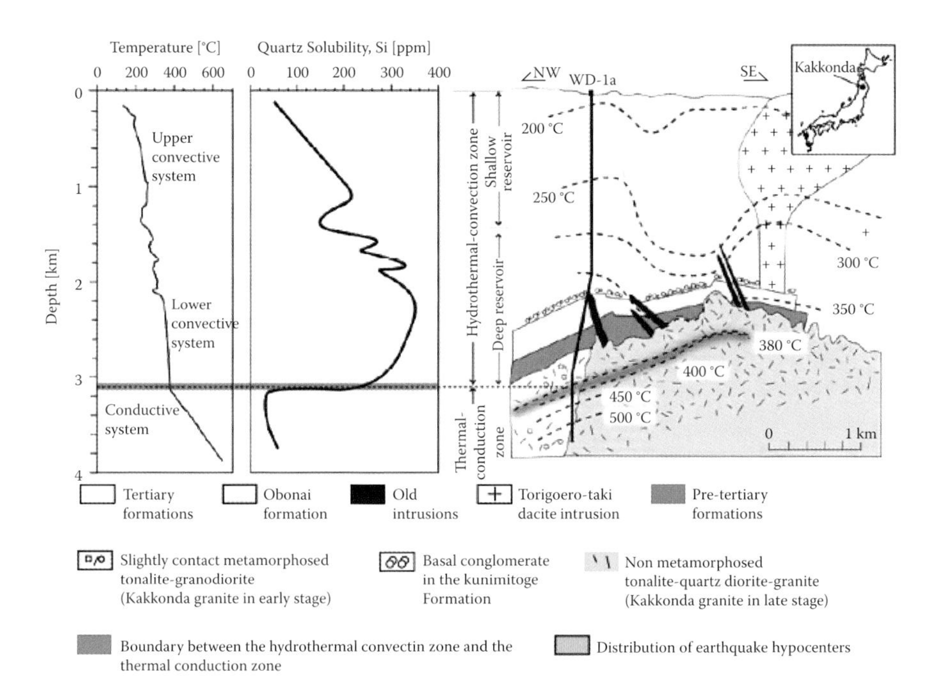

FIGURE 11.28 Location of Kakkonda geothermal field and temperature, quartz solubility, and geologic cross-section of the Kakkonda geothermal field. Note that the convective geothermal systems overlap with the distribution of microearthquake hypocenters (blue-shaded region) indicative of brittle fracturing; earthquake hypocenters do not extend below brittle ductile transition. (Adapted from Muraoka, H. et al., *Scientific Drilling*, 17, 51–59, 2014.)

Following the magnitude 9.1 Tohuku earthquake and subsequent damage to the Fukushima nuclear power plant in 2011, the national energy outlook in Japan, however, changed drastically. Government, industry, and citizenry are now looking to develop alternative energy sources, including greater exploitation of geothermal energy. The JBBP builds upon the results of studies evaluating the deep geothermal resources at the Kakkonda geothermal field (Asanuma et al., 2012; Muraoka et al., 2014; Tamanyu and Fujimoto, 2005), where a 3.7-km borehole encountered shallow and deep convective geothermal systems in brittle rock that extended to a depth of 3.1 km and a temperature of 380°C. At 3.1 km, the brittle–ductile transition zone was penetrated and thereafter temperature steadily increased to over 500°C in ductile and thermally conductive rock (Figure 11.28). Notice in Figure 11.28 that silica solubility reaches a maximum at about 360°C and a depth of about 2.4 km and then gradually decreases to the brittle–ductile transition zone at a depth of 3.1 km, where solubility drops precipitously. This abrupt drop in silica solubility reflects significant quartz/silica precipitation, which, together with rock creep from the onset of plastic deformation, causes any open-space permeability to be largely obliterated. Therefore, the brittle–ductile transition zone serves as a mainly impermeable barrier, isolating the overlying convective geothermal system from an underlying conductive geothermal system.

The JBBP explores the possibility of developing EGS resources at the brittle–ductile transition zone and deeper levels for the following reasons:

1. Because hot and ductile rocks at depth are much more widespread than shallow conventional geothermal resources, problems with localized distribution of shallow conventional geothermal resources, such as drilling dry wells, are reduced.
2. In EGSs developed to date in brittle rocks, only about 50% or less of the injected fluid is returned, and significant make-up water is required; however, about 100% recovery can be expected from ductile zones.
3. Induced seismicity resulting from hydraulic stimulation and thermal contraction in brittle rock would likely be suppressed in ductile rock due to the small size of fractures formed from thermal contraction of injected fluid. Also, seismic energy would be attenuated due to the ductile nature of the rock.
4. More homogeneous rock properties and stress conditions allow simpler designs for developing geothermal resources. Because site characteristics are less unique, widely applied methodologies for design, development, and production can be utilized, saving time and expense.

The key components in JBBP are illustrated in Figure 11.29. The JBBP envisions developing two types of reservoirs. Type I spans the brittle–ductile transition zone in which artificially produced fractures from hydraulic stimulation and thermal contraction would connect with fractures associated with the overlying deep-seated reservoir in brittle rock. This reservoir would be at depths of about 3 km with temperatures ranging from 350° to 400°C. Because type I lies in the zone of retrograde quartz, solubility efforts would be needed to control silica precipitation and scaling. This is critical, as experiments have demonstrated that quartz precipitation at temperatures greater than 400°C can seal permeability on time scales as short as days or weeks (Muraoka et al., 2014). Type II would lie in fully ductile rocks at depths of about 4 km and temperatures of about 500°C (supercritical conditions). Here, the isolated fracture network would be produced by artificial stimulation with permeability being improved by the supercritical nature of the aqueous fluid phase. The type II reservoir, relative to type I, would have higher enthalpy, nearly full recovery of injected fluid, and reduced potential for induced seismicity. As with type I, potential self-sealing from quartz precipitation could be a serious problem and would have to be addressed, probably by chemical manipulation of injected and circulated fluid.

Drilling to below the brittle–ductile transition zone is scheduled for early 2017. Prior to that time, studies will continue to focus on the following:

1. Characterizing rock conditions in ductile rock, including water–rock interactions, thermal properties, and nature of pore water under ductile conditions
2. Further understanding of rock mechanics at ductile conditions via experimentation to examine how stress may create fractures under otherwise ductile conditions, along with numerical modeling
3. Drilling methodologies for deep, hot wells including suitable drill muds, cementing of casing under harsh conditions, and use of logging and monitoring tools

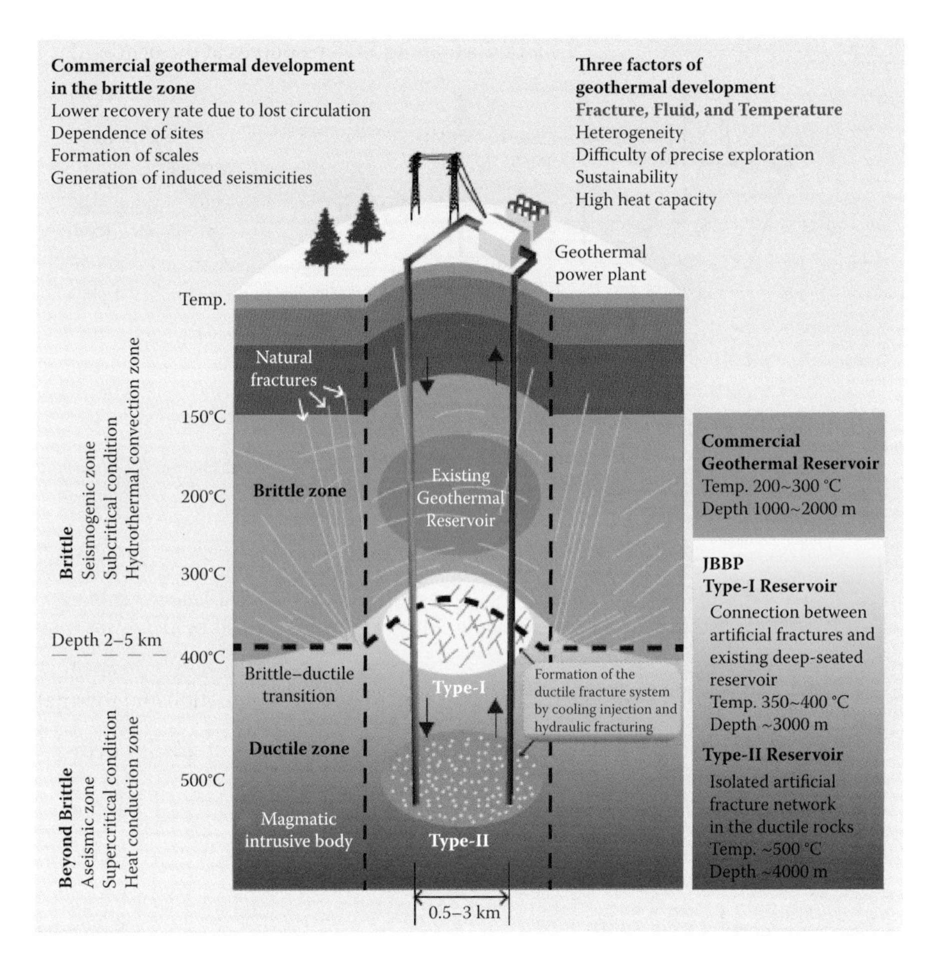

FIGURE 11.29 Diagram showing the two types of JBBP reservoirs that might underlie existing commercial geothermal systems in brittle rocks. (From JBBP (Japan Beyond-Brittle Project), http://www.icdp-online.org/fileadmin/icdp/projects/doc/jbbp/JBBP_Concept_poster_En.pdf.)

Hotter and Deeper Exploration Science, New Zealand

Similar to Iceland, New Zealand has abundant geothermal resources. Currently, the country has a geothermal power capacity of about 1080 MWe, which represents about 16% of the total installed capacity. Most of those resources are located in the Taupo Volcanic Zone on the North Island, one of the most volcanically active regions on the planet. All developed geothermal sites are relatively shallow (<3 km) and make use of conventional, convecting geothermal systems in brittle volcanic rocks having both fracture and primary permeability. The objective of the Hotter and Deeper Exploration Science (HADES) project is to explore the development of "hotter and deeper" systems below currently developed geothermal resources but, unlike JBBP, mainly above the brittle–ductile transition zone (Figure 11.30). Specific goals of the project are to

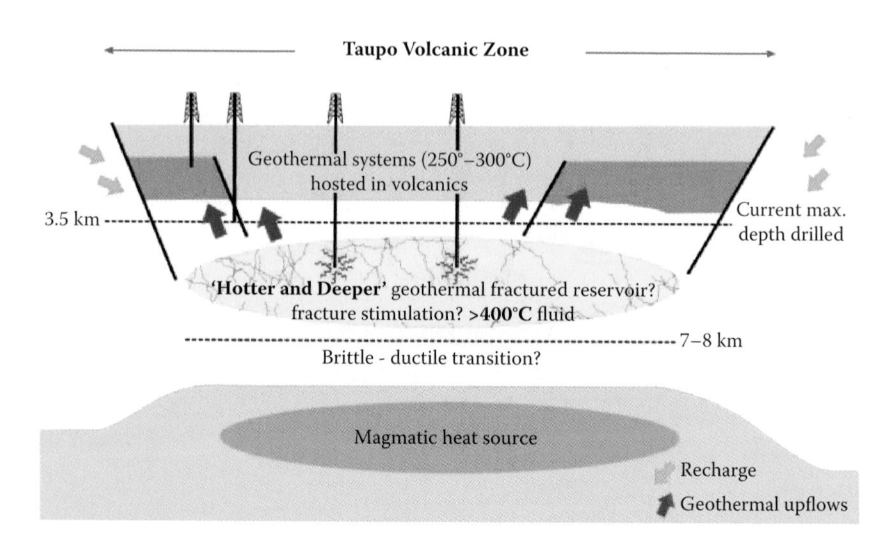

FIGURE 11.30 Schematic cross-sectional view of the Taupo Volcanic Zone showing the location of potential "hotter and deeper" geothermal resources located below 3.5 km and above the brittle–ductile transition zone (about 7 to 8 km deep) marked by the lower limit of shallow seismicity. This "hotter and deeper" region would be in the realm of supercritical fluids. (From Heise, W. et al., *Geophysical Research Letters*, 34, L14313, 2007.)

1. Provide incentives for developers, including reduced risks, to justify drilling deep exploration wells.
2. Image deep geothermal systems in the Taupo Volcanic Zone using detailed geophysical techniques including gravity, magnetotelluric resistivity surveys, and aeromagnetic studies.
3. Formulate predictive models for fluid behavior and mechanical rock properties in the envisioned deep target zones to help develop constraints on potential hydrofracturing to enhance permeability at deep levels (>4 km).
4. Drill a 4- to 5-km-deep exploratory well to assess the results of previous studies and with the goal of demonstrating feasibility to develop deep geothermal regions.

Currently developed conventional geothermal systems in New Zealand are developed in brittle rocks with sufficient permeability to support hydrothermal convection. Envisioned deep geothermal zones, below and spanning the brittle–ductile transition zone, would be heated mainly by conduction. Injection of cold water into this region is envisioned to thermally shock and fracture rock, creating improved permeability, stimulating convection, and offering the opportunity to withdraw heat to produce power. Enhanced permeability, however, may in part be negated by deposition of quartz.

Recent results from one of the largest three-dimensional magnetotelluric (MT) arrays, which are used to image deep zones (below 3 km) of resistivity anomalies, in the Taupo Volcanic Zone are consistent with sustained zones of convection extracting energy and volatiles from quasi-plastic rock at depths of 5 to 7 km (Lindsey et al.,

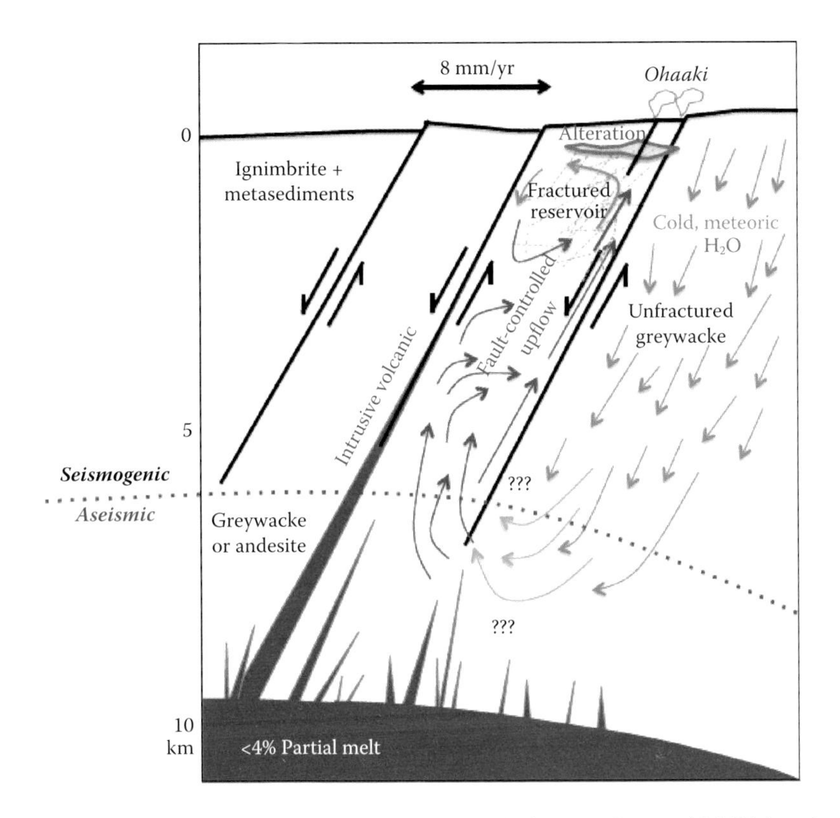

FIGURE 11.31 Conceptual model of deep fluid circulation, consistent with MT data, leading to geothermal exchange of energy and mass in the lower part of shallow geothermal systems, such as at Ohaaki shown here. The seismic/aseismic boundary coincides with the brittle–ductile transition zone. (Adapted from Lindsey, N.J. et al., *Geothermal Resources Council Transactions*, 38, 527–532, 2014.)

2014). Interpretation of the MT data suggests that hot brines migrate upward through localized fault zones leading to geothermal exchange of energy and mass to the lower parts of the shallow (<3 km deep) conventional geothermal systems (Figure 11.31).

SUMMARY

Enhanced and engineered geothermal systems have the potential to increase significantly the geothermal power output on the order of 100 times that of current production. If achieved, geothermally derived energy could provide about a third of the total U.S. power production, an increase largely due to the more widespread occurrences of hot rock (petra-heat) at depth than the current localized nature of currently producing conventional geothermal systems, which require the orchestration of unique geologic conditions for successful development.

Enhanced geothermal systems are typically developed in areas of existing conventional geothermal systems where certain wells are unproductive due to limited permeability or water availability. These wells are then stimulated through a combination of hydroshearing and more pressurized hydraulic fracturing to increase power production. This was successfully demonstrated at Desert Peak, Nevada, and is currently being explored in the Northwest Geysers geothermal field. Engineered geothermal systems focus on developing areas of hot rock where no conventional geothermal system exists. In this case, cold fluid is injected under varying amounts of pressure to hydroshear and thermally contract hot rock, inducing fractures and improved permeability. Engineered geothermal systems have been successfully developed at Insheim and Landau in Germany but they are typically small with installed capacity of about 5 MWe each. In the United States, a demonstration project for developing an engineered geothermal system is underway at Newberry Volcano in central Oregon. There, the operators have developed a thermally degradable zonal isolation material that temporarily plugs artificially stimulated fractures through hydroshearing. Then, during subsequent injection events, new fractures are formed that increase the size of the potential geothermal reservoir.

Also, instead of using water as the vehicle for extracting energy from hot rocks, some workers have suggested using supercritical CO_2. One advantage of using supercritical CO_2 is higher fluid transport rates due to the reduced viscosity of CO_2 compared to that of water. A second advantage is the limited solubility of dissolved minerals in supercritical CO_2 such that self-sealing of fractures from mineral deposition is minimized; however, obtaining a steady supply of CO_2 for injection at an economical cost remains problematical.

Deep and hot sedimentary aquifers occurring at depths of 3 to 4 km with temperatures of contained fluids ranging from about 150°C to as much as 200°C have an order of magnitude larger surface and target area compared to developed structurally controlled geothermal reservoirs. Due to their large size and potential high flow rates, geothermal power plants tapping such reservoirs could produce an order of magnitude more power (200 to 300 MWe) than typical structurally controlled reservoirs developed using binary technology (20 to 30 MWe). As such, development of deep, hot stratigraphic reservoirs could serve as a bridge between current development of conventional convective geothermal systems and realization of engineered geothermal systems.

Finally, the development of supercritical aqueous fluid systems remains experimental but prospective. Advantages include the high enthalpy and improved flow rates of supercritical fluids that can increase well power output by 5 to 10 times that of a well producing subcritical fluids. If realized, tapping of supercritical fluids would reduce the costs of drilling significantly and have less of an environmental impact, as fewer roads and drill pads would be required to develop the field. Nonetheless, some problems still remain to be resolved:

- Potential degradation of equipment due to the harsh conditions requiring expensive alloys for well casing
- Reduction in permeability due to the transition from brittle to ductile behavior at supercritical conditions for silicic or siliciclastic rocks
- Potential mineral deposition of quartz and other minerals that have retrograde solubility at the temperatures and pressures of supercritical conditions

This last problem can lead to self-sealing in the reservoir or development of scale in wellbores, thus potentially negating, at least in part, the elevated enthalpy and transmissivity of supercritical fluids.

In spite of these potential hurdles, work continues to test for developing supercritical fluids in Iceland with the Iceland Deep Drilling Project (IDDP), in Japan with the Japan Beyond-Brittle Project (JBBP), and in New Zealand with the Hotter and Deeper Exploration Science (HADES) program. Geothermal energy plays a significant role in each of these countries, both for power generation and direct use; it is not surprising that these countries are on the vanguard of geothermal research and development as each has limited or no fossil fuel resources.

SUGGESTED PROBLEMS

1. You are presenting at a town hall meeting on developing an engineered geothermal system that would provide the community with clean electrical energy. However, an avid, and emotional, environmentalist objects to your project, saying that high volumes of water will be consumed during creation of the reservoir and considerable make-up water will be required during operation of the geothermal power plant. Moreover, according to this person, fluid injection will create undesirable earthquakes that will be a threat to people's safety and property. How would you respond to this person to allay his or her concerns using knowledge gained from this chapter?
2. The potential EGS resource is very large, more than an order of magnitude available from conventional geothermal resources. If you were an investor, what criteria would you use to determine where first to deploy this technology and why?

REFERENCES AND RECOMMENDED READING

Allis, R. and Moore, J. (2014). Can deep stratigraphic reservoirs sustain 100 MW power plants? *Geothermal Resources Council Transactions*, 38: 1009–1016.

Allis, R., Moore, J., Blackett, B., Gwynn, M., Kirby, S., and Sprinkel, D. (2011). The potential for basin-centered geothermal resources in the Great Basin. *Geothermal Resource Council Transactions*, 35(1): 683–688.

Allis, R., Blackett, B., Gwynn, M. et al. (2012). Stratigraphic reservoirs in the Great Basin—the bridge to development of enhanced geothermal systems in the U.S. *Geothermal Resources Council Transactions*, 36: 351–357.

Allis, R., Moore, J.N., Anderson, T., Deo, M., Kirby, S., Roehner, R., and Spencer, T. (2013). Characterizing the power potential of hot stratigraphic reservoirs in the western U.S. In: *Proceedings of the 38th Workshop on Geothermal Reservoir Engineering*, Stanford, CA, February 11–13 (http://www.geothermal-energy.org/pdf/IGAstandard/SGW/2013/Allis.pdf).

AltaRock. (2014). *Thermally Degradable Zonal Isolation Materials (TZIMs)*. Seattle, WA: AltaRock Energy (http://altarockenergy.com/technology/tzim/).

Asanuma, H., Muraoka, H., Tsuchiya, N., and Ito, H. (2012). The concept of the Japan Beyond-Brittle Project (JBBP) to develop EGS reservoirs in ductile zones. *Geothermal Resources Council Transactions*, 36: 359–364.

Benoit, W.R., Hiner, J.E., and Forest, R.T. (1982). *Discovery and Geology of the Desert Peak Geothermal Field: A Case History*. Reno: Nevada Bureau of Mines and Geology.

Boissier, F., Lopez, S., Desplan, A., and Lesueur, H. (2009). 30 years of exploitation of the geothermal resource in Paris Basin for district heating. *Geothermal Resource Council Transactions*, 33: 355–360.

Bradford, J., Ohren, M., Osborn, W.L., McLennan, J., Moore, J., and Podgorney, R. (2014). Thermal stimulation and injectivity testing at Raft River, ID, EGS site. In: *Proceedings of the 39th Workshop on Geothermal Reservoir Engineering*, Stanford, CA, February 24–26 (http://www.geothermal-energy.org/pdf/IGAstandard/SGW/2014/Bradford.pdf).

Breede, K., Dzebisashvili, K., Liu, X., and Falcone, G. (2013). A systematic review of enhanced (or engineered) geothermal systems; past, present and future. *Geothermal Energy*, 1: 4.

Brown, D.W. (2000). A hot dry rock geothermal energy concept utilizing supercritical CO_2 instead of water. In: *Proceedings of the 25th Workshop on Geothermal Reservoir Engineering*, Stanford, CA, January 24–26 (http://www.geothermal-energy.org/pdf/IGAstandard/SGW/2000/Brown.pdf).

Chabora, E., Zemach, E., Spielman, P. et al. (2012). Hydraulic stimulation of well 27-15, Desert Peak geothermal field, Nevada, USA. In: *Proceedings of the 37th Workshop on Geothermal Reservoir Engineering*, Stanford, CA, January 30–February 1 (http://www.geothermal-energy.org/pdf/IGAstandard/SGW/2012/Chabora.pdf).

ChemWiki. (2016). *Fundamentals of Phase Transitions*, http://chemwiki.ucdavis.edu/Core/Physical_Chemistry/Physical_Properties_of_Matter/States_of_Matter/Phase_Transitions/Fundamentals_of_Phase_Transitions.

Cladouhos, T.T., Petty, S., Nordin, Y., Moore, M., Grasso, K., Uddenberg, M., and Swyer, M.W. (2013). Improving geothermal project economics with multi-zone stimulation: results from the Newberry Volcano EGS demonstration. *Geothermal Resource Council Transactions*, 37: 133–140.

David, J. and Herzog, H. (2000). The cost of carbon capture. In: *Proceedings of the 5th International Conference on Greenhouse Gas Control Technologies*, Cairns, Australia, August 13–16 (http://sequestration.mit.edu/pdf/David_and_Herzog.pdf).

EERE. (2012). *What Is an Enhanced Geothermal System (EGS)?*, DOE/EE-0785. Washington, DC: Office of Energy Efficiency and Renewable Energy (http://www1.eere.energy.gov/geothermal/pdfs/egs_basics.pdf).

EIA. (2015). *Count of Electric Power Industry Power Plants by Sector, by Predominant Energy Sources within a Plant, 2004–2014*. Washington, DC: Energy Information Administration (https://www.eia.gov/electricity/annual/html/epa_04_01.html).

EIA. (2016). *How Much of U.S. Carbon Dioxide Emissions Are Associated with Electricity Generation?* Washington, DC: Energy Information Administration (http://www.eia.gov/tools/faqs/faq.cfm?id=77&t=11).

Elders, W.A. and Fridleifsson, G.O. (2010). The science program of the Iceland Deep Drilling Project (IDDP): a study of supercritical geothermal resources. In: *Proceedings of World Geothermal Congress 2010*, Bali, Indonesia, April 25–30 (http://www.geothermal-energy.org/pdf/IGAstandard/WGC/2010/3903.pdf).

Elders, W.A., Fridleifsson, G.O., and Albertsson, A. (2014). Drilling into magma and the implications of the Iceland Deep Drilling Project (IDDP) for high-temperature geothermal systems worldwide. *Geothermics*, 49: 111–118.

Faulds, J.E., Garside, L.J., and Oppliger, G.L. (2003). Structural analysis of the Desert Peak–Brady geothermal fields, northwestern Nevada: implications for understanding linkages between northeast-trending structures and geothermal reservoirs in the Humboldt structural zone. *Geothermal Resources Council Transactions*, 27: 859–864.

Finkenworth, M. (2011). *Cost and Performance of Carbon Dioxide Capture from Power Generation*. Paris: International Energy Agency (https://www.iea.org/publications/freepublications/publication/costperf_ccs_powergen.pdf).

Fournier, R.O. (1999). Hydrothermal processes related to movement of fluid from plastic into brittle rock in the magmatic-epithermal environment. *Economic Geology*, 94(8): 1193–1211.

Fournier, R.O. (2007). Hydrothermal systems and volcano geochemistry. In: *Volcano Deformation: Geodetic Monitoring Techniques* (Dzurisin, D., Ed.), pp. 323–342. Heidelberg: Springer-Verlag.

Freifeld, B. and Hawkes, D. (2011). Achieving carbon sequestration and geothermal energy production: a win-win! *ESD News and Events*, June 28, http://esd.lbl.gov/achieving-carbon-sequestration-and-geothermal-energy-production-a-win-win/.

Garcia, J., Walters, M., Beall, J. et al. (2012). Overview of the Northwest Geysers EGS demonstration project. In: *Proceedings of the 37th Workshop on Geothermal Reservoir Engineering*, Stanford, CA, January 30–February 1 (http://www.geothermal-energy.org/pdf/IGAstandard/SGW/2012/Garcia.pdf).

Heise, W., Bibby, H.M., Caldwell, T.G. et al. (2007). Melt distribution beneath a young continental rift: the Taupo Volcanic Zone, New Zealand. *Geophysical Research Letters*, 34: L14313.

Herzberger, P., Kolbel, T., and Munch, W. (2009). Geothermal resources in the German basins. *Geothermal Resource Council Transactions*, 33: 352–354.

Homuth, S. and Sass, I. (2014). Outcrop analogue vs. reservoir data: characteristics and controlling factors of physical properties of the Upper Jurassic geothermal carbonate reservoirs of the Molasse Basin, Germany. In: *Proceedings of the 39th Workshop on Geothermal Reservoir Engineering*, Stanford, CA, February 24–26 (http://www.geothermal-energy.org/pdf/IGAstandard/SGW/2014/Homuth.pdf).

IEA. (2006). *IEA Energy Technology Essentials—CO_2 Capture & Storage*. Paris: International Energy Agency (http://www.iea.org/techno/essentials1.pdf).

Kirby, S.M. (2012). *Summary of Compiled Permeability with Depth Measurements for Basin, Igneous, Carbonate, and Siliciclastic Rocks in the Great Basin and Adjoining Regions*. Salt Lake City: Utah Department of Natural Resources.

Lindsey, N.J., Bertrand, E.A., Caldwell, T.G., Gasperikova, E., and Newman, G.A. (2014). Imaging the roots of high-temperature geothermal systems using MT: results from the Taupo Volcanic Zone, New Zealand. *Geothermal Resource Council Transactions*, 38: 527–532.

Lopez, S., Hamm, V., Le Brun, M. et al. (2010). 40 years of Dogger aquifer management in Ile-de-France, Paris Basin, France. *Geothermics*, 39(4): 339–356.

MIT. (2006). *The Future of Geothermal Energy: Impact of Enhanced Geothermal Systems (EGS) on the United States in the 21st Century*. Cambridge, MA: Massachusetts Institute of Technology (http://www1.eere.energy.gov/geothermal/egs_technology.html).

Muraoka, H., Asanuma, H., Tsuchiya, N., Ito, T., Mogi, T., and Ito, H. (2014). The Japan Beyond-Brittle Project. *Scientific Drilling*, 17: 51–59.

Nash, G.D. and Moore, J.N. (2012). Raft River EGS project: a GIS-centric review of geology. *Geothermal Resource Council Transactions*, 36: 951–958.

NREL. (2015). *Dynamic Maps, GIS Data, & Analysis Tools*. Washington, DC: National Renewable Energy Laboratory (http://www.nrel.gov/gis/geothermal.html).

Ormat. (2013). Success with Enhanced Geothermal Systems Changing the Future of Geothermal Power in the U.S. [press release]. Reno, NV: Ormat Technologies, Inc. (http://www.ormat.com/news/latest-items/success-enhanced-geothermal-systems-changing-future-geothermal-power-us).

Pruess, K. (2006). Enhanced geothermal systems (EGS) using CO_2 as working fluid—a novel approach for generating renewable energy with simultaneous sequestration of carbon. *Geothermics*, 35(4): 351–367.

Pruess, K. (2007). On production behavior of enhanced geothermal systems with CO_2 as a working fluid. *Energy Conversion and Management*, 49: 1446–1454.

SMU. (2016). Southern Methodist University Geothermal Laboratory website, http://www.smu.edu/dedman/academics/programs/geothermallab.

Tamanyu, S. and Fujimoto, K. (2005). Hydrothermal and heat source model for the Kakkonda geothermal field, Japan. In: *Proceedings of World Geothermal Congress 2005*, Antalya, Turkey, April 24–29 (http://www.geothermal-energy.org/pdf/IGAstandard/WGC/2005/0915.pdf).

Tester, J.W., Anderson, B., Batchelor, A. et al. (2006). *The Future of Geothermal Energy: Impact of Enhanced Geothermal Systems (EGS) on the United States in the 21st Century*. Cambridge: Massachusetts Institute of Technology (http://www1.eere.energy.gov/geothermal/egs_technology.html).

Union of Concerned Scientists. (2016). *Environmental Impacts of Coal Power: Air Pollution*. Cambridge, MA: Union of Concerned Scientists (http://www.ucsusa.org/clean_energy/coalvswind/c02c.html#.VLb4yfbQfcs).

USEPA. (2015). *Assessment of the Potential Impacts of Hydraulic Fracturing for Oil and Gas on Drinking Water Resources: Executive Summary*. Washington, DC: U.S. Environmental Protection Agency, Office of Research and Development (https://www.epa.gov/sites/production/files/2015-07/documents/hf_es_erd_jun2015.pdf).

Wikipedia. (2015). *Upper Rhine Plain*, https://en.wikipedia.org/wiki/Upper_Rhine_Plain.

Zemach, E., Drakos, P., Robertson-Tait, A., and Lutz, S.J. (2010). Feasibility evaluation of an "in-field" EGS project at Desert Peak, Nevada, USA. In: *Proceedings of World Geothermal Congress 2010*, Bali, Indonesia, April 25–30 (http://www.geothermal-energy.org/pdf/IGAstandard/WGC/2010/3159.pdf).

Zierenberg, R.A., Schiffman, P., Barfod, G.H. et al. (2013). Composition and origin of rhyolite melt intersected by drilling in the Krafla geothermal field, Iceland. *Contributions to Mineralogy and Petrology*, 165(2): 327–347.

12 Future Considerations of Geothermal Energy

KEY CHAPTER OBJECTIVES

- Examine promising projections of current geothermal assessments for the United States (Tester et al., 2006; Williams and DeAngelo, 2008) in light of current empirical findings.
- Discuss how geothermal is a renewable resource that may or may not be sustainable depending on how it is developed.
- Review encouraging aspects for future development of geothermal energy.
- Consider some key challenges facing further development of geothermal energy.

INTRODUCTION

About 95% of the Earth's mass is at a temperature greater than 500°C, and this heat energy constitutes a huge resource for helping satisfy the energy needs of society. Earth's flow of heat from its interior to the surface is estimated at 47 TW (Davies and Davies, 2010), which is about 20 times the total world power generation in 2012 (EIA, 2013a). At the end of 2015, worldwide geothermal power production was a little more than 12 GWe, which represents just a tiny fraction (0.025%) of Earth's heat flow, or about 0.5% of world power generation. If only 0.1% of Earth's heat energy could be tapped and utilized, it would amount to about 10% of the planet's power generation. Thus, harnessing geothermal energy for the benefit of society has tremendous potential to move from what is more or less a cottage industry at present for most regions of the planet toward becoming a major source of clean, baseload energy. In their U.S. Geological Survey assessment of domestic geothermal resources, Williams et al. (2008) indicated that an additional ~6000 MWe could be obtained from identified conventional geothermal systems and an estimated 30,000 MWe from undiscovered conventional geothermal resources. If realized, these sources would represent about a tenfold increase in the current U.S. installed geothermal power capacity of about 3500 MWe. In addition, these same researchers indicated that if engineered geothermal systems (EGSs) are included then the geothermal power potential could grow to 345,000 MWe (at 95% probability), which would represent about 35% of the total installed U.S. power generation (EIA, 2013b).

These promising projections, however, must be tempered with current empirical findings. The above assessments are based on volumetric heat-in-place calculations combined with Monte Carlo simulations.* This approach includes estimates of reservoir volumes, temperature, permeabilities, and surface heat flow measurements. For ease of calculations and model simulations, the hydrothermal systems are treated as static, not dynamic, entities and largely ignore the effects of injection to maintain reservoir pressure and to potentially cool the modeled system. Therefore, assessments based on volumetric heat-in-place calculations and modeling typically overestimate resources as evidenced empirically from developed projects to date (Benoit, 2013). As an example, the Blue Mountain binary geothermal power plant in northern Nevada was built for a capacity of 49.5 MWe based on volumetric heat-in-place calculations; however, after 6 years of production it is now producing less than 30 MWe and is projected to be producing about 15 MWe by 2020, according to the former operator, Nevada Geothermal Power (NGP, 2012). This rapid decline in power has resulted from thermal breakthrough of injection wells probably positioned too close to production wells and an overestimation of the original geothermal resource at Blue Mountain. Based on similar fault-controlled, producing geothermal systems elsewhere in Nevada, the mile-long fault-controlled geothermal reservoir at Blue Mountain would support about 12 to 18 MWe, not the 49.5 MWe indicated by volumetric heat-in-place calculations (Benoit, 2013).

In light of empirical data from operating power plants, assessments of domestic geothermal potential should be considered cautiously. Nonetheless, with continued advances in exploration techniques for finding blind undiscovered conventional geothermal systems and bringing online some of the most promising engineered geothermal systems, geothermal power output could reasonably increase from its current small fraction of a percent (~0.35%) to several percent of the nation's produced power by the middle of this century. Whether or not this magnitude of increase in geothermal power output is achieved, however, depends probably less on continual advances in understanding of and discovering new geothermal systems than on political and economic factors (as discussed below) that are major drivers of energy resource development and management. To understand how geothermal energy will evolve and the role it will play with time, the roles of renewable and sustainable development, as well as encouraging indicators and potential obstacles for geothermal development, must be considered.

RENEWABLE VS. SUSTAINABLE IN DEVELOPMENT OF GEOTHERMAL ENERGY

The terms *renewable* and *sustainable* are often interchanged, but they are different. Renewable refers to resources that can be replaced through natural processes within a time frame beneficial to societal needs. Such resources include harvesting timber, growing crops, or recovering salt through evaporation of seawater. Being renewable is thus an intrinsic characteristic of the resource (Axelsson, 2010). Being sustainable,

* *Monte Carlo simulations* are computerized mathematical algorithms that provide a range of forecasts of outcomes and probabilities for a given course of action. In the case of geothermal power potential, the input of parameters, such as heat flow, temperature, and size, yield a minimum output for a most likely condition to a maximum predicted value of geothermal power potential.

on the other hand, generally refers to the development of resources that meet today's needs without compromising the use of those resources by future generations. From that standpoint, even fossil fuel sources of energy, although finite, can be sustainable through conservation techniques and thus available for future generations. Fossil fuel resources, however, are not renewable as their development takes tens of thousands to millions of years; thus, they cannot be renewed on time scales of human needs. In this sense, then, sustainable can refer to how a resource is used by society, whether renewable or nonrenewable (Axelsson, 2010).

Geothermal energy, like solar and wind, is certainly renewable in that heat rising to the surface from the Earth's interior is ongoing and unyielding. External solar energy can be harnessed directly to produce electricity and heat (solar PV and solar thermal) and indirectly in the form of wind energy, as uneven heating of the planet drives global wind patterns. However, extraction of Earth's internal heat energy from geothermal systems can be rendered unsustainable because the transfer of heat to run a geothermal power plant generally exceeds the Earth's natural ability to sustain the reservoir's temperature. This is because heat transfer via convection is more efficient than conduction, and the former is how we currently harvest conventional geothermal systems to produce power, both electrical and thermal. A classic example of unsustainable geothermal production is The Geysers, which was producing about 1800 MWe in the late 1980s and early 1990s before the steam pressure dropped precipitously due to fluid removal without adequate recharge. On hot summer days, almost all of the produced steam is lost through the evaporative cooling towers (M. Walters, pers. comm., 2015). Now that treated wastewater is imported to The Geysers from Santa Rosa and Lake County, recharge has largely been restored and electrical production has stabilized at about 800 MWe. Nonetheless, the temperature of the main producing steam reservoir is declining slowly at about 1 to 2°F per year (M. Walters, pers. comm., 2015). This slow decline in temperature spurs the ongoing effort to enhance or at least maintain current production levels by producing from the hotter, but less permeable Northwest Geysers region. So, although geothermal heat is renewable, careful management of the resource is required to make it sustainable and available for future generations.

A possible way a geothermal field can be made at least more sustainable might be to rotate where the field is produced. In this way, when a producing area begins to cool, another part of the field can be "turned on" to allow the produced region to recharge, not unlike crop rotation where part of a field is left fallow to restore nutrients. Even then, a producing geothermal system eventually may have to be "turned off" for a few decades to allow it to be renewed by natural heat flow. But, when one geothermal field is "turned off" another could be "turned on," similar to well rotation in a given field but on a larger scale. Modeling estimates on recovery times for a geothermal system to recharge after production range from being equal to the duration of production to about three to four times the length of production (Axelsson, 2010; Bromley et al., 2006; O'Sullivan et al., 2010). For example, considering a 100-year production history for the Wairakei geothermal system in New Zealand, O'Sullivan et al. (2010) developed an empirical formula for the time of recovery: Recovery = (PR – 1) × (Duration of extraction), where PR is the production ratio and is defined as produced energy flow divided by the natural energy flow. For Wairakei, the average production ratio is 3.8, and using the above formula results in a recovery period of

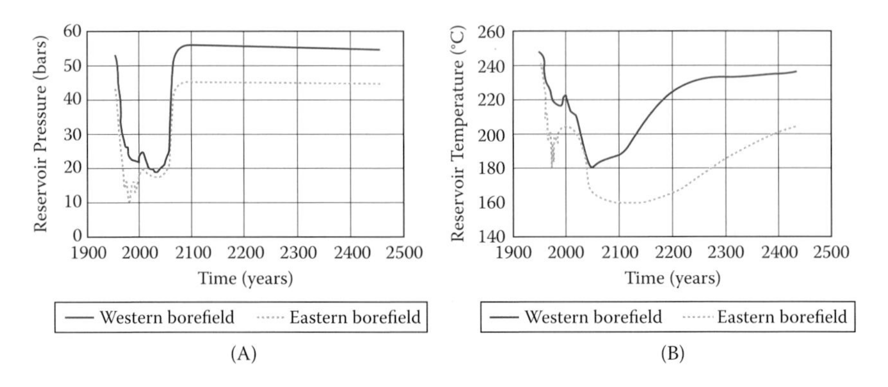

FIGURE 12.1 (A) Measured and modeled pressure decline and recovery and (B) temperature decline and recovery for Wairakei geothermal system. Note that the pressure recovery is much more rapid than temperature recovery due to the high permeability of the geothermal reservoir. (From O'Sullivan, M. et al., *Geothermics*, 39(4), 314–320, 2010.)

about 300 years. At Wairakei, the pressure recovery occurs much faster than temperature recovery (Figure 12.1). Although the outcome of applying the above formula for recovery time suggests about a 300-year recovery period for Wairakei after 100 years of production, about 85% of the temperature recovery occurs within a 200-year period (Figure 12.1B). This is still not within a useful time frame. For this reason, some geothermal experts argue that geothermal energy, although environmentally benign, is not fully sustainable (at least for some producing geothermal systems) as heat is being extracted faster than it is being restored, and a given producing geothermal system today may not be available for the next generation (S. Arnorsson, pers. comm., 2014).

Other studies modeling pressure and temperature recovery indicate shorter periods of reservoir recharge after shutdown in which recuperation periods are about equal to productions periods (Figure 12.2) (Bromley et al., 2006; Rybach, 2003). Furthermore, Rybach (2003) observed that a doublet system (one injection and one production well) operating cyclically in a production/recovery mode in a 10-year cycle produced more energy than 20- or 40-year cycles over a 160-year period. This latter result suggests that shorter operation rather than longer periods of operation, followed by recovery periods of comparable duration, yield more energy in the long run. Thus, when operated on a periodic basis of production followed by recovery, geothermal plants are renewable and sustainable.

SUMMARY OF ENCOURAGING INDICATORS

Geothermal as a renewable and sustainable resource of energy, if carefully managed, has many attractive attributes.

Baseload and High Capacity Factor

Unlike solar and wind, geothermal power is a baseload source of energy and not dependent on the sun shining or the wind blowing. Baseload power means the power plant produces all the power it can, with some variations due to condensing

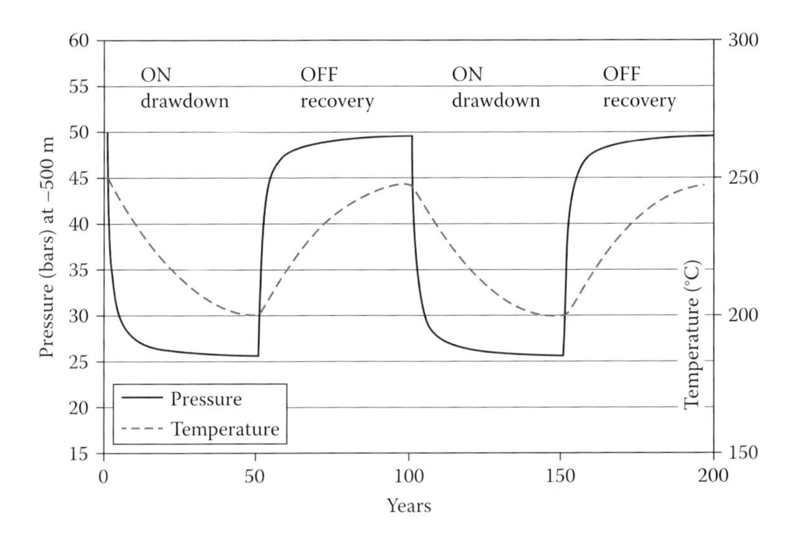

FIGURE 12.2 Cyclic utilization of geothermal systems in which production is followed by an equally long period of recuperation. Note the rapid recovery of system mass (pressure) compared to the slower recovery of temperature. (From Bromley, C.J. et al., in *Proceedings of International Solar Energy Society Renewable Energy Conference (RE2006)*, Chiba, Japan, October 9–13, 2006.)

efficiencies, and is basically available at all times of the day, 7 days a week, year in, year out. Furthermore, capacity factors* for geothermal power plants are >70% and not uncommonly >90%. For example, the capacity factor for the Dixie Valley geothermal plant (68 MWe total gross installed capacity) in central Nevada over a 25-year period is close to 99% (Benoit and Stock, 2015). By contrast, capacity factors for wind farms averaged about 33% in 2013 and 2014 and about 26% for solar photovoltaic facilities (EIA, 2016a). Capacity factors for hydropower, which is the other principal baseload renewable power source, are typically greater than 50%, reflecting external conditions such as drought and reservoir level for flood control and water storage. Indeed, in 2013 and 2014, the high capacity factors for geothermal power plants exceeded those of natural gas power plants (average about 50%) and those of coal-fired power plants (average about 60%) and rivaled those of nuclear power plants (average about 90%) (EIA, 2016a).

Environmental Aspects

As detailed in Chapter 9, the environmental elements of developing geothermal energy are outstanding. Greenhouse gas emissions are either nonexistent or very minor, and every megawatt of geothermal power produced displaces burning about 4350 tons of coal and 87 billion cubic feet of natural gas per year. The burning of this much coal

* *Capacity factor* is the ratio of actual produced energy to the potential output if operated at the full nameplate capacity for a given period of time.

and natural gas, using data published by the U.S. Energy Information Administration, adds about 12,000 tons and 7000 tons of CO_2 to the atmosphere per megawatt, respectively. Most if not all of these carbon emissions can be saved by geothermal power generation. Furthermore, geothermal direct use, including ground-source heat pumps, for cooling and space heating further displaces the use of fossil fuels. Geothermal heat pumps in particular, because of their widespread applicability, have the potential to greatly reduce power needs for air conditioning and the burning of natural gas for space heating and heated water supply. The U.S. Department of Energy's Office of Energy Efficiency and Renewable Energy estimated that 23 sponsored ground-source heat pump demonstration projects saved an estimated 153.6 GWh of energy per year and 9000 tons of CO_2 from being emitted. The Peppermill Resort Hotel and Casino in Reno, Nevada, is heated entirely by geothermal fluids and saves between $2M and $2.5M in natural gas costs per year along with the associated reduction of CO_2 (Dean Parker, pers. comm., 2015).

Geothermal power plants have one of the smallest footprints for energy produced—geothermal power plants consume just about a tenth of the land space per unit of power produced compared to coal (including the mine to supply the coal) and solar thermal facilities, about an eighth of the land space compared to a solar photovoltaic facility of comparable nameplate capacity, and less than a quarter of the land space of a wind farm for the same nameplate power capacity (Kagel et al., 2007). Moreover, the land used for geothermal development remains open for other uses if need be, such as agriculture in the Imperial Valley of California and free-range cattle grazing around some geothermal plants and well fields in Nevada.

Fuel Sources Not Needed and Low Operating Costs

The Earth serves as the boiler for geothermal power plants or thermal bank for direct use applications; no external fuel sources are needed to boil or heat the water. Geothermal power plants are immune to changes in availability or price fluctuations of fuel sources, such as coal, natural gas, or uranium. The tradeoff is that conventional geothermal resources are not everywhere available for power generation, such as a fossil-fuel-fired power plant. Direct use fluids, however, are more widespread, and geothermal heat pumps can be utilized virtually everywhere. However, the use of enhanced geothermal systems (EGSs), as noted below and discussed in the preceding chapter, could significantly expand and open new regions for developing geothermal power. Moreover, the high capacity factors of geothermal power, independence of external fuel sources, and generally low operating and maintenance costs keep the cost of geothermal power competitive with fossil-fuel-powered sources. Currently, operation and maintenance costs for geothermal varies from $0.01 to $0.03/kWh and from $0.024 to $0.04 for fossil-fuel-fired power plants (EIA, 2013c). The slightly higher costs of the latter mainly reflect added costs of fuel but also the higher operating conditions of temperature and pressure of fossil-fuel-fired power plants, which stress equipment and increase maintenance costs. The levelized* cost of geothermal

* *Levelized cost* is minimum price power can be sold in order to break even. Levelized costs include the expense of building and financing a power plant, the cost of operations and maintenance, and cost of fuel (none for geothermal).

is higher and ranges from \$0.04 to \$0.12/kWh. For comparison, wind power has comparably low levelized costs as geothermal due to its low initial capital expenditures and absence of a fuel cycle, despite its relatively low capacity factor. Solar power is somewhat more expensive, when price support or tax breaks are excluded, because the upfront capital expenditure is more than wind, especially for concentrating solar power. All in all, geothermal power is price competitive with fossil-fuel-fired plants and is comparable to or less expensive than wind- and solar-produced power.

Emerging Technologies and Geologic Settings

An area of fruitful development in enhanced geothermal systems (EGSs) is the use of hydroshear with thermally degradable packing material to increase the size of artificially produced geothermal reservoirs. These techniques are being explored at the Newberry Volcano EGS demonstration project in central Oregon by AltaRock Energy, Inc. It is also one of Department of Energy's sites for their Frontier Observatory for Research in Geothermal Energy (FORGE). Newberry Volcano appears to host one of the most potent heat reservoirs in the western United States, but geothermal power development has been hampered by poor permeability. The combined use of hydroshear and the thermally degradable packing material may allow creation of a sufficiently large volume of permeable rock to support a commercial geothermal power facility.

Also as chronicled in the last chapter, drilling into supercritical systems has the potential to increase power output per well by 5 to 10 times compared to a well tapping a subcritical reservoir. This means that fewer wells would need to be drilled, which would lower the overall costs. Development of supercritical systems is being actively explored in Iceland as part of the Iceland Deep Drilling Project.* For this to become reality, however, some important hurdles need to be overcome. These include the high cost of deep drilling, which requires specialized valves, extra-heavy-duty blow-out preventers, and corrosion-resistant casing. Also, elevated temperature and pressure (>374°C and >220 bars) are difficult on equipment, and the potential increase in mineral scaling, in particular silica, could result in partially clogged wellbores and could reduce the high buoyancy to viscous forces of supercritical fluids.

Deep, hot sedimentary aquifers may serve as a bridge on the way to developing true engineered geothermal projects like Newberry Volcano or utilizing supercritical geothermal fluids as proposed by Allis et al. (2012). As described in Chapter 11, deep (3 to 4 km) stratigraphic reservoirs in areas of elevated regional heat flow, such as in northeastern Nevada and northwest Utah, may be attractive geothermal reservoirs as their temperatures range from about 150°C to more than 200°C. Using data from oil and gas drilling, these deep stratigraphic reservoirs, especially in carbonate rocks, appear to have good permeabilities of 50 to 100 millidarcys (Allis et al., 2012). A key question is whether the costs of deeper drilling will permit commercial development. Considering that stratigraphic reservoirs cover areas two to four orders of magnitude larger than currently producing structurally

* A good project overview is available at www.iddp.is.

controlled reservoirs in the Great Basin, the extra expense of deep drilling could be justified, as these reservoirs have the potential for sustaining 100-MWe power plants (Allis and Moore, 2014).

As noted in Chapter 7, coproduced geothermal water from oil fields offer the potential for development, especially considering that no additional wells must be drilled. Temperatures of coproduced geothermal water from oil wells range from about 90°C at Teapot Dome in Wyoming to more than 150°C at depths of 4 km in the Williston Basin in the north-central United States (Gosnold et al., 2013). A limiting factor of utilizing these "waste" waters for power production is that pumping rates are low to maximize oil output. However, when these fields are tapped out of oil, pumping rates could be increased to support a potential binary-style geothermal power plant of modest size (15 to 30 MWe).

For the future, EGS development has the potential to increase geothermal power production by one to two orders of magnitude (Tester et al., 2006; Williams et al., 2008). Part of the reason for this large increase is that hot rock is considerably more widespread than conventional convecting geothermal systems already geologically primed for development. Although hot dry rock (petra-heat) is more widespread than conventional convecting geothermal systems, its development will require deep expensive drilling of wells and potentially copious amounts of added water, which could pose a problem during times of drought or in regions where water availability is limited.

Potential Flexible Load Provider

Although geothermal has been traditionally a baseload power source, the increasing addition of intermittent power sources, such as wind and solar, to the power grids allows utilities to meet mandated renewable portfolio standards quickly and economically. As a result, the new normal is for flexible power sources that can ramp up or down to accommodate the fluctuations of added wind and solar power to the grid. Because of the nature of geothermal energy, it is typically most efficiently operated under baseload conditions as throttling wells can compromise flow and temperature and can lead to condensation in piping, scaling, and corrosion. Nonetheless, flexible geothermal operation is possible, as demonstrated by the 38-MWe Puna Geothermal Venture plant in Hawaii, which provides 16 MWe of flexible delivery or dispatch ranging from 22 to 38 MWe (Nordquist et al., 2013). This is accomplished by using diverter values that can route heat and fluid around the binary fluid turbines during periods of lower power demand (Figure 12.3). Power at Puna can be ramped up or down at a rate of 2 MWe/minute while maintaining a spinning reserve* of 3 MWe. The Puna geothermal facility represents the first modern geothermal facility that became a provider of both baseload and flexible dispatchable power in 2012.

During the 1980s and early 1990s, The Geysers in California in part operated in flexible mode responding to the needs of one of its customers. The practice was curtailed, however, due to a combination of low demand for flexible power and the less expensive costs of supplying flexible power from dams and fossil-fuel-fired power

* *Spinning reserve* is the power that is ready to respond immediately in case a generator or transmission line fails unexpectedly.

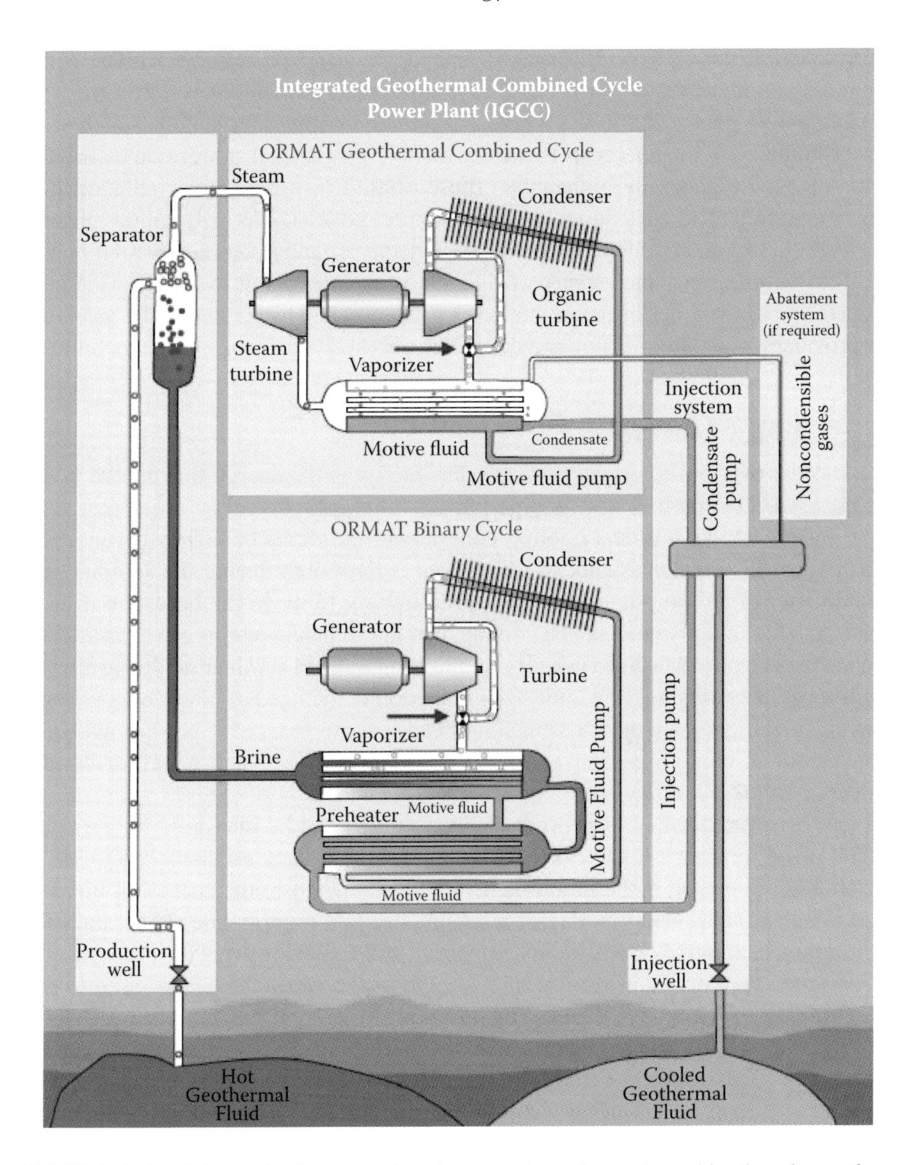

FIGURE 12.3 Schematic diagram of an integrated geothermal combined cycle geothermal plant and bottoming binary plant as used at the Puna geothermal facility in Hawaii. To achieve a flexible mode of dispatchable power, diverter valves are placed before the motive fluid turbines (noted at tip of red arrows). See text for discussion. (Adapted from Ormat, *Integrated Combined Cycle Units: Geothermal Power Plants*, Ormat Technologies, Inc., Reno, NV, 2016.)

plants. Indeed, geothermal plants can be retrofitted or built to provide both baseload and dispatchable power, but because of the associated extra costs to do so it becomes more difficult to keep the cost of flexible geothermal power competitive. What is needed is a pricing structure that provides a premium value for geothermal plants to deliver flexible power. The California Independent System Operator (CAISO), which monitors and regulates power needs on the California grid, is now very desirous of flexible power plants due to the increasing addition of intermittent solar and wind power being added to the grid. However, flexible geothermal power comes or will come at a price generally higher than those offered by solar- and wind-generated power (due in part to current uneven price support policies). So, will utilities change from purchasing relatively inexpensive power from ramping natural gas-fired power plants to flexible, albeit more expensive, geothermal power plants to offset the increasing intermittent sources of renewable power? Unless the answer is yes, the incentive for providers of geothermal power to make the necessary investment appears limited.

Future Role of Geothermal Heat Pumps

A too-often overlooked aspect of renewable energy is the energy that can be saved or not used. Geothermal heat pumps fall into this category; they are a very efficient means for heating and cooling commercial buildings or homes because the Earth acts like a thermal bank in which heat is deposited during the summer and withdrawn during the winter. Power is used mainly to move the heat that already exists, not for actually heating and cooling. Geothermal heat pumps can significantly reduce the use of power-intensive air-conditioning units in summer or the burning of natural gas or heating oil in winter. This reduction in the need of power or consumption of fossil fuels is a win–win situation. It can serve to reduce the power imbalance of intermittent wind- and solar-generated power as discussed above and further the reduction of greenhouse gas emissions.

The energy efficiency of geothermal heat pumps can be measured by the coefficient of performance (COP) for heating and by the energy efficiency ratio (EER) for cooling. The COP is the ratio of delivered heat energy to the energy required to move the heat, consisting mainly of a compressor and pumps. The COP values for geothermal heat pumps are typically between 3 and 5, meaning that the energy delivered to the space or water to be heated is three to five times the energy used to run the pump. For comparison, the most energy-efficient natural-gas-fired furnaces have a COP of about 0.95. On the cooling side, the EER is the ratio of the cooling capacity to the power input; the higher the rating, the greater the energy efficiency. The most energy-efficient conventional central air conditioners have an EER of about 15, whereas the most efficient closed-loop geothermal heat pumps have an EER of about 30, and some open-loop* geothermal heat pumps have EER values as high as 45 to 50. This means that geothermal heat pumps consume about a half to a third of the power used by the most efficient conventional air conditioners. Increased development of geothermal heat pumps across the globe could save considerable energy now used for heating and cooling.

* Closed- and open-loop geothermal heat pumps reflect the configuration of the in-ground piping to circulate fluid for depositing or withdrawing heat from the Earth. In a closed-loop arrangement, the fluid continuously circulates between the piping in the ground and the heat pump inside the building; no matter is exchanged, only the heat. In an open-loop system, both matter and heat are exchanged. In this case, a lake or pond or groundwater is used as a heat source or sink, such as tying into an existing domestic well.

CHALLENGES TO DEVELOPMENT

Although geothermal energy has considerable positive attributes, several obstacles stand in the way for greater use of this resource. These include land-use restrictions, such as the ongoing struggle of balancing development and wildlife management (e.g., sage grouse concerns in the western United States), the high upfront capital costs and risk associated with drilling production wells, low current costs of fossil fuels (particularly natural gas), short-term focus of investors for capital development, expense of transmission lines from remote locations, availability of water for EGS projects, and political vacillations resulting in come-and-go incentive policies such as tax credits and loan-guarantee programs and their ever-changing conditions and qualifications.

Risk, High Upfront Costs, and Short-Term Investor Focus

Finding and developing geothermal resources, particularly for power production, are expensive and entail risk. Currently, the total cost for putting a geothermal power plant in production, including initial exploration and development drilling, ranges from $3000/kWe to more than $7000/kWe. This cost range is sensitive to the temperature of the resource, geologic conditions such as the depth of the resource and flow rates of geothermal fluids, success or difficulty of drilling, and existing infrastructure such as proximity to existing transmission lines (Glassley, 2015). For example, a 30-MWe binary plant utilizing a moderate temperature resource of 165° to 175°C would cost about $120M and can take 5 to 7 years to realize from initial exploration to beginning of production.* The return on the investment for such an enterprise is also about 5 to 7 years. For comparison, the U.S. Energy Information Administration (EIA, 2013) reported that a conventional natural gas combined-cycle or combustion turbine power plant costs less than $1000/kWe to build or about a third of the cost per kWe of building a modest-size binary geothermal plant. Although a 300-MWe natural-gas-fired power plant would cost about $300M or about three times the cost of a 30-MWe binary geothermal plant, the former would deliver ten times the power and would take about one-third the time to bring to fruition, and its location is not dependent on geologic factors. A large natural-gas-fired power plant, by taking advantage of economies of scale, can sell its power at a very modest price, potentially underpricing power sold by a smaller geothermal power plant.

Nonetheless, a smaller binary geothermal power plant can stay competitive due mainly to its high capacity factor of about 90%, compared to about 50% for the natural-gas-fired plant, and its immunity to the price volatility of natural gas. Also, of course, there is the environmental benefit of zero greenhouse gas emissions for a binary geothermal plant. The 1749 natural gas-fired power plants in the United States produced 500 million metric tonnes of CO_2 in 2014, yielding an average of about 300,000 tonnes of emitted CO_2 per power plant per year, according to data compiled by the U.S. Energy Information Administration (EIA, 2015, 2016b). If the

* Ormat, however, developed the first 30-MWe phase of the now 72-MWe McGinness Hills near Austin, Nevada, in just under 4 years from initial exploration to transmission of power.

United States were to adopt a carbon tax at, say, a value of about $40 per tonne of CO_2 emitted (which is about the average for the eight countries that have adopted a carbon tax, according to the World Bank), that would add about $20 billion dollars to the cost of producing about 400 GWe from natural gas or about $50 per kWe of added cost.

Also unlike building a gas-fired power plant, where produced power is essentially guaranteed, development of geothermal energy has much more risk because geologic conditions vary considerably and expensive techniques, in particular drilling, are needed to help discover and evaluate the resource. Drilling production and injection wells currently costs from about $3M to more than $8M per well depending on geologic conditions and depth of the prospective resource. Indeed, about a third of the cost in developing a geothermal power facility is in exploration and defining the resource through drilling, a process that can cost a few tens of millions of dollars with the potential outcome that the resource is inadequate to warrant further development. Another potential risk is the fast-tracking of a project in order to receive federal or local government tax credits or loan guarantees. As a result, corners may be cut to meet incentive deadlines and the resource can be inaccurately characterized or improperly developed. For example, placing injection wells too close to production wells can lead to thermal breakthrough and reduced power generation within a few months or years from the beginning of power production. When this happens, investors become particularly skittish about supporting other worthwhile geothermal projects.

As noted above, it may take two to three times as long (5 to 7 years) to put a geothermal power plant online, beginning with initial exploration, relative to that of a gas-fired power plant. A comparable amount of time to pay off the debt is then required before profit is realized. Many of today's financial investors, indeed much of our economy, focus on short- rather than long-term returns on investment. Moreover, with the Great Recession at the beginning of this century still in the back of investors' minds, today's venture capitalists are particularly risk averse, which is a problem for an industry that entails risk even though substantial rewards can be found in the long term—both financially and climatologically. Going forward, as exploration techniques and expertise improve in the discovery of blind conventional (naturally convecting) geothermal systems and potentially bringing engineered geothermal systems online, the risks and time frame for geothermal development will decline. This should attract additional private and public financial investments.

Water Availability

Although much is touted about how development of EGSs could greatly expand geothermal power development, a key question is the availability of water to support these projects. Much of the potential for EGSs comes from the semi-arid western United States, where hot rock at depth can be found over relatively large areas, but surface or shallow groundwater to be injected and heated can be limited. As an example of the water demand, the cooling towers for the Dixie Valley geothermal plant in central Nevada evaporate about 1600 gallons per minute of water (D. Benoit, pers. comm., 2015); a comparable amount of water would then have to be made up in a water-cooled EGS power facility. Furthermore, at Dixie Valley the dual-flash power plant requires more than 500 kg/sec (~10,000 gpm) of fluid at 250°C to generate about 55 MWe. Tracer tests indicate return times ranging from 30 days to 150 days (Benoit and Stock, 2015). If we assume a simple average of 90 days, then the volume of water in the producing reservoir at Dixie Valley would be about 10,000 gal/min × 60 min/hr × 24 hr/day × 90 days, or 1.3 billion gallons of water (5 million m^3 or 4000 acre-ft). To make an EGS system comparable in size to Dixie Valley, about 4000 acre-ft of water would have to be injected to create a critical mass of fluid to be heated and produced from. Some water need could be reduced by using air-cooled binary geothermal binary plants, but these plants typically produce less power than water-cooled flash plants, which could impact the economic viability due to the added costs of developing an EGS power facility.

Political Whims and Governmental Regulations

The United States, unlike many other developed countries, lacks a unified national energy policy, in particular a policy that helps reduce the use of carbon-based fuels. Price supports, tax incentives, and loan guarantee programs come and go, making long-range planning for the development of geothermal resources complicated. The influence of governmental programs, both federal and state, on the growth of geothermal power output is illustrated in Figure 12.4. The large growth from less than

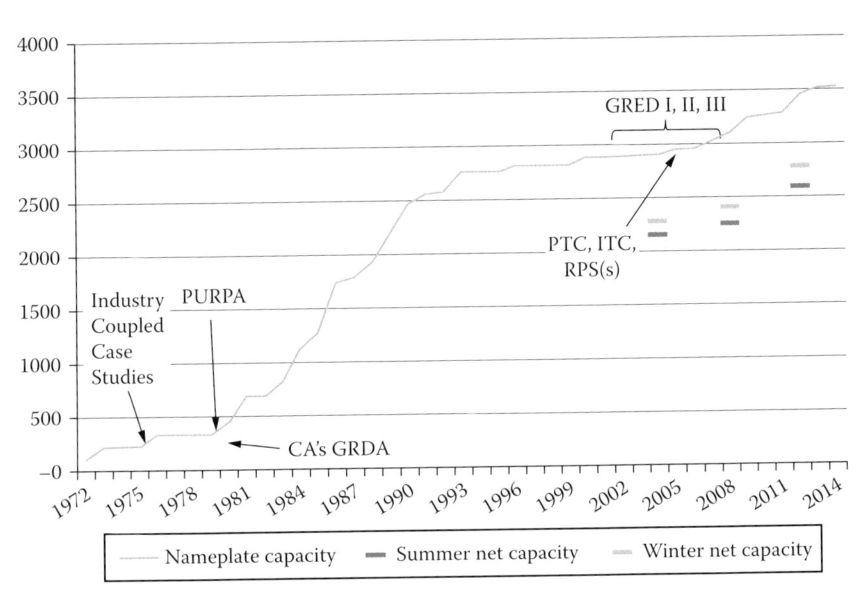

FIGURE 12.4 Growth of U.S. geothermal power (in MWe) with time. Abbreviations: GRDA, Geothermal Resource and Development Assessment, adopted in 1980; GRED, Geothermal Resource and Definition Program; ITC, investment tax credit; PTC, production tax credit; PURPA, Public Utilities Regulatory Policy Act of 1978; RPS(s), stated adopted renewable portfolio standards. See text for discussion. (From Matek, B., *2015 Annual U.S. & Global Geothermal Power Production Report*, Geothermal Energy Association, Washington, DC, 2015.)

400 MWe to 2750 MWe from 1980 to 1993 was stimulated in large part by the Public Utilities Regulatory Policy Act (PURPA) of 1978, as part of the National Energy Act that was engendered by the 1973 energy crisis. The focus of the act was to promote energy conservation and development of domestic energy sources, especially renewable energy, with the goal of reducing energy demand and increasing domestic supply of energy. Also, California's Geothermal Grant and Loan Program (or Geothermal Resource Development Account [GRDA]) began in 1980 and provided significant financial help for the expansion of power produced from The Geysers during the 1980s. Also during the early 1980s, the California Public Utilities Commission developed Standard Offer #4 (essentially a feed-in tariff), which required major utilities to purchase power from qualifying third parties offering renewable energy. Utilities purchased the renewable energy at an escalating price for the first 10 years of operation, before returning to short-term avoided costs (wholesale costs of conventional or fossil fuel power) for the remaining length of the contract. This allowed developers of geothermal energy to more easily secure financing and sell power at attractive rates (above those for natural gas and coal). More recently, the adoption of renewable portfolio standards by states and the federal government's investment in production tax credits and loan guarantee programs as part of the America Reinvestment Act of 2009 resulted in construction of 38 new geothermal power plants since 2005. Power output increased from less than 3000 MWe in 2006 to slightly more than 3500 MWe by the end of 2015, or about a 20% increase in ten years.

Since 2013, however, domestic growth of geothermal power has largely stalled. This is due in part to limited growth in demand for new power but also legislative uncertainty about production and investment tax credits and more stringent rules of federally backed loan guarantee programs. This has resulted in the reduction of geothermal projects in many states. For example, in geothermally prospective Nevada, the number of geothermal exploratory and development projects was scaled back from 45 in 2014 to 23 in 2015, and a similar significant decrease in projects has also occurred in California (Figure 12.5). However, according to a report by the Geothermal Energy Association (Matek, 2015), about 500 MWe of confirmed geothermal power projects stand in abeyance seeking power purchase agreements (PPAs) with various utility companies. Some of the delay stems from utilities seeking the least expensive sources of renewable power to satisfy their renewable portfolio standards, typically solar and wind, which currently enjoy tax incentives or price supports not available to geothermal. Of course, this has led to a higher proportion of intermittent power, which is why independent system operators such as CAISO now want geothermal to be a flexible source of power. This can be done, but at the cost of retrofitting geothermal plants, thus making geothermal power more expensive than what utilities currently want to pay—hence the conundrum.

New geothermal growth opportunities may lie ahead, spurred again by governmental actions. For instance, in his 2015 State of the State address, Governor Jerry Brown of California announced his plan to raise California's RPS goal to 50% by 2030. He also signed into law A.B. 2363, which in part stipulates that the California Public Utilities Commission is to develop integration costs of different renewable power technologies, including a possible surcharge on solar and wind power, because of their intermittent deliverability to the grid. If so, this would make geothermal

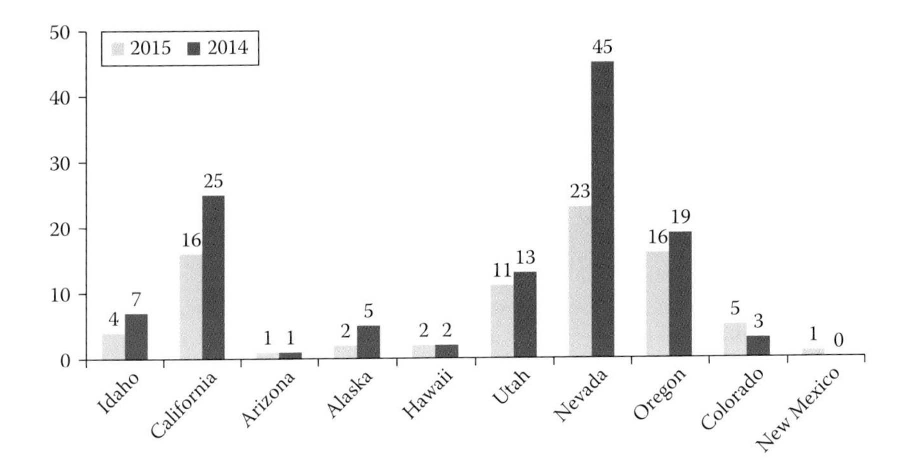

FIGURE 12.5 Change in number of geothermal projects by state from 2014 to 2015. Note the significant reduction in projects in the two leading states of geothermal power production, California and Nevada. See text for discussion that helps to explain these changes. (From Matek, B., *2015 Annual U.S. & Global Geothermal Power Production Report*, Geothermal Energy Association, Washington, DC, 2015.)

power more competitively priced. Furthermore, on the federal level, the budget of the Geothermal Technologies Office (GTO) of the Department of Energy has steadily increased over the last few years, reaching $55M for 2014–2015. For fiscal year 2015–2016, the GTO is requesting a 75% increase to $96M to help fund research and development in EGSs, including the newly developed Frontier Observatory for Research in Geothermal Energy (FORGE) and hydrothermal play fairway analysis. Money from these programs will be distributed to both research organizations and industry to advance development of geothermal resources.

A final question mark that may impact future geothermal development, mainly in the Great Basin of Nevada, eastern Oregon, and western Utah, is the newly adopted regulations for protecting the greater sage grouse on federally managed lands. Impacts of these new regulations are still being studied. Current indications are that new regulations will most likely lengthen the time of permitting and the time from exploration to putting power on the grid if a site is found viable. This will probably further stress the short-term focus of many investors.

FINAL ASSESSMENT

Further geothermal growth depends on many factors. Is there a national will to develop geothermal energy on par with other forms of non-carbon-producing alternative energy sources, including wind and solar? Or, will the *status quo* of currently inexpensive natural gas and market forces be the main drivers of energy policy? Will federal and state governments encourage further growth of the geothermal industry, such as providing investment and production tax credits and loan guarantee programs to reduce and encourage capital availability? Legislators and the public need to be better informed of the benefits of geothermal energy. The various attributes discussed have included the following:

- Cascading temperature applications include high-temperature resources used for power generation, moderate-temperature direct use applications for space and water heating, and low-temperature geoexchange or geothermal heat pumps used for both heating and cooling of buildings.
- Geothermal has a small carbon footprint for the energy produced, or saved in the case of geothermal heat pumps.
- Geothermal is a baseload power source that can be designed to provide flexible power to help offset the intermittent deliverability of solar and wind power.

In the end, development of geothermal power entails risk that is lacking in other renewable energy technologies due to the variable geologic conditions in which geothermal systems occur. This risk, however, can be reduced with continued improvements in exploration and assessment techniques, a better understanding of the geologic processes within active hydrothermal systems, more accurate geochemical and geophysical measuring and imaging techniques, and superior drilling technologies. Also. incentives offered by state and federal governmental agencies can help reduce risk and encourage private capital investment. Emerging technologies in the areas of harnessing supercritical geothermal systems; deep, hot sedimentary aquifers; and enhanced geothermal systems have the potential to reduce risk and costs, expand the geographic applicability of developing geothermal resources, and significantly increase the contribution of geothermal energy production. As the share of geothermal and other renewable energy sectors increases, our carbon footprint will diminish and our environment will benefit.

REFERENCES AND RECOMMENDED READING

Allis, R. and Moore, J. (2014). Can deep stratigraphic reservoirs sustain 100 MW power plants? *Geothermal Resources Council Transactions*, 38: 1009–1016.

Allis, R., Blackett, B., Gwynn, M. et al. (2012). Stratigraphic reservoirs in the Great Basin—the bridge to development of enhanced geothermal systems in the U.S. *Geothermal Resources Council Transactions*, 36: 351–357.

Axelsson, G. (2010). Sustainable geothermal utilization; case histories, definitions, research issues and modelling. *Geothermics*, 39(4): 283–291.

Benoit, D. (2013). An empirical injection limitation in fault-hosted basin and range geothermal systems. *Geothermal Resources Council Transactions*, 37: 887–894.

Benoit, D. and Stock, D. (2015). A case history of the Dixie Valley geothermal field. *Geothermal Resource Council Transactions*, 39: 3–11.

Bromley, C.J., Rybach, L., Mongillo, M.A., and Matsunaga, I. (2006). Geothermal resources: utilisation strategies to promote beneficial environmental effects and to optimize sustainability. In: *Proceedings of International Solar Energy Society Renewable Energy Conference (RE2006)*, Chiba, Japan, October 9–13.

Davies, J.H. and Davies, D.R. (2010). Earth's surface heat flux. *Solid Earth*, 1(1): 5–24.

EIA. (2013a). *International Energy Statistics*. Washington, DC: U.S. Energy Information Administration (http://www.eia.gov/cfapps/ipdbproject/IEDIndex3.cfm?tid=2&pid=2&aid=12).

EIA. (2013b). *Electricity Generating Capacity*. Washington, DC: U.S. Energy Information Administration (https://www.eia.gov/electricity/capacity/).

EIA. (2013c). *Electric Power Annual Report 2013*. Washington, DC: U.S. Energy Information Administration.

EIA. (2015). *Count of Electric Power Industry Power Plants by Sector, by Predominant Energy Sources within a Plant, 2004–2014*. Washington, DC: U.S. Energy Information (https://www.eia.gov/electricity/annual/html/epa_04_01.html).

EIA. (2016a). *Electric Power Monthly*. Washington, DC: U.S. Energy Information Administration (https://www.eia.gov/electricity/monthly/epm_table_grapher.cfm?t=epmt_6_07_b).

EIA. (2016b). *How Much of U.S. Carbon Dioxide Emissions Are Associated with Electricity Generation?* Washington, DC: U.S. Energy Information (http://www.eia.gov/tools/faqs/faq.cfm?id=77&t=11).

Glassley, W.E. (2015). *Geothermal Energy: Renewable Energy and the Environment*, 2nd ed. Boca Raton, FL: CRC Press.

Gosnold, W.D., Barse, K., Bubach, B. et al. (2013). Co-produced geothermal resources and EGS in the Williston Basin. *Geothermal Resources Council Transactions*, 37: 721–726.

Kagel, A., Bates, D., and Gawall, K. (2007). *A Guide to Geothermal Energy and the Environment*. Washington, DC: Geothermal Energy Association, pp. 20–60 (http://geo-energy.org/pdf/reports/AGuidetoGeothermalEnergyandtheEnvironment10.6.10.pdf).

Matek, B. (2015). *2015 Annual U.S. & Global Geothermal Power Production Report*. Washington, DC: Geothermal Energy Association (http://geo-energy.org/reports/2015/2015%20Annual%20US%20%20Global%20Geothermal%20Power%20Production%20Report%20Draft%20final.pdf).

NGP. (2012). Update on Projected Availability of Geothermal Resource for the Blue Mountain Geothermal Field [press release]. Vancouver, BC: Nevada Geothermal Power, Inc.

Nordquist, J., Buchanan, T., and Kaleikini, M. (2013). Automatic generation control and ancillary services. *Geothermal Resources Council Transactions*, 37: 761–766.

Ormat. (2016). *Integrated Combined Cycle Units: Geothermal Power Plants*. Reno, NV: Ormat Technologies, Inc. (http://www.ormat.com/solutions/Geothermal_Integrated_Combined_Cycle).

O'Sullivan, M., Yeh, A., and Mannington, W. (2010). Renewability of geothermal resources. *Geothermics*, 39(4): 314–320.

Rybach, L. (2003). Geothermal energy; sustainability and the environment. *Geothermics*, 32(4–6): 463–470.

Tester, J.W., Anderson, B., Batchelor, A. et al. (2006). *The Future of Geothermal Energy: Impact of Enhanced Geothermal Systems (EGS) on the United States in the 21st Century*. Cambridge: Massachusetts Institute of Technology (http://www1.eere.energy.gov/geothermal/egs_technology.html).

Williams, C.F. and DeAngelo, J. (2008). Mapping geothermal potential in the western United States. *Geothermal Resources Council Transactions*, 32: 155–161.

Williams, C.F., Reed, M.J., and Mariner, R.H. (2008). *A Review of Methods Applied by the U.S. Geological Survey in the Assessment of Identified Geothermal Resources*, USGS Open-File Report 2008-1296. Reston, VA: U.S. Geological Survey (http://pubs.usgs.gov/of/2008/1296/).

Index

D

E

F

G

H

N

O

P

T

U

V